High School Calculus: A New Perspective

For Any High School Curriculum, Including IB, A-Level, AP, and More

Spring Seeds

Published by Spring Seeds
Print ISBN: 979-8-89571-040-1

Contents

CHAPTER 1 PREPARATORY KNOWLEDGE 1

1.1. SOME MATHEMATICAL CONCEPTS 1

1.2. SETS 3

1.3. SET OF REAL NUMBERS 11

1.4. FUNCTIONS 16

CHAPTER 2 LIMITS 37

2.1. LIMITS OF FUNCTIONS AND VARIABLES 37

2.2. COMPUTATION RULES OF LIMITS 40

2.3. TWO IMPORTANT LIMITS 44

CHAPTER 3 DERIVATIVES AND DIFFERENTIALS 52

3.1. TWO EXAMPLES OF THE DERIVATIVE CONCEPT 52

3.2. CONCEPT OF THE DERIVATIVE 55

3.3. DIFFERENTIALS 57

3.4. FUNDAMENTAL DERIVATIVE FORMULAE AND COMPUTATION RULES 62

3.5. HIGHER DERIVATIVES 85

CHAPTER 4 DIFFERENTIAL MEAN VALUE THEOREM, L'HÔPITAL'S RULE, AND CONTINUITY AND DIFFERENTIABILITY OF FUNCTIONS 92

4.1. MEAN VALUE THEOREM 92

4.2. L'HÔPITAL'S RULE 99

4.3. CONTINUITY AND DIFFERENTIABILITY OF FUNCTIONS 114

CHAPTER 5 APPLICATIONS OF DERIVATIVES AND CURVE SKETCHING 132

5.1. TANGENTS AND NORMALS 132

5.2. DETERMINING MONOTONIC INTERVALS OF FUNCTIONS 140

5.3. EXTREME VALUES OF FUNCTIONS 143

5.4. MAXIMUM AND MINIMUM APPLICATIONS OF EXTREME VALUES 155

5.5. CONCAVITY AND POINTS OF INFLECTION OF CURVES 171

5.6. CURVE SKETCHING 184

5.6.1. Asymptotes of a curve 184
5.6.2. Sketching graphs of functions 191
5.6.3. Sketching the gradient function 201
5.7. RELATED RATES 209

CHAPTER 6 INDEFINITE INTEGRALS 219

6.1. CONCEPT AND PROPERTIES OF INDEFINITE INTEGRALS AND BASIC INTEGRATION FORMULAE 219

6.2. INTEGRATION BY SUBSTITUTION 226

6.3. INTEGRATION BY PARTS 246

6.4. INTEGRATION OF RATIONAL FUNCTIONS 256

CHAPTER 7 DEFINITE INTEGRALS 269

7.1. DEFINITION AND CALCULATION FORMULA OF THE DEFINITE INTEGRAL 269

7.2. INTEGRATION BY SUBSTITUTION AND INTEGRATION BY PARTS OF DEFINITE INTEGRALS 279

7.3. MEAN VALUE THEOREM FOR DEFINITE INTEGRALS AND FORMULA FOR FINDING THE DERIVATIVE OF A DEFINITE INTEGRAL 290

7.4. APPLICATIONS OF DEFINITE INTEGRALS 296

7.4.1. Area between two curves 296
7.4.2.* Area in polar coordinates 312
7.4.3. Volume of solids of revolution 317
7.4.4.* Volume of a solid whose area of parallel cross-sections is known . 331
7.4.5.* Arc length of a plane curve 333
7.4.6. Kinematics problems 338
7.5. IMPROPER INTEGRAL 351

7.6.* APPROXIMATE CALCULATION OF DEFINITE INTEGRALS 357

CHAPTER 8 FIRST ORDER DIFFERENTIAL EQUATIONS............369

8.1. SOME CONCEPTS OF FIRST ORDER DIFFERENTIAL EQUATIONS....................369

8.2. SEPARABLE FIRST ORDER DIFFERENTIAL EQUATIONS 371

8.3. HOMOGENEOUS DIFFERENTIAL EQUATIONS..381

8.4. FIRST ORDER LINEAR DIFFERENTIAL EQUATIONS......................................391

8.5. EULER'S METHOD 399

8.6.* SOLVING DIFFERENTIAL EQUATIONS USING PICARD'S METHOD.................412

CHAPTER 9 INFINITE SERIES...416

9.1. CONCEPT OF INFINITE SERIES 416

9.2. FUNDAMENTAL PROPERTIES OF INFINITE SERIES 420

9.3. SERIES OF NONNEGATIVE TERMS..426

9.4. SERIES OF ARBITRARY TERMS, ABSOLUTE CONVERGENCE.........................440

9.5. POWER SERIES...449

9.6. TAYLOR POLYNOMIAL, TAYLOR FORMULA, TAYLOR SERIES AND MACLAURIN SERIES 459

9.7. POWER SERIES EXPANSION OF FUNCTIONS...474

9.8. APPLICATIONS OF POWER SERIES..486

9.9. EXTENSION OF THE BINOMIAL THEOREM FOR ANY REAL NUMBER λ.........500

ANSWERS TO EXERCISES...509

This page is intentionally left blank.

Preface

The first chapter of this book gives the preparatory knowledge on which calculus depends. The second chapter explains the concept of limit and its definition from a dynamic point of view. With this basic knowledge, further study of calculus becomes concrete and straightforward.

We have written this calculus textbook to use in our teaching, and it has received high praise from students. Students found the concepts and theories in this book to be very clear and logical, and thus, learning calculus became very easy, which is easier than learning other mathematics courses. Students also achieved excellent grades in the study of calculus. The teaching effect is remarkable. It can be said that this book has completely changed the difficult situation of teaching calculus and made learning calculus easy.

Calculus is one of the outstanding achievements of the human intellect. It is a powerful tool for solving complex problems in mathematics, physics, economics, the social sciences, engineering, and the biological sciences, among other fields. The importance of calculus is obvious. Therefore, we have written this textbook in the hope that all calculus students will gain a true understanding and mastery of the subject and that it will provide strong support for further development in related fields in the future.

To make this textbook fully adaptable, it contains essential contents applicable to different high school curricula, as well as contents marked with * that can be selected by various high schools.

Spring Seeds

2024

This page is intentionally left blank.

Chapter 1 Preparatory knowledge

1.1. Some mathematical concepts

1. The multiplication property: any number multiplied by zero is equal to zero.

That is, $0 \cdot a = 0$

Cross-multiplying gives

$$\frac{0}{0} = a \quad \text{(any number)}$$

Further $a = \frac{0}{0} = 0 \cdot \frac{1}{0} = 0 \cdot \infty = \frac{1}{\infty} \cdot \infty = \frac{\infty}{\infty} = \frac{c-c}{0} = \frac{c}{0} - \frac{c}{0} = \infty - \infty$

Thus $a = \frac{0}{0} = 0 \cdot \infty = \frac{\infty}{\infty} = \infty - \infty$

$a \neq \pm\infty$ is defined. Its expression is called the '**indeterminate form**'.

where $\frac{c}{0} = \pm\infty$ (infinity), $c \in \mathrm{R}$, $c \neq 0$

Infinity ∞ is a dynamic infinite number. It is undefined. That is, $\infty \notin \mathrm{R}$. It follows that a fraction $\frac{c}{0}$ $(c \neq 0)$ is undefined when the denominator is equal to zero.

2. A property of intervals

In any interval $[a, b]$, regardless of its length, there are infinitely many points. That is, there are infinitely many values $\{x_1, x_2, \ldots\}$ within the interval, as shown in Figure 1.1.

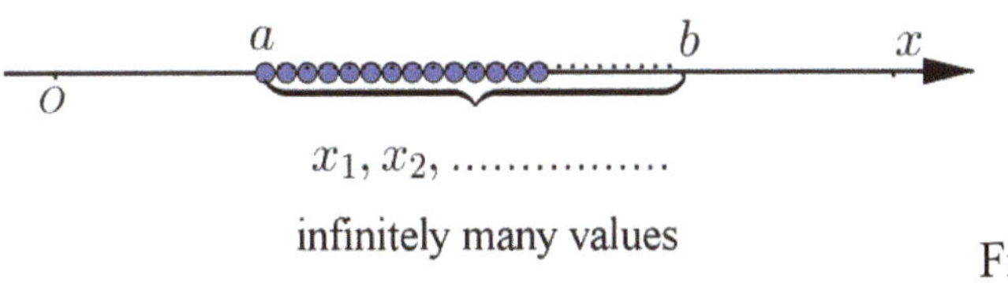

Figure 1.1

3. Geometric meaning of zero

(1)

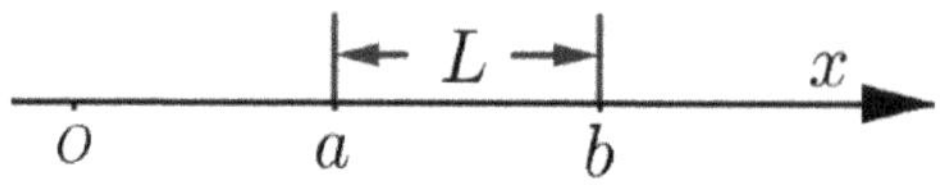

Figure 1.2

As shown in Figure 1.2, the length $L = b - a$, a, b are two points on the graph. When $b = a$

$$L = b - a = a - a = 0$$

As shown in Figure 1.3, $L = 0$ implies that two points coincide into one point. Hence, 0 represents the length of a single point geometrically. A point is a visible entity.

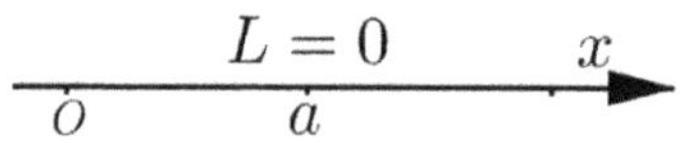

Figure 1.3

(2) As shown in Figure 1.4,

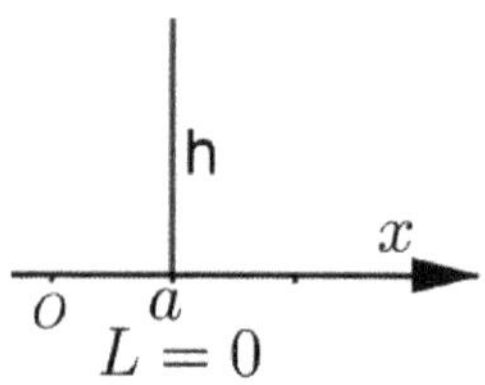

Figure 1.4

the area of the line segment is $\quad S = h \cdot L = h \cdot (a - a) = 0$

where h is the height, $L = (a - a) = 0$ represents the length of a single point.

Although the numerical value for the area of the line segment, S, is zero, the segment is a geometrically visible entity.

A thorough grasp of these properties is key to understanding calculus, which was originally created by combining geometric figures. On geometric figures, the shortest length is a point, the smallest area is a point or area of a line segment, and the smallest volume is a point or volume of a line segment. Although their numerical values are all zero, they are all geometrically visible entities.

1.2. Sets

I. Concept of sets

Sets are a fundamental concept in mathematics, playing a crucial role in modern mathematics.

We often study collections of objects, such as a class of students, a batch of products, or all positive integers. Such collections are regarded as sets.

Generally speaking, a set is a collection of all things possessing a common attribute, or a gathering of well-defined objects. The things or objects that constitute a set are called the **elements** of the set.

Here are some examples of sets:

Example 1. People born in Paris on May 1, 2025

Example 2. Colour TVs, refrigerators, computers

Example 3. The roots of the equation $x^2 - 7x + 12 = 0$

Example 4. All even numbers

Example 5. All points on the line $x + y - 1 = 0$

A set consisting of a finite number of elements is called a **finite set** (for example, Examples 1, 2, and 3). A set consisting of an infinite number of elements is called an **infinite set** (for example, Examples 4 and 5).

Generally, we can use uppercase letters A, B, C, ... to denote sets, and lowercase letters a, b, c, ... to denote set elements. If a is an element of set A, we write $a \in A$, which is read as 'a belongs to A' or 'a is in A'. If a is not an element of set A, we write $a \notin A$, read as 'a does not belong to A' or 'a is not in A'.

For example, if Q denotes the set of all rational numbers, then $0.8 \in Q$, $\sqrt{2} \notin Q$.

The sets we are discussing here are characterised by determinacy; that is, whether a given element belongs to a particular set is definite – it is either 'yes' or 'no', with no third possibility.

II. Representing sets

1. **Listing Method**: List all elements of the set in any order, enclosed in curly braces {}.

Example 1. The set A consisting of four elements a, b, c, and d can be represented as: $A = \{a, b, c, d\}$

Example 2. The set A consisting of the roots of $x^2 - 7x + 12 = 0$ can be represented as: $A = \{3, 4\}$

When representing a set using the listing method, all elements of the set must be listed; no elements may be omitted or repeated.

2. **Descriptive Method**: Let P(a) be a condition or rule pertaining to a, and A be the set of all elements a that satisfy P(a). Then it is denoted by

$$A = \{a \mid P(a)\}$$

Example 3. Let A be the set of roots of the equation $x^2 - 7x + 12 = 0$, which can be represented as: $A = \{x \mid x^2 - 7x + 12 = 0\}$

Example 4. Let A be the set of all even numbers, which can be represented as:

$$A = \{x \mid x = 2n, \text{n is an integer.}\}$$

A set and the relationships between sets can be represented graphically using **Venn diagrams**. A Venn diagram uses a simple region in the plane to represent a set, as shown in Figure 1.5. Elements within a set are represented by points within that region.

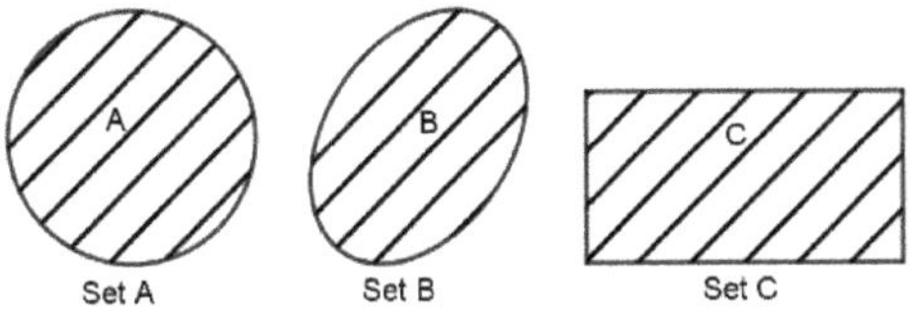

Figure 1.5

III. Universal set and empty set

The set consisting of all the things under study is called the **universal set**, denoted by U. The universal set is relative; a set may be universal under certain conditions but not under others. For example, if the context is limited to positive integers, then the set of all positive integers is the universal set. If the context includes both positive and negative integers, then the set of all positive integers is not the universal set. The set containing no elements is called the **empty set**, denoted by ∅.

Example 1. The set of real roots of the equation $x^2 + 1 = 0$ is the empty set ∅.

Example 2. The set of intersection points of two parallel lines in a plane is the empty set ∅.

Note: Neither {0} nor {∅} is an empty set; the former contains the element '0', while the latter has the empty set '∅' as its element.

IV. Subsets

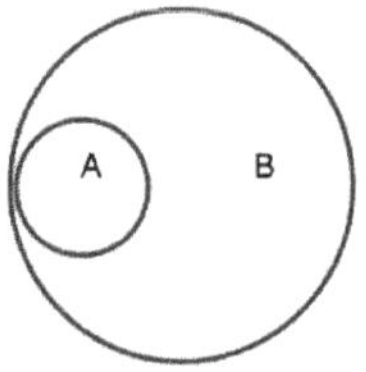

Figure 1.6

Definition 1.1

If every element of set A is also an element of set B – that is, if $a \in A$, then $a \in B$ – then A is called a **subset** of B, denoted by $A \subset B$ or $B \supset A$, and read as 'A is contained in B' or 'B contains A', as shown in Figure 1.6.

Example 1. Let N denote the set of all natural numbers and Q denote the set of all rational numbers. Then,

$$N \subset Q$$

Example 2. Let $A = \{1, 2, 3, 4, 5\}$ and $B = \{1, 3, 5\}$. Then,

$$B \subset A$$

Definition 1.2

Let A and B be sets. If $A \subset B$ and $B \subset A$, i.e., $A \subset B \subset A$, then A and B are said to be equal, denoted by $A = B$.

Example 3. Let $A = \{x \mid x \text{ is an integer greater than 1 and less than 4.}\} = \{2, 3\}$

$$B = \{x \mid x^2 - 5x + 6 = 0\} = \{2, 3\}$$

Then $A = B$

V. Operations on sets

We have previously introduced the concept of sets; we shall now define operations on sets. Like numerical operations, these are derived from practical applications and reflect the relationships between quantities in the real world.

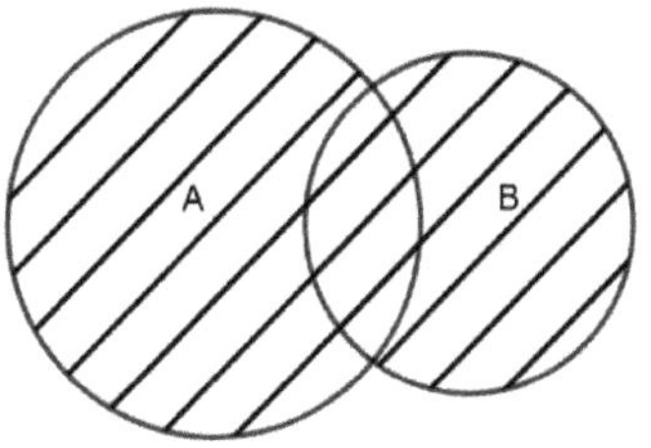

Figure 1.7. $A \cup B$

Definition 1.3

Let A and B be sets. The set consisting of all elements of A and B is called the **union** of A and B. It is denoted by $A \cup B$, as shown in Figure 1.7; that is, $A \cup B = \{x \mid x \in A \text{ or } x \in B\}$

The union of the sets has the following properties:

(1) $A \subset (A \cup B)$, $B \subset (A \cup B)$

(2) For any set A, the following hold.

$$A \cup \varnothing = A, \qquad A \cup U = U, \qquad A \cup A = A$$

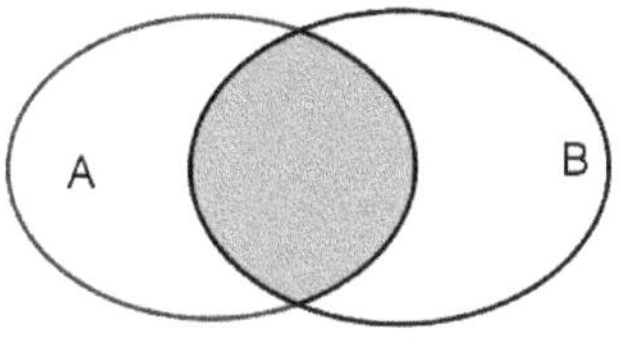

Figure 1.8. $A \cap B$

Definition 1.4

Let A and B be sets. The set consisting of all the common elements of A and B is called the **intersection** of A and B. It is denoted by $A \cap B$, as shown in the shaded region of Figure 1.8; that is, $A \cap B = \{x \mid x \in A \text{ and } x \in B\}$

The intersection of the sets has the following properties:

(1) $(A \cap B) \subset A$, $(A \cap B) \subset B$

(2) For any set A, the following hold.

$$A \cap \varnothing = \varnothing, \qquad A \cap U = A, \qquad A \cap A = A$$

Example 1. Let $A = \{1, 2, 3, 4\}$, $B = \{3, 4, 5, 6\}$. Then

$$A \cup B = \{1, 2, 3, 4, 5, 6\}, \qquad A \cap B = \{3, 4\}$$

Example 2. Let $A = \{a, b, c\}$, $B = \{a, b\}$. Then

$$A \cup B = \{a, b, c\}, \qquad A \cap B = \{a, b\}$$

Example 3. Let A be the set of people in a company who speak German, and B be the set of people who speak Japanese. Then

$A \cup B$ represents the set of people who speak either German or Japanese.

$A \cap B$ represents the set of people who speak both German and Japanese.

Example 4. Let $A = \{x \mid -1 \le x \le 1\}$, $B = \{x \mid x > 0\}$. Then

$$A \cup B = \{x \mid x \ge -1\}, \qquad A \cap B = \{x \mid 0 < x \le 1\}$$

Example 5. If A is the set of odd numbers and B is the set of even numbers, then

$$A \cup B = \{x \mid x \text{ is either odd or even}\}, \qquad A \cap B = \varnothing$$

If $A \cap B = \varnothing$, then A and B are said to be **disjoint**, as shown in Figure 1.9.

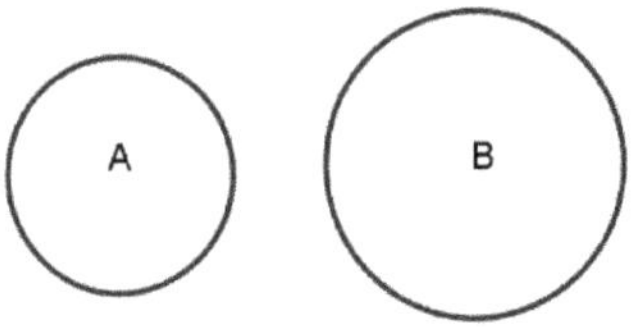

Figure 1.9

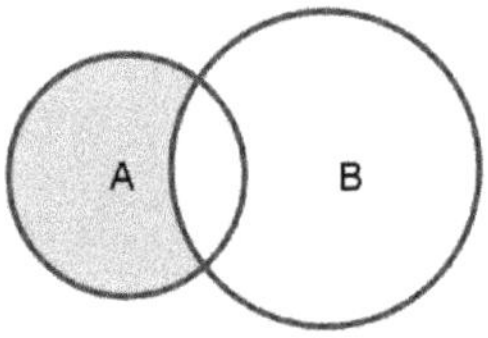

Figure 1.10

Definition 1.5

Let A and B be sets. The set of all elements that belong to A but not to B is called the **difference** between A and B, denoted by $A - B$ or $A \backslash B$, as shown in the shaded region of Figure 1.10; that is,

$$A - B = \{x \mid x \in A \text{ and } x \notin B\}$$

Example 6. If $A = \{1, 2, 3, 4\}$, $B = \{1, 3, 5, 7\}$, then

$$A - B = \{2, 4\}$$

Definition 1.6

The set of all elements in the universal set U that are not elements of A is called the **complement** of A, denoted by A' or $\overline{A}$, as shown in Figure 1.11; that is,

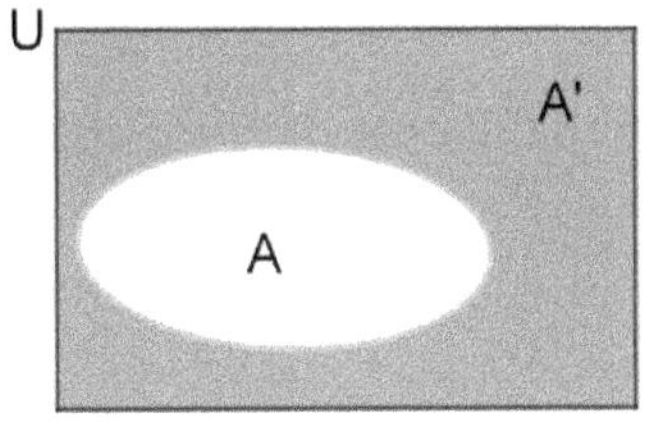

Figure 1.11

$A' = \{x \mid x \in U \text{ and } x \notin A\}$

Properties of the complement:

$A \cup A' = U$, $\quad A \cap A' = \varnothing$

Example 7. Let the set of students taking the exam be the universal set U. If A represents the set of students who passed, then A' represents the set of students who failed. $\quad A \cup A' = U$

VI. Laws of set operations (proof omitted)

(1) Commutative Law (i) $A \cup B = B \cup A$

(ii) $A \cap B = B \cap A$

(2) Associative Law (i) $(A \cup B) \cup C = A \cup (B \cup C)$

(ii) $(A \cap B) \cap C = A \cap (B \cap C)$

(3) Distributive Law (i) $(A \cup B) \cap C = (A \cap C) \cup (B \cap C)$

(ii) $(A \cap B) \cup C = (A \cup C) \cap (B \cup C)$

(4) De Morgan's Law (i) $(A \cup B)' = A' \cap B'$

(ii) $(A \cap B)' = A' \cup B'$

Example 1. Suppose students' exam grades are divided into four categories: excellent, good, average, and poor. Let A be the set of students with an excellent grade and B be the set of students with a good grade. Verify that De Morgan's Law (i) $(A \cup B)' = A' \cap B'$ holds.

Solution. Let $A \cup B$ be the set of students with an excellent or good grade. Therefore, $(A \cup B)'$ is the set of students with an average or poor grade. Let A' be the set of students with a good, average, or poor grade. Let B' be the set of students with an excellent, average, or poor grade. Therefore, A' ∩ B' is the set of students with an average or poor grade. Hence, (A ∪ B)' = A' ∩ B'.

Example 2. Prove using the set operation laws that

$$(A \cap B) \cup (A' \cap B) = B$$

Proof. By the distributive law (i), we have

$$(A \cap B) \cup (A' \cap B) = (A \cup A') \cap B = B$$

VII. Cartesian product of sets

The elements of a set do not involve any consideration of order. For example, {a, b} and {b, a} refer to the same set; that is, {a, b} = {b, a}. However, sometimes it is necessary to study cases where elements must be arranged according to a prescribed order. Arranging two elements x and y in a sequential order to form the pair (x, y) is called an ordered set of elements. (x, y) and (y, x) are different ordered sets of elements; that is, $(x, y) \neq (y, x)$.

For the ordered sets of elements (x_1, y_1) and (x_2, y_2), they are said to be equal if and only if $x_1 = x_2$ and $y_1 = y_2$.

An ordered set consisting of two elements, (x_1, x_2), is called an ordered pair. An ordered set consisting of three elements, (x_1, x_2, x_3), is called an ordered triple. An ordered set consisting of n elements, $(x_1, x_2, \ldots, x_n)$, is called an ordered n-tuple.

Definition 1.7

Let A and B be sets. For $x \in A$ and $y \in B$, the set of all ordered pairs (x, y) is called the Cartesian product of the sets A and B, denoted by $A \times B$, i.e.

$$A \times B = \{(x, y) \mid x \in A, y \in B\}$$

Example 1. Let $A = \{1, 2, 3, 4\}$, $B = \{2, 3\}$, then

$$A \times B = \{(1, 2), (1, 3), (2, 2), (2, 3), (3, 2), (3, 3), (4, 2), (4, 3)\}$$

Example 2. Let $A = \{a, b\}$. Then

$$A \times A = \{(a, a), (a, b), (b, a), (b, b)\}$$

Example 3. Let R be the set of all real numbers. Then the coordinate plane of the Cartesian coordinate system can be denoted by

$$R \times R = \{(x, y) \mid x \in R, y \in R\}$$

Example 4. Let $A = \{x \mid 0 \leq x \leq 2\}$, $B = \{y \mid 0 \leq y \leq 1\}$, then

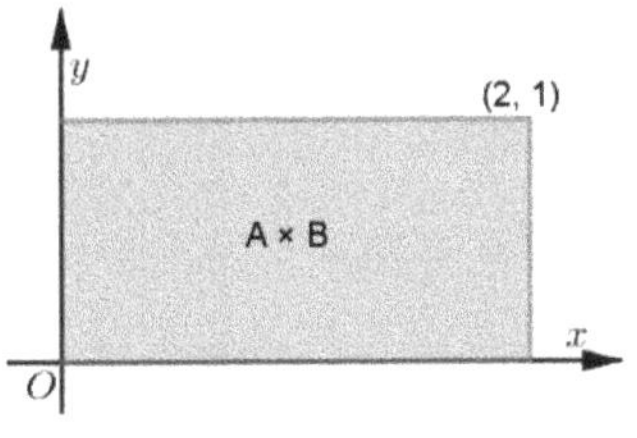

Figure 1.12

$$A \times B = \{(x, y) \mid 0 \leq x \leq 2,\ 0 \leq y \leq 1\}$$

It represents the rectangular region shown in Figure 1.12 in the Cartesian coordinate system.

Similarly, we can define

$$A \times B \times C = \{(x, y, z) \mid x \in A, y \in B, z \in C\}$$

Example 5. Let $A = \{0, 1\}$, $B = \{1, 2\}$, $C = \{3\}$, then we have

$$A \times B \times C = \{(0, 1, 3), (0, 2, 3), (1, 1, 3), (1, 2, 3)\}$$

1.3. Set of real numbers

I. Real numbers and the number line

Human understanding of numbers has developed gradually. First came the natural numbers, which then expanded to the rational numbers (i.e. positive and negative integers, positive and negative fractions, and zero). Further advancement led to the irrational numbers (for example, $\sqrt{2}$ and π are both irrational numbers). Rational numbers can be expressed in the form $\frac{p}{q}$, while irrational numbers cannot be expressed in the form $\frac{p}{q}$, where p and q are integers, and $q \neq 0$.

Fractions can be represented as terminating decimals or infinite recurring decimals. Conversely, terminating decimals or infinite recurring decimals can also be represented as fractions. In contrast, irrational numbers are infinite non-recurring decimals and cannot be represented as fractions.

Consider a horizontal straight line. On this line, choose a point O, called the origin O. Specify a positive direction (conventionally taken to the right of the origin O).

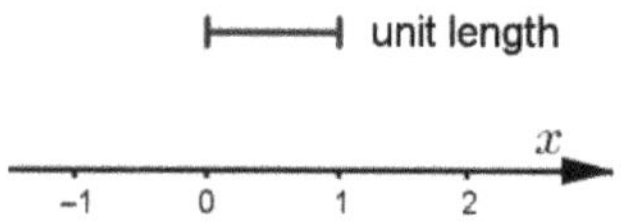

Figure 1.13

Next, define a length equal to 1, called the unit length. This straight line with an origin, a positive direction, and a unit length is called the number line, as shown in Figure 1.13.

For any rational number $\frac{p}{q}$, there is a corresponding point on the number line such that the ratio of the distance from the origin to this point to the unit length is equal to $\frac{p}{q}$. This point is called a rational point. It is the geometric representation of a rational number. $\frac{p}{q}$ is called the coordinate of this rational point. Conversely, every rational point on the number line corresponds to a rational number.

Given any two rational numbers a and b (a < b), there exists at least one rational number c between a and b such that $a < c < b$, for example, $c = \frac{a+b}{2}$. Similarly, there exists at least one rational number d between a and c such that $a < d < c$. By deduction, no matter how small the difference between the rational numbers a and b, there will always be infinitely many rational numbers between a and b. This is the density property of rational numbers. Because every rational number must correspond to a rational point on the number line, there are always infinitely many rational points between any two rational points on the number line. That is, the rational points are everywhere dense on the number line.

Although rational points are everywhere dense on the number line, they do not completely fill the line. For example, a square with a side length of one unit has a diagonal length of $\sqrt{2}$ units. It can be proven that $\sqrt{2}$ is not a rational number. Therefore, a point with the coordinate $\sqrt{2}$ on the number line is not a rational point. There are infinitely many such points and they are also everywhere dense on the number line. For example, points with coordinates $\sqrt{2}+1$, $\sqrt{2}+0.1$, $\sqrt{3}$, $\sqrt[3]{7}$, π are not rational points. Therefore, besides rational points, there are infinitely many 'gaps' on the number line. The points in these gaps are called irrational points, and the numbers corresponding to irrational points are called irrational numbers.
Rational and irrational numbers are collectively called real numbers. Real numbers fill the number line completely, leaving no gaps. This is known as the continuity of real numbers. Consequently, every real number must be the coordinate of a point on the number line. Conversely, the coordinate of every point on the number line represents a real number. This means that there is a one-to-one correspondence between all real numbers and all points on the number line. From now on, all numbers we study are real numbers. For simplicity, we often do not distinguish between a real number and its corresponding point on the number line; that is, point a and real number a are considered identical.

II. Absolute value

We often use the concept of the absolute value of real numbers when studying various problems. The definition and properties of the absolute value of real numbers are introduced below.

Definition 1.8

The absolute value of a real number x, denoted by $|x|$, is defined as

$$|x| = \begin{cases} x, & x \geq 0 \\ -x, & x < 0 \end{cases}$$

The geometric meaning of $|x|$: $|x|$ represents the distance between the point x on the number line (regardless of whether x is to the left or right of the origin) and the origin O.

The absolute value and its operations have the following properties (proof omitted).

(1) $|x|^2 = x^2$ (2) $|x| \geq 0$ (3) $|-x| = |x|$ (4) $-|x| \leq x \leq |x|$

(5) If $\mathrm{a} > 0$, then the following two sets are equal.

$$\{x \mid |x| < \mathrm{a}\} = \{x \mid -\mathrm{a} < x < \mathrm{a}\}$$

where the inequality is: $|x| < \mathrm{a} \Leftrightarrow -\mathrm{a} < x < \mathrm{a}$

(6) If $\mathrm{b} > 0$, then the following two sets are equal.

$$\{x \mid |x| > \mathrm{b}\} = \{x \mid x < -\mathrm{b}\} \cup \{x \mid x > \mathrm{b}\}$$

where the inequality is: $|x| > \mathrm{b} \Leftrightarrow x < -\mathrm{b}$ or $x > \mathrm{b}$

(7) $|x + y| \leq |x| + |y|$ (8) $|x - y| \geq |x| - |y|$

(9) $|xy| = |x| \cdot |y|$ (10) $\left|\dfrac{x}{y}\right| = \dfrac{|x|}{|y|}$, $y \neq 0$

III. Intervals

Let a and b be real numbers and a < b.

Figure 1.14

(1) The set of all real numbers satisfying the inequality $a < x < b$ is called an **open interval** with endpoints a and b, denoted as (a, b) or]a, b[, as shown in Figure 1.14. That is, $(a, b) = \{x \mid a < x < b\}$

Figure 1.15

(2) The set of all real numbers satisfying the inequality $a \le x \le b$ is called a **closed interval** with endpoints a and b, denoted as [a, b], as shown in Figure 1.15. That is, $[a, b] = \{x \mid a \le x \le b\}$

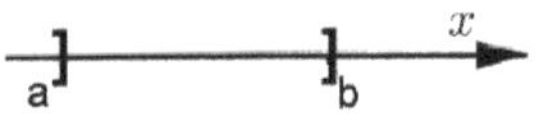

Figure 1.16

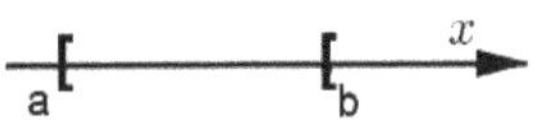

Figure 1.17

(3) The set of all real numbers satisfying the inequality $a < x \le b$ (or $a \le x < b$) is called a **half-open interval** with endpoints a and b, denoted]a, b] (or [a, b[), as shown in Figures 1.16 and 1.17, respectively. That is,

$$]a, b] = \{x \mid a < x \le b\}$$

$$[a, b[= \{x \mid a \le x < b\}$$

The above three types of intervals are finite intervals. The difference between the right endpoint b and the left endpoint a of a finite interval, $\Delta x = b - a$, is called the **length** of the interval. There are also the following types of infinite intervals.

(4) $(a, +\infty) = \{x \mid a < x\}$, $[a, +\infty[= \{x \mid a \le x\}$

(5) $(-\infty, b) = \{x \mid x < b\}$, $]-\infty, b] = \{x \mid x \le b\}$

(6) $(-\infty, +\infty) = \{x \mid -\infty < x < +\infty\}$

IV. Neighbourhood

By the properties of absolute value, the set of real numbers

$$\{x \mid |x - x_0| < \delta ,\ \delta > 0\} = \{x \mid x_0 - \delta < x < x_0 + \delta ,\ \delta > 0\}$$

where the inequality is: $|x - x_0| < \delta \quad \Rightarrow \quad -\delta < x - x_0 < \delta \quad \Rightarrow \quad x_0 - \delta < x < x_0 + \delta$

on the number line is an open interval, $(x_0 - \delta, x_0 + \delta)$, centred at the point x_0 with length 2δ. This is called the **neighbourhood** of the point x_0, where x_0 is called the **centre** of the neighbourhood, and δ is called the **radius** of the neighbourhood, as shown in Figure 1.18. For example, for $|x - 5| < 0.5$, it is the neighbourhood centred at the point $x_0 = 5$ with a radius of 0.5, which is the open interval (4.5, 5.5).

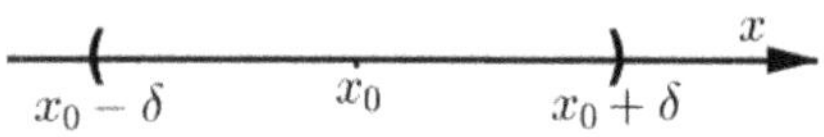

Figure 1.18

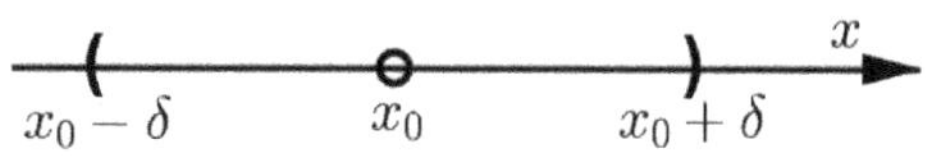

Figure 1.19

The following sets are also frequently used in calculus:

$$\{x \mid 0 < |x - x_0| < \delta ,\ \delta > 0\} = \{x \mid x_0 - \delta < x < x_0 + \delta ,\ x \neq x_0 ,\ \delta > 0\}$$

where the inequality is:

$$0 < |x - x_0| < \delta \quad \Rightarrow \quad 0 < |x - x_0| ,\ |x - x_0| < \delta \quad \Rightarrow \quad x \neq x_0 ,\ x_0 - \delta < x < x_0 + \delta$$

Here, removing the point x_0 from the neighbourhood of the point x_0, i.e., $x \neq x_0$, the set of remaining points, namely the set $(x_0 - \delta, x_0) \cup (x_0, x_0 + \delta)$, is called the **hollow neighbourhood** centred at x_0 with radius δ, as shown in Figure 1.19. For example, $0 < |x - 1| < 2$, which is the hollow neighbourhood centred at $x_0 = 1$ with radius 2:

$$(-1, 1) \cup (1, 3)$$

1.4. Functions

I. Concept of a function

Elementary mathematics studies primarily invariant quantities, whereas advanced mathematics studies varying quantities. Variables and functions are the fundamental objects of study in calculus.

1. Definition 1.9

If D is a non-empty set of real numbers, and there exists a rule of correspondence f such that for every $x \in D$, there is a definite real number y corresponding to it, then y is called a function of x, denoted by

$$y = f(x), \qquad x \in D$$

where the variable x is called the **independent variable**, and the variable y is called the **dependent variable** or function. The dependency relationship between the dependent variable y and the independent variable x is usually called a functional relationship. The set D (i.e., the set of values for the independent variable x) is called the **domain** of the function, which can also be denoted as D(f). The set of all function values is called the **range** of the function f(x), denoted by Z or Z(f), i.e.,

$$Z = \{y \mid y = f(x),\ x \in D\}$$

The notation for a function can be chosen arbitrarily. Besides the commonly used 'f', other English or Greek letters can be used, such as φ, h, g and F. The function would then be written as $\varphi(x)$, h(x), g(x), F(x), etc. Sometimes, the notation for the dependent variable is used directly to represent the function, i.e., the function is written as $y = y(x)$. In this case, y on the left side of the equal sign represents the function value, and y on the right side represents the rule of correspondence.

2. Representing the domain D and range Z of a function

Generally, there are three methods to represent the domain D and range Z of a function.

(1) Set notation

$$D = \{x \mid f(x)\}, \qquad Z = \{y \mid y = f(x),\ x \in D\}$$

(2) Interval notation

$$D = [a, b] \qquad \text{or} \qquad x \in [a, b]$$

If the domain is the set of all real numbers R, then

$$D = R = (-\infty, +\infty) \qquad \text{or} \qquad x \in R$$

(3) Inequality notation

$$D = \{x \mid a \le x \le b\} \qquad \text{or} \qquad a \le x \le b$$

In secondary school mathematics, we became familiar with many functional relationships. To study a function, we must know the rule of correspondence between the independent and dependent variables, as well as the function's domain; otherwise, the function is not determined. However, by convention, we often only provide the rule of correspondence without specifying the domain. For example, given $y = \sqrt{25 - x^2}$, it is clear that for every $x \in [-5, 5]$, there is a definite real number y that corresponds to it. Therefore, the domain of the function is $D = [-5, 5]$.

Example 1. Determine the domain of the function $y = \dfrac{1}{\log(3x-2)}$.

Solution. The function can only take a definite real number when $3x - 2 > 0$ and $3x - 2 \ne 1$, i.e. $x > \dfrac{2}{3}$ and $x \ne 1$. Therefore, the domain of the function

$y = \dfrac{1}{\log(3x-2)}$ is $D = (\dfrac{2}{3}, 1) \cup (1, +\infty)$.

Example 2. Determine the domain of the function $y = \arcsin\dfrac{x-1}{5} + \dfrac{1}{\sqrt{25-x^2}}$.

Solution. $\left|\frac{x-1}{5}\right| \le 1$ and $x^2 < 25$

That is, $|x-1| \le 5$ and $|x| < 5$

$-4 \le x \le 6$ and $-5 < x < 5$

Hence, $-4 \le x < 5$

Therefore, the domain of the given function is D = [-4, 5[.

3. Multivalued functions

The definition of a function requires that for every $x \in$ D, there is a definite y-value corresponding to it. However, we often encounter another type of relationship, for example, $y = \pm\sqrt{25 - x^2}$, where for every $x \in [-5, 5]$, there are two corresponding y-values. This does not conform to the previous function definition. It should be said that it is not a functional relationship. But for convenience, we call the relationship where multiple y-values correspond to each x-value in a non-empty set D a **multivalued function**. Consequently, the functional relationship in the previous function definition can be called a **single-valued function**. For example, for the multivalued function $y = \pm\sqrt{25 - x^2}$, we can divide it into two single-valued functions, $y = \sqrt{25 - x^2}$ and $y = -\sqrt{25 - x^2}$, as shown in Figure 1.20.

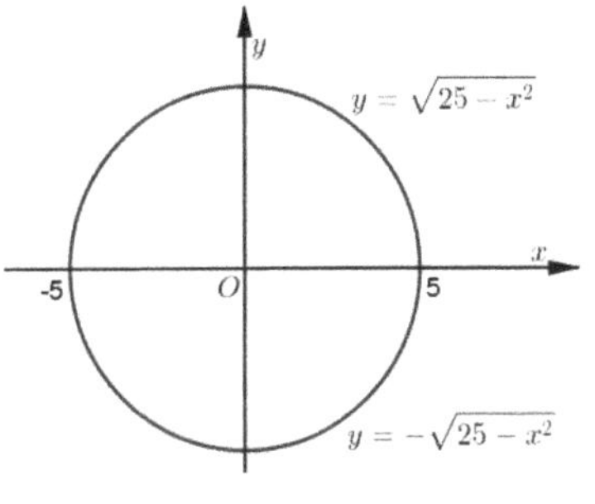

Figure 1.20

Graphically, they represent the two single-valued functions above and below the x-axis, respectively.

4. Use of function notation f(x)

In the function f(x), 'f' represents the rule of correspondence in the functional relationship. For each $x \in D$, according to the rule f, there is a definite y value corresponding to it. We can replace x in f(x) with a specific numerical value in D, a letter representing a numerical value, or a mathematical expression. For example, let $f(x) = x^2$, then we have:

$$f(2) = 2^2 = 4, \qquad f(a) = a^2, \qquad f(x+1) = (x+1)^2 = x^2 + 2x + 1$$

$$f(-\frac{1}{y}) = (-\frac{1}{y})^2 = \frac{1}{y^2}, \qquad f[f(x)] = (x^2)^2 = x^4$$

II. Representing functions

1. Three common methods for representing functions: analytic method, tabular method, and graphical method

(1) Analytic method

An equation is used to express that one variable is a function of another variable. This equation is called the analytic expression (or functional relationship) of the function. For example, $y = \sqrt{x+1}$, $S = \pi r^2$, $V = \frac{4}{3}\pi r^3$ etc.

(2) Tabular method

Tables can be used to represent one variable as a function of another. For example, the table below shows the functional relationship between the lake depth h and water volume V.

Lake depth h meter	0	5	10	15	20	30	35
Water volume V (10^4 m^3)	0	20	40	90	275	437.5	650

(3) Graphical Method

By taking a value of the independent variable x and the corresponding value of the function y as the x-coordinate and y-coordinate of a point, respectively, a point can be plotted in a Cartesian coordinate system. The set of all these points is called the graph of the function. The graphical method uses graphs to represent one variable as a function of another variable.

For example, a weather station uses an automatic temperature recording device to plot a curve of temperature changes throughout the day, as shown in Figure 1.21. This graph vividly illustrates the relationship between the variables y and x. Through this graph, for each specific time value x, there is a corresponding specific temperature value y. For example, when $x = 4$ hours, $y = 2°C$; when $x = 14$ hours, $y = 12°C$.

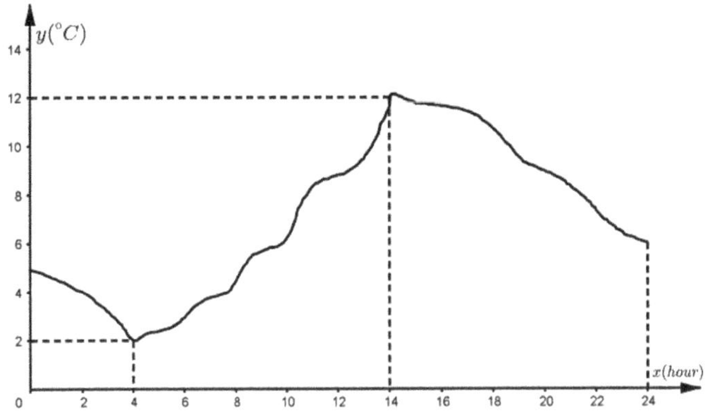

Figure 1.21

The three methods described above can be used in combination. However, in some scientific experiments or measurements, although we may obtain data, we may not be able to express the functional relationship analytically, as in the examples above. In such cases, we can only use tables or graphs to record the obtained data. Therefore, in practice, the tabular and graphical methods have significant practical significance. We should keep this in mind.

2. Graph of a function

In a Cartesian coordinate system, the independent variable varies along the horizontal axis, and the dependent variable varies along the vertical axis. The set of points in the plane

$$\{(x, y) \mid y = f(x),\ x \in D\}$$

is then the graph of the function $y = f(x)$ defined on D.

Example. Sketch the graph of the function $y = \frac{1}{8}x^3$.

Solution. Step 1. Tabulate the values. Take some values within the domain of x, calculate the corresponding y values, and list them in the table below.

x	…	-4	-3	-2	-1	0	1	2	3	4	…
$y = \frac{1}{8}x^3$	…	-8	-3.38	-1	-0.13	0	0.13	1	3.38	8	…

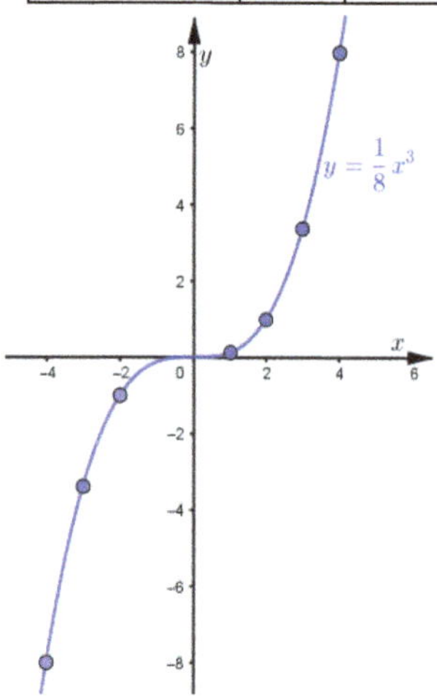

Figure 1.22

Step 2. Plot the points in the coordinate system based on the corresponding values in the table.

Step 3. Connect the plotted points with a smooth curve in the increasing order of the independent variable to form the graph of the function $y = \frac{1}{8}x^3$ (Figure 1.22)

3. Piecewise functions

Some functions, for different values of the independent variable x within their domain, cannot be represented by a single mathematical expression, but require two or more expressions. These functions are called **piecewise functions**. For example,

$$y = |x| = \begin{cases} x, & x \geq 0 \\ -x, & x < 0 \end{cases} \quad \text{(Figure 1.23)} \qquad y = \begin{cases} x+2, & x > 2 \\ 2x-4, & x \leq 2 \end{cases} \quad \text{(Figure 1.24)}$$

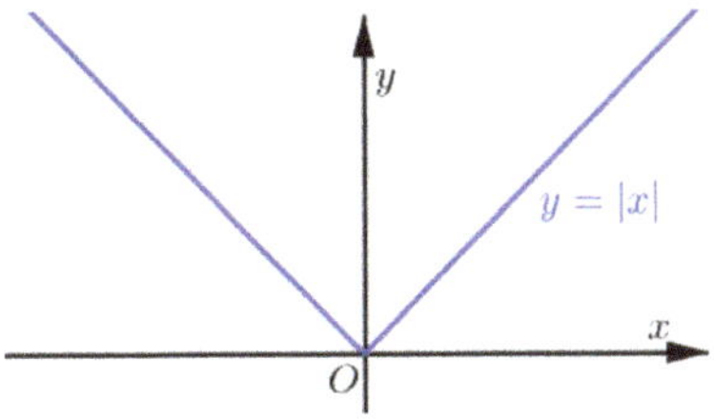

Figure 1.23

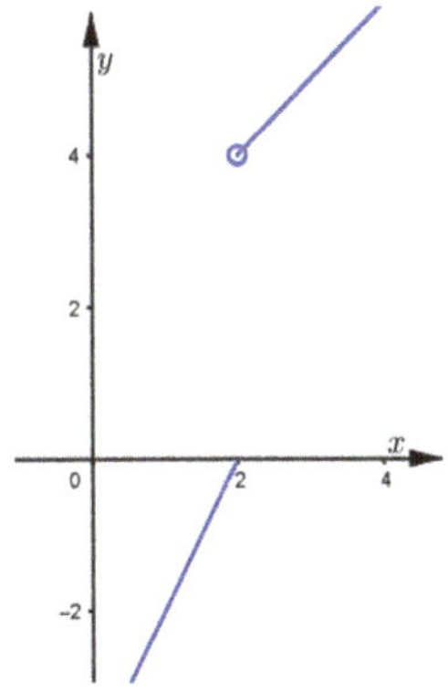

Figure 1.24

Note: A piecewise function uses multiple formulae combined to represent a single function, not several separate functions.

To evaluate such a function, we first determine the interval to which the x-value belongs, then identify the corresponding formula for that interval, and finally substitute the x-value into it to obtain the function value.

For example, $$f(x) = \begin{cases} x+3, & -3 \leq x < -1 \\ 5, & -1 \leq x \leq 1 \\ \sqrt{x}, & 1 < x < 9 \end{cases}$$

If we want to evaluate f(4), f(0), f(-2), and f(9), the steps are as follows:

f(4) represents the value of the function when $x = 4$. Since the x-value lies within the interval $1 < x < 9$, we substitute $x = 4$ into the formula $\sqrt{x}$. Hence, f(4) = $\sqrt{4} = 2$.

f(0) represents the value of the function when $x = 0$. Since the x-value lies within the interval $-1 \leq x \leq 1$, Therefore, f(0) = 5.

f(-2) represents the value of the function when $x = -2$. Since the x-value lies within the interval $-3 \leq x < -1$, we substitute $x = -2$ into the formula $x + 3$. Hence, f(-2) = 1.

f(9) represents the value of the function when $x = 9$. However, none of the three formulae correspond to this x-value. Therefore, f(9) is undefined, or in other words, f(9) does not exist.

To sketch the graph of a piecewise function, we need to plot each section separately. Take care to distinguish the difference between 'open' and 'closed' endpoints. The graph of this function is shown in Figure 1.25.

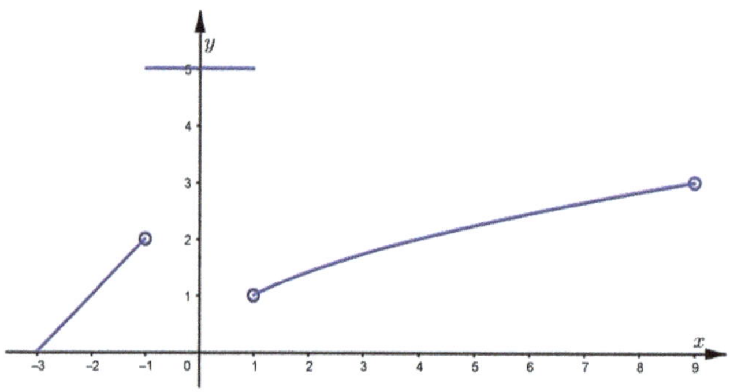

Figure 1.25

4. Implicit functions

Some functions have their dependent variable directly expressed by an expression of the independent variable; these functions are called **explicit functions**. Examples include $y = x^2$, $y = \sqrt{a^2 - x^2}$, $y = \log_a(3x+1)$, and so on. However, for some functions, the rule of correspondence between the dependent variable and the independent variable is expressed by $F(x, y) = 0$; these functions are called **implicit functions**. Examples include $Ax + By + C = 0$, $xy = 1$, $x^2 + y^2 = r^2$, etc.

III. Several simple properties of functions

1. Parity of functions

Definition 1.10

Given a function $y = f(x)$

• if, for all $x \in D$, $f(-x) = f(x)$, then $f(x)$ is called an **even function**;

• if, for all $x \in D$, $f(-x) = -f(x)$, then $f(x)$ is called an **odd function**.

For an even function, since $f(-x) = f(x)$, if the point $P(x, f(x))$ lies on the graph of the function, then the point $P'(-x, f(x))$, which is symmetrical to P about the y-axis, also lies on the graph of the function. Therefore, the graph of an even function is symmetrical about the y-axis, as shown in Figure 1.26.

For an odd function, since $f(-x) = -f(x)$, if the point $Q(x, f(x))$ lies on the graph of the function, then the point $Q'(-x, -f(x))$, which is symmetrical to Q about the origin, also lies on the graph of the function. Therefore, the graph of an odd function is symmetrical about the origin, as shown in Figure 1.27.

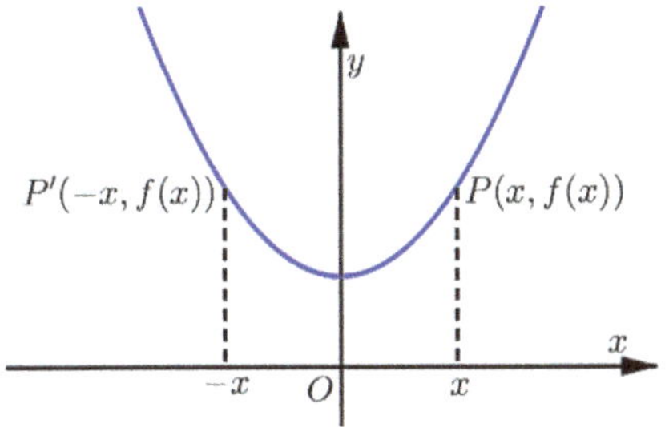

Figure 1.26

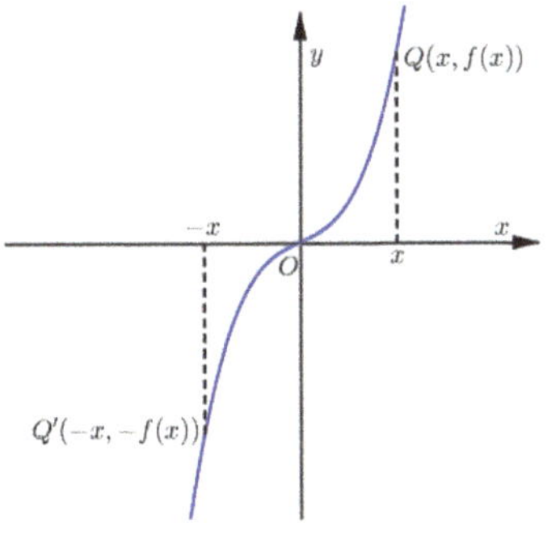

Figure 1.27

Example 1. Determine the parity of $y = x^4 - 2x^2$.

Solution. Since $f(-x) = (-x)^4 - 2(-x)^2 = x^4 - 2x^2 = f(x)$,

$y = x^4 - 2x^2$ is an even function, as shown in Figure 1.28.

Example 2. Determine the parity of $y = \dfrac{1}{x}$.

Solution. Since $f(-x) = \frac{1}{-x} = -\frac{1}{x} = -f(x)$,

$y = \frac{1}{x}$ is an odd function, as shown in Figure 1.29.

Example 3. Determine the parity of $y = x^3 + 1$.

Solution. Since $f(-x) = (-x)^3 + 1 = -x^3 + 1$, which is neither equal to $f(x) = x^3 + 1$ nor to $-f(x) = -x^3 - 1$, the function $y = x^3 + 1$ is neither an even function nor an odd function, as shown in Figure 1.30.

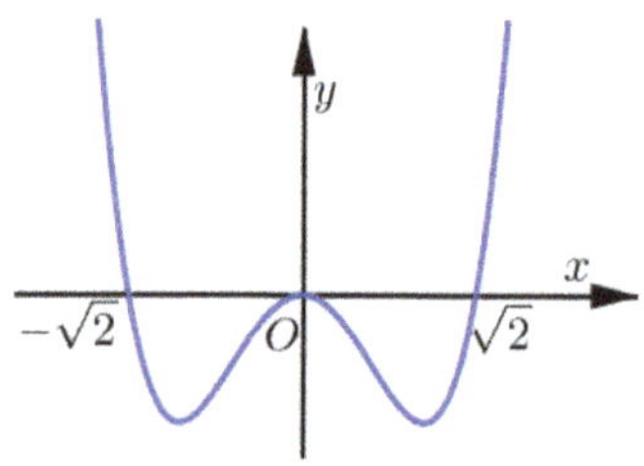

Figure 1.28

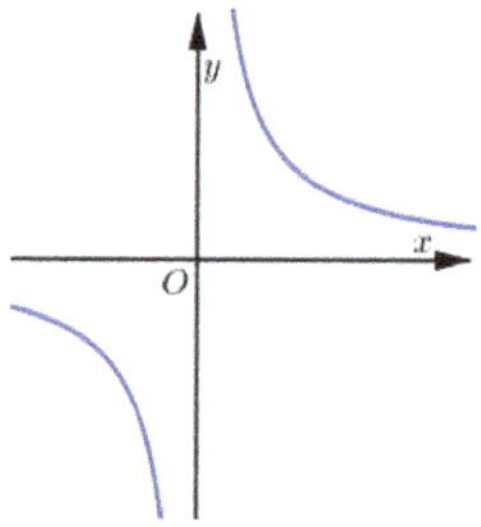

Figure 1.29

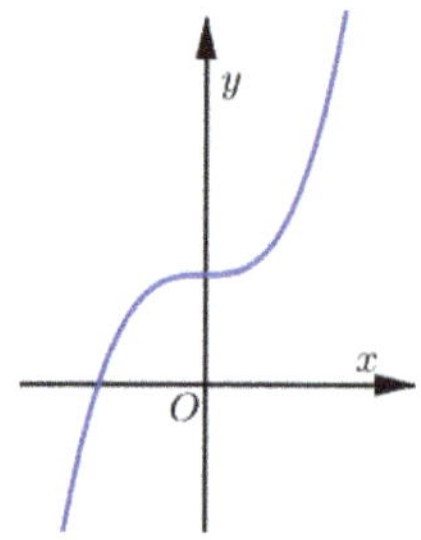

Figure 1.30

2. Periodicity of functions

Definition 1.11

For a function $y = f(x)$, if there exists a positive number a such that

$$f(x) = f(x + a)$$

always holds, then $f(x)$ is called a **periodic function**. The smallest positive number a that satisfies this equation is called the **period** of the function.

For example, $y = \sin x$ is a periodic function with a period of 2π.

3. Monotonicity of functions

Definition 1.12

If for any two points x_1 and x_2 in the interval (a, b), when $x_1 < x_2$, $f(x_1) < f(x_2)$, then the function is said to be monotonically increasing on the interval (a, b). If when $x_1 < x_2$, $f(x_1) > f(x_2)$, then the function is said to be monotonically decreasing on the interval (a, b).

The graph of a monotonically increasing function gradually rises along the positive x-axis, as shown in Figure 1.31. The graph of a monotonically decreasing function gradually falls along the positive x-axis, as shown in Figure 1.32.

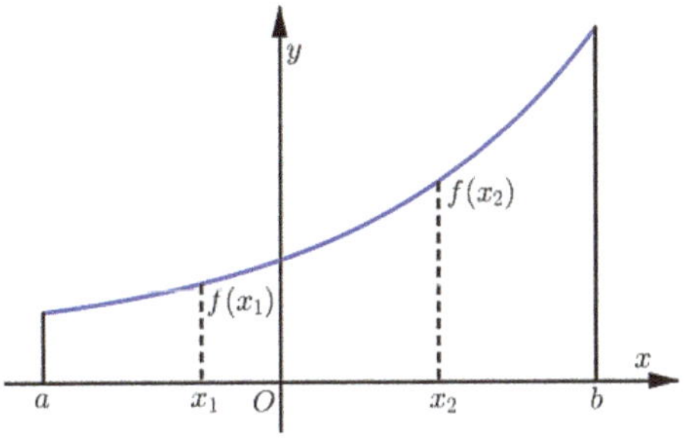

Figure 1.31

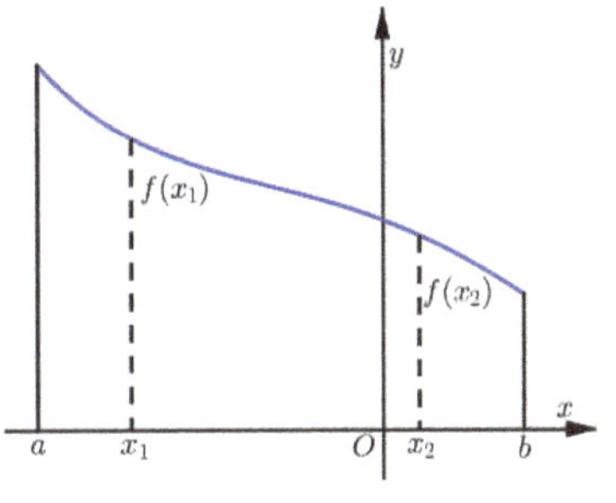

Figure 1.32

Example 1. Determine the monotonicity of the function $y = x^3$.

Solution. For any x_1 and x_2, if $x_1 < x_2$, then $x_1^3 < x_2^3$. Therefore, $f(x_1) < f(x_2)$, so $y = x^3$ is monotonically increasing on (-∞, +∞).

Example 2. Determine the monotonicity of the function $y = 2x^2 + 1$.

Solution. For any x_1 and x_2, we have

$$f(x_2) - f(x_1) = (2x_2^2 + 1) - (2x_1^2 + 1) = 2(x_2^2 - x_1^2)$$

On the interval]-∞, 0], if $x_1 < x_2$, then $x_1^2 > x_2^2$, thus $2(x_2^2 - x_1^2) < 0$. Therefore, $f(x_2) - f(x_1) < 0$, i.e., $f(x_2) < f(x_1)$. Hence, $y = 2x^2 + 1$ is monotonically decreasing on $]-\infty, 0]$.

On the interval $[0, +\infty[$, if $x_1 < x_2$, then $x_1^2 < x_2^2$, thus $2(x_2^2 - x_1^2) > 0$. Therefore, $f(x_2) - f(x_1) > 0$, i.e., $f(x_2) > f(x_1)$. Hence, $y = 2x^2 + 1$ is monotonically increasing on $[0, +\infty[$.

Therefore, on $(-\infty, +\infty)$, the function $y = 2x^2 + 1$ is not a monotonic function.

4. Boundedness of functions

Definition 1.13

Let the function $y = f(x)$ be defined on the interval (a, b). If there exists a positive number M such that $| f(x) | \le M$ for all $x \in (a, b)$, then the function $f(x)$ is said to be **bounded** on (a, b). If no such positive number M exists, then the function $f(x)$ is said to be **unbounded** on (a, b).

For example, the function $y = \sin x$ is bounded on $(-\infty, +\infty)$ because $| \sin x | \le 1$ for any real number x. The function $y = \frac{1}{x}$ is unbounded on (0, 2) and bounded on $[1, +\infty[$.

IV. Inverse functions and composite functions

1. Inverse functions

Definition 1.14

Let $y = f(x)$ be a function defined on D with a range Z(f). If for every $y \in Z(f)$, there is a definite $x \in D$ that satisfies $y = f(x)$, defined by the rule of correspondence f^{-1}, then this function $x = f^{-1}(y)$ defined on Z(f) is called the inverse function of $y = f(x)$, or they are said to be inverse functions of each other.

For the function $y = \mathrm{f}(x)$, x is the independent variable, y is the dependent variable, the domain is D, and the range is Z(f).

For the function $x = \mathrm{f}^{-1}(y)$, y is the independent variable, x is the dependent variable, the domain is Z(f), and the range is D.

Conventionally, we use x to represent the independent variable and y to represent the dependent variable. Therefore, we rewrite $x = \mathrm{f}^{-1}(y)$ as a function with x as the independent variable and y as the dependent variable: $y = \mathrm{f}^{-1}(x)$. In this case, we say that $y = \mathrm{f}^{-1}(x)$ is the inverse function of $y = \mathrm{f}(x)$.

The relationship between $y = \mathrm{f}(x)$ and $y = \mathrm{f}^{-1}(x)$ is that x and y are interchanged, so their graphs are symmetrical about the line $y = x$, as shown in Figure 1.33.

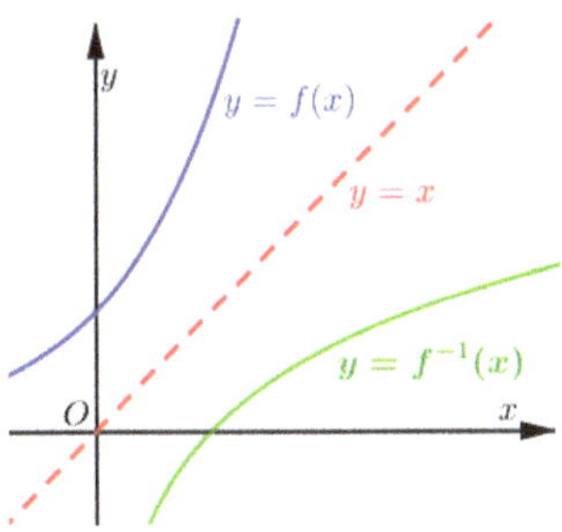

Figure 1.33

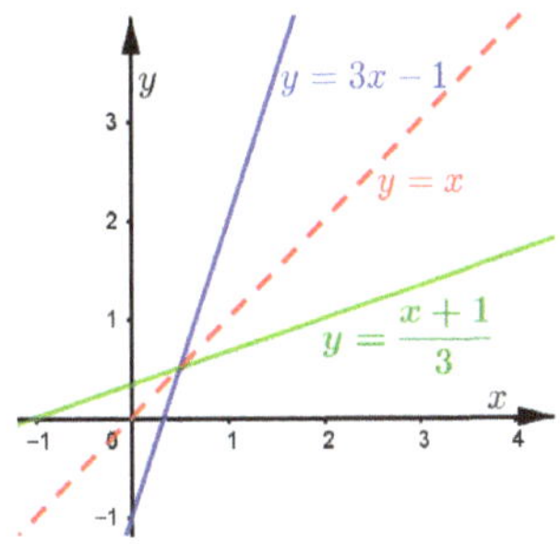

Figure 1.34

Example. Find the inverse function of $y = 3x - 1$.

Solution. From $y = \mathrm{f}(x) = 3x - 1$, we can find that

$$x = \mathrm{f}^{-1}(y) = \frac{y+1}{3}$$

Interchanging x and y in the above equation, we obtain the inverse function of $y = 3x - 1$ as $y = \dfrac{x+1}{3}$, as shown in Figure 1.34. □

If a function has an inverse function, it must be a one-to-one function. For example, on (-∞, +∞) , $y = x^2$ is not a one-to-one function, so it has no inverse function. However, on (0, +∞), $y = x^2$ has the inverse function $y = \sqrt{x}$, and on (-∞, 0), $y = x^2$ has the inverse function $y = -\sqrt{x}$.

2. Composite functions

Definition 1.15

Let the domain of the function $y = \mathrm{f}(x)$ be D(f). If the range of the function $u = u(x)$ is Z(u), and $Z(u) \cap D(f)$ is not an empty set, then $y = \mathrm{f}(u)$ is called a **composite function**. x is the independent variable, y is the dependent variable, and u is the intermediate variable.

Example 1. Given $y = \mathrm{f}(u) = \sqrt{u}$, $u = u(x) = a - x^2$. Examine whether $y = \mathrm{f}(u)$ is a composite function when $a = 1$ and $a = -1$.

Solution. (1) When $a = 1$, we have

$$y = \sqrt{u}\ ,\quad u = 1 - x^2$$

The domain of the function $y = \sqrt{u}$ is $0 \le u < \infty$, that is, D(f) = [0, +∞[.

The range of the function $u = 1 - x^2$ is Z(u) =]-∞, 1].

$$Z(u) \cap D(f) = [0, 1] \neq \varnothing \text{ (empty set)}$$

Therefore, $y = \mathrm{f}(u) = \sqrt{1-x^2}$ is a composite function.

(2) When $a = -1$, we have

$$y = \sqrt{u}\ , \qquad u = -1 - x^2$$

$$D(f) = [0, +\infty[\qquad Z(u) =]-\infty, -1]$$

$$Z(u) \cap D(f) = \varnothing \text{ (empty set)}$$

Therefore, $y = f(u) = \sqrt{-1-x^2}$ is not a composite function.

Example 2. Find the domain of the composite function $y = \arcsin\frac{2x-1}{3}$.

Solution. $y = \arcsin u$, $u = \frac{2x-1}{3}$

To find the domain, we require $|u| \le 1$, which means $\left|\frac{2x-1}{3}\right| \le 1$

Therefore, $-1 \le x \le 2$

Thus, the domain of $y = \arcsin\frac{2x-1}{3}$ is D = [-1, 2]. □

Using the concept of a composite function, a relatively complex function can be seen as a composite of several simpler functions, which facilitates the study of functions.

Example 3. The function $y = e^{\sqrt{x^2+1}}$ can be seen as a composite of the following three functions: $y = e^u$, $u = \sqrt{v}$, $v = x^2 + 1$

V. Elementary functions

The following functions are called **basic elementary functions**:

(1) Constant function: $y = c$

(2) Power function: $y = x^a$ (a is any real number)

(3) Exponential function: $y = a^x$ $(a > 0, a \ne 1)$

(4) Logarithmic function: $y = \log_a x$ $(a > 0, a \ne 1)$

(5) Trigonometric functions:

$y = \sin x$, $y = \cos x$, $y = \tan x$, $y = \cot x$, $y = \sec x$, $y = \operatorname{cosec} x$

(6) Inverse trigonometric functions:

$y = \arcsin x$, $y = \arccos x$, $y = \arctan x$, $y = \operatorname{arccot} x$, $y = \operatorname{arcsec} x$, $y = \operatorname{arccosec} x$

All functions formed by a finite number of four basic arithmetic operations and compositions of basic elementary functions are collectively called elementary functions. For example,

$$y = \sqrt{1-x^2}, \quad y = \sin^2 x, \quad y = \sqrt{\cot\frac{x}{2}}$$

are elementary functions. The vast majority of functions discussed in this course are elementary functions.

VI. Simple combinations and transformations of function graphs

This section introduces only a few simple and commonly used methods for constructing related function graphs from known function graphs.

1. Superposition

Given the graphs of $y = f(x)$ and $y = g(x)$, to sketch the graph of $y = f(x) + g(x)$, simply add the y-coordinates of the two graphs at the same x-value. For example, given the graphs of $y = x$ and $y = \frac{1}{x}$, we can sketch the graph of $y = x + \frac{1}{x}$, as shown in Figure 1.35.

2. Reflection

Given the graph of $y = f(x)$, to sketch the graph of $y = -f(x)$, change the sign of the y-coordinate of each point on the graph of $f(x)$. If the graph is above the x-axis, reflect it downwards. If the graph is below the x-axis, reflect it upwards. This creates a graph that is a reflection of $f(x)$ in the x-axis. For example, given the graph of $y = x^2 - 1$, we can sketch the graph of $y = -(x^2 - 1)$, as shown in Figure 1.36.

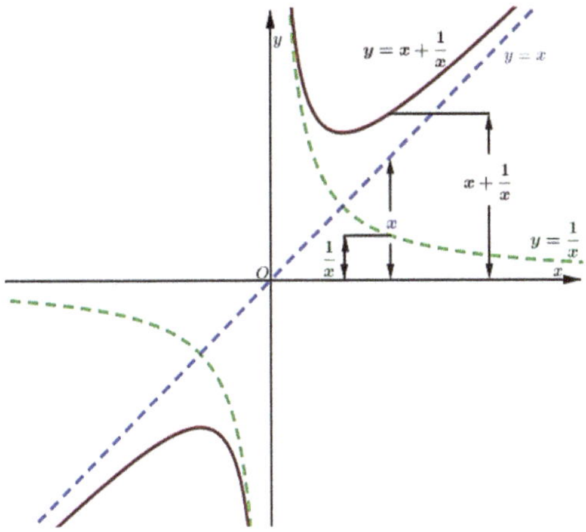

Figure 1.35

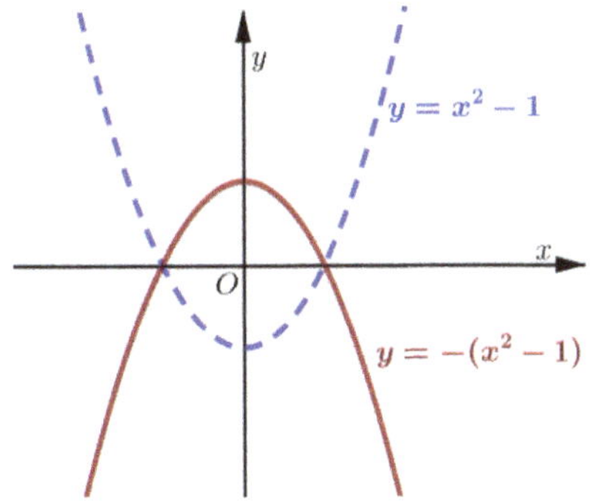

Figure 1.36

3. Stretch

Given the graph of $y = \text{f}(x)$, we can sketch the graph of $y = \text{kf}(x)$ (where k is a non-zero constant). When $\text{k} > 1$, stretch the graph of $\text{f}(x)$ vertically by a scale factor of k by multiplying the y-coordinate of the graph of $\text{f}(x)$ by a factor of k at each x-value. When $0 < \text{k} < 1$, compress the graph of $\text{f}(x)$ vertically by a scale factor of $\frac{1}{\text{k}}$. When $\text{k} < 0$, the graph is both stretched and reflected. For example, given the graph of $y = x^2$, we can sketch the graphs of $y = 2x^2$, $y = \frac{1}{4}x^2$, $y = -\frac{1}{2}x^2$, as shown in Figure 1.37.

4. Translation

Given the graph of $y = \text{f}(x)$, we can sketch the graph of $y = \text{f}(x) + \text{c}$ (where c is a constant). When $\text{c} > 0$, shift the graph of $\text{f}(x)$ upwards by a vertical distance c. When $\text{c} < 0$, shift the graph of $\text{f}(x)$ downwards by a vertical distance $|\text{c}|$. For example, given the graph of $y = x^3$, we can sketch the graphs of $y = x^3 + 1$ and $y = x^3 - 2$, as shown in Figure 1.38.

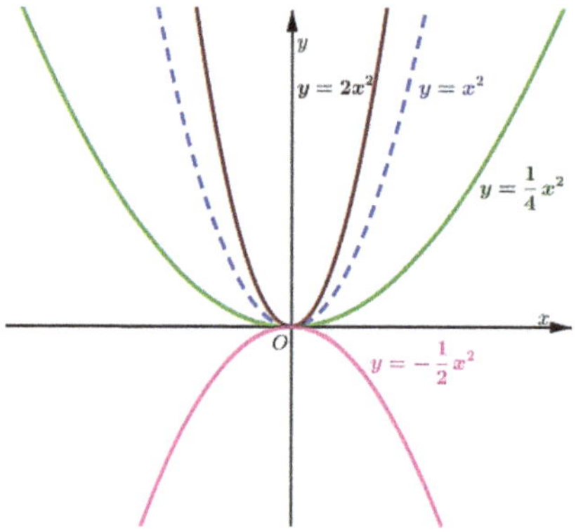

Figure 1.37

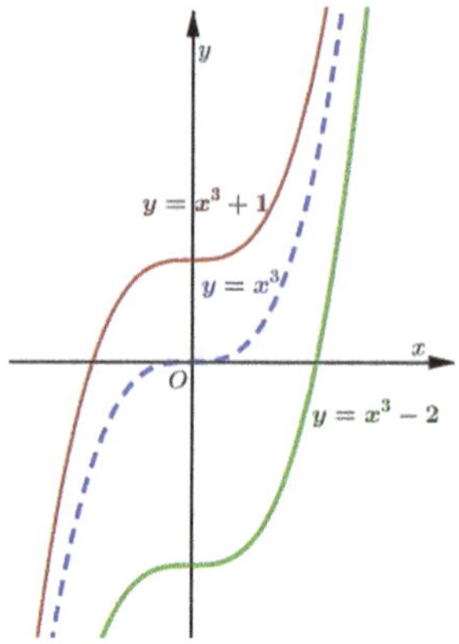

Figure 1.38

Exercise 1

1. Represent the following sets using the descriptive method.

a. The set of all real numbers greater than 5;

b. The set of all points inside the circle $x^2 + y^2 = 25$ (excluding the circumference);

c. The set of the intersection points of the parabola $y = x^2$ and the line $x - y = 0$.

2. Represent the following sets using the listing method.

a. The set of roots of the equation $x^2 - 8x + 15 = 0$;

b. The set of intersection points of the parabola $y = x^2$ and the line $x - y = 0$;

c. The set $\{x \mid |x-1| \leq 5$ and x is an integer.$\}$.

3. Which of the following sets are empty sets?

$A = \{x \mid x + 1 = 0\}$, $B = \{x \mid x^2 + 1 = 0$, x is a real number.$\}$, $C = \{x \mid x > 1$ and $x < 0\}$

4. Let $A = \{1, 2, 3\}$, $B = \{1, 3, 5\}$, $C = \{2, 4, 6\}$. Find

a. $A \cup B$ **b**. $A \cap B$ **c**. $A \cup B \cup C$ **d**. $A \cap B \cap C$ **e**. $A - B$

5. If $A = \{x \mid 3 < x < 5\}$, $B = \{x \mid x > 4\}$, find

a. $A \cup B$ **b.** $A \cap B$ **c.** $A - B$

6. If $U = \{1, 2, 3, 4, 5, 6\}$, $A = \{1, 2, 3\}$, $B = \{2, 4, 6\}$, find

a. A' **b.** B' **c.** $A' \cup B'$ **d.** $A' \cap B'$

7. If $A = \{a, b, c, d\}$, $B = \{a, b, c\}$, find $A \times B$.

8. If $A = \{a, b, c, d\}$, $B = \{c, d, e\}$, $C = \{d, e, f\}$, verify that

$$A \cap (B \cup C) = (A \cap B) \cup (A \cap C).$$

9. Prove using laws of set operations that: $X \cup (X \cap Y)' \cup Y - U$

10. Solve the following inequalities.

a. $x^2 < 9$ **b.** $|x - 4| < 7$ **c.** $0 < (x - 2)^2 < 4$

11. Let $f(x) = x^2 - 3x + 1$, find $f(0)$, $f(1)$, $f(2)$, $f(-x)$, $f(x + 1)$.

12. Find the domain of the following functions.

a. $y = \sqrt{9 - x^2}$ **b.** $y = \dfrac{1}{1 - x^2} + \sqrt{x + 2}$ **c.** $y = \dfrac{-5}{x^2 + 4}$

d. $y = \arcsin \dfrac{x - 1}{2}$ **e.** $y = \log(x + 3)$ **f.** $y = \log_2(\log_3 x)$

13. Let $f(x) = \dfrac{x}{1 - x}$, find $f[f(x)]$, $f\{f[f(x)]\}$.

14. If $f(x) = x^5 - 2x^3 + 3x$, prove that $f(-x) = -f(x)$.

15. If $f(x) = \dfrac{e^{-x} - 1}{e^{-x} + 1}$, prove that $f(-x) = -f(x)$.

16. If $f(x) = \dfrac{1-x^2}{\cos x}$, prove that $f(-x) = f(x)$.

17. Determine the domain of the following functions and sketch their graphs.

a. $f(x) = \begin{cases} x+3, & -3 \le x \le 0 \\ x^2+3, & 0 \le x \le 1 \end{cases}$ b. $f(x) = \begin{cases} \sqrt{1-x^2}, & |x| \le 1 \\ x-1, & 1 < |x| < 2 \end{cases}$

18. Determine the parity of the following functions.

a. $y = \dfrac{1}{x^2}$ b. $y = \tan x$ c. $y = \dfrac{a^x + a^{-x}}{2}$

d. $y = x + \sin x$ e. $y = xe^x$ f. $y = |x| + \cos x$

19. Find the inverse function of each of the following functions.

a. $y = 2x + 1$ b. $y = 2^x$ c. $y = \dfrac{x+2}{x-2}$

d. $y = x^3 + 2$ e. $y = 1 + \log(x + 2)$ f. $y = 3 + \dfrac{x}{1+x}$

20. If $y = u^2$, $u = \log_a x$, express y as a function of x.

21. If $y = \sqrt{u}$, $u = 2 + v^2$, $v = \cos x$, express y as a function of x.

22. If $f(x) = 3x^3 + 2x$, $\varphi(t) = \log(1 + t)$, find $f[\varphi(t)]$.

23. Discuss whether $y = \log(a - \sin x)$ is a composite function for $a = 2$, $a = 0.5$ and $a = -2$, respectively. If it is a composite function, find its domain.

24. Sketch the graphs of the following functions from the graph of $y = 2^x$.

a. $y = 3 \times 2^x$ b. $y = 2^x + 4$ c. $y = -2^x$

25. Sketch the graphs of the following functions from the graph of $y = \sin x$.

a. $y = \sin 2x$ b. $y = 2\sin 2x$ c. $y = 1 - 2\sin 2x$

Chapter 2 Limits

2.1. Limits of functions and variables

The concept and theory of limits are mathematical theories from a dynamic point of view. They form the foundation of calculus. The two fundamental concepts in calculus, differentiation and integration, are based on the concept of the limit.

The approach symbol $x \to x_0$ indicates the change process of the independent variable x approaching the limit value x_0.

from the left of x_0 $\quad x \to x_0 \quad$ Or $\quad x_0 \leftarrow x \quad$ from the right of x_0

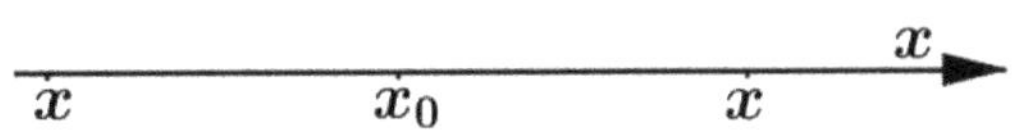

Figure 2.1

As shown in Figure 2.1, at the beginning, x takes value in the direction of x_0. In the end, x takes a value that is infinitely close to x_0. That is, at the end of the change process,

$$x = x_0 + a \qquad \text{or} \qquad x \approx x_0$$

where a is an infinitesimal.

An infinitesimal a is a variable that approaches zero, that is, $a \to 0$. At the end of the change process

$$a \approx 0 \qquad \Leftrightarrow \qquad 0 < |a| \le 0.05$$

Using the above notion of the dynamic approach, we now provide an equally dynamic definition for the limit of a function:

If a function $f(x)$ is defined in a neighbourhood of x_0, when x approaches infinitely close to x_0, the corresponding functional value of $f(x)$ will approach infinitely close to a certain constant A, then A is called the limit of the function $f(x)$ as $x \to x_0$, denoted by

$$\lim_{x \to x_0} f(x) = A$$

If $f(x)$ is a polynomial function or a fractional function whose denominator's limit is not zero as $x \to x_0$, then we have

$$\lim_{x \to x_0} f(x) = f(x_0)$$

The limit symbol, $\lim_{x \to x_0} (\bullet)$, means that the independent variable x in the expression $(\bullet)$ will finally take the value of $x = x_0$ in the change process when x approaches x_0.

For the change process of x as $x \to x_0$, the corresponding change process of $f(x)$ is $f(x) \to A$, this is generally denoted by

$$f(x) \to A \qquad (x \to x_0)$$

In summary

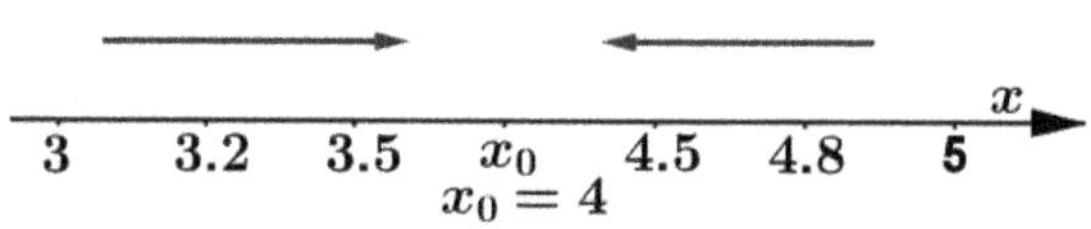

Figure 2.2

The approach symbol $x \to x_0$ and the limit symbol $\lim_{x \to x_0} (\bullet)$ mean that at the beginning, x takes value in the direction of x_0, either from the left of x_0 $(x < x_0)$ or from the right of x_0 $(x > x_0)$, as shown in Figure 2.2.

$$x = 3 \ldots 3.2 \ldots 3.5 \ldots \; x_0 - a, \; x_0$$

or

$$x = 5 \ldots 4.8 \ldots 4.5 \ldots \; x_0 + a, \; x_0$$

At the end of the change process

$$x = x_0 \pm a \approx x_0 \qquad \text{(for } x \to x_0\text{)}$$

$$x = x_0 \qquad \text{(for } \lim_{x \to x_0} (\bullet)\text{)}$$

With the above concept of limits, we can now apply limit method to prove the indeterminate form $\frac{0}{0} = a$.

Proof. Let $\lim_{x \to 0} x = 0$

then $a \cdot \lim_{x \to 0} x = a \cdot 0 = 0$ $\qquad a$ is a random number.

Dividing both sides by $\lim_{x \to 0} x$ gives

$$a \cdot \frac{\lim_{x \to 0} x}{\lim_{x \to 0} x} = \frac{0}{\lim_{x \to 0} x}$$

That is, $a \cdot 1 = \frac{0}{0}$

Hence, $a = \frac{0}{0} = 0 \cdot \infty = \frac{\infty}{\infty} = \infty - \infty$

Note: Above, we have given the concepts and definitions of approach and limit from a dynamic point of view. They should be easy to understand and grasp. However, in order to have a more comprehensive understanding of the nature of the limit, we will provide the definition of the limit from a static point of view below:

If, for any given positive number $\varepsilon > 0$, there always exists a positive number $\delta > 0$ such that whenever $0 < |x - x_0| < \delta$, the inequality

$$|f(x) - A| < \varepsilon$$

always holds, then the function $f(x)$ is said to have a limit at constant A when x approaches x_0. This is denoted by

$$\lim_{x \to x_0} f(x) = A \qquad \text{or} \qquad f(x) \to A \qquad (x \to x_0)$$

Here, the limit and approach have the same definition. However, their properties and calculation results are different. Care must be taken to distinguish the differences and do not confuse them. In addition, this definition only conforms to mathematical logic but lacks operability and is difficult to understand.

2.2. Computation rules of limits

As $x \to x_0$, the limits of the functions $f(x)$ and $g(x)$ are A and B, respectively.

That is, $\lim_{x \to x_0} f(x) = A, \qquad \lim_{x \to x_0} g(x) = B$

Then (1) $\lim_{x \to x_0} \left[f(x) \pm g(x) \right] = \lim_{x \to x_0} f(x) \pm \lim_{x \to x_0} g(x) = A \pm B$

(2) $\lim_{x \to x_0} \left[f(x) \times g(x) \right] = \lim_{x \to x_0} f(x) \times \lim_{x \to x_0} g(x) = A \times B$

If $g(x) = b$ is a constant, then

$\lim_{x \to x_0} bf(x) = b \lim_{x \to x_0} f(x) = bA$

(3) $\lim_{x \to x_0} \frac{f(x)}{g(x)} = \frac{\lim_{x \to x_0} f(x)}{\lim_{x \to x_0} g(x)} = \frac{A}{B} \qquad B \neq 0$

- In the calculations above, the variable x is a continuous variable. For a discontinuous integer variable n, except $n \to \infty$, $\lim_{n \to \infty}(\bullet)$, there is no other computation of limit.

Example 1. Find $\lim_{x\to 2}\frac{x^2-4}{x-2}$.

Solution. As $x\to 2$, $x\neq 2$, the common factor $(x-2)$ can be cancelled from the numerator and the denominator.

Thus $$\lim_{x\to 2}\frac{x^2-4}{x-2}=\lim_{x\to 2}\frac{(x+2)(x-2)}{x-2}=\lim_{x\to 2}(x+2)=2+2=4$$

(At the end of the change process, $x = 2$.)

Example 2. Find $\lim_{x\to 3}\frac{x^2-9}{x^3-27}$.

Solution. Because $\lim_{x\to 3}(x^3-27)=0$, the computation rule (3) cannot be used directly. However, as $x\to 3$, $x\neq 3$. Both the numerator and denominator have a factor $(x-3)$. Factorising the numerator and the denominator, we have

$$\lim_{x\to 3}\frac{x^2-9}{x^3-27}=\lim_{x\to 3}\frac{(x-3)(x+3)}{(x-3)(x^2+3x+9)}=\lim_{x\to 3}\frac{x+3}{x^2+3x+9}=\frac{3+3}{3^2+3\times 3+9}=\frac{2}{9}$$

Example 3. Find $\lim_{x\to 2}\frac{2x^2+x-5}{3x+1}$.

Solution. Since $\lim_{x\to 2}(3x+1)=3\cdot 2+1=7\neq 0$

thus $$\lim_{x\to 2}\frac{2x^2+x-5}{3x+1}=\frac{\lim_{x\to 2}(2x^2+x-5)}{\lim_{x\to 2}(3x+1)}=\frac{2\times 2^2+2-5}{3\times 2+1}=\frac{5}{7}$$

Example 4. Find $\lim_{x\to 1}(3x^2-2x+4)$.

Solution. $\lim_{x\to 1}(3x^2 - 2x + 4) = \lim_{x\to 1} 3x^2 - \lim_{x\to 1} 2x + \lim_{x\to 1} 4 = 3\cdot 1^2 - 2\cdot 1 + 4 = 5$

Example 5. Find $\lim_{x\to 0}\frac{x}{\sqrt{1+x}-1}$.

Solution. $$\lim_{x\to 0}\frac{x}{\sqrt{1+x}-1} = \lim_{x\to 0}\frac{x\left(\sqrt{1+x}+1\right)}{\left(\sqrt{1+x}-1\right)\left(\sqrt{1+x}+1\right)}$$

$$= \lim_{x\to 0}\frac{x\left(\sqrt{1+x}+1\right)}{x} = \lim_{x\to 0}\left(\sqrt{1+x}+1\right) = 2$$

Example 6. Find $\lim_{x\to\infty}\frac{3x^3+x^2+4}{5x^3-3x-7}$.

Solution. Dividing both the numerator and the denominator by x^3, we obtain

$$\lim_{x\to\infty}\frac{3x^3+x^2+4}{5x^3-3x-7} = \lim_{x\to\infty}\frac{3+\frac{1}{x}+\frac{4}{x^3}}{5-\frac{3}{x^2}-\frac{7}{x^3}} = \frac{3+0+0}{5-0-0} = \frac{3}{5}$$

Example 7. Find $\lim_{x\to\infty}\frac{x^5-4x^2}{6x^2+3x+2}$.

Solution. Dividing both the numerator and the denominator by x^5, we obtain

$$\lim_{x\to\infty}\frac{x^5-4x^2}{6x^2+3x+2} = \lim_{x\to\infty}\frac{1-\frac{4}{x^3}}{\frac{6}{x^3}+\frac{3}{x^4}+\frac{2}{x^5}} = \frac{1-0}{0+0+0} = \infty$$

Because $\infty \notin$ R, we say that the limit does not exist.

Example 8. Find $\lim_{x \to 0} \sin \frac{1}{x}$.

Solution. $\lim_{x \to 0} \sin \frac{1}{x} = \sin \frac{1}{0} = \sin \infty$

Because ∞ is dynamic, $\sin\infty$ takes values dynamically in the interval [-1, 1]. We say that it oscillates without limit. That is, the limit does not exist.

Example 9. Find $\lim_{x \to \infty} \frac{1}{x} \sin x$.

Solution. $\lim_{x \to \infty} \frac{1}{x} \sin x = \frac{1}{\infty} \sin \infty = 0 \cdot \sin \infty = 0$

Example 10. Find $\lim_{n \to \infty} \frac{3n+1}{n}$.

Solution. $\lim_{n \to \infty} \frac{3n+1}{n} = \lim_{n \to \infty} (3 + \frac{1}{n}) = 3 + \frac{1}{\infty} = 3 + 0 = 3$

Example 11. Find $\lim_{n \to \infty} \frac{1+(-1)^n}{2}$.

Solution. $\lim_{n \to \infty} \frac{1+(-1)^n}{2} = \frac{1+(-1)^\infty}{2} = 0 \text{ or } 1$

Because ∞ is dynamic, the limit sometimes takes 0 and sometimes 1. We say that it oscillates without limit. That is, the limit does not exist.

2.3. Two important limits

In the following, we will use two principles that limit exists to establish two important limits. With these two important limits, many problems in finding the limits of functions can be solved. And it also provides the necessary foundation for deriving the standard derivative formulae in the subsequent chapter.

a) Two principles that limit exists

Principle 1

Theorem. If, in the change process $x \to x_0$, the three variables $f(x)$, $g(x)$, and $h(x)$ always have a relationship

$$g(x) \le f(x) \le h(x)$$

and

$$\lim_{x \to x_0} g(x) = \lim_{x \to x_0} h(x) = A$$

then

$$\lim_{x \to x_0} f(x) = A$$

This theorem holds generally in mathematics. For example,

$$5 \le x \le 5 \xleftrightarrow{\textit{definite} \quad \textit{result}} x = 5$$

Principle 2

Consider a sequence $u_n = f(n)$. If, for all positive integers n, the inequality

$$f(n) < f(n+1)$$

always holds, then $f(n)$ is a monotonically increasing sequence.

If, for all positive integers n, the inequality

$$f(n) > f(n+1)$$

always holds, then $f(n)$ is said to be a monotonically decreasing sequence.

If there exist two constants m and M (m < M), for all positive integers n, the inequality

$$m \leq f(n) \leq M$$

always holds, then $f(n)$ is a bounded sequence.

Theorem. If a sequence $u_n = f(n)$ is monotonic and has a boundary, then the limit

$$\lim_{n\to\infty} f(n) = A$$

will definitely exist, where A is a constant.

b) Two important limits

(1) By applying principle 1, we can derive the first important limit.

$$\lim_{x\to 0}\frac{\sin x}{x} = 1$$

Proof. Consider a unit circle, as shown in Figure 2.3.

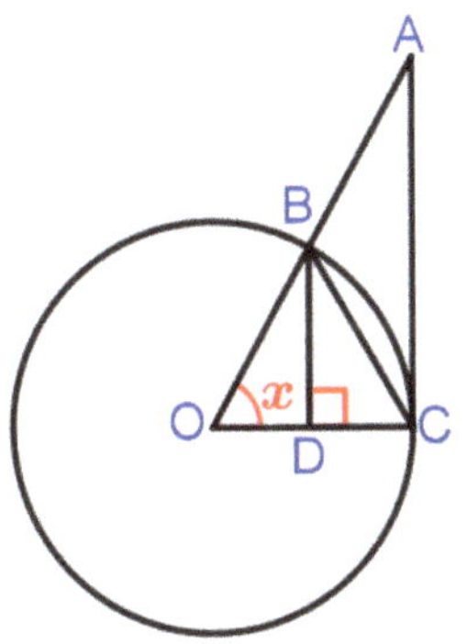

Figure 2.3

Let the central angle $B\hat{O}C = x \quad (0 < x < \frac{\pi}{2})$.

Then area of ΔOBC < area of sector OBC < area of ΔOAC

Since area of $\Delta OBC = \frac{1}{2} OC \cdot BD = \frac{1}{2}\cdot 1 \cdot \sin x = \frac{1}{2}\sin x$

$$\text{area of sector OBC} = \frac{1}{2}\text{OC}\cdot\text{OB}\cdot B\hat{O}C = \frac{1}{2}\cdot 1\cdot 1\cdot x = \frac{1}{2}x$$

$$\text{area of }\Delta\text{OAC} = \frac{1}{2}\text{OC}\cdot\text{AC} = \frac{1}{2}\cdot 1\cdot \tan x = \frac{1}{2}\tan x$$

thus $$\frac{1}{2}\sin x < \frac{1}{2}x < \frac{1}{2}\tan x$$

Dividing the inequality by $\frac{1}{2}\sin x$, we obtain

$$1 < \frac{x}{\sin x} < \frac{1}{\cos x}$$

That is, $$\cos x < \frac{\sin x}{x} < 1$$

Since the inequality holds when x is replaced by $-x$, it remains valid for all x in the interval $0 < |x| < \frac{\pi}{2}$.

And $$\lim_{x\to 0}\cos x = 1, \qquad \lim_{x\to 0} 1 = 1$$

By principle 1, we obtain the limit

$$\lim_{x\to 0}\frac{\sin x}{x} = 1$$ □

Example 1. Find $\lim_{x\to 0}\frac{\tan x}{x}$.

Solution. $$\lim_{x\to 0}\frac{\tan x}{x} = \lim_{x\to 0}\frac{\sin x}{x\cos x} = \lim_{x\to 0}\frac{\sin x}{x}\lim_{x\to 0}\frac{1}{\cos x} = 1$$

Example 2. Find $\lim_{x\to 0}\frac{\tan 3x}{\sin 6x}$.

Solution. $\lim\limits_{x\to0}\frac{\tan 3x}{\sin 6x}=\lim\limits_{x\to0}\frac{\sin 3x}{\sin 6x\cos 3x}=\lim\limits_{x\to0}\frac{\sin 3x}{\sin 6x\cos 3x}\times\frac{18x}{18x}$

$$=\lim_{x\to0}\frac{\sin 3x}{\sin 6x\cos 3x}\times\frac{6x}{3x}\times\frac{3}{6}=\lim_{x\to0}(\frac{\sin 3x}{3x}\times\frac{6x}{\sin 6x}\times\frac{1}{\cos 3x}\times\frac{3}{6})=\frac{3}{6}=\frac{1}{2}$$

Example 3. Find $\lim\limits_{x\to0}\frac{1-\cos x}{x^2}$.

Solution. $\lim\limits_{x\to0}\frac{1-\cos x}{x^2}=\lim\limits_{x\to0}\frac{2\sin^2\frac{x}{2}}{x^2}=\lim\limits_{x\to0}\frac{2\sin^2\frac{x}{2}}{2^2\times\frac{x^2}{2^2}}=\frac{2}{4}\lim\limits_{x\to0}(\frac{\sin\frac{x}{2}}{\frac{x}{2}})^2=\frac{1}{2}$

(2) By applying principle 2, we can derive the second important limit.

$$\lim_{n\to\infty}(1+\frac{1}{n})^n=e$$

where e is a constant. It is an irrational number, and its value is

$$e=2.7182818284590\ldots$$

Proof. Let $f(n)=(1+\frac{1}{n})^n$.

First, we show that $f(n)$ is monotonically increasing. Applying the binomial theorem, we obtain

$$f(n)=(1+\frac{1}{n})^n$$

$$=1+1+\frac{n(n-1)}{2!}\cdot\frac{1}{n^2}+\frac{n(n-1)(n-2)}{3!}\cdot\frac{1}{n^3}+\cdots+\frac{n(n-1)(n-2)\cdots2\cdot1}{n!}\cdot\frac{1}{n^n}$$

$$=1+1+\frac{1}{2!}(1-\frac{1}{n})+\frac{1}{3!}(1-\frac{1}{n})(1-\frac{2}{n})+\cdots+\frac{1}{n!}(1-\frac{1}{n})(1-\frac{2}{n})\cdots(1-\frac{n-1}{n})$$

Similarly, we obtain

$$f(n+1)=(1+\frac{1}{n+1})^{n+1}$$

$$=1+1+\frac{1}{2!}(1-\frac{1}{n+1})+\frac{1}{3!}(1-\frac{1}{n+1})(1-\frac{2}{n+1})+\cdots$$

$$+\frac{1}{n!}(1-\frac{1}{n+1})(1-\frac{2}{n+1})\cdots(1-\frac{n-1}{n+1})+\frac{1}{(n+1)!}(1-\frac{1}{n+1})(1-\frac{2}{n+1})\cdots(1-\frac{n}{n+1})$$

Comparing the two expansions, the first two terms are identical. However, from the third term onwards, each term in $f(n+1)$ is greater than the corresponding term in $f(n)$. Furthermore, the expansion of $f(n+1)$ contains one more positive term at the end. Therefore,

$$f(n)<f(n+1) \qquad (n=1, 2, 3, \ldots)$$

$f(n)$ is a monotonically increasing sequence.

In the following, we show that $f(n)$ is bounded. In the expansion of $f(n)$, all numbers in the brackets are less than 1. Consequently

$$f(n)<1+1+\frac{1}{2!}+\frac{1}{3!}+\ldots+\frac{1}{k!}+\ldots+\frac{1}{n!}$$

Since $\frac{1}{2^{k-1}}\geq\frac{1}{k!} \qquad (k\geq 2)$

thus $f(n)<1+1+\frac{1}{2}+\frac{1}{2^2}+\ldots+\frac{1}{2^{k-2}}+\ldots+\frac{1}{2^{n-1}}=1+\frac{1-\frac{1}{2^n}}{1-\frac{1}{2}}=3-\frac{1}{2^{n-1}}<3$

That is, $f(n)<3$

It can be seen from here that no matter how big the value n is, $f(n)$ is always less than 3. That is, $f(n)$ is a bounded sequence. Therefore, by principle 2, we obtain the limit.

$$\lim_{n\to\infty}(1+\frac{1}{n})^n = e$$

It can be shown that for a continuous independent variable x, we also have

$$\lim_{x\to\infty}(1+\frac{1}{x})^x = e$$

This formula can also be written in the form

$$\lim_{a\to 0}(1+a)^{\frac{1}{a}} = e$$

A logarithm with base e is called a natural logarithm. The natural logarithm of x is denoted by $\ln x$. In mathematics, we frequently employ the natural logarithmic functions, $y = \ln x$ and exponential functions with base e, $y = e^x$.

Example 4. Find $\lim_{x\to\infty}(\frac{x+4}{x})^x$.

Solution. $\lim_{x\to\infty}(\frac{x+4}{x})^x = \lim_{x\to\infty}(1+\frac{4}{x})^x = \lim_{x\to\infty}(1+\frac{1}{(\frac{x}{4})})^{x\cdot\frac{4}{4}} = \lim_{x\to\infty}[(1+\frac{1}{(\frac{x}{4})})^{\frac{x}{4}}]^4 = e^4$

Example 5. Find $\lim_{x\to\infty}(1-\frac{1}{x})^x$.

Solution. $\lim_{x\to\infty}(1-\frac{1}{x})^x = \lim_{x\to\infty}[(1+\frac{1}{(-x)})^{-x}]^{-1} = e^{-1}$

Exercise 2

1. Find the following limits.

a. $\lim_{x\to 3}\left(5+x^2-x\right)$

b. $\lim_{x\to 4}\frac{x^2-16}{x-4}$

c. $\lim_{x\to 1}\frac{x-1}{\sqrt{x+1}}$

d. $\lim_{x\to 1}\frac{x+1}{x^2-x-2}$

e. $\lim_{x\to 2}\frac{\sqrt{x+7}-3}{x-2}$

f. $\lim_{x\to\infty}\frac{x^3+x^2-4x+1}{2x^3+x+9}$

g. $\lim_{x\to 3}\frac{x^2-2x-3}{x-3}$

h. $\lim_{x\to\infty}\frac{2x+5}{x}$

i. $\lim_{x\to 0}\frac{x}{\sqrt{x+4}-2}$

j. $\lim_{x\to 0}\sin(x+\pi)$

k. $\lim_{x\to -5}(3x+14)^{99}$

l. $\lim_{x\to -3}4$

2. Find the following limits.

a. $\lim_{x\to -1}\sqrt{8-x}$

b. $\lim_{x\to 2}\sqrt{x^2+12}$

c. $\lim_{x\to 4}\frac{\sqrt{x}-2}{x-2}$

d. $\lim_{x\to 2}\frac{x^3-8}{x-2}$

e. $\lim_{x\to 9}\frac{9-x}{3-\sqrt{x}}$

f. $\lim_{x\to 0}\frac{4x}{(3+x)^2-9}$

g. $\lim_{x\to -1}\frac{x^2+5x+6}{x^2+3x+2}$

h. $\lim_{x\to 1}\frac{2x^2-10x+8}{3x^2-3x}$

3. Find the following limits.

a. $\lim_{x\to 3}\frac{x^4-81}{x^3-3x^2}$

b. $\lim_{x\to 3}\frac{x^3-3x^2+12x-4}{x^3-4x}$

c. $\lim_{x\to 2}\frac{x-2}{\sqrt{x+2}-2}$

d. $\lim_{x\to\infty}\frac{3x+1}{3x^2+x-4}$

e. $\lim_{x\to -2}\frac{x^2+3x+2}{x^2-x-6}$

f. $\lim_{x\to -3}\frac{\sin\frac{\pi x}{6}}{x+1}$

g. $\lim_{x\to 0}\dfrac{\sqrt{x+4}-2}{x}$

h. $\lim_{x\to -8}\dfrac{\sqrt{1-x}-3}{2+\sqrt[3]{x}}$

4. Find the following limits.

a. $\lim_{x\to\infty}\dfrac{x^5-x^4+3}{4x^5-3x^3+2x+6}$

b. $\lim_{x\to 3}\dfrac{x^3-3x^2+x-3}{x-3}$

c. $\lim_{x\to 1}\dfrac{x^2-1}{(x-1)\sqrt{x^2+1}}$

d. $\lim_{x\to 4}\dfrac{2-\sqrt{x}}{3-\sqrt{2x+1}}$

e. $\lim_{x\to 3}\dfrac{2x^3+5x^2-39x+18}{x^3+3x^2-13x-15}$

f. $\lim_{x\to -1}\dfrac{x^3+6x^2+11x+6}{x^3-4x^2-x+4}$

5. Find the following limits.

a. $\lim_{x\to 0}\dfrac{\sin 2x}{\sin 5x}$

b. $\lim_{x\to\infty}\left(1+\dfrac{3}{x}\right)^{3x}$

c. $\lim_{x\to 0}\dfrac{\tan 3x}{\tan x}$

d. $\lim_{x\to 0}\dfrac{x}{\sin 5x}$

e. $\lim_{x\to 0}\dfrac{1-\cos 4x}{x^2}$

f. $\lim_{x\to\infty}\left(\dfrac{5x+2}{5x}\right)^{x}$

g. $\lim_{x\to 0}(1-2x)^{\frac{1}{x}}$

h. $\lim_{x\to 0}\dfrac{\tan^3 x-\sin^3 x}{x^3}$

Chapter 3 Derivatives and differentials

Before studying this chapter, we define the **change in a function** (or **function increment**). Let the variable x change from its initial value x_1 to its final value x_2. The difference between the final value and the initial value is $x_2 - x_1$. It is called the **change in the variable** x, denoted by $\Delta x = x_2 - x_1$.

For the function $y = f(x)$, when the independent variable x changes from x_1 to $x_2 = x_1 + \Delta x$, the corresponding change in the function y is Δy, as shown in Figure 3.1, this is expressed as

$$\Delta y = y_2 - y_1 = f(x_2) - f(x_1) = f(x_1 + \Delta x) - f(x_1)$$

Note: The change can be positive or negative.

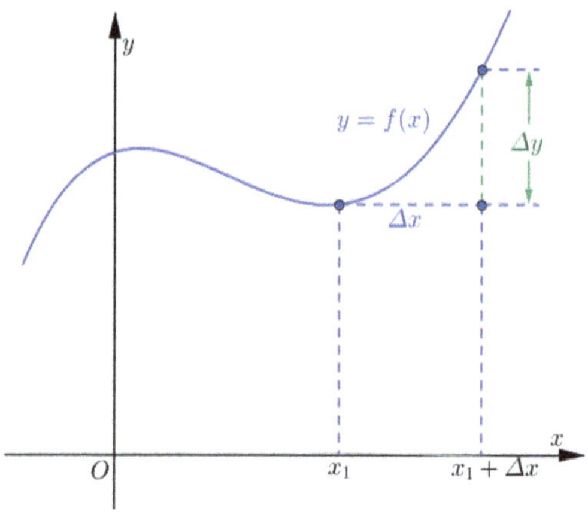

Figure 3.1

3.1. Two examples of the derivative concept

When solving real-world problems, we often need to understand more than just the functional relationship between variables; we also need to study the rate of change. Examples include the velocity of a moving object, the rate of population growth in a city, or the pace of a country's economic development. To address these problems, we introduce the concept of the derivative to better describe how these quantities change. Let us consider two practical examples.

(1) Instantaneous velocity of non-uniform linear motion

If s denotes the displacement travelled by an object in linear motion from an initial time to time t, then s is a function of time t, denoted by $s = f(t)$.

We now consider the velocity of the object at the specific time $t = t_0$, denoted by $v(t_0)$. When the time changes from t_0 to $t_0 + \Delta t$, as shown in Figure 3.2,

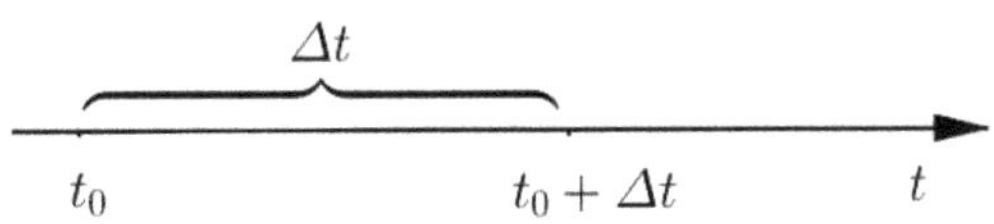

$\Delta t = (t_0 + \Delta t) - t_0$

Figure 3.2

the displacement that the object travels over the time interval Δt is

$$\Delta s = f(t_0 + \Delta t) - f(t_0)$$

If the object undergoes uniform motion, its velocity, $v(t)$, does not change with time. In this case

$$v(t_0) = \frac{\Delta s}{\Delta t} = \frac{f(t_0 + \Delta t) - f(t_0)}{\Delta t} = \text{constant}$$

Here, $v(t_0)$ is a constant. It is the velocity of the object at time t_0, which is identical to the velocity of the object at any time t. That is, $v(t) = v(t_0) = \text{constant}$.

However, when the object moves with varying velocity, its velocity $v(t)$ changes over time. In this case, the ratio $\frac{\Delta s}{\Delta t}$ represents the average velocity, denoted by $\overline{v}$, over the time interval from t_0 to $t_0 + \Delta t$:

$$\overline{v} = \frac{\Delta s}{\Delta t} = \frac{f(t_0 + \Delta t) - f(t_0)}{\Delta t}$$

When Δt is very small, $\bar{v}$ can be used to approximate the velocity of the object at time t_0, denoted by $v(t_0)$. As Δt becomes smaller, the approximation becomes more accurate. As $\Delta t \to 0$, if the limit $\lim_{\Delta t \to 0} \frac{\Delta s}{\Delta t}$ exists, it is called the instantaneous velocity of the object at time t_0. That is,

$$v(t_0) = \lim_{\Delta t \to 0} \frac{\Delta s}{\Delta t} = \lim_{\Delta t \to 0} \frac{f(t_0 + \Delta t) - f(t_0)}{\Delta t}$$

(2) Tangent

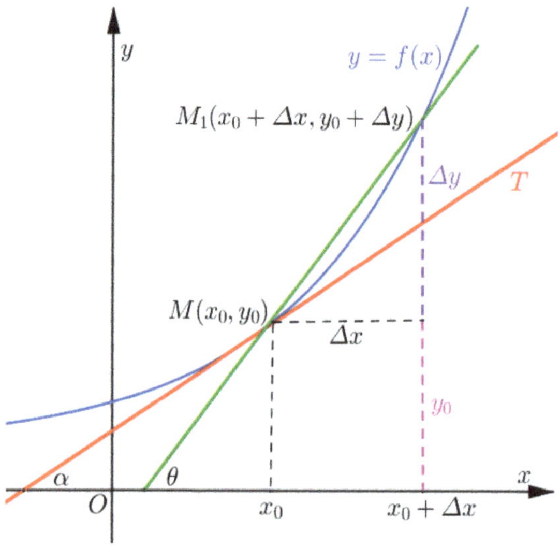

Figure 3.3

Let $M(x_0, y_0)$ be a fixed point on the curve $y = f(x)$. And let $M_1(x_0 + \Delta x, y_0 + \Delta y)$ be another point on the same curve. The position of M_1 is determined by Δx, making it a moving point. The line connecting M and M_1 is called a chord. Let its angle of inclination (the angle between the chord MM_1 and the x-axis) be θ.

From Figure 3.3, it can be seen that the gradient of the chord MM_1 is

$$\tan\theta = \frac{\Delta y}{\Delta x} = \frac{f(x_0 + \Delta x) - f(x_0)}{\Delta x}$$

As $\Delta x \to 0$, the moving point M_1 approaches the fixed point M along the curve. Consequently, the chord MM_1 changes and approaches its limiting position, represented by the line MT. We define the line MT as the tangent to the curve at the point M. It follows that θ approaches the angle of inclination α of the tangent MT.

That is, the gradient of the tangent MT is

$$\tan\alpha = \lim_{\Delta x \to 0} \tan\theta = \lim_{\Delta x \to 0} \frac{\Delta y}{\Delta x} = \lim_{\Delta x \to 0} \frac{f(x_0 + \Delta x) - f(x_0)}{\Delta x}$$

Although the two examples above have different practical meanings, speaking from the abstract numeric relationship, their mathematical nature is the same. They can both be generalised as the calculation of the ratio of the change in the function to the change in the independent variable, and the determination of the limit as the change in the independent variable approaches zero. The limit of this specific ratio $\frac{0}{0}$ is called the **derivative** of the function.

3.2. Concept of the derivative

Definition of the derivative

Let the function $y = f(x)$ be defined in a neighbourhood of x_0. When the independent variable x undergoes an increment Δx at the point x_0, the function $f(x)$ undergoes a corresponding increment.

$$\Delta y = f(x_0 + \Delta x) - f(x_0)$$

If the limit of $\frac{\Delta y}{\Delta x}$ exists as $\Delta x \to 0$, that is,

$$\lim_{\Delta x \to 0} \frac{\Delta y}{\Delta x} = \lim_{\Delta x \to 0} \frac{f(x_0 + \Delta x) - f(x_0)}{\Delta x} \qquad \text{exists}$$

then this limit is called the **derivative** or **differential** of the function $f(x)$ at x_0. This can be denoted by

$$f'(x_0) \qquad y'\big|_{x=x_0} \qquad \left.\frac{dy}{dx}\right|_{x=x_0} \qquad \left.\frac{df(x)}{dx}\right|_{x=x_0}$$

The ratio $\frac{\Delta y}{\Delta x} = \frac{f(x_0 + \Delta x) - f(x_0)}{\Delta x}$ represents the average speed of change in the function $f(x)$ as the independent variable x changes from x_0 to $x_0+\Delta x$, which is called the average rate of change in the function. The derivative $f'(x_0)=\lim\limits_{\Delta x\to 0}\frac{\Delta y}{\Delta x}$ represents the instantaneous rate of change of the function at x_0. This is simply referred to as the rate of change of the function at x_0.

Example. Find the derivative of the function $y=x^2$ at $x=2$.

Solution. When x changes from 2 to 2 + Δx, the increment in the function is

$$\Delta y=f(2+\Delta x)-f(2)=(2+\Delta x)^2-2^2=4\Delta x+(\Delta x)^2$$

Therefore, $$\frac{\Delta y}{\Delta x}=4+\Delta x$$

$$f'(2)=\lim_{\Delta x\to 0}\frac{\Delta y}{\Delta x}=\lim_{\Delta x\to 0}(4+\Delta x)=4$$ □

If the function $f(x)$ has a derivative at x_0, we say that the function $f(x)$ is **differentiable** at x_0 or that the derivative $f'(x_0)$ exists. Otherwise, we say that the function $f(x)$ is not differentiable at x_0 or that the derivative $f'(x_0)$ does not exist. If $f'(x_0)=\lim\limits_{\Delta x\to 0}\frac{\Delta y}{\Delta x}=\pm\infty$, in this case, it cannot be treated that $f'(x_0)$ exists because $\pm\infty\notin$ R, hence, the derivative $f'(x_0)$ does not exist. However, for convenience, we sometimes say that the derivative of function at x_0, $f'(x_0)$, is infinity. If the function $f(x)$ is differentiable at every point within the interval (a, b), then we say that the function $f(x)$ is differentiable on (a, b).

If $f(x)$ is differentiable on (a, b), then at each point x in the interval (a, b), there is a corresponding derivative value. Consequently, a new function can be defined, which is called the derivative function of the function $y = f(x)$ with respect to x on (a, b). This is often referred to simply as the derivative, and is denoted by

$$y' \qquad f'(x) \qquad \frac{dy}{dx} \qquad \frac{df(x)}{dx}$$

That is, $$y' = f'(x) = \frac{dy}{dx} = \frac{df(x)}{dx} = \lim_{\Delta x \to 0} \frac{\Delta y}{\Delta x}$$

According to the definition of the derivative, the two examples in Section 3.1 can be stated as follows:

(1) Instantaneous velocity is the derivative of displacement s with respect to time t, that is,

$$v(t) = s' = \frac{ds}{dt}$$

(2) The gradient of the tangent MT to the curve $y = f(x)$ at x is the derivative of the ordinate (the y-coordinate) of the curve with respect to the abscissa (the x-coordinate).

$$\tan\alpha = f'(x) = \frac{dy}{dx}$$

3.3. Differentials

Above, we use $\frac{dy}{dx}$ to denote the derivative, that is,

$$\frac{dy}{dx} = \lim_{\Delta x \to 0} \frac{\Delta y}{\Delta x}$$

This equality is a proportion.

Proof. Let $f(x)$ be a differentiable function of x and its derivative is

$$f'(x) = \lim_{\Delta x \to 0} \frac{\Delta y}{\Delta x}$$

From $\Delta y = \frac{\Delta y}{\Delta x} \cdot \Delta x$ we can obtain

$$\lim_{\Delta x \to 0} \Delta y = \lim_{\Delta x \to 0} \frac{\Delta y}{\Delta x} \cdot \Delta x = \lim_{\Delta x \to 0} \frac{\Delta y}{\Delta x} \cdot \lim_{\Delta x \to 0} \Delta x = f'(x) \cdot 0 = 0$$

That is, $$f'(x) = \lim_{\Delta x \to 0} \frac{\Delta y}{\Delta x} = \frac{\lim_{\Delta x \to 0} \Delta y}{\lim_{\Delta x \to 0} \Delta x} = \frac{0}{0}$$

Hence, $$\frac{dy}{dx} = \lim_{\Delta x \to 0} \frac{\Delta y}{\Delta x} = \frac{\lim_{\Delta x \to 0} \Delta y}{\lim_{\Delta x \to 0} \Delta x} = \frac{0}{0}$$

That is, $\frac{dy}{dx} = \frac{\lim_{\Delta x \to 0} \Delta y}{\lim_{\Delta x \to 0} \Delta x}$ is a proportion.

The derivative $f'(x)$ is a limit of the $\frac{0}{0}$ type.

Definition of the differential

Because $\lim_{\Delta x \to 0} \frac{\Delta y}{\Delta x} = f'(x)$, if $\Delta x \to 0$, then we have

$$\frac{\Delta y}{\Delta x} \to f'(x) \quad \text{or} \quad [\frac{\Delta y}{\Delta x} - f'(x)] \to 0$$

Let $\frac{\Delta y}{\Delta x} - f'(x) = \text{a}$ $\qquad \text{a} \to 0$

Then $\Delta y = f'(x)\,\Delta x + \text{a}\Delta x$

That is, $\Delta y - \text{a}\Delta x = f'(x)\,\Delta x$

Let $dy = \Delta y - \text{a}\Delta x,$ $\qquad dx = \Delta x$

Then $dy = f'(x)\,dx,$ $\qquad \frac{dy}{dx} = f'(x)$

Since $dx = \Delta x \to 0$

thus dx is an infinitesimal, $0 < |dx| \leq 0.05$.

The value of dy is determined by the value of $f'(x)\, dx$.

Generally, both dy and dx are infinitesimals.

dy is called the differential of the function $f(x)$, and dx is called the differential of the independent variable x.

Having introduced the concept of the differential, we can see that $\frac{dy}{dx}$ represents the quotient of the differential of the function dy and the differential of the independent variable dx. Consequently, the derivative $f'(x)$ is also known as the quotient of differentials. Since problems of finding the differential $dy = f'(x)\, dx$ is essentially the problems of finding derivatives, methods for finding derivatives and differentials are called differentiation methods.

Example 1. Find the differential dy of the function $y = x^2$ when x changes from 1 to 1.01.

Solution. Because $\frac{dy}{dx} = \frac{dx^2}{dx} = \lim_{\Delta x \to 0} \frac{(x+\Delta x)^2 - x^2}{\Delta x} = \lim_{\Delta x \to 0}(2x + \Delta x) = 2x$

it follows that $dy = 2xdx$

From the given condition

$$dx = \Delta x = 1.01 - 1 = 0.01$$

Here, $dx = \Delta x$ signifies that the differential dx of the independent variable x is equal to its increment Δx. Therefore, the differential of the function is

$$dy = 2xdx = 2 \cdot 1 \cdot (0.01) = 0.02$$

Geometric significance of the differential and its application

Sketch the graph of the function $y = f(x)$ on the xy–plane, as shown in Figure 3.4.

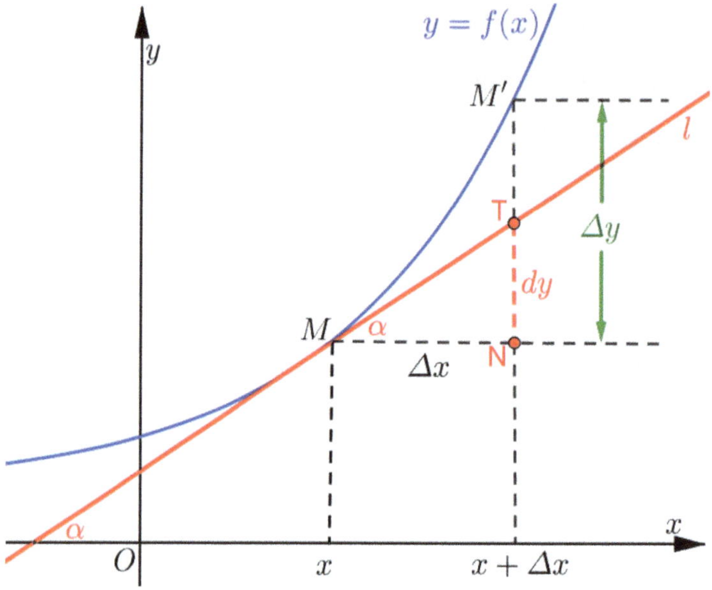

Figure 3.4

Let $M(x, y)$ be a fixed point on the curve. Draw the tangent l to the curve at the point M. The gradient of this tangent l is

$$\tan\alpha = f'(x)$$

When the independent variable x undergoes an increment Δx, we obtain another point on the curve, $M'(x + \Delta x, y + \Delta y)$. From Figure 3.4, it can be seen that

$$\mathrm{N}M = \Delta x \qquad \mathrm{N}M' = \Delta y \qquad \tan\alpha = \frac{\mathrm{TN}}{\mathrm{N}M}$$

That is, $$\mathrm{TN} = \mathrm{N}M{\cdot}\tan\alpha = f'(x)\cdot\Delta x = dy \qquad (\Delta x = dx)$$

Therefore, the differential dy of the function $y = f(x)$ represents the increment in the ordinate (the y-coordinate) of the tangent l at the point $M(x, y)$. In the graph, the segment $\mathrm{T}M'$ represents the difference between Δy and dy.

That is, $$\mathrm{T}M' = \mathrm{e} = \Delta y - dy, \qquad \text{where} \qquad \mathrm{e} = \mathrm{a}\Delta x$$

When $|\Delta x|$ is very small, e is also very small and can be neglected.

That is, $\Delta y - dy = e \approx 0$

Therefore, the formula for calculating an approximation using differentials is

$$\Delta y \approx dy = f'(x)\,\Delta x \qquad (\Delta x = dx)$$

Since $\Delta y = f(x+\Delta x) - f(x)$

it can be written as

$$f(x+\Delta x) - f(x) \approx f'(x)\,\Delta x$$

Alternatively, $f(x+\Delta x) \approx f(x) + f'(x)\,\Delta x$

Example 2. Find an approximation for the value of $\sqrt[3]{1.02}$.

Solution. This problem can be approached by finding an approximation for the value of the function $f(x) = \sqrt[3]{x}$ at $x = 1.02$.

Since $f'(x) = \dfrac{1}{3\sqrt[3]{x^2}}$ (Differentiation process is omitted.)

thus $f(x+\Delta x) \approx f(x) + f'(x)\,\Delta x = \sqrt[3]{x} + \dfrac{1}{3\sqrt[3]{x^2}} \cdot \Delta x$

Let $x = 1$ and $\Delta x = 0.02$ (This satisfies the requirement that Δx is very small.)

Therefore, $f(1+0.02) = f(1.02) = \sqrt[3]{1.02} \approx \sqrt[3]{1} + \dfrac{1}{3\sqrt[3]{1^2}} \cdot 0.02 \approx 1.0067$

That is, $\sqrt[3]{1.02} \approx 1.0067$

Example 3. A spherical ball has an outer diameter of 10 cm. The thickness of the spherical shell is $\dfrac{1}{20}$ cm. Find an approximation for the value of the volume of the spherical shell.

Solution. The radius of the ball is $r = \dfrac{10}{2} = 5$.

The thickness of the shell of the ball is $dr = \Delta r = -\frac{1}{20}$.

The volume of the ball is

$$V = f(r) = \frac{4}{3}\pi r^3$$

The derivative is $f'(r) = 4\pi r^2$ (Differentiation process is omitted.)

The volume of the spherical shell is ΔV, then

$$\Delta V \approx dV = f'(r)dr = 4\pi r^2 \cdot dr = 4\pi \cdot 5^2 \cdot (-\frac{1}{20}) = -5\pi$$

That is, the required volume of the spherical shell is

$$|\Delta V| = 5\pi \text{ cm}^3$$

3.4. Fundamental derivative formulae and computation rules

From the definition of the derivative, the steps for finding a derivative can be summarised as follows:

Step 1. Determine the increment in the function, Δy, corresponding to the increment Δx in the independent variable x.

$$\Delta y = f(x+\Delta x) - f(x)$$

Step 2. Form the ratio of the increments.

$$\frac{\Delta y}{\Delta x} = \frac{f(x+\Delta x) - f(x)}{\Delta x}$$

Step 3. Calculate the limit of $\frac{\Delta y}{\Delta x}$ as $\Delta x \to 0$, that is,

$$y' = f'(x) = \lim_{\Delta x \to 0} \frac{f(x+\Delta x) - f(x)}{\Delta x}$$

In the definition of derivative, we have already clearly explained the essence of the concept of derivative. The method above for finding the derivative $f'(x)$ of the function $f(x)$ is based directly on the definition (i.e. differentiating from first principles). However, if, for every function, we have to find its derivative directly according to the definition, it can be cumbersome and complex. We therefore seek to find some fundamental formulae and computation rules to simplify the computation in finding derivatives.

a) Derivative of a constant

Function $y = \mathrm{c}$ (c is a constant.)

$$\Delta y = f(x+\Delta x) - f(x) = \mathrm{c} - \mathrm{c} = 0$$

Since $\Delta y = 0$ for any value of Δx, it follows that the ratio $\dfrac{\Delta y}{\Delta x} = 0$.

Therefore, $$y' = \mathrm{c}' = \lim_{\Delta x \to 0} \frac{\Delta y}{\Delta x} = \lim_{\Delta x \to 0} 0 = 0$$

That is, $\mathrm{c}' = 0$

The derivative of a constant is equal to 0.

b) Derivatives of power functions

Function $y = x^n$ (n is a positive integer.)

From the binomial theorem, we know that

$$\Delta y = (x + \Delta x)^n - x^n$$

$$= [x^n + nx^{n-1}\Delta x + \frac{n(n-1)}{2}x^{n-2}(\Delta x)^2 + \ldots + (\Delta x)^n] - x^n$$

$$= nx^{n-1}\Delta x + \frac{n(n-1)}{2}x^{n-2}(\Delta x)^2 + \ldots + (\Delta x)^n$$

Therefore, $$\frac{\Delta y}{\Delta x} = nx^{n-1} + \frac{n(n-1)}{2}x^{n-2}\Delta x + ... + (\Delta x)^{n-1}$$

$$y' = \lim_{\Delta x \to 0} \frac{\Delta y}{\Delta x} = nx^{n-1}$$

That is, $$(x^n)' = nx^{n-1}$$

Later in this chapter, it will be shown that when n is any real number a, this formula remains valid. That is, for the function

$$y = x^a \qquad y' = ax^{a-1} \qquad a \in R$$

c) Derivatives of sums and differences

If $u = u(x)$ and $v = v(x)$ are differentiable functions of x, then their sum (or difference) $y = u \pm v$ is also a differentiable function of x. Moreover,

$$y' = (u \pm v)' = u' \pm v'$$

Proof. Consider y as a function of two independent variables, u and v.

That is, $$y = f(u, v) = u \pm v$$

Therefore, $$\Delta y = f(u + \Delta u, v + \Delta v) - f(u, v)$$

$$= [(u + \Delta u) \pm (v + \Delta v)] - (u \pm v) = \Delta u \pm \Delta v$$

$$\frac{\Delta y}{\Delta x} = \frac{\Delta u}{\Delta x} \pm \frac{\Delta v}{\Delta x}$$

Hence, $$y' = \lim_{\Delta x \to 0} \frac{\Delta y}{\Delta x} = \lim_{\Delta x \to 0} \frac{\Delta u}{\Delta x} \pm \lim_{\Delta x \to 0} \frac{\Delta v}{\Delta x} = u' \pm v'$$

That is, $$(u \pm v)' = u' \pm v'$$

The formula can be applied to the sums and differences of a finite number of functions. That is,

$$(u_1 + u_2 + \ldots + u_n)' = u_1' + u_2' + \ldots + u_n'$$

Example 1. Find the derivative of the function $y = x^3 - x^2$.

Solution. $$y' = (x^3 - x^2)' = (x^3)' - (x^2)' = 3x^{3-1} - 2x^{2-1} = 3x^2 - 2x$$

d) Derivative of a product

If both $u = u(x)$ and $v = v(x)$ are differentiable functions of x, then their product $y = uv$ is also a differentiable function of x. Moreover,

$$y' = (uv)' = uv' + u'v$$

Proof. Consider y as a function of two independent variables, u and v, that is,

$$y = f(u, v) = uv$$

Hence, $$\Delta y = f(u + \Delta u, v + \Delta v) - f(u, v)$$

$$= (u + \Delta u)(v + \Delta v) - uv = u\Delta v + v\,\Delta u + \Delta u \Delta v$$

$$\frac{\Delta y}{\Delta x} = u\frac{\Delta v}{\Delta x} + v\frac{\Delta u}{\Delta x} + \frac{\Delta u}{\Delta x}\Delta v$$

As $\Delta x \to 0$, values of u and v do not change, as u and v depend on x, rather than Δx. Furthermore, since v is differentiable, it is necessarily continuous. It follows that

$$\lim_{\Delta x \to 0} \Delta v = 0.$$

Therefore, $$y' = \lim_{\Delta x \to 0} \frac{\Delta y}{\Delta x} = u \lim_{\Delta x \to 0} \frac{\Delta v}{\Delta x} + v \lim_{\Delta x \to 0} \frac{\Delta u}{\Delta x} + \lim_{\Delta x \to 0} \frac{\Delta u}{\Delta x} \lim_{\Delta x \to 0} \Delta v$$

$$= uv' + vu' + u' \cdot 0$$

$$= uv' + u'v$$

That is, $$(uv)' = uv' + u'v$$

Especially when $u = \mathrm{c}$ (c is a constant)

$$y' = (\mathrm{c}v)' = \mathrm{c}v'$$

That is, a constant factor can be moved outside of the derivative operator.

This formula can be generalised to the product of a finite number of functions. That is, if $y = u_1 u_2 \cdots u_n$, then

$$(u_1 u_2 \cdots u_n)' = u_1' u_2 \cdots u_n + u_1 u_2' \cdots u_n + \ldots + u_1 u_2 \cdots u_{n-1} u_n'$$

Example 2. Find the derivative of the function $y = (1 + 2x)(3x^3 - 2x^2)$.

Solution.

$$\begin{aligned} y' &= (1 + 2x)(3x^3 - 2x^2)' + (1 + 2x)'(3x^3 - 2x^2) \\ &= (1 + 2x)[(3x^3)' - (2x^2)'] + [1' + (2x)'](3x^3 - 2x^2) \\ &= (1 + 2x)[3(x^3)' - 2(x^2)'] + [0 + 2(x)'](3x^3 - 2x^2) \\ &= (1 + 2x)(9x^2 - 4x) + 2(3x^3 - 2x^2) \\ &= 24x^3 - 3x^2 - 4x \end{aligned}$$

e) Derivative of a quotient

If u and v are differentiable functions of x and $v \neq 0$, then the quotient $y = \dfrac{u}{v}$ is also a differentiable function of x. Moreover,

$$y' = \left(\frac{u}{v}\right)' = \frac{u'v - uv'}{v^2}$$

Proof. Consider y as a function of two independent variables, u and v. That is,

$$y = f(u, v) = \frac{u}{v}$$

Hence,

$$\begin{aligned} \Delta y &= f(u + \Delta u, v + \Delta v) - f(u, v) \\ &= \frac{u + \Delta u}{v + \Delta v} - \frac{u}{v} = \frac{\Delta u \cdot v - u \cdot \Delta v}{v(v + \Delta v)} \end{aligned}$$

$$\frac{\Delta y}{\Delta x} = \frac{\frac{\Delta u}{\Delta x} \cdot v - u \cdot \frac{\Delta v}{\Delta x}}{v(v+\Delta v)}$$

Since as $\Delta x \to 0$, values of u and v do not change while $\Delta v \to 0$.

Therefore, $$y' = \lim_{\Delta x \to 0} \frac{\Delta y}{\Delta x} = \frac{\lim_{\Delta x \to 0} \frac{\Delta u}{\Delta x} \cdot v - u \cdot \lim_{\Delta x \to 0} \frac{\Delta v}{\Delta x}}{v(v + \lim_{\Delta x \to 0} \Delta v)} = \frac{u'v - uv'}{v^2}$$

That is, $$(\frac{u}{v})' = \frac{u'v - uv'}{v^2}$$

Especially when $u = c$ (c is a constant)

$$(\frac{c}{v})' = c(\frac{1}{v})' = c \cdot \frac{1'v - 1 \cdot v'}{v^2} = -c\frac{v'}{v^2}$$

This formula demonstrates that the formula $y' = (x^n)' = nx^{n-1}$ is also valid when n is a negative integer.

In fact, when $n = -m$ ($m > 0$), that is, n is a negative integer,

$$y = x^n = x^{-m} = \frac{1}{x^m}$$

therefore, $$y' = (x^{-m})' = (\frac{1}{x^m})' = -\frac{(x^m)'}{(x^m)^2} = -\frac{mx^{m-1}}{x^{2m}} = -mx^{-m-1}$$

Example 3. Find the derivative of the function $y = \frac{x^4}{3} - \frac{4}{x^3}$.

Solution. $$y' = (\frac{x^4}{3} - \frac{4}{x^3})' = (\frac{x^4}{3})' - (\frac{4}{x^3})' = \frac{1}{3}(x^4)' - 4(x^{-3})'$$

$$= \frac{4}{3}x^3 + 12x^{-4} = \frac{4}{3}x^3 + \frac{12}{x^4}$$

Example 4. Find the derivative of the function $y=\dfrac{x^2-1}{x^2+1}$.

Solution. $$y'=(\frac{x^2-1}{x^2+1})'=\frac{(x^2-1)'(x^2+1)-(x^2-1)(x^2+1)'}{(x^2+1)^2}$$

$$=\frac{(2x)(x^2+1)-(x^2-1)(2x)}{(x^2+1)^2}=\frac{4x}{(x^2+1)^2}$$

f) Derivatives of logarithm functions

Function $y=\log_a x \qquad (a>0,\ a\neq 1)$

From $\Delta y=\log_a(x+\Delta x)-\log_a x=\log_a(1+\dfrac{\Delta x}{x})$

we obtain $\dfrac{\Delta y}{\Delta x}=\dfrac{1}{\Delta x}\log_a(1+\dfrac{\Delta x}{x})=\dfrac{1}{x}\cdot\dfrac{x}{\Delta x}\log_a(1+\dfrac{\Delta x}{x})=\dfrac{1}{x}\log_a(1+\dfrac{\Delta x}{x})^{\frac{x}{\Delta x}}$

Let $b=\dfrac{\Delta x}{x}$. Then as $\Delta x\to 0,\ b\to 0$.

Thus $\displaystyle\lim_{\Delta x\to 0}\frac{\Delta y}{\Delta x}=\lim_{\Delta x\to 0}\frac{1}{x}\log_a(1+\frac{\Delta x}{x})^{\frac{x}{\Delta x}}=\frac{1}{x}\log_a\lim_{b\to 0}(1+b)^{\frac{1}{b}}$

$$=\frac{1}{x}\log_a e=\frac{1}{x\ln a} \qquad (\log_a b=\frac{\log_c b}{\log_c a})$$

where $\displaystyle\lim_{b\to 0}(1+b)^{\frac{1}{b}}=e$ is an important limit.

Therefore, $y'=(\log_a x)'=\dfrac{1}{x}\log_a e=\dfrac{1}{x\ln a}$

Especially when $a=e$

$$y'=(\log_e x)'=(\ln x)'=\frac{1}{x}\log_e e=\frac{1}{x}\ln e=\frac{1}{x}, \quad \text{where} \quad \ln e=1$$

That is, $y'=(\ln x)'=\dfrac{1}{x}$

This is the frequently used derivative formula for natural logarithms.

g) Derivatives of trigonometric functions

(1) Derivative of $y = \sin x$

From $$\Delta y = \sin(x + \Delta x) - \sin x = 2\cos(x + \frac{\Delta x}{2})\sin\frac{\Delta x}{2}$$

we obtain $$\frac{\Delta y}{\Delta x} = 2\cos(x + \frac{\Delta x}{2})\frac{\sin\frac{\Delta x}{2}}{\Delta x} = \cos(x + \frac{\Delta x}{2})\frac{\sin\frac{\Delta x}{2}}{\frac{\Delta x}{2}}$$

From the continuity of $\cos x$, we have

$$\lim_{\Delta x \to 0} \cos(x + \frac{\Delta x}{2}) = \cos x$$

From the limit formula $\lim_{x \to 0}\frac{\sin x}{x} = 1$, we have

$$\lim_{\Delta x \to 0} \frac{\sin\frac{\Delta x}{2}}{\frac{\Delta x}{2}} = 1$$

Hence, $$y' = \lim_{\Delta x \to 0} \cos(x + \frac{\Delta x}{2})\frac{\sin\frac{\Delta x}{2}}{\frac{\Delta x}{2}} = \lim_{\Delta x \to 0} \cos(x + \frac{\Delta x}{2}) \lim_{\Delta x \to 0} \frac{\sin\frac{\Delta x}{2}}{\frac{\Delta x}{2}} = \cos x$$

That is, $$(\sin x)' = \cos x$$

Similarly, we can show that

(2) Derivative of $y = \cos x$

$$(\cos x)' = -\sin x$$

(3) Derivative of $y = \tan x$

$$y' = (\tan x)' = \left(\frac{\sin x}{\cos x}\right)' = \frac{(\sin x)'\cos x - \sin x(\cos x)'}{\cos^2 x}$$

$$= \frac{\cos x\cos x - \sin x(-\sin x)}{\cos^2 x} = \frac{\cos^2 x + \sin^2 x}{\cos^2 x} = \frac{1}{\cos^2 x} = \sec^2 x$$

That is, $(\tan x)' = \sec^2 x$

It is not difficult to derive the following:

(4) Derivatives of $y = \cot x, y = \sec x, y = \csc x$

$$(\cot x)' = -\frac{1}{\sin^2 x} = -\csc^2 x$$

$$(\sec x)' = \left(\frac{1}{\cos x}\right)' = \sec x \cdot \tan x$$

$$(\csc x)' = \left(\frac{1}{\sin x}\right)' = -\csc x \cdot \cot x$$

Example 5. Find the derivative of the function $y = 2\sqrt{x} \cdot \sin x + \cos x \cdot \ln x$.

Solution. $y' = (2\sqrt{x} \cdot \sin x + \cos x \cdot \ln x)'$

$$= (2\sqrt{x} \cdot \sin x)' + (\cos x \cdot \ln x)'$$

$$= 2(\sqrt{x})'\sin x + 2\sqrt{x}(\sin x)' + (\cos x)'\ln x + \cos x(\ln x)'$$

$$= 2\cdot\frac{1}{2\sqrt{x}}\sin x + 2\sqrt{x}\cos x - \sin x\ln x + \frac{1}{x}\cos x$$

$$= \left(\frac{1}{\sqrt{x}} - \ln x\right)\sin x + \left(2\sqrt{x} + \frac{1}{x}\right)\cos x$$

h) Derivatives of composite functions (The Chain Rule)

To find a derivative easily, we usually treat a function $f(x)$ as a composite function,

that is, $f(x) \quad \Leftrightarrow \quad f(u), \quad u = u(x)$

For example,

$$y = \sqrt{x+1} = \sqrt{u}, \qquad u = x + 1$$

On the other hand, by the multiplication rule, we have

$$a = a \cdot 1 = a \cdot 1 \cdot 1 \cdot 1 \cdot \ldots$$

Hence, we get the derivative formula for composite functions, that is, the Chain Rule.

$$y' = \frac{dy}{dx} = \frac{dy}{dx} \cdot \frac{du}{du} = \frac{dy}{dx} \cdot \frac{du}{du} \cdot \frac{dv}{dv} \cdot \ldots$$

$$= \frac{dy}{du} \cdot \frac{du}{dx} = \frac{dy}{du} \cdot \frac{du}{dv} \cdot \frac{dv}{dx} \cdot \ldots$$

where $\dfrac{du}{du} = 1, \qquad \dfrac{du}{du} \cdot \dfrac{dv}{dv} \cdot \ldots = 1 \cdot 1 \cdot \ldots = 1.$

Example 6. Find the derivative of $y = (1 + 2x)^{30}$.

Solution. **Method 1** (write down the substitution)

Let $u = 1 + 2x$ then $y = (1 + 2x)^{30} = u^{30}$

Therefore, $y' = \dfrac{dy}{dx} = \dfrac{dy}{dx} \cdot \dfrac{du}{du} = \dfrac{dy}{du} \cdot \dfrac{du}{dx} = 30u^{30-1} \cdot 2$

where $\dfrac{dy}{du} = 30u^{30-1}, \qquad \dfrac{du}{dx} = 2.$

Hence, $y' = 60(1 + 2x)^{29}$

Method 2 (direct method without writing down the substitution)

$$y' = \frac{dy}{dx} = \frac{d(1+2x)^{30}}{dx} = \frac{d(1+2x)^{30}}{dx} \cdot \frac{d(1+2x)}{d(1+2x)} = \frac{d(1+2x)^{30}}{d(1+2x)} \cdot \frac{d(1+2x)}{dx}$$

$$= 30(1 + 2x)^{30-1} \cdot 2 = 60(1 + 2x)^{29}$$

Example 7. Find the derivative of the function $y = \ln(\sin x)$.

Solution. $y' = \frac{dy}{dx} = \frac{d\ln(\sin x)}{dx} \cdot \frac{d\sin x}{d\sin x} = \frac{d\ln(\sin x)}{d\sin x} \cdot \frac{d\sin x}{dx} = \frac{1}{\sin x} \cdot \cos x = \cot x$

Example 8. Find the derivative of the function $y = \cos(nx)$.

Solution. $y' = \frac{dy}{dx} = \frac{d\cos(nx)}{dx} \cdot \frac{d(nx)}{d(nx)} = \frac{d\cos(nx)}{d(nx)} \cdot \frac{d(nx)}{dx} = -n\sin(nx)$

Example 9. Find the derivative of the function $y = (\frac{x}{2x+1})^n$.

Solution. $y' = \frac{dy}{dx} = \frac{d(\frac{x}{2x+1})^n}{dx} \cdot \frac{d\frac{x}{2x+1}}{d\frac{x}{2x+1}} = \frac{d(\frac{x}{2x+1})^n}{d\frac{x}{2x+1}} \cdot \frac{d\frac{x}{2x+1}}{dx}$

$$= n(\frac{x}{2x+1})^{n-1} \cdot \frac{x'(2x+1) - x(2x+1)'}{(2x+1)^2}$$

$$= n(\frac{x}{2x+1})^{n-1} \cdot \frac{2x+1-2x}{(2x+1)^2} = \frac{nx^{n-1}}{(2x+1)^{n+1}}$$

Example 10. Find the derivative of the function $y = \sin(x^2+4)$.

Solution. $y' = \frac{dy}{dx} = \frac{d\sin(x^2+4)}{d(x^2+4)} \cdot \frac{d(x^2+4)}{dx} = \cos(x^2+4) \cdot 2x$

Example 11. Find the derivative of the function $y = \left[\ln(2x)\right]^3$.

Solution. $$y'=\frac{dy}{dx}=\frac{d\left[\ln(2x)\right]^3}{d\ln(2x)}\cdot\frac{d\ln(2x)}{d2x}\cdot\frac{d2x}{dx}$$

$$=3\left[\ln(2x)\right]^2\cdot\frac{1}{2x}\cdot 2=\frac{3\left[\ln(2x)\right]^2}{x}$$

Example 12. Find the derivative of the function $y=\cos\left[\ln(3x^2+6)^5\right]$.

Solution. $$y'=\frac{dy}{dx}=\frac{d\cos\left[\ln(3x^2+6)^5\right]}{d\ln(3x^2+6)^5}\cdot\frac{d\ln(3x^2+6)^5}{d(3x^2+6)^5}\cdot\frac{d(3x^2+6)^5}{d(3x^2+6)}\cdot\frac{d(3x^2+6)}{dx}$$

$$=-\sin\left[\ln(3x^2+6)^5\right]\cdot\frac{1}{(3x^2+6)^5}\cdot 5(3x^2+6)^4\cdot 6x$$

$$=-\frac{10x\sin\left[\ln(3x^2+6)^5\right]}{x^2+2}$$

Example 13. Find the derivative of the function $y=\sqrt{\cos\sqrt{x}}$.

Solution. $$y'=\frac{dy}{dx}=\frac{d\left(\cos\sqrt{x}\right)^{\frac{1}{2}}}{d\cos\sqrt{x}}\cdot\frac{d\cos\sqrt{x}}{d\sqrt{x}}\cdot\frac{d\sqrt{x}}{dx}$$

$$=\frac{1}{2}\left(\cos\sqrt{x}\right)^{-\frac{1}{2}}\cdot(-\sin\sqrt{x})\cdot\frac{1}{2}x^{-\frac{1}{2}}=-\frac{\sin\sqrt{x}}{4\sqrt{x\cos\sqrt{x}}}$$

Example 14. Find the derivative of the function $y=\sin\left[x^2\cos(3x)\right]$.

Solution. $$y'=\frac{dy}{dx}=\frac{d\sin\left[x^2\cos(3x)\right]}{dx^2\cos(3x)}\cdot\frac{dx^2\cos(3x)}{dx}$$

$$= \cos\left[x^2 \cos(3x)\right] \cdot \left[(x^2)'\cos(3x) + x^2 \cdot \frac{d\cos(3x)}{d3x} \cdot \frac{d3x}{dx}\right]$$

$$= \cos\left[x^2 \cos(3x)\right] \cdot \left[2x\cos(3x) - 3x^2 \sin(3x)\right] \qquad \square$$

By applying the Chain Rule for composite functions, we can also obtain the derivative formulae of a product and a quotient.

$$y = uv \qquad u = u(x) \qquad v = v(x)$$

Take logarithms of both sides.

$$\ln y = \ln(uv) = \ln u + \ln v$$

Differentiate both sides of the equation with respect to x.

$$\frac{d\ln y}{dx} = \frac{d\ln u}{dx} + \frac{d\ln v}{dx}$$

$$\frac{d\ln y}{dy} \cdot \frac{dy}{dx} = \frac{d\ln u}{du} \cdot \frac{du}{dx} + \frac{d\ln v}{dv} \cdot \frac{dv}{dx}$$

Hence, $$\frac{1}{y} \cdot \frac{dy}{dx} = \frac{1}{u} \cdot \frac{du}{dx} + \frac{1}{v} \cdot \frac{dv}{dx}$$

Substitute $y = uv$ and rearrange.

$$\frac{dy}{dx} = v\frac{du}{dx} + u\frac{dv}{dx}$$

Therefore, the derivative formula for a product is

$$y' = (uv)' = vu' + uv'$$

Similarly, we can also derive the derivative formula for a quotient, which is omitted here.

Generally speaking, apart from elementary functions, for functions $f(x)$ of all other forms, we can always use the Chain Rule for composite functions to find their derivatives.

i) Derivatives of implicit functions

Suppose the equation $P(x, y) = 0$ defines y as a function of x, where the function y is differentiable. The method of finding derivatives of composite functions (the Chain Rule) can also be applied to find derivatives of implicit functions y with respect to x.

Example 15. Given the equation $x^2 + y^2 - r^2 = 0$, where y is a function of x. Find the derivative y'.

Solution. Differentiate both sides with respect to x.

$$\frac{d(x^2)}{dx} + \frac{d(y^2)}{dx} - \frac{d(r^2)}{dx} = \frac{d0}{dx}$$

$$2x + \frac{d(y^2)}{dy} \cdot \frac{dy}{dx} - 0 = 0$$

$$2x + 2y \cdot \frac{dy}{dx} = 0$$

Divide both sides by 2.

$$x + y \cdot \frac{dy}{dx} = 0$$

$$y \cdot \frac{dy}{dx} = -x$$

$$\frac{dy}{dx} = -\frac{x}{y}$$

That is, $$y' = \frac{dy}{dx} = -\frac{x}{y}$$

Example 16. Given the equation $y^2 = 2px$, where y is a function of x. Find the derivative y'.

Solution. Differentiate both sides with respect to x.

$$\frac{d(y^2)}{dx} = \frac{d(2px)}{dx}$$

$$\frac{d(y^2)}{dy} \cdot \frac{dy}{dx} = 2p$$

$$2y \cdot \frac{dy}{dx} = 2p$$

$$\frac{dy}{dx} = \frac{p}{y}$$

Therefore, $y' = \dfrac{dy}{dx} = \dfrac{p}{y}$

Example 17. Given the equation $y = x\ln y$, where y is a function of x. Find the derivative y'.

Solution. Differentiate both sides with respect to x.

$$y' = (x)'\ln y + x(\ln y)'$$

$$= \ln y + x \cdot \frac{d(\ln y)}{dx}$$

$$= \ln y + x \cdot \frac{d(\ln y)}{dy} \cdot \frac{dy}{dx}$$

$$= \ln y + x \cdot \frac{1}{y} \cdot y'$$

Rearrange. $y' - \dfrac{x}{y} y' = \ln y$

$$y'\left(\frac{y-x}{y}\right) = \ln y$$

Hence, $y' = \dfrac{y \ln y}{y - x}$

Example 18. Given the equation $x^2 y^2 + 6xy + y^2 = 3$, where y is a function of x. Find the derivative y'.

Solution. Differentiate both sides with respect to *x*.

$$(x^2y^2)'+(6xy)'+(y^2)'=3'$$

$$\frac{d\left(x^2y^2\right)}{dx}+\frac{d\left(6xy\right)}{dx}+\frac{d(y^2)}{dx}=\frac{d3}{dx}$$

$$\frac{dx^2}{dx}\cdot y^2+x^2\cdot\frac{dy^2}{dy}\cdot\frac{dy}{dx}+6\cdot\frac{dx}{dx}\cdot y+6x\cdot\frac{dy}{dx}+\frac{dy^2}{dy}\cdot\frac{dy}{dx}=0$$

$$2xy^2+2x^2y\frac{dy}{dx}+6y+6x\frac{dy}{dx}+2y\frac{dy}{dx}=0$$

$$(2x^2y+6x+2y)\frac{dy}{dx}=-2xy^2-6y$$

$$y'=\frac{dy}{dx}=\frac{-2xy^2-6y}{2x^2y+6x+2y}=\frac{-xy^2-3y}{x^2y+3x+y}$$

j) Derivatives of inverse functions

If the function $y=f(x)$ has a non-zero derivative $f'(x)$ at the point *x*, and its inverse function $x=f^{-1}(y)$ is continuous at the corresponding point, then the derivative $\left[f^{-1}(y)\right]'$ exists and is given by

$$\left[f^{-1}(y)\right]'=\frac{1}{f'(x)} \quad \text{or} \quad f'(x)=\frac{1}{[f^{-1}(y)]'} \quad \text{or} \quad f'(x)\cdot\left[f^{-1}(y)\right]'=1$$

That is, $\dfrac{dx}{dy}=\dfrac{1}{\frac{dy}{dx}}$ or $\dfrac{dy}{dx}=\dfrac{1}{\frac{dx}{dy}}$ or $\dfrac{dy}{dx}\cdot\dfrac{dx}{dy}=1$

Proof. Since $\dfrac{dy}{dy}\cdot\dfrac{dx}{dx}=1\cdot1=1$

thus $\dfrac{dy}{dx}\cdot\dfrac{dx}{dy}=1$ that is, $f'(x)\cdot\left[f^{-1}(y)\right]'=1$

k) Derivatives of inverse trigonometric functions

(1) Derivative of function $y = \arcsin x \quad (-1 < x < 1)$

Since the inverse function of $y = \arcsin x \quad (-1 < x < 1)$ is

$$x = \sin y \qquad (-\frac{\pi}{2} < y < \frac{\pi}{2})$$

thus
$$\frac{dx}{dy} = \frac{d\sin y}{dy} = \cos y = \sqrt{1-\sin^2 y} = \sqrt{1-x^2}$$

$$y' = (\arcsin x)' = \frac{dy}{dx} = \frac{1}{\frac{dx}{dy}} = \frac{1}{\sqrt{1-x^2}}$$

That is,
$$(\arcsin x)' = \frac{1}{\sqrt{1-x^2}} \qquad (-1 < x < 1)$$

Similarly, it can be shown that

(2)
$$(\arccos x)' = -\frac{1}{\sqrt{1-x^2}} \qquad (-1 < x < 1)$$

$$(\arctan x)' = \frac{1}{1+x^2}$$

$$(\operatorname{arccot} x)' = -\frac{1}{1+x^2}$$

Example 19. Find the derivative of the function $y = \arcsin(3x^2)$.

Solution.
$$y' = \frac{dy}{dx} = \frac{d\arcsin(3x^2)}{dx}\cdot\frac{d(3x^2)}{d(3x^2)} = \frac{d\arcsin(3x^2)}{d(3x^2)}\cdot\frac{d(3x^2)}{dx}$$

$$= \frac{1}{\sqrt{1-(3x^2)^2}}\cdot 6x = \frac{6x}{\sqrt{1-9x^4}}$$

Example 20. Find the derivative of the function $y = \arctan\frac{1}{x}$.

Solution. $y' = \frac{dy}{dx} = \frac{d\arctan\frac{1}{x}}{dx} \cdot \frac{d\frac{1}{x}}{d\frac{1}{x}} = \frac{d\arctan\frac{1}{x}}{d\frac{1}{x}} \cdot \frac{d\frac{1}{x}}{dx} = \frac{1}{1+(\frac{1}{x})^2} \cdot (-\frac{1}{x^2}) = -\frac{1}{1+x^2}$

l) Derivatives of exponential functions

Exponential function $y = a^x \quad (a > 0,\ a \neq 1)$

Take logarithms of both sides.

$$\ln y = x\ln a$$

Differentiate both sides with respect to x.

$$\frac{1}{y}y' = \ln a$$

That is, $\quad y' = y\ln a = a^x\ln a$

Therefore, $\quad y' = (a^x)' = a^x\ln a$

Especially when $a = e$, we have

$$(e^x)' = e^x\ln e = e^x \cdot 1 = e^x$$

That is, $\quad (e^x)' = e^x \qquad$ This is a frequently used derivative formula.

Example 21. Find the derivative of the function $y = a^{-x}$.

Solution. $\quad y' = \frac{dy}{dx} = \frac{d(a^{-x})}{dx} \cdot \frac{d(-x)}{d(-x)} = \frac{d(a^{-x})}{d(-x)} \cdot \frac{d(-x)}{dx} = a^{-x}\ln a \cdot (-1) = -a^{-x}\ln a$

Example 22. Find the derivative of the function $y = e^{ax^2+bx+c}$.

Solution. **Method 1** Take logarithms of both sides.

$$\ln y = (ax^2 + bx + c)\ln e$$

$$\ln y = ax^2 + bx + c$$

Differentiate both sides with respect to x.

$$\frac{1}{y}y' = 2ax + b$$

$$y' = (2ax + b)y$$

Therefore, $\quad y' = (2ax + b)\,e^{ax^2+bx+c}$

Method 2

$$y' = \frac{de^{ax^2+bx+c}}{d(ax^2+bx+c)} \cdot \frac{d(ax^2+bx+c)}{dx} = e^{ax^2+bx+c} \cdot (2ax+b) = (2ax+b)\,e^{ax^2+bx+c}$$

Example 23. Given the equation $e^y = xy$, where y is a function of x. Find the derivative y'.

Solution. Differentiate both sides with respect to x.

$$\frac{de^y}{dx} = x'y + xy'$$

$$e^y y' = y + xy'$$

Rearranging gives $\quad y' = \dfrac{y}{e^y - x}$

m) Differentiate with the method of taking logarithms (Logarithmic Differentiation)

In deriving the derivative formula for the exponential function above, we first take logarithms of both sides of $y = a^x$. We then differentiate both sides with respect to x to find the derivative y'. This technique is called differentiating with the method of taking logarithms (or logarithmic differentiation). It is a particularly useful method in finding the derivatives of some functions. The following examples illustrate this method further.

Example 24. Find the derivative of the function $y = x^x$.

Solution. This function is neither a power function nor an exponential function; it is a composite exponential function (where both the base and the exponent are variables). Consequently, we cannot directly apply the standard derivative formulae for power or exponential functions. Instead, we differentiate this by using the method of logarithmic differentiation.

Take logarithms of both sides.

$$\ln y = x \ln x$$

Differentiate both sides with respect to x.

$$\frac{1}{y} y' = x' \ln x + x(\ln x)'$$

$$= \ln x + x \cdot \frac{1}{x}$$

$$= \ln x + 1$$

Therefore, $y' = y(\ln x + 1) = x^x(\ln x + 1)$

Example 25. Find the derivative of the function $y = \sqrt{\dfrac{(x-1)(x-2)}{(x-3)(x-4)}}$.

Solution. Differentiating this directly using the Chain Rule would be unnecessarily complex. Consequently, we first take logarithms of both sides.

$$\ln y = \frac{1}{2}[\ln(x-1) + \ln(x-2) - \ln(x-3) - \ln(x-4)]$$

Differentiate both sides with respect to x.

$$\frac{1}{y} y' = \frac{1}{2}\left(\frac{1}{x-1} + \frac{1}{x-2} - \frac{1}{x-3} - \frac{1}{x-4}\right)$$

Therefore, $y' = \frac{1}{2}y(\frac{1}{x-1} + \frac{1}{x-2} - \frac{1}{x-3} - \frac{1}{x-4})$

Example 26. Find the derivative of the power function $y = x^a$ $(a \in \mathrm{R})$.

Solution. Take logarithms of both sides.

$$\ln y = a \ln x$$

Differentiate both sides with respect to x.

$$\frac{1}{y}y' = a \cdot \frac{1}{x}$$

Hence, $y' = a \cdot \frac{y}{x} = a \cdot \frac{x^a}{x} = ax^{a-1}$

Therefore, $y' = (x^a)' = ax^{a-1}$

This is the frequently used derivative formula for power functions.

• Derivative formulae

For the ease of reference and use, the standard derivative formulae are listed below:

(1) $(c)' = 0$ (c is a constant.)

(2) $(u \pm v)' = u' \pm v'$

(3) $(uv)' = u'v + uv'$

(4) $(cu)' = cu'$ (c is a constant.)

(5) $(\frac{u}{v})' = \frac{u'v - uv'}{v^2}$ $(v \neq 0)$

(6) $\frac{dy}{dx} = \frac{dy}{du} \cdot \frac{du}{dx} = f'(u)u'(x)$ where $y = f(u)$, $u = u(x)$

(7) $\frac{dx}{dy} = \frac{1}{\frac{dy}{dx}}$ or $\left[f^{-1}(y)\right]' = \frac{1}{f'(x)}$ $(f'(x) \neq 0)$

(8) $(x^a)' = ax^{a-1}$

(9) $(\log_a x)' = \frac{1}{x}\log_a e$ $(a > 0, a \neq 1)$

(10) $(\ln x)' = \dfrac{1}{x}$

(11) $(a^x)' = a^x \ln a$ $\qquad (a > 0)$

(12) $(e^x)' = e^x$

(13) $(\sin x)' = \cos x$

(14) $(\cos x)' = -\sin x$

(15) $(\tan x)' = \dfrac{1}{\cos^2 x} = \sec^2 x$

(16) $(\cot x)' = -\dfrac{1}{\sin^2 x} = -\csc^2 x$

(17) $(\sec x)' = \sec x \cdot \tan x$

(18) $(\csc x)' = -\csc x \cdot \cot x$

(19) $(\arcsin x)' = \dfrac{1}{\sqrt{1-x^2}}$ $\qquad (-1 < x < 1)$

(20) $(\arccos x)' = -\dfrac{1}{\sqrt{1-x^2}}$ $\qquad (-1 < x < 1)$

(21) $(\arctan x)' = \dfrac{1}{1+x^2}$

(22) $(\text{arccot}\, x)' = -\dfrac{1}{1+x^2}$

• Differential formulae

From $dy = f'(x)\,dx$, we can see that to find the differential dy, we simply calculate the derivative $f'(x)$ and multiply it by dx.

Based on the differentiation rules for the sum, product, and quotient of functions, we can derive the corresponding differential formulae for the sum, product and quotient of functions. For example, using the product rule for the function $y = uv$

$$y' = uv' + u'v$$

and multiplying both sides by dx, we obtain the differential formula:

$$dy = d(uv) = udv + vdu$$

The differential formulae are summarised as follows:

(1) $dc = 0$ (c is a constant.)

(2) $d(u \pm v) = du \pm dv$

(3) $d(uv) = udv + vdu$

(4) $d(cu) = cdu$ (c is a constant.)

(5) $d(\frac{u}{v}) = \frac{vdu - udv}{v^2}$ $(v \neq 0)$

(6) $d(x^a) = ax^{a-1}dx$

(7) $d(\log_a x) = \frac{1}{x}\log_a e dx$ $(a > 0, a \neq 1)$

(8) $d(\ln x) = \frac{1}{x}dx$

(9) $d(a^x) = a^x \ln a dx$ $(a > 0)$

(10) $d(e^x) = e^x dx$

(11) $d(\sin x) = \cos x dx$

(12) $d(\cos x) = -\sin x dx$

(13) $d(\tan x) = \frac{1}{\cos^2 x}dx = \sec^2 x dx$

(14) $d(\cot x) = -\frac{1}{\sin^2 x}dx = -\csc^2 x dx$

(15) $d(\sec x) = \sec x \cdot \tan x dx$

(16) $d(\csc x) = -\csc x \cdot \cot x dx$

(17) $d(\arcsin x) = \frac{1}{\sqrt{1-x^2}}dx$ $(-1 < x < 1)$

(18) $d(\arccos x) = -\frac{1}{\sqrt{1-x^2}}dx$ $(-1 < x < 1)$

(19) $d(\arctan x) = \frac{1}{1+x^2}dx$

(20) $d(\text{arccot}\, x) = -\frac{1}{1+x^2}dx$

3.5. Higher derivatives

At the beginning of this chapter, we discussed the instantaneous velocity of an object moving in a straight line with varying speed. If the displacement of the object is given by the equation $s = f(t)$, then the instantaneous velocity of the object at time t is the derivative of s with respect to t, that is,

$$v = s' = f'(t)$$

Velocity $v = f'(t)$ is also a function of time t. Its derivative with respect to time t is called the instantaneous acceleration a at time t.

$$a = v' = (s')' = s''$$

This is known as the second derivative (or the second order derivative) of s with respect to t.

For example, the motion equation for an object in free fall is

$$s = \frac{1}{2}gt^2$$

Hence, the instantaneous velocity is

$$v = s' = (\frac{1}{2}gt^2)' = gt$$

The instantaneous acceleration is

$$a = s'' = (gt)' = g$$

In general, if the derivative $f'(x)$ of the function $f(x)$ is differentiable at the point x, then the derivative of $f'(x)$ at the point x is called the second derivative (or second order derivative) of the function $f(x)$ at the point x, denoted by

$$f''(x) \qquad y'' \qquad \text{or} \qquad \frac{d^2y}{dx^2}$$

Similarly, the derivative of the second derivative $y'' = f''(x)$ is called the third derivative (or the third order derivative) of the function $y = f(x)$, denoted by

$$f'''(x) \qquad y''' \qquad \text{or} \qquad \frac{d^3y}{dx^3}$$

In general, we define the nth derivative of the function $y = f(x)$ as the derivative of the $(n-1)$th derivative of the function $y = f(x)$. That is,

$$\left[y^{(n-1)}\right]' = y^{(n)} \qquad (n = 2, 3, 4, \ldots)$$

denoted by $\quad f^{(n)}(x) \qquad y^{(n)} \qquad \text{or} \qquad \dfrac{d^ny}{dx^n}$

All derivatives of order two and above are collectively known as higher derivatives (or higher order derivatives). The values of derivatives of all orders for the function $f(x)$, evaluated at the point $x = x_0$, are denoted by

$$f'(x_0) \qquad f''(x_0) \qquad \cdots \qquad f^{(n)}(x_0)$$

alternatively, $\quad y'\big|_{x=x_0} \qquad y''\big|_{x=x_0} \qquad \cdots \qquad y^{(n)}\big|_{x=x_0}$

Example 1. Find derivatives of all orders for the function $y = x^4$.

Solution. $\quad y' = 4x^3 \quad y'' = 12x^2 \quad y''' = 24x \quad y^{(4)} = 24$

$$y^{(5)} = y^{(6)} = \ldots = 0$$

Example 2. Find the nth derivative for the function $y = e^x$.

Solution. Since $(e^x)' = e^x$, the function does not change after differentiation, it follows that $\quad y^{(n)} = e^x$

Example 3. Find the nth derivative for the function $y = \sin x$.

Solution.
$$y' = (\sin x)' = \cos x = \sin(x + \frac{\pi}{2})$$

$$y'' = [\sin(x + \frac{\pi}{2})]' = \cos(x + \frac{\pi}{2}) = \sin(x + 2 \cdot \frac{\pi}{2})$$

$$y''' = [\sin(x + 2 \cdot \frac{\pi}{2})]' = \cos(x + 2 \cdot \frac{\pi}{2}) = \sin(x + 3 \cdot \frac{\pi}{2})$$

$$\cdots$$

Generally, we have

$$y^{(n)} = (\sin x)^{(n)} = \sin(x + n \cdot \frac{\pi}{2})$$

Similarly, we can prove that

$$(\cos x)^{(n)} = \cos(x + n \cdot \frac{\pi}{2})$$

Exercise 3

1. Use the definition of the derivative to find the following derivatives.

a. $f(x) = x^2$ **b.** $f(x) = x^3 + x$

c. $f(x) = 1 - 2x^2$ **d.** $f(x) = \frac{1}{x}$

e. $f(x) = \sqrt{x}$ **f.** $f(x) = \frac{1}{x^2}$

2. Find the derivatives of the following functions.

a. $y = 2x^3 - x + 6$ **b.** $y = \frac{3}{x^3} + \frac{x^3}{3}$

c. $y = 5\sqrt{x} + \frac{2}{x} - \sqrt[5]{x^3}$

d. $y = \frac{x-1}{x+1}$

e. $y = (2x-3)\sqrt{x}$

f. $y = \frac{\sin x}{x^2}$

g. $y = \cos x \ln x$

h. $y = \tan(5x) + \sin(3x) - \sin^2 x$

i. $y = \sqrt{\cos(5x)}$

j. $y = \frac{5\ln x}{1+\cos x}$

k. $y = \frac{x}{\sin x + \cos x}$

l. $y = \sec(2x-3)\ln(3x+2)$

3. Find the derivatives of the following functions.

a. $y = \sin(3x) + \sin^3(2x)$

b. $y = \sin(x^3) + \log_2(x^2+1)$

c. $y = \cos(7x-5) - \sin(2x+3)$

d. $y = (2-x)(2+x^3)^2$

e. $y = (1-\sin x)^2$

f. $y = \tan^3(\frac{x}{3}) + \sin(\cos\sqrt{x})$

g. $y = \sqrt{\frac{1+x^2}{1-x}}$

h. $y = \tan(5x) + \left[\ln x^2\right]^3$

i. $y = \sqrt{\cos(3x+5)}$

j. $y = \sec^5(2x+1)$

k. $y = \ln\sqrt{3x+4} + \sqrt{\ln(3x+4)}$

l. $y = (\ln x)^5 + \frac{1}{\cos^4(2x)}$

4. Find the derivatives of the following functions.

a. $y = \tan\frac{5x}{4}\cos^3\frac{x}{2}$

b. $y = \sqrt[3]{\ln\frac{2x}{5}}$

c. $y = \ln(x\cos x)$

d. $y = (9+x^4)\sqrt{5-x}$

e. $y = (2+3x^2)\sqrt{1+4x^3}$

f. $y = \log_3(1+x^3)$

g. $y = \ln\frac{1-\sqrt{x}}{1+\sqrt{x}}$

h. $y = \tan^3(\log_3(4x+5))$

i. $y = \ln^3(\tan(2x+3))$

j. $y = \cos(2x^3 \ln x)$

k. $y = \dfrac{\cos x}{\sqrt{1-x^2}}$

l. $y = x^6 \sqrt[5]{x^7 - 2}$

5. Find the derivatives of the following functions.

a. $y = \sqrt[3]{\dfrac{1}{1-x^2}}$

b. $y = \dfrac{1}{\tan^2(3x)}$

c. $y = \sin(x^2 \cos(3x))$

d. $y = \dfrac{\ln(\cos(2x))}{\sin^2 x}$

e. $y = \sin^3[\ln x \cos(5x)]$

f. $y = \dfrac{\ln^5 x}{\sec(5x)}$

g. $y = (x^4 - x^3 + 2x)^7$

h. $y = (\dfrac{x-3}{2x+1})^4$

i. $y = (x^5 + 3x^2)^{\frac{3}{4}}$

j. $y = \sqrt{\cos ec(5x)}$

k. $y = \sec(\dfrac{3x}{1+x^2})$

l. $y = \ln[\sin^2(5x+2)]$

6. Find $\dfrac{dy}{dx}$ of the following.

a. $x^3 + y^3 = 5$

b. $2xy - x^2 y = 9$

c. $x^3(2x - 3y) = 4$

d. $e^{xy} + y \ln x - \cos(3x) = 1$

e. $\sin(x^2 + y^2) = 3x$

f. $y = 1 + xe^y$

g. $3xy^3 + x^2 y - y^2 = 7$

h. $\sqrt{x + y^2} = x$

i. $\ln(x^2 + y) = x^4 y - \cos x$

j. $e^y + 8xy + x^2 = 7$

k. $x^y = y^x$

l. $\dfrac{1}{x^2} + \dfrac{1}{y} = -8$

7. Find the derivatives of the following functions.

a. $y = \dfrac{\sin x}{e^{5x}}$ **b.** $y = 3^{\sqrt{x}} \arctan x^2$

c. $y = 5^{\sqrt{\cos(2x)}}$ **d.** $y = \dfrac{\arcsin(2x)}{\sin x + \cos x}$

e. $y = \ln(xe^{3x})$ **f.** $y = (\arccos \dfrac{x}{3})^4$

g. $y = e^{4x+3} \sin(e^{x^2})$ **h.** $y = \cos(2x^3 e^{-x})$

i. $y = \dfrac{\arccos x}{\sqrt{1-x^2}}$ **j.** $y = e^{\sin^2(3x)} \cdot \cos(e^{-2x})$

k. $y = \arctan^2(\dfrac{1}{2x})$ **l.** $y = \sin^3[e^{2x}\cos(5x)]$

m. $y = \dfrac{\ln^5 x}{e^{2x+3}}$ **n.** $y = \sin(2^{\sin(2x)})$

o. $y = e^{2-\cos^3(3x+5)}$ **p.** $y = \ln(2^{\sqrt{x}} - e^{-2x})$

q. $y = \arctan(\sqrt{x^5 - x})$ **r.** $y = \arccos \dfrac{x-1}{x+2}$

s. $y = e^{\arcsin\sqrt{2x}}$ **t.** $y = \cos \dfrac{\arcsin x}{5}$

u. $y = 3^{4^{x^3}}$ **v.** $y = 10^{3+x^4}$

w. $y = \sqrt[3]{\arctan \sqrt[3]{x}}$ **z.** $y = \arcsin(\cos^2 x)$

8. Find the derivatives of the following functions.

a. $y = x^{\cos x}$ **b.** $y = x\sqrt{\dfrac{1+x}{1-x}}$

c. $y = (\ln x)^x$ **d.** $y = \sqrt[3]{\dfrac{x^2+3}{x+2}}$

e. $y = (\tan x)^{\sin x}$ **f.** $y = (x^3+2)^{\frac{1}{4}} \cdot \sin^5 x \cdot 3^{x+2}$

g. $y = x^{x^{x^3}}$

h. $y = (\sin x)^{\cos^2 x}$

i. $y = \dfrac{\sqrt{4x+3}}{5^{x^3}\cos(2x)}$

j. $y = (2x+3)^5 \cdot (4x-1)^7 \cdot (x^3+2)^8$

9. Find the first and second derivatives of the following.

a. $y = \ln(1+x^3)$

b. $y = x^4 + 5x + 8$

c. $y = xe^{-2x}$

d. $y\sin x - \cos(x-y) = 0$

e. $e^x + \sin(x^2) = y$

f. $y^3 + 2\ln y = x^3$

g. $x + \arctan y = y$

h. $x^2 + y^2 = e^{\arctan y}$

10. Given that $xy - \sin(\pi y^2) = 0$, find the values for $y'|_{x=0,y=1}$ and $y''|_{x=0,y=1}$.

11. Find the differential, *dy*, for each of the following functions.

a. $y = 2x^4$

b. $y = \dfrac{x}{1-x^2}$

c. $y = e^{-x}\cos x$

d. $y = e^{\sin x}$

e. $xy = 5$

f. $y = 1 + xe^y$

12. Find an approximation for each of the following using differentials.

a. $\sqrt[5]{0.95}$

b. $\sqrt[3]{8.02}$

c. $\ln(1.01)$

d. $e^{0.05}$

13. Find the *n*th derivative of each of the following functions.

a. $y = 3^x$

b. $y = \ln(1+x)$

c. $y = e^{5x}$

d. $y = (2+x)^m$ where m is a constant.

Chapter 4 Differential Mean Value Theorem, L'Hôpital's rule, and Continuity and Differentiability of Functions

4.1. Mean Value Theorem

1. The maximum and minimum theorem for continuous functions

If a function $f(x)$ is continuous on the closed interval $[a, b]$, then it will definitely have a maximum and a minimum on this interval.

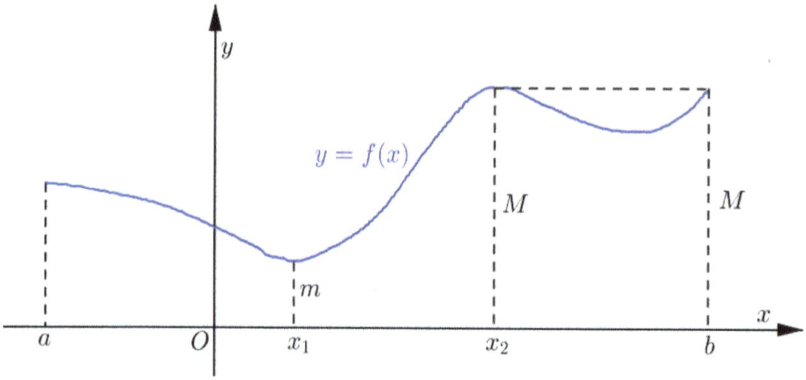

Figure 4.1.1

Figure 4.1.1 shows that a function $f(x)$ is continuous on the closed interval $[a, b]$. The function attains its minimum value, m, at x_1, and its maximum value, M, at x_2 and b.

2. Mean Value Theorem

(1) Rolle's Theorem

If the function $f(x)$ satisfies the following conditions:

(i) it is continuous on the closed interval $[a, b]$;

(ii) it is differentiable on the open interval (a, b);

(iii) it takes the same value at both endpoints, that is, $f(b) = f(a)$;

then there exists at least one point $c \in (a, b)$ such that $f'(c) = 0$.

Geometric significance of Rolle's Theorem

As shown in Figure 4.1.2, the continuous function $f(x)$ takes the same value at points A and B. Therefore, on the interval (a, b), there exists at least one point c such that the tangent to the function curve at the point c is parallel to the x-axis.

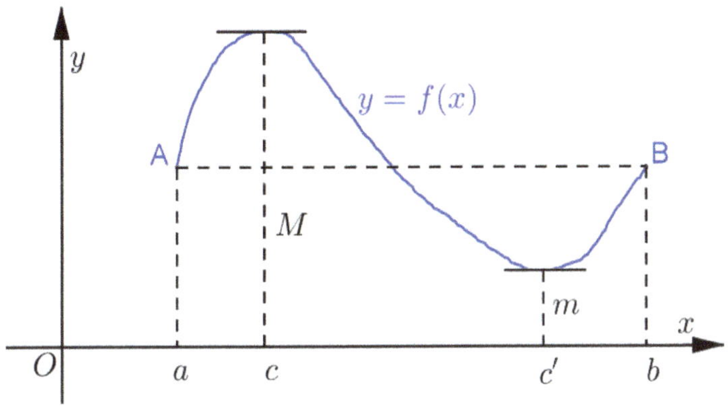

Figure 4.1.2

(2) Lagrange's Mean Value Theorem

If the function $f(x)$ satisfies the following conditions:

(i) it is continuous on the closed interval $[a, b]$;

(ii) it is differentiable on the open interval (a, b);

then there exists at least one point $c \in (a, b)$ such that $f'(c) = \dfrac{f(b) - f(a)}{b - a}$.

That is, $$f(b) = f(a) + f'(c)(b - a)$$

Geometric significance of Lagrange's Mean Value Theorem

As shown in Figure 4.1.3, the function $f(x)$ is continuous on the interval $[a, b]$. Therefore, on the interval (a, b), there exists at least one point c such that the tangent to the function curve at the point c is parallel to the chord AB.

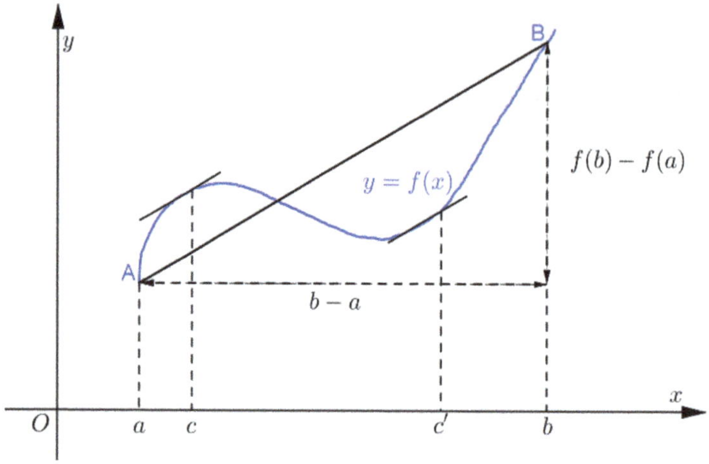

Figure 4.1.3

It is easy to see that Rolle's Theorem is a special case of Lagrange's Mean Value Theorem when $f(a) = f(b)$.

Proof of Lagrange's Mean Value Theorem

From Figure 4.1.3, the equation of the chord AB is

$$y - f(a) = \frac{f(b) - f(a)}{b - a}(x - a)$$

That is,
$$y = f(a) + \frac{f(b) - f(a)}{b - a}(x - a)$$

Let
$$g(x) = f(x) - y$$

$$= f(x) - \left[f(a) + \frac{f(b) - f(a)}{b - a}(x - a) \right]$$

where $f(x)$ is the function, and y is the equation of the chord AB.

Since
$$g(a) = g(b) = 0$$

it follows from Rolle's Theorem that there exists at least one point $c \in (a, b)$ such that

$$g'(c) = f'(c) - \frac{f(b) - f(a)}{b - a} = 0$$

That is, $$f'(c) = \frac{f(b) - f(a)}{b - a} \qquad c \in (a,\ b)$$

Alternatively, it can be written as

$$f'(c) = \frac{\Delta y}{\Delta x} \qquad \Delta x = x_2 - x_1 \qquad c \in (x_1,\ x_2)$$ □

Comparison between the derivatives $f'(x)$ and $f'(c)$

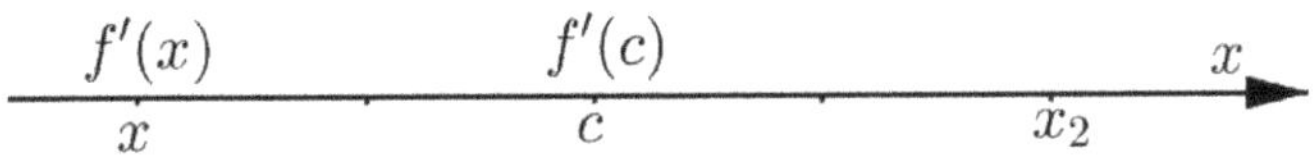

Figure 4.1.4

(1) If $\Delta x = x_2 - x,$ then

$f'(x) = \lim\limits_{\Delta x \to 0} \frac{\Delta y}{\Delta x}$, $x \in [x,\ x_2]$ is the initial endpoint of the interval.

$f'(c) = \frac{\Delta y}{\Delta x}$, $c \in (x,\ x_2)$ is a value lying between the two endpoints of the interval.

(2) The limit $$\lim_{\Delta x \to 0} f'(c) = \lim_{\Delta x \to 0} \frac{\Delta y}{\Delta x} = f'(x)$$

Since as $\Delta x \to 0 \quad \Leftrightarrow \quad x_2 \to x, \quad c \to x$

hence, $$\lim_{\Delta x \to 0} f'(c) = \lim_{c \to x} f'(c) = f'(x)$$

For the limit $\lim\limits_{\Delta x \to 0} f'(c)$, at the end of the change process, $c = x$.

Therefore, $$\lim_{\Delta x \to 0} f'(c) \cdot \Delta x = f'(x) \cdot \lim_{\Delta x \to 0} \Delta x = \lim_{\Delta x \to 0} \Delta y$$

(3) Although the definitions of derivatives $f'(x)$ and $f'(c)$ are different, they are both derivatives.

For example, $f(x) = x^2 \qquad f'(x) = 2x \qquad f'(c) = 2c$

$f(x) = \sin x \qquad f'(x) = \cos x \qquad f'(c) = \cos c$

Example 1. $f(x) = \sin x,\ 0 \le x \le \dfrac{\pi}{2}$. Find a value of c such that

$f'(c) = \dfrac{f(b)-f(a)}{b-a}$ (In other words, find a value of c that satisfies the Mean Value Theorem.)

Solution. Here $a = 0,\ b = \dfrac{\pi}{2},\ f(0) = \sin 0 = 0,\ f(\dfrac{\pi}{2}) = \sin\dfrac{\pi}{2} = 1,\ f'(x) = \cos x$

Hence, from $f'(c) = \dfrac{f(b)-f(a)}{b-a}$ we have

$$\cos c = \frac{1-0}{\frac{\pi}{2}-0} = \frac{2}{\pi}$$

Therefore, $c \approx 0.881$ radians $\quad (0.881 \in (0, \dfrac{\pi}{2}))$

Example 2. Prove that

$$\arctan x_2 - \arctan x_1 \le x_2 - x_1 \qquad (x_1 < x_2)$$

Proof. Let $f(x) = \arctan x$. The function $f(x)$ satisfies the conditions of Lagrange's Mean Value Theorem.

Since $f'(x) = (\arctan x)' = \dfrac{1}{1+x^2}$

it follows that $f'(c) = \dfrac{1}{1+c^2}$

Therefore, we have

$$\arctan x_2 - \arctan x_1 = \frac{1}{1+c^2}(x_2 - x_1) \qquad c \in (x_1, x_2)$$

Since $\dfrac{1}{1+c^2} \le 1$

Hence, we have $\arctan x_2 - \arctan x_1 \le x_2 - x_1$

Example 3. Consider $f(x) = x^7 + 5x + 1$ on the interval [-2, 2]. Use Rolle's Theorem to prove that the equation $f(x) = 0$ has exactly one real root on this interval.

Solution. Since $f(-1) = -5 < 0$ and $f(0) = 1 > 0$, there is a root of $f(x)$ on the interval (-1, 0).

Suppose there are two zeros of $f(x)$ on [-2, 2]. That is, there exist a, b on the interval [-2, 2] such that $f(a) = f(b) = 0$.

Since $f(x)$ is differentiable and continuous on this interval, it follows from Rolle's Theorem that there exists a point c such that $f'(c) = 0$.

However, $f'(x) = 7x^6 + 5 > 0$ for all c.

Therefore, $f(x)$ cannot have more than one zero on the interval [-2, 2].

Consequently, the equation $f(x) = 0$ has exactly one real root on the interval [-2, 2].

Exercise 4.1

1. Find the value(s) of c on the given intervals whose existence is guaranteed by Rolle's Theorem.

a. $f(x) = 2x^2 - 5x - 3 \quad x \in [-0.5, 3]$

b. $f(x) = \dfrac{2}{1+x^2} \quad x \in [-3, 3]$

c. $f(x) = e^{x^2} - 1 \quad x \in [-1, 1]$

d. $f(x) = \cos x \quad x \in [0, 2\pi]$

2. Find the value(s) of c on the given intervals whose existence is guaranteed by Lagrange's Mean Value Theorem.

a $f(x) = 3x^3 - 4x^2 + 2x - 5 \quad x \in [1, 3]$

b. $f(x) = \arctan x \quad x \in [0, 1]$

c. $f(x)=\sqrt{16-x^2}$ $\quad x\in[-2,2]$

d. $f(x)=x^{\frac{3}{5}}$ $\quad x\in[0,32]$

3. Prove that the equation $x^3+2x^2+2x-5=0$ has exactly one real root.

4. Suppose that $f(x)$ is continuous on the interval [0, 4] and differentiable on (0, 4). If $f(0)=-6$ and $|f'(x)|\le 8$, find the bounds for $f(x)$ on the interval [0, 4].

5. Suppose that $f(x)$ is differentiable everywhere. If $f(1)=-4$ and $3<f'(x)<7$ for $1<x<6$, show that $11<f(x)<31$ for $1<x<6$.

6. Use Lagrange's Mean Value Theorem to prove that

$$|\tan y-\tan x|\ge|y-x| \qquad x,y\in(-\frac{\pi}{2},\frac{\pi}{2})$$

7. Let $a>b>0$. Show that $\dfrac{a-b}{a}<\ln\dfrac{a}{b}<\dfrac{a-b}{b}$.

8. Prove that when $x>1$, $e^x>ex$.

9. Apply the Mean Value Theorem to the function $f(x)=\sqrt[3]{x}$ on the interval [26, 27] to find bounds for the value of $\sqrt[3]{26}$. Give the answer correct to five decimal places.

10. Apply the Mean Value Theorem to the function $f(x)=\sqrt{x}$ on the interval [8, 9] to find bounds for the value of $\sqrt{8}$. Give the answer correct to five decimal places.

11. Apply the Mean Value Theorem to the function $f(x)=\ln x$ on the interval $[e^2, 8]$

to show that $2 < \ln 8 < \dfrac{8}{e^2} + 1.$

4.2. L'Hôpital's Rule

For limits of ratios of two infinitesimals (or two infinite quantities), some may exist, and some may not. For example, $\lim\limits_{x \to 0} \dfrac{\sin x}{x}$ is the limit of a ratio of two infinitesimals, and this limit exists.

$$\lim_{x \to 0} \frac{\sin x}{x} = 1 \qquad (\frac{0}{0} \text{ type})$$

Although $\lim\limits_{x \to 0} \dfrac{\sin x}{x^2}$ is also the limit of a ratio of two infinitesimals, this limit does not exist.

$$\lim_{x \to 0} \frac{\sin x}{x^2} = \lim_{x \to 0} \frac{1}{x} \cdot \frac{\sin x}{x} = \lim_{x \to 0} \frac{1}{x} \cdot \lim_{x \to 0} \frac{\sin x}{x} = \infty \cdot 1 = \infty \qquad (\frac{0}{0} \text{ type})$$

Consequently, we refer to such limits as '**indeterminate forms**' of the $\dfrac{0}{0}$ type.

Another indeterminate form is the $\dfrac{\infty}{\infty}$ type. Previously, we were limited to evaluating limits of only certain indeterminate forms. We will now deduce L'Hôpital's rule, a method for evaluating limits of indeterminate forms.

Theorem 4.1

If the functions $f(x)$ and $g(x)$ satisfy the following conditions:

(1) $\lim\limits_{x \to x_0} f(x) = \lim\limits_{x \to x_0} g(x) = 0$;

(2) they are differentiable in a neighbourhood of the point x_0;

(3) $\lim_{x \to x_0} \frac{f'(x)}{g'(x)} = A$ (or ∞), where $g'(x) \neq 0$.

then we have

$$\lim_{x \to x_0} \frac{f(x)}{g(x)} = \lim_{x \to x_0} \frac{f'(x)}{g'(x)} = A \text{ (or } \infty)$$

Proof. Since $\lim_{x \to x_0} f(x) = \lim_{x \to x_0} g(x) = 0$

thus $\lim_{x \to x_0} f(x) = f(x_0) = 0$, $\lim_{x \to x_0} g(x) = g(x_0) = 0$

Hence, $\lim_{x \to x_0} f'(x) = f'(x_0) = \lim_{\Delta x \to 0} \frac{f(x) - f(x_0)}{\Delta x} = \lim_{\Delta x \to 0} \frac{f(x) - 0}{\Delta x} = \lim_{\Delta x \to 0} \frac{f(x)}{\Delta x}$

Similarly, we have

$$\lim_{x \to x_0} g'(x) = g'(x_0) = \lim_{\Delta x \to 0} \frac{g(x) - g(x_0)}{\Delta x} = \lim_{\Delta x \to 0} \frac{g(x) - 0}{\Delta x} = \lim_{\Delta x \to 0} \frac{g(x)}{\Delta x}$$

Hence, $\lim_{x \to x_0} \frac{f'(x)}{g'(x)} = \lim_{x \to x_0} \frac{\frac{f(x)}{\Delta x}}{\frac{g(x)}{\Delta x}} = \lim_{x \to x_0} \frac{f(x)}{g(x)}$

That is, $\lim_{x \to x_0} \frac{f(x)}{g(x)} = \lim_{x \to x_0} \frac{f'(x)}{g'(x)}$ □

Using this theorem, we can evaluate the limits of indeterminate forms of the $\frac{0}{0}$ type.

If $\lim_{x \to x_0} \frac{f'(x)}{g'(x)}$ remains an indeterminate form of the $\frac{0}{0}$ type, and the functions $f'(x)$ and $g'(x)$ also satisfy the conditions that $f(x)$ and $g(x)$ must satisfy in the theorem, then we can apply L'Hôpital's rule repeatedly. That is,

$$\lim_{x \to x_0} \frac{f(x)}{g(x)} = \lim_{x \to x_0} \frac{f'(x)}{g'(x)} = \lim_{x \to x_0} \frac{f''(x)}{g''(x)} = \ldots = \lim_{x \to x_0} \frac{f^{(n)}(x)}{g^{(n)}(x)} = A \text{ (or } \infty)$$

If we cannot determine the limit of $\lim\limits_{x\to x_0}\dfrac{f'(x)}{g'(x)}$ or we can determine that it oscillates and has no limit, then L'Hôpital's rule cannot be applied. In such cases, we must use alternative methods to evaluate the limit of indeterminate forms of the $\dfrac{0}{0}$ type.

Example 1. Find $\lim\limits_{x\to 3}\dfrac{x^4-81}{x-3}$.

Solution. Since $\lim\limits_{x\to 3}\left(x^4-81\right)=3^4-81=0$

$$\lim_{x\to 3}\left(x-3\right)=3-3=0$$

thus $\lim\limits_{x\to 3}\dfrac{x^4-81}{x-3}$ $\quad \dfrac{0}{0}$ type

we can apply L'Hôpital's rule to evaluate this limit.

$$\lim_{x\to 3}\frac{x^4-81}{x-3}=\lim_{x\to 3}\frac{(x^4-81)'}{(x-3)'}=\lim_{x\to 3}\frac{4x^3-0}{1-0}=4\times 3^3=108$$

That is, $\lim\limits_{x\to 3}\dfrac{x^4-81}{x-3}=108$

Example 2. Find $\lim\limits_{x\to 0}\dfrac{1-e^x}{x^2-x}$.

Solution. Since $\lim\limits_{x\to 0}\dfrac{1-e^x}{x^2-x}$ $\quad \dfrac{0}{0}$ type

thus $\lim\limits_{x\to 0}\dfrac{1-e^x}{x^2-x}=\lim\limits_{x\to 0}\dfrac{(1-e^x)'}{(x^2-x)'}=\lim\limits_{x\to 0}\dfrac{-e^x}{2x-1}=1$

That is, $\lim\limits_{x\to 0}\dfrac{1-e^x}{x^2-x}=1$

Example 3. Find $\lim\limits_{x\to 1}\frac{\ln x}{x^2-1}$.

Solution. Since $\lim\limits_{x\to 1}\frac{\ln x}{x^2-1}$ $\quad \frac{0}{0}$ type

thus $$\lim_{x\to 1}\frac{\ln x}{x^2-1}=\lim_{x\to 1}\frac{(\ln x)'}{(x^2-1)'}=\lim_{x\to 1}\frac{\frac{1}{x}}{2x}=\frac{1}{2}$$

That is, $$\lim_{x\to 1}\frac{\ln x}{x^2-1}=\frac{1}{2}$$

Example 4. Find $\lim\limits_{x\to 0}\frac{x-\sin x}{x^3}$.

Solution. Since $\lim\limits_{x\to 0}\frac{x-\sin x}{x^3}$ $\quad \frac{0}{0}$ type

thus $$\lim_{x\to 0}\frac{x-\sin x}{x^3}=\lim_{x\to 0}\frac{(x-\sin x)'}{(x^3)'}=\lim_{x\to 0}\frac{1-\cos x}{3x^2}$$

Because $\lim\limits_{x\to 0}\frac{1-\cos x}{3x^2}$ $\quad \frac{0}{0}$ type

we can continue to apply L'Hôpital's rule and obtain the derivatives of the numerator and denominator.

Hence, $$\lim_{x\to 0}\frac{1-\cos x}{3x^2}=\lim_{x\to 0}\frac{(1-\cos x)'}{(3x^2)'}=\lim_{x\to 0}\frac{\sin x}{6x}=\frac{1}{6}$$ Use $\lim\limits_{x\to 0}\frac{\sin x}{x}=1$.

Alternatively, continue to apply L'Hôpital's rule and obtain the derivatives of the numerator and denominator.

$$\lim_{x\to 0}\frac{\sin x}{6x}=\lim_{x\to 0}\frac{(\sin x)'}{(6x)'}=\lim_{x\to 0}\frac{\cos x}{6}=\frac{\cos 0}{6}=\frac{1}{6}$$

Therefore, $$\lim_{x\to 0}\frac{x-\sin x}{x^3}=\frac{1}{6}$$

Example 5. Find $\lim_{x\to 0}\frac{\ln(1+x)}{x^2}$.

Solution. Since $\lim_{x\to 0}\frac{\ln(1+x)}{x^2}$ $\quad \frac{0}{0}$ type

thus
$$\lim_{x\to 0}\frac{\ln(1+x)}{x^2}=\lim_{x\to 0}\frac{[\ln(1+x)]'}{(x^2)'}=\lim_{x\to 0}\frac{\frac{1}{1+x}}{2x}=\lim_{x\to 0}\frac{1}{2x(1+x)}=\infty$$

That is,
$$\lim_{x\to 0}\frac{\ln(1+x)}{x^2}=\infty$$

Example 6. Find $\lim_{x\to 0}\frac{x^2\sin\frac{1}{x}}{\sin x}$.

Solution. Since $\lim_{x\to 0}\frac{x^2\sin\frac{1}{x}}{\sin x}$ $\quad \frac{0}{0}$ type

thus
$$\lim_{x\to 0}\frac{x^2\sin\frac{1}{x}}{\sin x}=\lim_{x\to 0}\frac{(x^2\sin\frac{1}{x})'}{(\sin x)'}=\lim_{x\to 0}\frac{2x\sin\frac{1}{x}-\cos\frac{1}{x}}{\cos x}$$

$$=\frac{2\cdot 0\cdot\sin\frac{1}{0}-\cos\frac{1}{0}}{\cos 0}=\frac{0-\cos\infty}{1}=-\cos\infty$$

That is,
$$\lim_{x\to 0}\frac{x^2\sin\frac{1}{x}}{\sin x}=-\cos\infty$$

However, since infinity is a changing value with no end so the value of cos∞ oscillates between -1 and 1 (that is, change occurs repeatedly on the interval [-1, 1]) and has no limit. Consequently, L'Hôpital's rule cannot be applied in this instance. Nevertheless, the limit of the original expression does exist, and we can evaluate it using the following method.

$$\lim_{x\to 0}\frac{x^2\sin\frac{1}{x}}{\sin x}=\lim_{x\to 0}\frac{x}{\sin x}\cdot x\sin\frac{1}{x}=\lim_{x\to 0}\frac{x\sin\frac{1}{x}}{\frac{\sin x}{x}}=\frac{0}{1}=0$$

That is, $$\lim_{x\to 0}\frac{x^2\sin\frac{1}{x}}{\sin x}=0$$ □

It can be shown that for limits of indeterminate forms of the $\frac{\infty}{\infty}$ type, the following theorem holds.

Theorem 4.2

If the functions $f(x)$ and $g(x)$ satisfy the following conditions:

(1) $\lim\limits_{x\to x_0} f(x)=\lim\limits_{x\to x_0} g(x)=\infty$;

(2) they are differentiable in a neighbourhood of the point x_0;

(3) $\lim\limits_{x\to x_0}\frac{f'(x)}{g'(x)}=A$ (or ∞), where $g'(x)\neq 0$,

then we have

$$\lim_{x\to x_0}\frac{f(x)}{g(x)}=\lim_{x\to x_0}\frac{f'(x)}{g'(x)}=A \text{ (or } \infty)$$

Proof. **Method 1**

Let $$h(x)=\frac{1}{f(x)}$$

then $$h'(x)=\frac{d\frac{1}{f(x)}}{dx}=\frac{d[f(x)]^{-1}}{df(x)}\cdot\frac{df(x)}{dx}=-\frac{1}{[f(x)]^2}\cdot f'(x)$$

Let $$s(x) = \frac{1}{g(x)}$$

then $$s'(x) = \frac{d\frac{1}{g(x)}}{dx} = \frac{d[g(x)]^{-1}}{dg(x)} \cdot \frac{dg(x)}{dx} = -\frac{1}{[g(x)]^2} \cdot g'(x)$$

Since $$\lim_{x \to x_0} f(x) = \lim_{x \to x_0} g(x) = \infty$$

thus $$\lim_{x \to x_0} h(x) = \lim_{x \to x_0} \frac{1}{f(x)} = \frac{1}{\infty} = 0$$

$$\lim_{x \to x_0} s(x) = \lim_{x \to x_0} \frac{1}{g(x)} = \frac{1}{\infty} = 0$$

That is, $$\lim_{x \to x_0} h(x) = \lim_{x \to x_0} s(x) = 0$$

By Theorem 4.1, we have

$$\lim_{x \to x_0} \frac{s(x)}{h(x)} = \lim_{x \to x_0} \frac{s'(x)}{h'(x)}$$

Hence, $$\lim_{x \to x_0} \frac{f(x)}{g(x)} = \lim_{x \to x_0} \frac{s(x)}{h(x)} = \lim_{x \to x_0} \frac{s'(x)}{h'(x)} = \lim_{x \to x_0} \left(-\frac{g'(x)}{[g(x)]^2}\right)\left(-\frac{[f(x)]^2}{f'(x)}\right)$$

$$= \lim_{x \to x_0} \frac{[f(x)]^2}{[g(x)]^2} \cdot \frac{g'(x)}{f'(x)}$$

That is, $$\lim_{x \to x_0} \frac{f(x)}{g(x)} = \lim_{x \to x_0} \frac{[f(x)]^2}{[g(x)]^2} \cdot \frac{g'(x)}{f'(x)}$$

Cross-multiplying, we obtain

$$\lim_{x \to x_0} \frac{f'(x)}{g'(x)} = \lim_{x \to x_0} \frac{f(x)}{g(x)}$$

That is, $$\lim_{x \to x_0} \frac{f(x)}{g(x)} = \lim_{x \to x_0} \frac{f'(x)}{g'(x)}$$

Method 2

Since $$\frac{\infty}{\infty} = \frac{1}{\infty} \cdot \infty = 0 \cdot \infty = 0 \cdot \frac{1}{0} = \frac{0}{0}$$

thus $$\lim_{x\to x_0}\frac{f(x)}{g(x)}=\frac{\infty}{\infty}=\frac{0}{0}=\lim_{x\to x_0}\frac{f'(x)}{g'(x)}$$

That is, $$\lim_{x\to x_0}\frac{f(x)}{g(x)}=\lim_{x\to x_0}\frac{f'(x)}{g'(x)}$$ □

- Limits of the $\frac{0}{0}$ type and the $\frac{\infty}{\infty}$ type will have only two possible results: A (or ∞). Consequently, we know that the derivative of a function $f(x)$

  $$f'(x_0)=\lim_{\Delta x\to 0}\frac{\Delta y}{\Delta x}=\frac{0}{0}=A\ \ (\text{or }\infty)$$ has only two possible results.

Example 7. Find $\lim_{x\to\frac{\pi}{2}}\frac{\tan 5x}{\tan x}$.

Solution. Since $\lim_{x\to\frac{\pi}{2}}\frac{\tan 5x}{\tan x}$ $\quad\frac{\infty}{\infty}$ type

then $$\lim_{x\to\frac{\pi}{2}}\frac{\tan 5x}{\tan x}=\lim_{x\to\frac{\pi}{2}}\frac{(\tan 5x)'}{(\tan x)'}=\lim_{x\to\frac{\pi}{2}}\frac{5\sec^2 5x}{\sec^2 x}=\lim_{x\to\frac{\pi}{2}}\frac{\frac{5}{\cos^2 5x}}{\frac{1}{\cos^2 x}}=5\lim_{x\to\frac{\pi}{2}}\frac{\cos^2 x}{\cos^2 5x}$$

Since $\lim_{x\to\frac{\pi}{2}}\frac{\cos^2 x}{\cos^2 5x}$ $\quad\frac{0}{0}$ type

we can continue to apply L'Hôpital's rule to obtain the derivatives of the numerator and denominator, therefore,

$$5\lim_{x\to\frac{\pi}{2}}\frac{\cos^2 x}{\cos^2 5x}=5\lim_{x\to\frac{\pi}{2}}\frac{(\cos^2 x)'}{(\cos^2 5x)'}=5\lim_{x\to\frac{\pi}{2}}\frac{2\cos x\cdot(-\sin x)}{2\cos 5x\cdot(-5\sin 5x)}=\lim_{x\to\frac{\pi}{2}}\frac{\sin 2x}{\sin 10x}$$

Since $\lim_{x\to\frac{\pi}{2}}\frac{\sin 2x}{\sin 10x}$ $\quad\frac{0}{0}$ type we can continue to apply L'Hôpital's rule to obtain the derivatives of the numerator and denominator.

$$\lim_{x\to\frac{\pi}{2}}\frac{\sin 2x}{\sin 10x}=\lim_{x\to\frac{\pi}{2}}\frac{(\sin 2x)'}{(\sin 10x)'}=\lim_{x\to\frac{\pi}{2}}\frac{2\cos 2x}{10\cos 10x}=\frac{1}{5}\cdot\frac{\cos 2\cdot\frac{\pi}{2}}{\cos 10\cdot\frac{\pi}{2}}=\frac{1}{5}$$

That is, $\lim_{x\to\frac{\pi}{2}}\frac{\tan 5x}{\tan x}=\frac{1}{5}$

Example 8. Find $\lim_{x\to\infty}\frac{\ln x}{x^b}$ $(b>0)$.

Solution. Since $\lim_{x\to\infty}\frac{\ln x}{x^b}$ $\frac{\infty}{\infty}$ type

thus $$\lim_{x\to\infty}\frac{\ln x}{x^b}=\lim_{x\to\infty}\frac{(\ln x)'}{(x^b)'}=\lim_{x\to\infty}\frac{\frac{1}{x}}{bx^{b-1}}=\lim_{x\to\infty}\frac{1}{bx^b}=0$$

That is, $\lim_{x\to\infty}\frac{\ln x}{x^b}=0$

Example 9. Find $\lim_{x\to\infty}\frac{e^{2x}}{x^2}$.

Solution. Since $\lim_{x\to\infty}\frac{e^{2x}}{x^2}$ $\frac{\infty}{\infty}$ type

thus $$\lim_{x\to\infty}\frac{e^{2x}}{x^2}=\lim_{x\to\infty}\frac{(e^{2x})'}{(x^2)'}=\lim_{x\to\infty}\frac{2e^{2x}}{2x}=\lim_{x\to\infty}\frac{(2e^{2x})'}{(2x)'}=\lim_{x\to\infty}\frac{4e^{2x}}{2}=\infty$$

That is, $\lim_{x\to\infty}\frac{e^{2x}}{x^2}=\infty$ □

Please note: L'Hôpital's rule can only be applied directly to evaluate limits of indeterminate forms of the $\frac{0}{0}$ type and the $\frac{\infty}{\infty}$ type. For other indeterminate forms, such as $\infty - \infty$, $0 \cdot \infty$, 1^∞, 0^0, ∞^0 etc., an appropriate transformation is required to convert them into the $\frac{0}{0}$ type or the $\frac{\infty}{\infty}$ type first before L'Hôpital's rule can be applied.

Example 10. Find $\lim\limits_{x \to 0^+}\left(\frac{1}{x}-\frac{1}{\sin x}\right)$.

Solution.

$$\lim_{x \to 0^+}\left(\frac{1}{x}-\frac{1}{\sin x}\right) \qquad \infty - \infty \text{ type}$$

$$= \lim_{x \to 0^+}\frac{\sin x - x}{x \sin x} \qquad \frac{0}{0} \text{ type}$$

$$= \lim_{x \to 0^+}\frac{(\sin x - x)'}{(x \sin x)'} = \lim_{x \to 0^+}\frac{\cos x - 1}{\sin x + x\cos x} \qquad \frac{0}{0} \text{ type}$$

$$= \lim_{x \to 0^+}\frac{(\cos x - 1)'}{(\sin x + x\cos x)'} = \lim_{x \to 0^+}\frac{-\sin x}{\cos x + \cos x - x\sin x}$$

$$= \frac{-\sin 0}{\cos 0 + \cos 0 - 0 \cdot \sin 0} = \frac{0}{2} = 0$$

Example 11. Find $\lim\limits_{x \to \infty} x\left(\frac{\pi}{2} - \arctan x\right)$.

Solution.

$$\lim_{x \to \infty} x\left(\frac{\pi}{2} - \arctan x\right) \qquad \infty \cdot 0 \text{ type}$$

$$= \lim_{x \to \infty}\frac{\frac{\pi}{2} - \arctan x}{\frac{1}{x}} \qquad \frac{0}{0} \text{ type}$$

$$= \lim_{x\to\infty} \frac{(\frac{\pi}{2} - \arctan x)'}{(\frac{1}{x})'} = \lim_{x\to\infty} \frac{-\frac{1}{1+x^2}}{-\frac{1}{x^2}} = \lim_{x\to\infty} \frac{x^2}{1+x^2} \qquad \frac{\infty}{\infty} \text{ type}$$

$$= \lim_{x\to\infty} \frac{(x^2)'}{(1+x^2)'} = \lim_{x\to\infty} \frac{2x}{2x} = 1$$

Example 12. Find $\lim_{x\to 1} x^{\frac{1}{1-x}}$.

Solution. $\lim_{x\to 1} x^{\frac{1}{1-x}}$ $\qquad 1^\infty$ type

$$= \lim_{x\to 1} e^{\ln x^{\frac{1}{1-x}}} = \lim_{x\to 1} e^{\frac{\ln x}{1-x}} = e^{\lim_{x\to 1}\frac{\ln x}{1-x}}$$

For $\lim_{x\to 1} \frac{\ln x}{1-x}$ $\qquad \frac{0}{0}$ type

$$= \lim_{x\to 1} \frac{(\ln x)'}{(1-x)'} = \lim_{x\to 1} \frac{\frac{1}{x}}{-1} = -1$$

Therefore, $\lim_{x\to 1} x^{\frac{1}{1-x}} = e^{-1}$

Note: In this example, we apply the formula.

$$N = a^{\log_a N}$$

Proof. Since $\log_a N = \log_a N \cdot \log_a a \qquad (\log_a a = 1)$

$$= \log_a a^{\log_a N}$$

thus $N = a^{\log_a N}$ □

Example 13. Find $\lim_{x\to 0} x^x$.

Solution. $\lim_{x\to 0} x^x$ $\qquad$ 0^0 type

$$= \lim_{x\to 0} e^{\ln x^x} = \lim_{x\to 0} e^{x\ln x} = \lim_{x\to 0} e^{\frac{\ln x}{\frac{1}{x}}} = e^{\lim_{x\to 0}\frac{\ln x}{\frac{1}{x}}}$$

For $\lim_{x\to 0} \frac{\ln x}{\frac{1}{x}}$ $\qquad$ $\frac{-\infty}{\infty}$ type

$$= \lim_{x\to 0} \frac{(\ln x)'}{(\frac{1}{x})'} = \lim_{x\to 0} \frac{\frac{1}{x}}{-\frac{1}{x^2}} = \lim_{x\to 0}(-x) = 0$$

therefore, $\lim_{x\to 0} x^x = e^{\lim_{x\to 0}\frac{\ln x}{\frac{1}{x}}} = e^0 = 1$

Example 14. Find $\lim_{x\to 0}(1+x)^{\frac{1}{x}}$.

Solution. $\lim_{x\to 0}(1+x)^{\frac{1}{x}}$ $\qquad$ 1^∞ type

$$= \lim_{x\to 0} e^{\ln(1+x)^{\frac{1}{x}}} = \lim_{x\to 0} e^{\frac{\ln(1+x)}{x}} = e^{\lim_{x\to 0}\frac{\ln(1+x)}{x}} = e^{\lim_{x\to 0}\frac{\frac{1}{1+x}}{1}} = e^1 = e$$

Example 15. Find $\lim_{x\to 0}\left(\frac{1}{x} - \frac{1}{e^x - 1}\right)$.

Solution. $\lim_{x\to 0}\left(\frac{1}{x} - \frac{1}{e^x - 1}\right)$ $\qquad$ $\infty - \infty$ type

$$= \lim_{x \to 0} \frac{e^x - 1 - x}{x\left(e^x - 1\right)} \qquad \frac{0}{0} \text{ type}$$

$$= \lim_{x \to 0} \frac{e^x - 1}{e^x - 1 + xe^x} \qquad \frac{0}{0} \text{ type}$$

$$= \lim_{x \to 0} \frac{e^x}{e^x + e^x + xe^x} = \frac{e^0}{e^0 + e^0 + 0 \cdot e^0} = \frac{1}{2}$$

Example 16. Find $\lim_{x \to 0^+} x^p \ln x \quad (p > 0)$.

Solution. $\lim_{x \to 0^+} x^p \ln x \quad 0 \cdot -\infty$ type

$$= \lim_{x \to 0^+} \frac{\ln x}{\frac{1}{x^p}} \qquad \frac{-\infty}{\infty} \text{ type}$$

$$= \lim_{x \to 0^+} \frac{\frac{1}{x}}{\frac{-p}{x^{p+1}}} = \lim_{x \to 0^+} \frac{x^p}{-p} = 0$$

Example 17. Find $\lim_{x \to 0^+} \left(\ln \frac{1}{x}\right)^x$.

Solution. $\lim_{x \to 0^+} \left(\ln \frac{1}{x}\right)^x \qquad \infty^0$ type

$$= \lim_{x \to 0^+} e^{\ln\left(\ln \frac{1}{x}\right)^x} = e^{\lim_{x \to 0^+} x \ln\left(\ln \frac{1}{x}\right)}$$

For $\lim_{x \to 0^+} x \ln\left(\ln \frac{1}{x}\right) \qquad 0 \cdot \infty$ type

$$= \lim_{x\to 0^+} \frac{\ln\left(\ln\frac{1}{x}\right)}{\frac{1}{x}} \qquad \frac{\infty}{\infty} \text{ type}$$

$$= \lim_{x\to 0^+} \frac{\dfrac{d\ln\left(\ln\frac{1}{x}\right)}{d\ln\frac{1}{x}} \cdot \dfrac{d\ln\frac{1}{x}}{d\frac{1}{x}} \cdot \dfrac{d\frac{1}{x}}{dx}}{\dfrac{d\frac{1}{x}}{dx}} = \lim_{x\to 0^+} \frac{1}{\ln\frac{1}{x}} \cdot x \qquad 0 \cdot 0 \text{ type}$$

$$= \lim_{x\to 0^+} \frac{x}{-\ln x} = 0$$

therefore, $\lim_{x\to 0^+} \left(\ln\frac{1}{x}\right)^x = e^0 = 1$

Example 18. Find $\lim_{x\to 0^+} \left(\frac{1}{x} - \ln x\right)$.

Solution. $\lim_{x\to 0^+} \left(\frac{1}{x} - \ln x\right)$ $\qquad \infty - (-\infty)$ type

$= \infty$

Example 19. Find $\lim_{x\to -\infty} \left(x + x^3\right)$.

Solution. $\lim_{x\to -\infty} \left(x + x^3\right)$ $\qquad (-\infty) - \infty$ type

$= -\infty$

Note: Limits of the types $\frac{0}{\infty}$, $\infty \cdot \infty$, $\infty + \infty$, $\infty - (-\infty)$ are not indeterminate forms.

Exercise 4.2

1. Evaluate the following using L'Hôpital's rule.

a. $\lim_{x\to 0}\frac{\sin 5x}{x}$

b. $\lim_{x\to 1}\frac{x^3-3x^2+2}{x^3-x^2-x+1}$

c. $\lim_{x\to 0}\frac{e^x-e^{-x}}{x}$

d. $\lim_{x\to 3}\frac{\sqrt{x+1}-2}{x-3}$

e. $\lim_{x\to \frac{\pi}{2}^+}\frac{\ln(x-\frac{\pi}{2})}{\tan x}$

f. $\lim_{x\to 0}\frac{7^x-2^x}{3x}$

2. Evaluate the following using L'Hôpital's rule.

a. $\lim_{x\to \pi}\frac{\sin 3x}{\tan 5x}$

b. $\lim_{x\to 0}\frac{\sin(2x)-2x}{x^4}$

c. $\lim_{x\to 0^+}\frac{\ln(\tan 5x)}{\ln(\tan 4x)}$

d. $\lim_{x\to 0}\frac{e^x-e^{-x}-2x}{\sin(x^2)}$

e. $\lim_{x\to 0}\frac{1-\cos x^2}{x^2\sin(x^2)}$

f. $\lim_{x\to 0}\frac{\arcsin(4x)}{\arctan(3x)}$

3. Evaluate the following limits.

a. $\lim_{x\to 0}(\frac{1}{e^x-1}-\frac{1}{x})$

b. $\lim_{x\to \infty}4xe^{-3x}$

c. $\lim_{x\to \infty}x\sin\frac{1}{x}$

d. $\lim_{x\to 0^+}(\cot x-\cos ecx)$

e. $\lim_{x\to \infty}e^{-3x}\ln 5x$

f. $\lim_{x\to 0^+}x\ln x$

4. Evaluate the following limits.

a. $\lim_{x\to \infty}x^{\frac{1}{x^2}}$

b. $\lim_{x\to 0}x^{1-\cos 3x}$

4.3. Continuity and differentiability of functions

4.3.1. Left and right limits

1.

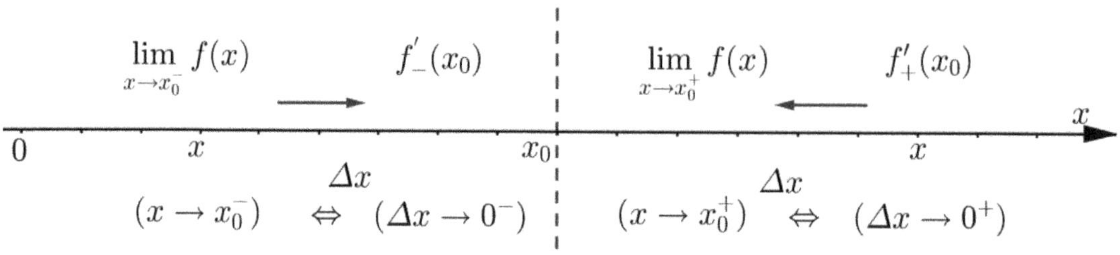

Figure 4.3.1

If x approaches x_0 from the left $(x < x_0)$ or from the right $(x > x_0)$, we can use the superscripts '–' and '+' on the limit points x_0 or 0 to denote the direction. That is,

$$x \to x_0^- \qquad \Delta x \to 0^- \qquad x \to x_0^+ \qquad \Delta x \to 0^+$$

$$(x \to x_0) = (x \to x_0^-) \quad \text{or} \quad (x \to x_0^+)$$

The left limit of the function $f(x)$ at $x = x_0$ is $\lim\limits_{x \to x_0^-} f(x)$

The right limit of the function $f(x)$ at $x = x_0$ is $\lim\limits_{x \to x_0^+} f(x)$

The left derivative of the function $f(x)$ at $x = x_0$ is

$$f_-'(x_0) = \lim_{\Delta x \to 0^-} \frac{\Delta y}{\Delta x} = \lim_{\Delta x \to 0^-} \frac{f(x_0 + \Delta x) - f(x_0)}{\Delta x}$$

The right derivative of the function $f(x)$ at $x = x_0$ is

$$f_+'(x_0) = \lim_{\Delta x \to 0^+} \frac{\Delta y}{\Delta x} = \lim_{\Delta x \to 0^+} \frac{f(x_0 + \Delta x) - f(x_0)}{\Delta x}$$

2. The relationship between the derivative, the left derivative and the right derivative

Since $\quad (x \to x_0) = (x \to x_0^-) \quad \text{or} \quad (x \to x_0^+)$

thus $$\lim_{\Delta x_0 \to 0} \frac{\Delta y}{\Delta x_0} = \lim_{\Delta x_0 \to 0^-} \frac{\Delta y}{\Delta x_0} \quad \text{or} \quad \lim_{\Delta x_0 \to 0^+} \frac{\Delta y}{\Delta x_0}$$

That is, $$f'(x_0) = f_-'(x_0) \quad \text{or} \quad f_+'(x_0)$$

- If the function $f(x)$ is differentiable at the point $x = x_0$, it follows from the definition of the limit that

  $$f_-'(x_0) = f_+'(x_0) = \text{A}$$

  then $$f'(x_0) = f_-'(x_0) = f_+'(x_0) = \text{A}$$

  That is, $f'(x_0) = \text{A}$ (one value)

  Therefore, the derivative $f'(x_0)$ exists.

- If $$f_-'(x_0) \neq f_+'(x_0)$$

  then $f'(x_0) = f_-'(x_0)$ or $f_+'(x_0)$ (two different values)

  which implies that the derivative $f'(x_0)$ does not exist.

  Therefore, the function $f(x)$ is not differentiable at the point $x = x_0$.

3. The definition of the limit is

$$\lim_{x \to x_0} f(x) = \text{A} \qquad \text{(one value)}$$

From the definition of the limit, it follows that

if $\lim_{x \to x_0} f(x) = \text{A}_1, \text{A}_2, ..., \text{A}_n$ (n different values, $n \geq 2$)

and this does not satisfy the definition of a limit as a number A; therefore $\lim_{x \to x_0} f(x)$ does not exist.

Note: Explanation of the definitions of a function, a limit and a derivative

function $f(x) = y$, limit $\lim_{x \to x_0} f(x) = \text{A}$, derivative $f'(x) = y'$

These three definitions imply that for every real number x, there is one definite real number (y, A, y', respectively) corresponding to it, thus, all of them are single-value functions.

The expressions for the function $f(x)$, the limit $\lim_{x \to x_0} f(x)$ and the derivative $f'(x)$ at the same time are also expressions of operation. If these operations yield only one result (y, A, y', respectively) – that is, if they are single-value functions – then they exist. Otherwise, they do not exist.

For instance,

let $\quad f(x) = x^2 + 3$

then $\quad f(2) = 2^2 + 3 = 7 \quad$ one result of operation: 7

Corresponding to $x = 2$, y has only one value 7.

This satisfies the definition of a single-value function.

Therefore, $f(2)$ exists.

However, suppose that

$f(2) = 2^2 + 3 = 7$ or 5 $\quad$ two different results of operation: 7 or 5

Corresponding to $x = 2$, y has two different values, 7 or 5.

This does not satisfy the definition of a single-value function.

Therefore, $f(2)$ does not exist.

4.3.2. Continuity and differentiability of piecewise functions

A piecewise function $f(x)$ consists of n different expressions, that is, n different sub-functions (also known as branches). Consequently, it has n corresponding derivatives. Therefore, at a common endpoint, $x = x_0$ of neighbouring intervals, the two sub-functions are not necessarily continuous, nor do they necessarily share the same derivative. A piecewise function typically takes the following form:

$$\text{a piecewise function} \quad f(x) = \begin{cases} x-1, & x \le 0 \\ 2x, & 0 < x \le 1 \\ x^2+1, & 1 < x \le 2 \\ \frac{1}{2}x+4, & x > 2 \end{cases} = \begin{cases} g(x), & x \le 0 \\ h(x), & 0 < x \le 1 \\ u(x), & 1 < x \le 2 \\ v(x), & x > 2 \end{cases}$$

1. Continuity of a piecewise function $f(x)$

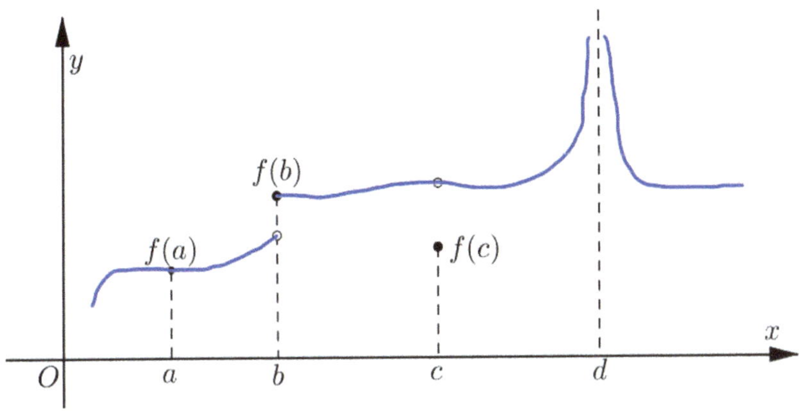

Figure 4.3.2

As shown in Figure 4.3.2, the function is discontinuous at *b*, *c* and *d*.

To determine whether a function $f(x)$ is continuous at $x = x_0$, we may check whether the function $f(x)$ is defined at $x = x_0$. However, this judgment method is not generally valid. For example, it fails to determine the continuity at b in Figure 4.3.2. Therefore, another method is required.

Definition 4.1

Let the function $y = f(x)$ be defined in a neighbourhood of the point $x = x_0$. If, as the change in the independent variable *x*, Δx, approaches zero at $x = x_0$, the change in the function, Δy, also approaches zero – that is,

$$\lim_{\Delta x \to 0} \Delta y = 0$$

– then we say that the function $f(x)$ is continuous at $x = x_0$.

We can use the judgement formula $\lim_{\Delta x \to 0} \Delta y = 0$ to determine whether the piecewise function $f(x)$ is continuous at $x = x_0$.

Since $$\lim_{\Delta x \to 0} \Delta y = \lim_{x \to x_0} \left[f(x) - f(x_0) \right] = 0$$

rearranging, we obtain

$$\lim_{x \to x_0} f(x) = f(x_0)$$

Since $$\lim_{x \to x_0} f(x) = \lim_{x \to x_0^-} f(x) = \lim_{x \to x_0^+} f(x)$$

thus $$\lim_{x \to x_0} f(x) = \lim_{x \to x_0^-} f(x) = \lim_{x \to x_0^+} f(x) = f(x_0)$$

Therefore,

$$\lim_{\Delta x \to 0} \Delta y = 0 \Leftrightarrow \lim_{x \to x_0} f(x) = f(x_0) \Leftrightarrow \lim_{x \to x_0} f(x) = \lim_{x \to x_0^-} f(x) = \lim_{x \to x_0^+} f(x) = f(x_0)$$

Hence, the judgement formula $\lim_{\Delta x \to 0} \Delta y = 0$ can be rearranged to the following form.

$$\lim_{x \to x_0} f(x) = \lim_{x \to x_0^-} f(x) = \lim_{x \to x_0^+} f(x) = f(x_0)$$

- If $\lim_{x \to x_0^-} f(x) = \lim_{x \to x_0^+} f(x) = \text{A}$ (same value)

  then $\lim_{x \to x_0} f(x) = \lim_{x \to x_0^-} f(x) = \lim_{x \to x_0^+} f(x) = \text{A}$

  That is, $\lim_{x \to x_0} f(x) = \text{A}$

  If $f(x_0) = \text{A}$

  then $\lim_{x \to x_0} f(x) = \text{A} = f(x_0)$ $\quad (\lim_{\Delta x \to 0} \Delta y = 0)$

  Therefore, the piecewise function $f(x)$ is continuous at $x = x_0$.

- If $\lim_{x \to x_0^-} f(x) \neq \lim_{x \to x_0^+} f(x)$

  then $\lim_{x \to x_0} f(x)$ does not exist.

  $\lim_{x \to x_0} f(x) \neq f(x_0)$ $\quad (\lim_{\Delta x \to 0} \Delta y \neq 0)$

  Therefore, the piecewise function $f(x)$ is discontinuous at $x = x_0$.

2. Differentiability of a piecewise function $f(x)$

(1) If the piecewise function $f(x)$ is continuous at $x = x_0$ and satisfies that

$$f_-'(x_0) = f_+'(x_0)$$

then $\quad f'(x_0) = f_-'(x_0) = f_+'(x_0)$

Consequently, the derivative $f'(x_0)$ exists.

Therefore, the piecewise function $f(x)$ is differentiable at $x = x_0$.

If $\quad f_-'(x_0) \neq f_+'(x_0)$

then the derivative $f'(x_0)$ does not exist.

Therefore, the piecewise function $f(x)$ is not differentiable at $x = x_0$.

(2) If the piecewise function $f(x)$ is discontinuous at $x = x_0$, then regardless of whether the left or right derivatives are equal or not, the piecewise function $f(x)$ is not differentiable at $x = x_0$.

4.3.3. Continuity and differentiability of single-expression functions

1. Properties of single-expression functions

If $\lim_{x \to x_0} f(x) = A$ (a constant), then we always have the identity

$$\lim_{x \to x_0} f(x) = \lim_{x \to x_0^-} f(x) = \lim_{x \to x_0^+} f(x) = A$$

If the derivative $f'(x)$ exists, then we always have the identity

$$f'(x) = f_-'(x) = f_+'(x)$$

These identities hold because the expressions for the left and right limits are identical; furthermore, at the end of the change process, the left and right limits have

$$x = x_0^- = x_0^+ = x_0, \quad \Delta x = 0^- = 0^+ = 0$$

Therefore, when we evaluate the limit $\lim_{x \to x_0} f(x)$ or the derivative $f'(x)$, we do not need to consider the left and right limits, regardless of whether the function is continuous or not, This distinguishes them from piecewise functions.

2. Methods for determining the continuity and differentiability of a single-expression function $f(x)$ at $x = x_0$

(1) If $f(x)$ is a polynomial function or a rational function whose denominator is non-zero at $x = x_0$, it follows from the calculation rules of limits that

$$\lim_{x \to x_0} f(x) = f(x_0)$$

That is, the function $f(x)$ is continuous at $x = x_0$.

(2) If when $x = x_0$, $f(x_0) = \frac{0}{0}, \pm 0 \cdot \infty, \pm \frac{\infty}{\infty} = a$ (a is a random number), then by L'hôpital's rule we have

$$\lim_{x \to x_0} f(x) = A \quad (\text{or } \pm\infty)$$

Since a random number $a \neq A$ (or $\pm\infty$)

thus $\lim_{x \to x_0} f(x) \neq f(x_0)$

That is, $f(x)$ is neither continuous nor differentiable at $x = x_0$.

(3) If $f(x_0) = \dfrac{a}{0}$ $(a \neq 0)$,

then $\lim_{x \to x_0} f(x) = f(x_0) = \pm\infty$ (not defined)

That is, $f(x)$ is neither continuous nor differentiable at $x = x_0$.

3. The Theorem on the relationship between the differentiability and continuity of a single-expression function

If a function $f(x)$ is differentiable at $x = x_0$, then $f(x)$ is necessarily continuous at $x = x_0$. However, if $f(x)$ is not differentiable at $x = x_0$, we cannot determine whether $f(x)$ is continuous or not at $x = x_0$. The proof is as follows.

- If $f(x)$ is differentiable at $x = x_0$, then $f'(x_0) = A$ (a constant).

  Since the judgement formula is $\lim_{\Delta x \to 0} \Delta y = f'(x_0) \cdot \lim_{\Delta x \to 0} \Delta x = A \cdot 0 = 0$

  therefore, $f(x)$ is continuous at $x = x_0$.

- If $f(x)$ is not differentiable at $x = x_0$, then $f'(x_0) = \pm\infty$.

  Since the judgement formula is

  $$\lim_{\Delta x \to 0} \Delta y = f'(x_0) \cdot \lim_{\Delta x \to 0} \Delta x = \pm\infty \cdot 0 = a \quad (\text{a random number})$$

  hence, it is impossible to determine if $f(x)$ is continuous at $x = x_0$.

In summary

> If a function $f(x)$ satisfies the following conditions:
>
> 1. $f(x)$ is defined at $x = x_0$;
> 2. the limit $\lim_{x \to x_0} f(x)$ exists;
> 3. $\lim_{\Delta x \to 0} \Delta y = 0$ that is, $\lim_{x \to x_0} f(x) = f(x_0)$,
>
> then the function $f(x)$ is continuous at $x = x_0$. Otherwise, $x = x_0$ is a point of discontinuity of the function $f(x)$ (where the function is neither continuous nor differentiable).

Example 1. Determine the continuity and differentiability of $y = \frac{1}{x}$ at $x = 0$.

Solution. Since $y = \frac{1}{x} = \frac{1}{0} = \infty$, thus $y = \frac{1}{x}$ undefined at $x = 0$. $y = \frac{1}{x}$ is neither continuous nor differentiable at $x = 0$.

Example 2. Determine whether $f(x) = \frac{x^2 + 2}{x + 3}$ is continuous at $x = 1$.

Solution. $$f(1) = \frac{1^2 + 2}{1 + 3} = \frac{3}{4}$$

Since the denominator of $f(1)$ is $x + 3 = 1 + 3 = 4 \neq 0$, by the calculation rules of limits, we have

$$\lim_{x \to 1} f(x) = f(1) = \frac{3}{4}$$

Therefore, the function $f(x)$ is continuous at $x = 1$.

Example 3. Determine if $f(x)=\dfrac{x^2-4}{x-2}$ is continuous at $x=2$.

Solution. Since $f(x)=\dfrac{x^2-4}{x-2}=\dfrac{(x+2)(x-2)}{x-2}=x+2$

thus $\qquad f(x)=x+2$

Since $f(x)$ is a polynomial function, we have

$$\lim_{x\to 2} f(x)=f(2)=4$$

Hence, $f(x)$ is continuous at $x=2$.

Note: This example demonstrates that if the expression for a function $f(x)$ is a fraction that can be further simplified, then the function $f(x)$ is not necessarily a rational function. This is an important distinction to observe.

Example 4. Determine if $f(x)=\dfrac{e^x-1}{x^2-x}$ is continuous at $x=0$.

Solution. $\qquad f(0)=\dfrac{e^0-1}{0^2-0}=\dfrac{0}{0}=a \qquad$ (a random number)

Since $f(0)=\dfrac{0}{0}$, by L'hôpital's rule,

$$\lim_{x\to x_0} f(x)=A \quad (\text{or } \pm\infty)$$

thus $\qquad \lim_{x\to 0} f(x)\neq f(0)$

Therefore, $f(x)$ is not continuous at $x=0$.

Example 5. The derivative of $f(x)$ at $x=1$ is $f'(1)=3$. Determine whether $f(x)$ is continuous at $x=1$.

Solution. Since $f'(1)=3$

the judgement formula is $\lim\limits_{\Delta x\to 0}\Delta y = f'(1)\cdot \lim\limits_{\Delta x\to 0}\Delta x = 3\cdot 0 = 0$

thus, $f(x)$ is continuous at $x=1$.

Example 6. Determine the continuity and differentiability of $y=f(x)=|x|$ at $x=0$.

Solution. $y=f(x)=|x|=\begin{cases} x, & x\ge 0 \\ -x, & x<0 \end{cases}$

Since $\lim\limits_{x\to 0^-} f(x) = \lim\limits_{x\to 0^-} -x = 0$

$\lim\limits_{x\to 0^+} f(x) = \lim\limits_{x\to 0^+} x = 0$

then $\lim\limits_{x\to 0^-} f(x) = \lim\limits_{x\to 0^+} f(x) = 0$ and $f(0)=0$

therefore, $\lim\limits_{x\to 0} f(x) = 0 = f(0)$

Hence, $f(x)$ is continuous at $x=0$.

Since

$$f_-'(0) = \lim_{\Delta x\to 0^-}\frac{f(0+\Delta x)-f(0)}{\Delta x} = \lim_{\Delta x\to 0^-}\frac{-(0+\Delta x)-0}{\Delta x} = \lim_{\Delta x\to 0^-}\frac{-\Delta x}{\Delta x} = \lim_{\Delta x\to 0^-} -1 = -1$$

$$f_+'(0) = \lim_{\Delta x\to 0^+}\frac{f(0+\Delta x)-f(0)}{\Delta x} = \lim_{\Delta x\to 0^+}\frac{(0+\Delta x)-0}{\Delta x} = \lim_{\Delta x\to 0^+}\frac{\Delta x}{\Delta x} = \lim_{\Delta x\to 0^+} 1 = 1$$

thus $f_-'(0) \ne f_+'(0)$.

The derivative $f'(0)$ does not exist.

Therefore, $f(x)$ is not differentiable at $x=0$.

Example 7. $f(x) = \begin{cases} x-1, & x<0 \\ 0, & x=0 \\ x+1, & x>0 \end{cases}$, determine if $f(x)$ is continuous at $x=0$.

Solution.

Since $\lim\limits_{x\to 0^-} f(x) = \lim\limits_{x\to 0^-}(x-1) = 0-1 = -1$

$\lim\limits_{x\to 0^+} f(x) = \lim\limits_{x\to 0^+}(x+1) = 0+1 = 1$

thus $\lim\limits_{x\to 0^-} f(x) \neq \lim\limits_{x\to 0^+} f(x)$

$\lim\limits_{x\to 0} f(x)$ does not exist.

Therefore, $f(x)$ is discontinuous at $x = 0$, as shown in Figure 4.3.3.

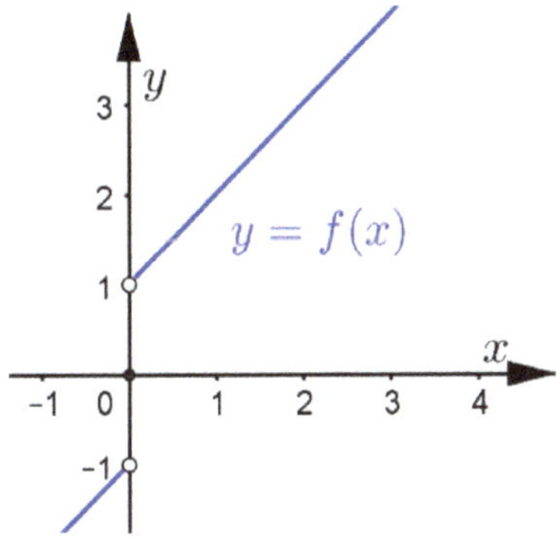

Figure 4.3.3

Example 8. $f(x) = \begin{cases} ax^3 - 1, & x > 2 \\ 3 + x^2, & x \leq 2 \end{cases}$, determine the value of a such that $f(x)$ is continuous at $x = 2$.

Solution. When $x = 2$, $f(2) = 3 + 2^2 = 7$

If $f(x)$ is continuous at $x = 2$

then $\lim\limits_{x\to 2^+} f(x) = \lim\limits_{x\to 2^+}(ax^3 - 1) = a\cdot 2^3 - 1 = 7$

That is, $a\cdot 8 - 1 = 7$ Hence, $a = 1$

Example 9. Determine whether $f(x) = \begin{cases} -1-x, & -1 \leq x \leq 0 \\ -1+x, & 0 < x < 1 \end{cases}$ is differentiable at $x = 0$.

Solution. Since $\lim\limits_{x \to 0^-} f(x) = \lim\limits_{x \to 0^-}(-1-x) = -1-0 = -1$

$$\lim_{x \to 0^+} f(x) = \lim_{x \to 0^+}(-1+x) = -1+0 = -1$$

then $\lim\limits_{x \to 0^-} f(x) = \lim\limits_{x \to 0^+} f(x) = -1$ and $f(0) = -1$

Therefore, $\lim\limits_{x \to 0} f(x) = -1 = f(0)$

Hence, $f(x)$ is continuous at $x = 0$.

Since

$$f_-'(0) = \lim_{\Delta x \to 0^-} \frac{f(0+\Delta x) - f(0)}{\Delta x} = \lim_{\Delta x \to 0^-} \frac{(-1-\Delta x)-(-1)}{\Delta x} = \lim_{\Delta x \to 0^-} -1 = -1$$

$$f_+'(0) = \lim_{\Delta x \to 0^+} \frac{f(0+\Delta x) - f(0)}{\Delta x} = \lim_{\Delta x \to 0^+} \frac{(-1+\Delta x)-(-1)}{\Delta x} = \lim_{\Delta x \to 0^+} 1 = 1$$

thus $f_-'(0) \neq f_+'(0)$.

The derivative $f'(0)$ does not exist.

Therefore, $f(x)$ is not differentiable at $x = 0$.

Example 10. $f(x) = \begin{cases} x-1, & x \le 0 \\ 2x, & 0 < x \le 1 \\ x^2+1, & 1 < x \le 2 \\ \frac{1}{2}x+4, & x > 2 \end{cases}$, discuss the continuity and differentiability of $f(x)$ at $x = 0, x = 1$ and $x = 2$.

Solution.

(1) At $x = 0$

$$\lim_{x \to 0^-} f(x) = \lim_{x \to 0^-}(x-1) = 0-1 = -1$$

$$\lim_{x \to 0^+} f(x) = \lim_{x \to 0^+}(2x) = 2 \cdot 0 = 0$$

then $\lim\limits_{x \to 0^-} f(x) \neq \lim\limits_{x \to 0^+} f(x)$

$$\lim_{x\to 0} f(x) \text{ does not exist.}$$

Hence, $f(x)$ is neither continuous nor differentiable at $x = 0$.

(2) At $x = 1$

$$\lim_{x\to 1^-} f(x) = \lim_{x\to 1^-}(2x) = 2\cdot 1 = 2$$

$$\lim_{x\to 1^+} f(x) = \lim_{x\to 1^+}(x^2+1) = 1^2+1 = 2$$

then $\quad \lim_{x\to 1^-} f(x) = \lim_{x\to 1^+} f(x) = 2 \quad$ and $\quad f(1) = 2$

hence, $\quad \lim_{x\to 1} f(x) = 2 = f(1)$

Therefore, $f(x)$ is continuous at $x = 1$.

Since

$$f_-'(1) = \lim_{\Delta x\to 0^-} \frac{f(1+\Delta x)-f(1)}{\Delta x} = \lim_{\Delta x\to 0^-} \frac{2(1+\Delta x)-2}{\Delta x} = \lim_{\Delta x\to 0^-} \frac{2\Delta x}{\Delta x} = \lim_{\Delta x\to 0^-} 2 = 2$$

$$f_+'(1) = \lim_{\Delta x\to 0^+} \frac{f(1+\Delta x)-f(1)}{\Delta x} = \lim_{\Delta x\to 0^+} \frac{(1+\Delta x)^2+1-2}{\Delta x} = \lim_{\Delta x\to 0^+}(2+\Delta x) = 2+0 = 2$$

thus $\quad f_-'(1) = f_+'(1)$

$$f'(1) = f_-'(1) = f_+'(1)$$

The derivative $f'(1)$ exists.

Therefore, $f(x)$ is differentiable at $x = 1$.

(3) At $x = 2$

$$\lim_{x\to 2^-} f(x) = \lim_{x\to 2^-}(x^2+1) = 2^2+1 = 5$$

$$\lim_{x\to 2^+} f(x) = \lim_{x\to 2^+}\left(\frac{1}{2}x+4\right) = \frac{1}{2}\cdot 2+4 = 5$$

then $\quad \lim_{x\to 2^-} f(x) = \lim_{x\to 2^+} f(x) = 5 \quad$ and $\quad f(2) = 5$

thus $\quad \lim_{x\to 2} f(x) = 5 = f(2)$

Therefore, $f(x)$ is continuous at $x = 2$.

Since

$$f_-'(2) = \lim_{\Delta x \to 0^-} \frac{f(2+\Delta x) - f(2)}{\Delta x} = \lim_{\Delta x \to 0^-} \frac{(2+\Delta x)^2 + 1 - 5}{\Delta x} = \lim_{\Delta x \to 0^-} (4 + \Delta x) = 4 + 0 = 4$$

$$f_+'(2) = \lim_{\Delta x \to 0^+} \frac{f(2+\Delta x) - f(2)}{\Delta x} = \lim_{\Delta x \to 0^+} \frac{\frac{1}{2}(2+\Delta x) + 4 - 5}{\Delta x} = \lim_{\Delta x \to 0^+} (\frac{1}{2}) = \frac{1}{2}$$

thus $\quad f_-'(2) \neq f_+'(2)$.

The derivative $f'(2)$ does not exist.

Hence, $f(x)$ is not differentiable at $x = 2$. The result is shown in Figure 4.3.4.

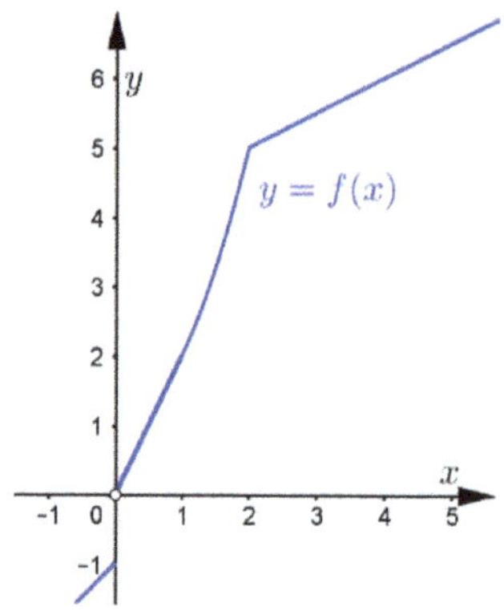

Figure 4.3.4

Example 11. $f(x) = \begin{cases} -\frac{x}{2} + 1, & x \le 0 \\ -x^2 + 6x + 1, & 0 < x < 2 \\ x^3 + 1, & x \ge 2 \end{cases}$, determine where the function $f(x)$ is differentiable.

Solution.

(1) At $x = 0$

$$\lim_{x \to 0^-} f(x) = \lim_{x \to 0^-} (-\frac{x}{2} + 1) = -\frac{0}{2} + 1 = 1$$

$$\lim_{x \to 0^+} f(x) = \lim_{x \to 0^+} (-x^2 + 6x + 1) = -0^2 + 6 \cdot 0 + 1 = 1$$

then $\quad \lim_{x \to 0^-} f(x) = \lim_{x \to 0^+} f(x) = 1 \quad$ and $\quad f(0) = 1$

hence, $\lim_{x \to 0} f(x) = 1 = f(0)$

Therefore, $f(x)$ is continuous $x = 0$.

Since

$$f_-'(0) = \lim_{\Delta x \to 0^-} \frac{f(0+\Delta x) - f(0)}{\Delta x} = (-\frac{x}{2}+1)'\Big|_{x=0} = -\frac{1}{2}\Big|_{x=0} = -\frac{1}{2}$$

$$f_+'(0) = \lim_{\Delta x \to 0^+} \frac{f(0+\Delta x) - f(0)}{\Delta x} = (-x^2+6x+1)'\Big|_{x=0} = (-2x+6)\Big|_{x=0} = -2 \cdot 0 + 6 = 6$$

thus $f_-'(0) \neq f_+'(0)$.

The derivative $f'(0)$ does not exist.

Therefore, $f(x)$ is not differentiable at $x = 0$.

(2) At $x = 2$

$$\lim_{x \to 2^-} f(x) = \lim_{x \to 2^-} (-x^2 + 6x + 1) = -2^2 + 6 \cdot 2 + 1 = 9$$

$$\lim_{x \to 2^+} f(x) = \lim_{x \to 2^+} (x^3 + 1) = 2^3 + 1 = 9$$

then $\lim_{x \to 2^-} f(x) = \lim_{x \to 2^+} f(x) = 9$ and $f(2) = 9$

hence, $\lim_{x \to 2} f(x) = 9 = f(2)$

Therefore, $f(x)$ is continuous at $x = 2$.

Since

$$f_-'(2) = \lim_{\Delta x \to 0^-} \frac{f(2+\Delta x) - f(2)}{\Delta x} = (-x^2+6x+1)'\Big|_{x=2} = (-2x+6)\Big|_{x=2} = -2 \cdot 2 + 6 = 2$$

$$f_+'(2) = \lim_{\Delta x \to 0^+} \frac{f(2+\Delta x) - f(2)}{\Delta x} = (x^3+1)'\Big|_{x=2} = (3x^2)\Big|_{x=2} = 3 \cdot 2^2 = 12$$

thus $f_-'(2) \neq f_+'(2)$.

The derivative $f'(2)$ does not exist.

Therefore, $f(x)$ is not differentiable at $x = 2$.

Hence, $f(x)$ is not differentiable at $x = 0$ and $x = 2$.

That is, $f(x)$ is differentiable at $x \in \mathrm{R},\ x \neq 0, x \neq 2$.

Exercise 4.3

1. $$f(x)=\begin{cases} x^2+1, & 0 \le x < 1 \\ 3x-1, & x \ge 1 \end{cases}$$

a. Determine whether the function $f(x)$ is continuous at $x = 1$.

b. Determine whether the function $f(x)$ is differentiable at $x = 1$.

2. $$f(x)=\begin{cases} \ln(1+x), & -1 < x \le 0 \\ \sqrt{1+x}-\sqrt{1-x}, & 0 < x < 1 \end{cases}$$

Discuss the continuity and differentiability of $f(x)$ at $x = 0$.

3. $$f(x)=\begin{cases} 1, & x \le 0 \\ 2x+1, & 0 < x \le 1 \\ x^2+2, & 1 < x \le 2 \\ x, & x > 2 \end{cases}$$

Discuss the continuity and differentiability of $f(x)$ at $x = 0, x = 1, x = 2$.

4. Determine whether the function $f(x)$ is differentiable.

$$f(x)=\begin{cases} x^3+2x^2+1, & x \le -1 \\ \left|x^2-x-2\right|, & -1 < x < 3 \\ 3x-5, & x \ge 3 \end{cases}$$

5. The function $f(x)$ is defined for $x \in \mathrm{R}$.

$$f(x)=\begin{cases} (x-2)^{\frac{5}{3}}, & x \le 2 \\ x^3+ax+b, & x > 2 \end{cases}$$

Find the values of a and b such that the function $f(x)$ is differentiable for all $x \in$ R.

6. The function $f(x)$ is defined for $x \in$ R.

$$f(x)=\begin{cases} e^{ax+b}, & x \le -1 \\ -ex, & x > -1 \end{cases}$$

Find the values of a and b such that the function $f(x)$ is differentiable for all $x \in$ R.

7. Given that

$$f(x)=\begin{cases} -x^3+2x^2+x-2, & x \ge 1 \\ -\sqrt{a-x}, & x < 1 \end{cases}$$

Find the value of a such that the function $f(x)$ is continuous at $x = 1$.

8. Show that $f(x) = \cos|x|$ is differentiable at $x = 0$.

9. Given that

$$f(x)=\begin{cases} 7x-3, & x \le -1 \\ -5x^2, & -1 < x < 1 \\ -10x+5, & x \ge 1 \end{cases}$$

show that $f(x)$ is not differentiable at x = -1 but differentiable at $x = 1$.

10. **a**. Find the value of a and b such that the function

$$f(x)=\begin{cases} \sqrt[3]{x-2}, & x \le 2 \\ x^2+ax+b, & 2 < x < 4 \\ -x+6, & x \ge 4 \end{cases}$$

is continuous everywhere.

b. Where is $f(x)$ differentiable?

Chapter 5 Applications of derivatives and curve sketching

5.1. Tangents and normals

The tangent to the function $f(x)$ at the point (x_0, y_0) is a straight line that touches the curve of the function $f(x)$ at the point (x_0, y_0), as shown in Figure 5.1.1. The gradient of the tangent, denoted by m_t, is equal to the value of the derivative $f'(x_0)$ of the function $f(x)$ at the point (x_0, y_0).

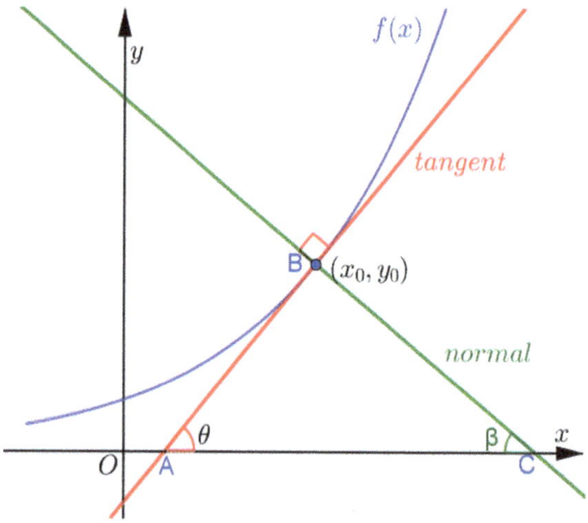

Figure 5.1.1

The normal to the function $f(x)$ at the point (x_0, y_0) is a straight line that passes through the point (x_0, y_0) and is perpendicular to the tangent at the point (x_0, y_0), as shown in Figure 5.1.1. Since the normal is perpendicular to the tangent, the product of the gradient of the normal, m_n, and the gradient of the tangent at that point, m_t, is equal to -1. Thus, the gradient of the normal is the negative reciprocal of the gradient of the tangent at that point.

$$m_n \cdot m_t = -1$$

$$m_n = -\frac{1}{m_t}$$

Proof. Figure 5.1.1 illustrates that

the gradient of the tangent is $\quad m_t = \tan\theta = \dfrac{BC}{AB} = \dfrac{1}{\dfrac{AB}{BC}} = \dfrac{1}{\tan\beta}$

the gradient of the normal is $\quad m_n = \tan(180^\circ - \beta) = -\tan\beta$

Therefore, $\quad m_n \cdot m_t = \tan(180^\circ - \beta)\cdot\tan\theta = -\tan\beta\cdot\dfrac{1}{\tan\beta} = -1$

That is, $\quad m_n \cdot m_t = -1$ □

The gradient can be denoted by the derivative.

$$m_t = \left.\frac{dy}{dx}\right|_{x=x_0}$$

$$m_n = -\left.\frac{dx}{dy}\right|_{x=x_0}$$

Example 1. Find the equation of the tangent to the curve $y = x^2 + 2x$ at the point (1, 3).

Solution. The gradient of the tangent at the point (1, 3) is

$$m_t = y'\big|_{x=1} = (x^2 + 2x)'\big|_{x=1} = (2x+2)\big|_{x=1} = 4$$

The equation of the tangent to the curve $y = x^2 + 2x$ at the point (1, 3) is

$$y = m_t x + C$$

$$y = 4x + C$$

$$3 = 4 + C$$

$$C = -1$$

Hence, the equation of the tangent to the curve $y = x^2 + 2x$ at the point (1, 3) is

$$y = 4x - 1$$

Example 2. Find the equation of the normal to the curve $y = x^3 + x^2 - 3x$ at the point (-1, 3). Find the coordinates of the point of intersection of this normal and the curve.

Solution. First, find the gradient of the tangent to the curve $y = x^3 + x^2 - 3x$ at the point (-1, 3).

$$m_t = y'\big|_{x=-1} = (x^3 + x^2 - 3x)'\big|_{x=-1} = (3x^2 + 2x - 3)\big|_{x=-1} = -2$$

Therefore, the gradient of the normal at the point is

$$m_n = -\frac{1}{m_t} = \frac{1}{2}$$

The equation of the normal to the curve $y = x^3 + x^2 - 3x$ at the point (-1, 3) is

$$y = m_n x + C$$

$$y = \frac{1}{2}x + C$$

$$3 = -\frac{1}{2} + C$$

$$C = \frac{7}{2}$$

That is, the equation of the normal at the point (-1, 3) to the curve is

$$y = \frac{1}{2}x + \frac{7}{2}$$

To find the intersection point of this normal and the curve, equate the equations of the normal and the curve.

$$x^3 + x^2 - 3x = \frac{1}{2}x + \frac{7}{2}$$

Using a calculator, we find that the coordinates of the point of intersection are

(-1.87, 2.56), (-1, 3), (1.87, 4.44)

Example 3. Given the equation $xy^3 - x^2y = 6$, where y is a function of x. Find the equation of the tangent and the normal to the curve at the point (3, 2).

Solution. Differentiate both sides of the equation with respect to x.

$$\frac{d(xy^3)}{dx} - \frac{d(x^2y)}{dx} = \frac{d6}{dx}$$

$$\frac{dx}{dx} \cdot y^3 + x \cdot \frac{dy^3}{dy} \cdot \frac{dy}{dx} - \left(\frac{dx^2}{dx} \cdot y + x^2 \cdot \frac{dy}{dx} \right) = 0$$

$$y^3 + 3y^2x \cdot \frac{dy}{dx} - 2xy - x^2 \cdot \frac{dy}{dx} = 0$$

$$(3y^2x - x^2)\frac{dy}{dx} = 2xy - y^3$$

$$y' = \frac{dy}{dx} = \frac{2xy - y^3}{3y^2x - x^2}$$

The gradient of the tangent to the curve at the point (3, 2) is

$$m_t = y'\Big|_{\substack{x=3 \\ y=2}} = \frac{4}{27}$$

Hence, the equation of the tangent to the curve at the point (3, 2) is

$$y = m_t x + C$$

$$y = \frac{4}{27}x + C$$

$$2 = \frac{4}{27} \cdot 3 + C$$

$$C = \frac{14}{9}$$

That is, the equation of the tangent is

$$y = \frac{4}{27}x + \frac{14}{9}$$

The gradient of the normal is

$$m_n = -\frac{1}{m_t} = -\frac{27}{4}$$

Therefore, the equation of the normal to the curve at the point (3, 2) is

$$y = m_n x + C$$

$$y = -\frac{27}{4}x + C$$

$$2 = -\frac{27}{4}\cdot 3 + C$$

$$C = \frac{89}{4}$$

Hence, the equation of the normal is

$$y = -\frac{27}{4}x + \frac{89}{4}$$

Example 4. Given the curve $(y+1)^3 = (2x-1)^2$, find the equation of the tangent to the curve that has a gradient of 1.

Solution. Differentiate both sides of the equation with respect to x.

$$\frac{d(y+1)^3}{d(y+1)}\cdot\frac{d(y+1)}{dx} = \frac{d(2x-1)^2}{d(2x-1)}\cdot\frac{d(2x-1)}{dx}$$

$$3(y+1)^2\frac{dy}{dx} = 4(2x-1)$$

$$\frac{dy}{dx} = \frac{4(2x-1)}{3(y+1)^2}$$

The gradient of the tangent to the curve is 1.

$$\frac{dy}{dx} = \frac{4(2x-1)}{3(y+1)^2} = 1$$

That is, $$2x - 1 = \frac{3}{4}(y+1)^2$$

Square both sides.

$$(2x-1)^2 = \frac{9}{16}(y+1)^4$$

Substituting $(y+1)^3 = (2x-1)^2$ gives

$$(y+1)^3 = \frac{9}{16}(y+1)^4$$

Divide both sides by $(y+1)^3$ and we obtain

$$1 = \frac{9}{16}(y+1)$$

$$y = \frac{7}{9}$$

Substituting into $(y+1)^3 = (2x-1)^2$ gives

$$x = \frac{91}{54}$$

That is, the point on the curve, whose gradient of the tangent is 1, is $(\frac{91}{54}, \frac{7}{9})$.

Therefore, the equation of the tangent at $(\frac{91}{54}, \frac{7}{9})$ is

$$y = m_t x + C$$

$$y = x + C$$

$$\frac{7}{9} = \frac{91}{54} + C$$

$$C = -\frac{49}{54}$$

That is, the equation of the tangent is

$$y = x - \frac{49}{54}$$

Exercise 5.1

1. Find the equations of the tangents and normals to each of the following curves at the specified point.

a. $y = 3x^2 + x, \quad x = 2$

b. $y = \dfrac{2}{x^2+1}, \quad x = 1$

c. $y = xe^{-x}, \quad x = 3$

d. $y = \dfrac{2x}{\cos x}, \quad x = \pi$

e. $y = \sqrt{4 - x^2}, \quad x = 1$

f. $y = x \ln x, \quad x = 5$

g. $y^3 + y^2 - 2xy = 5, \quad x = 1$

h. $ye^x + \ln y = 1, \quad x = 0$

i. $3xy + \sqrt{x^3 + y^3} = 9, \quad (1, 2)$

j. $\tan(xy) = x, \quad (1, \frac{\pi}{4})$

k. $x^2 + xy + y^2 = 27, \quad (3, 6)$

l. $(x^2 + y^2)^3 - 30xy^2 = 5, \quad (1, 2)$

m. $2e^x y + xe^y = 1, \quad (1, 0)$

n. $\dfrac{4x^2}{2x^2 - 3y^3} = -\dfrac{1}{2}y, \quad (1, 2)$

2. Find the equation of the tangent to the curve $y = x^3 - 4x^2 + x - 9$ at the point where it intersects the x-axis.

3. The normals to the curve $y = \sqrt{2x}$ at the points where $x = 1$ and $x = 5$ intersect at the point A. Find the coordinates of A.

4. The tangent to the curve $y = ax^2 + bx$ at the point where $x = 2$ has gradient 1. The curve passes through the point (3, 6). Find the values of a and b.

5. Find the equation of the tangent to the curve $x^3 + y^3 = 6xy$ that is parallel to the x-axis.

6. The tangent to the curve $y^2 = (1 - x)^3$ at the point (-8, -27) meets the curve again at the point B. Find the coordinates of the point B.

7. The tangent to the curve $y = x^2 + 2x$ at the point P is parallel to the line $y = 4x - 5$. Find the coordinates of the point P.

8. The curve $y = e^{1-x^2}$ intersects the line $x = -1$ at the point P. Find the equation of the tangent to the curve $y = e^{1-x^2}$ at the point P.

9. Given the equation $e^{2x+y} - \cos(xy) = e - 1$, where y is a function of x, find the equation of the normal to the curve at the point (0, 1).

10. Find the equation of the normal to the curve

$$x^2y^3 + xy^2 = 12$$

at the point where $x = 1$.

11. Find the equation of the tangent and the normal to the curve

$$\cos(3x) + 2\cos y + 2 = 0$$

at the point $(\frac{\pi}{3}, \frac{2\pi}{3})$.

12. The normal to the curve

$$\frac{x^2}{25} + \frac{y^2}{16} = 1$$

at the point $(4, \frac{12}{5})$ meets the x-axis at the point Q. The tangent to the curve at the same point meets the y-axis at R. Find the area of triangle ΔOQR.

5.2. Determining monotonic intervals of functions

In our study of functions, we have defined monotonic increasing or decreasing functions on specific intervals. Now, we will introduce a method of using the derivative to determine a function's monotonicity.

First, analyse this geometrically:

- If, on an interval (a,b), the gradient of the tangent at every point on the curve is positive ($\tan\theta = f'(x) > 0$), then the function's curve is rising. That is, the function $f(x)$ is monotonically increasing. As shown in Figure 5.2.1, as the value of x increases, the corresponding functional value $f(x)$ also increases.
- If, on an interval (a,b), the gradient of the tangent at every point on the curve is negative ($\tan\theta = f'(x) < 0$), then the function's curve is declining. That is, the function $f(x)$ is monotonically decreasing. As shown in Figure 5.2.2, as the value of x increases, the corresponding functional value $f(x)$ decreases.

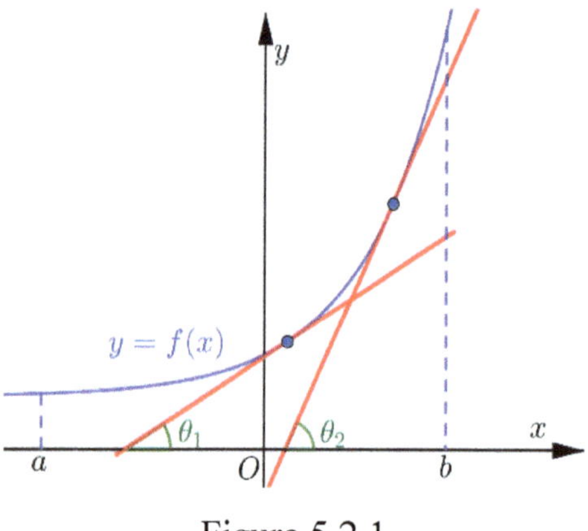

Figure 5.2.1

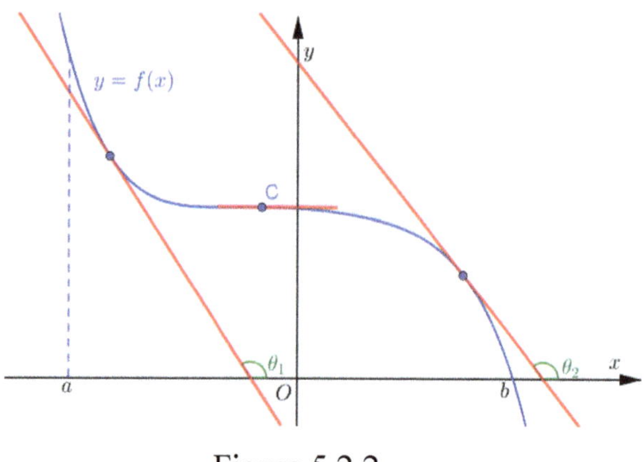

Figure 5.2.2

Theorem 5.1

Let the function $f(x)$ be differentiable on the interval (a,b).

- If $f'(x) > 0$ for all $x \in (a,b)$, then the function $f(x)$ is monotonically increasing on the interval (a,b).
- If $f'(x) < 0$ for all $x \in (a,b)$, then the function $f(x)$ is monotonically decreasing on the interval (a,b).

Proof. Take any two points, x_1 and x_2, in the interval (a,b). Let $x_1 < x_2$. $x_2 - x_1 = \Delta x > 0$. By Lagrange's Mean Value Theorem, we have

$$f'(c) = \frac{\Delta y}{\Delta x} \qquad c \in (x_1, x_2)$$

That is, $$f(x_2) - f(x_1) = f'(c) \cdot \Delta x$$

(1) If $f'(x) > 0$ for all $x \in (a,b)$, then let $x = c$, $f'(c) > 0$ and we have

$$f(x_2) - f(x_1) = f'(c) \cdot \Delta x > 0$$

$$f(x_2) > f(x_1)$$

Therefore, the function $f(x)$ increases monotonically on the interval (a,b).

(2) If $f'(x) < 0$ for all $x \in (a,b)$, then $f'(c) < 0$ and we have

$$f(x_2) - f(x_1) = f'(c) \cdot \Delta x < 0$$

$$f(x_2) < f(x_1)$$

Therefore, the function $f(x)$ decreases monotonically on the interval (a,b). □

Note: If on the interval (a,b), we have $f'(x) \geq 0$ (or $f'(x) \leq 0$), but $f'(x) = 0$ occurs only at isolated points (such as point C in Figure 5.2.2), then the function $f(x)$ is still considered monotonically increasing (or decreasing) on the interval (a,b).

Example 1. $f(x) = x^3 + 3x^2 - 4$. Find the values of x for which the function $f(x)$ is increasing.

Solution. When $f'(x) > 0$, the function $f(x)$ is increasing.

$$f'(x) = 3x^2 + 6x$$

$$3x^2 + 6x > 0$$

$$3x(x+2) > 0$$

$$x > 0 \quad \text{or} \quad x < -2$$

Example 2. $f(x) = 5x^2 + ax + b$ is decreasing for $x < -\frac{4}{10}$. Given that the function $f(x)$ passes through the point (1, 8), find the values of the constants a and b.

Solution. When $f'(x) < 0$, $f(x)$ is decreasing.

$$f'(x) = 10x + a$$

$$10x + a < 0$$

$$x < -\frac{a}{10}$$

Hence, $$-\frac{a}{10} = -\frac{4}{10}$$

$$a = 4$$

$f(x)$ passes through the point (1, 8).

$$f(1) = 5\cdot 1^2 + 4\cdot 1 + b = 8$$

$$b = -1$$

Therefore, $$f(x) = 5x^2 + 4x - 1$$

Example 3. Prove that $f(x) = \frac{1}{x}$ is a decreasing function on the interval $(0, \infty)$.

Proof. $f'(x) = (\frac{1}{x})' = -\frac{1}{x^2}$

On the interval $(0,\infty)$, $x^2 > 0$.

Thus $f'(x) < 0$

Therefore, $f(x) = \frac{1}{x}$ is a decreasing function on the interval $(0,\infty)$.

Exercise 5.2

1. Determine the intervals of increase and decrease for each of the following functions.

a. $y = x^4 - 2x^2 - 5$ **b.** $y = 2\sin x + \cos(2x)$ $(0 \le x \le 2\pi)$

c. $y = 2x^3 - 9x^2 + 12x - 3$ **d.** $y = \frac{x^2}{1+x}$

e. $y = x^2 + 5x + 6$ **f.** $y = x - \ln(1 + x^2)$

2. Find the interval on which $y = \frac{\ln x}{x}$ is a decreasing function.

3. Determine the interval on which the function $y = x\cos x - \sin x$ is increasing, for $0 \le x \le 2\pi$.

4. Determine the interval on which the function $y = \frac{x}{x^2 - 3x + 2}$ is increasing.

5.3. Extreme values of functions

Definition 5.1: Let the function $f(x)$ be defined in a δ neighbourhood of the point $x = x_0$, $(x_0 - \delta, x_0 + \delta)$, as shown in Figure 5.3.1.

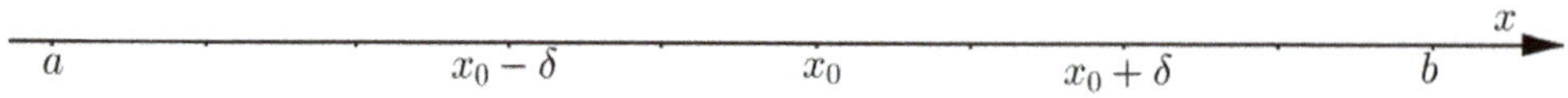

Figure 5.3.1

The left interval of x_0 is $(x_0 - \delta, x_0)$

The right interval of x_0 is $(x_0, x_0 + \delta)$

If, for all $x \in (x_0 - \delta, x_0) \cup (x_0, x_0 + \delta)$, $f(x) < f(x_0)$, then $f(x_0)$ is called a local maximum value of the function, and x_0 is called a local maximum point.

If, for all x in this neighbourhood, $f(x) > f(x_0)$, then $f(x_0)$ is called a local minimum value of the function, and x_0 is called a local minimum point.

The local maximum and local minimum values, along with their corresponding points x_0, are collectively called **extreme values** and **extremum points** (or **extreme points**). An extreme value is a local concept; it is only compared with the function values of all points in the immediate neighbourhood of the extreme point x_0. It is not necessarily the global maximum or minimum for the entire domain (a,b) of the function $f(x)$.

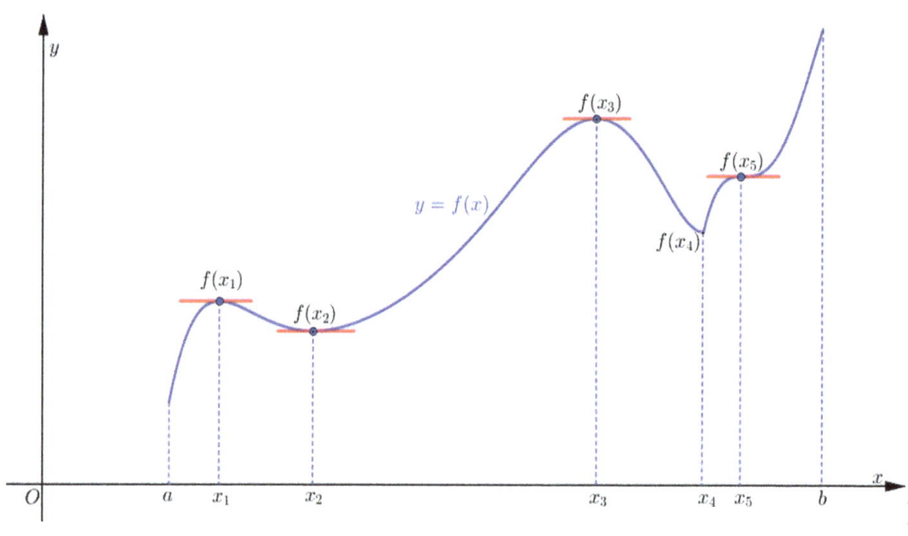

Figure 5.3.2

For the function $f(x)$ shown in Figure 5.3.2, the **local maxima** $f(x_1)$ and $f(x_3)$ are not the **absolute (or global) maximum** of the function $f(x)$ on the domain (a,b). Similarly, the **local minima** $f(x_2)$ and $f(x_4)$ are not the **absolute minimum** of the function $f(x)$ on the domain (a,b). Notably, the local maximum $f(x_1)$ is still smaller than the local minimum $f(x_4)$, that is, $f(x_1) < f(x_4)$.

As shown in Figure 5.3.2, at extreme points, if a tangent to the curve exists and has a definite gradient, then the tangent is parallel to the x-axis. In such cases, the gradient of the tangent is zero ($f'(x) = 0$).

However, even if the tangent at a certain point on the curve is parallel to the x-axis, that point is not necessarily an extremum point. For example, the point x_5 in Figure 5.3.2 is not an extremum point, even though the tangent to the curve at the point x_5 is parallel to the x-axis.

Based on the geometric observations above, we state the following theorem regarding the extreme values of functions.

Theorem 5.2 (Necessary Condition)
If, at the point x_0, the function $f(x)$ has an extreme value $f(x_0)$ and $f'(x_0)$ exists, then $f'(x_0) = 0$.

Proof. If $f(x_0)$ is a local maximum, then there exists some neighbourhood of x_0 such that for all x in that neighbourhood, we have

$$f(x_0 + \Delta x) < f(x_0)$$

That is,
$$f(x_0 + \Delta x) - f(x_0) < 0$$

On the left of x_0

$$\frac{f(x_0+\Delta x)-f(x_0)}{\Delta x}>0 \qquad \Delta x<0$$

On the right of x_0

$$\frac{f(x_0+\Delta x)-f(x_0)}{\Delta x}<0 \qquad \Delta x>0$$

Therefore, $$f_-'(x_0)=\lim_{\Delta x\to 0^-}\frac{f(x_0+\Delta x)-f(x_0)}{\Delta x}\geq 0$$

$$f_+'(x_0)=\lim_{\Delta x\to 0^+}\frac{f(x_0+\Delta x)-f(x_0)}{\Delta x}\leq 0$$

If $f'(x_0)$ exists, then we must have

$$f_-'(x_0)=f_+'(x_0)=0$$

Therefore, $$f'(x_0)=f_-'(x_0)=f_+'(x_0)=0$$

That is, $$f'(x_0)=0$$

Similarly, the proof for a local minimum is analogous and is therefore omitted. □

Note:

(1) Theorem 5.2 shows that $f'(x_0)=0$ is a necessary condition for the point x_0 to be an extreme point, but it is not a sufficient condition. For example, for $f(x)=x^3$, $f'(0)=0$, yet there is no extreme point at $x=0$, as shown in Figure 5.3.3.

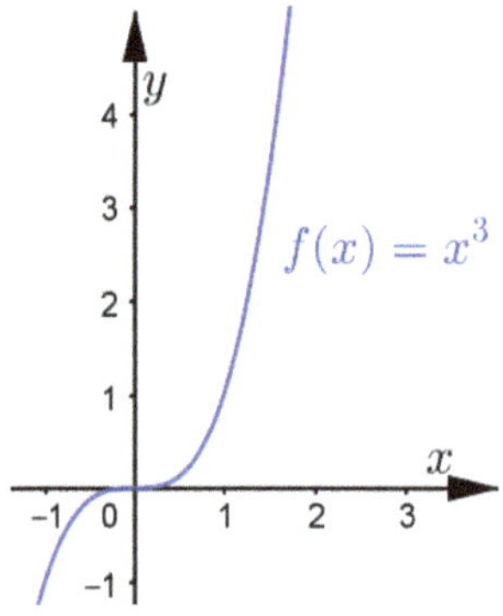

Figure 5.3.3

The points where $f'(x)=0$ are called the **stationary points** of the function. The stationary point may or may not be an extreme point of the function.

(2) Theorem 5.2 applies to functions that are differentiable at the point x_0.

- It is possible for a function $f(x)$ to have extreme values at a point where the derivative does not exist ($f'(x)=\infty$). For example, for $f(x)=x^{\frac{2}{3}}$, $f'(x)=\frac{2}{3}x^{-\frac{1}{3}}$, $f'(0)=\infty$, the derivative does not exist. However, at the point $x=0$, the function has a minimum $f(0)=0$, as shown in Figure 5.3.4.

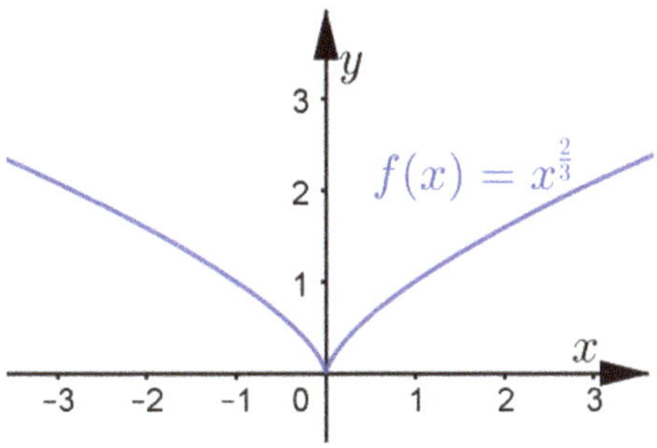

Figure 5.3.4

- It is also possible for a function $f(x)$ not to have an extreme value at a point where the derivative does not exist ($f'(x)=\infty$). For example, for $f(x)=x^{\frac{1}{3}}$, $f'(x)=\frac{1}{3}x^{-\frac{2}{3}}$, $f'(0)=\infty$, the derivative does not exist. However, at the point $x=0$, the function does not have an extreme value, as shown in Figure 5.3.5.

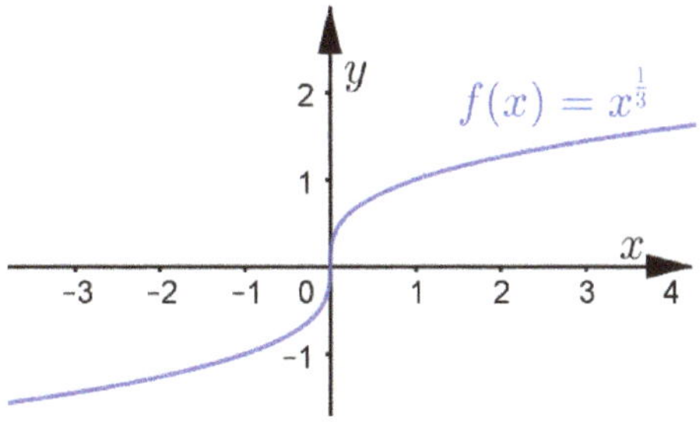

Figure 5.3.5

From the above, it can be seen that a stationary point ($f'(x)=0$) or a point where the derivative does not exist ($f'(x)=\infty$) is not necessarily an extreme point. However, an extreme point must be either a stationary point or a point where the derivative does not exist.

The following introduces the sufficient condition for the function $f(x)$ to have extreme values, that is, the method for determining the extreme values.

Theorem 5.3 First derivative test for determining extreme values

In a neighbourhood of the point $x=x_0$, $(x_0-\delta, x_0+\delta)$, let the function $f(x)$ be continuous and differentiable (but $f'(x_0)$ may not exist).

(1) If when $x\in(x_0-\delta, x_0)$, that is, $x<x_0$, on the left of x_0 there is $f'(x)>0$;

when $x\in(x_0, x_0+\delta)$, that is, $x>x_0$, on the right of x_0 there is $f'(x)<0$,

then the function $f(x)$ has a local maximum $f(x_0)$ at the point $x=x_0$, as shown in Figure 5.3.6.

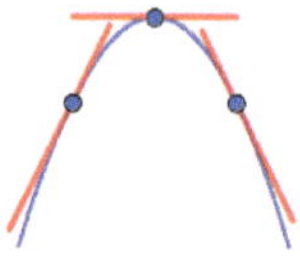

Figure 5.3.6

(2) If when $x\in(x_0-\delta, x_0)$, that is, $x<x_0$, on the left of x_0 there is $f'(x)<0$;

when $x\in(x_0, x_0+\delta)$, that is, $x>x_0$, on the right of x_0 there is $f'(x)>0$,

then the function $f(x)$ has a local minimum $f(x_0)$ at the point $x=x_0$, as shown in Figure 5.3.7.

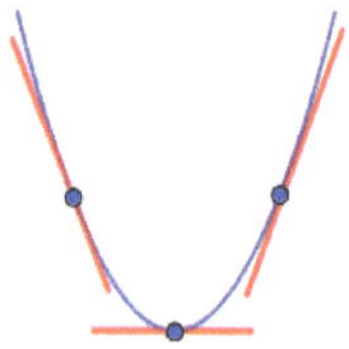

Figure 5.3.7

(3) If when $x \in (x_0 - \delta, x_0)$ and $x \in (x_0, x_0 + \delta)$, the sign of $f'(x)$ does not change, then the function does not have an extreme value at the point $x = x_0$.

Proof.

(i) When $x \in (x_0 - \delta, x_0)$, $f'(x) > 0$, then $f(x)$ is monotonically increasing on $(x_0 - \delta, x_0)$ so $f(x_0) > f(x)$;

When $x \in (x_0, x_0 + \delta)$, $f'(x) < 0$, then $f(x)$ is monotonically decreasing on $(x_0, x_0 + \delta)$ so $f(x_0) > f(x)$;

That is, to $x \in (x_0 - \delta, x_0) \cup (x_0, x_0 + \delta)$, there is always $f(x_0) > f(x)$. Hence, $f(x_0)$ is a local maximum of $f(x)$.

(ii) Similarly, we can prove (2), and here is omitted.

(iii) Since on the interval $(x_0 - \delta, x_0 + \delta)$, the sign of $f'(x)$ does not change, then there is always $f'(x) > 0$ or $f'(x) < 0$. Therefore, on the left and right sides of x_0, $f(x)$ increases or decreases monotonically; thus, it is impossible to have an extreme point at the point x_0.

Example 1. Find the monotonic intervals and extreme values of the function.

$$f(x) = (x+1)^3(x-2)^2$$

Solution. First, set $f'(x) = 0$ to find the *x*-values of the stationary points.

$$f'(x) = 3(x+1)^2(x-2)^2 + 2(x-2)(x+1)^3$$
$$= (x+1)^2(x-2)[3(x-2) + 2(x+1)]$$
$$= (x+1)^2(x-2)(5x-4) = 0$$
$$x = -1 \qquad x = 0.8 \qquad x = 2$$

These three points divide the interval $(-\infty, \infty)$ into four sub-intervals.

$$(-\infty,-1) \quad (-1,0.8) \quad (0.8,2) \quad (2,\infty)$$

We can summarize the sign of $f'(x)$ and the monotonicity of $f(x)$ on each interval in the following table.

x	$(-\infty,-1)$	-1	$(-1,0.8)$	0.8	$(0.8,2)$	2	$(2,\infty)$
$f'(x)$	$+$	0	$+$	0	$-$	0	$+$
$f(x)$	↗	0 no extreme value	↗	$\frac{26244}{3125}$ local maximum	↘	0 local minimum	↗

As shown in the table, the function $f(x)$ is monotonically increasing on the intervals $(-\infty,0.8)$ and $(2,\infty)$, while it is monotonically decreasing on the interval $(0.8,2)$.

At the point $x=0.8$, the function has a local maximum $f(0.8)=\dfrac{26244}{3125}=8.39808$, and at the point $x=2$, it has a local minimum $f(2)=0$, as shown in Figure 5.3.8.

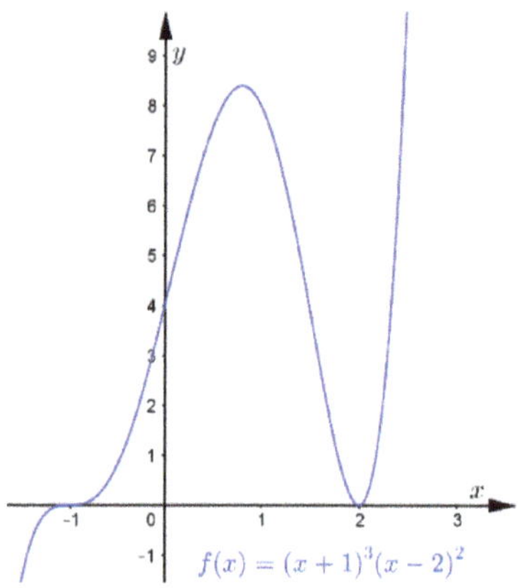

Figure 5.3.8

Example 2. Find the monotonic intervals and extreme values of the function.

$$f(x)=-3x^{\frac{4}{5}}+2x$$

Solution. First, find the derivative.

$$f'(x) = -\frac{12}{5}x^{-\frac{1}{5}} + 2$$

Set $\quad f'(x) = -\frac{12}{5}x^{-\frac{1}{5}} + 2 = 0, \qquad f'(x) = -\frac{12}{5}x^{-\frac{1}{5}} + 2 = \infty$

We find the stationary point $x = \frac{7776}{3125} = 2.48832$ and identify the discontinuity point of $f'(x)$ at $x = 0$. Thus, the function $f(x)$ can only have extreme values at these two points. These two points divide the interval $(-\infty, \infty)$ into three sub-intervals: $(-\infty, 0)$, $(0, 2.48832)$ and $(2.48832, \infty)$. We then summarize the behaviour of the function in the following table.

x	$(-\infty, 0)$	0	$(0, 2.48832)$	2.48832	$(2.48832, \infty)$
$f'(x)$	+	does not exist	−	0	+
$f(x)$	↗	0 local maximum	↘	−1.24416 local minimum	↗

As shown in the table, the function $f(x)$ is monotonically increasing on the intervals $(-\infty, 0)$ and $(2.48832, \infty)$, while the function is monotonically decreasing on the interval $(0, 2.48832)$. At the point $x = 0$, it has a local maximum $f(0) = 0$, and at the point $x = 2.48832$, it has a local minimum $f(2.48832) = -1.24416$, as shown in Figure 5.3.9.

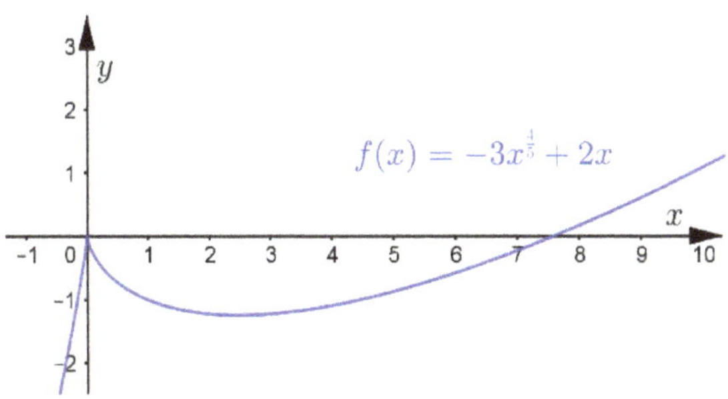

Figure 5.3.9

If the second derivative of the function $f(x)$ exists at the stationary point x_0, we have the following theorem.

Theorem 5.4 Second derivative test for determining extreme values

Let $f'(x_0)=0$, $f''(x_0)$ exists.

(1) If $f''(x_0)<0$, then $f(x_0)$ is a local maximum of $f(x)$.

(2) If $f''(x_0)>0$, then $f(x_0)$ is a local minimum of $f(x)$.

Proof. (1) From the definition of derivative, $f'(x_0)=0$ and $f''(x_0)<0$, we have

$$f''(x_0)=\lim_{\Delta x\to 0}\frac{f'(x)-f'(x_0)}{\Delta x}=\lim_{\Delta x\to 0}\frac{f'(x)-0}{\Delta x}=\lim_{\Delta x\to 0}\frac{f'(x)}{\Delta x}<0$$

From this, we obtain

$$\frac{f'(x)}{\Delta x}<0 \qquad (x\neq x_0)$$

When $x<x_0$, that is, $x-x_0=\Delta x<0$, then $f'(x)>0$.

When $x>x_0$, that is, $x-x_0=\Delta x>0$, then $f'(x)<0$.

Therefore, when $f''(x_0)<0$, $f(x_0)$ is a local maximum of $f(x)$.

Similarly, we can prove (2), which is omitted here. □

Note:

(1) Theorem 5.4 (Second Derivative Test) is only valid for a stationary point x_0, where $f'(x_0)=0$. It cannot be applied at points where the derivative does not exist ($f'(x_0)=\infty$). Whether the discontinuity point x_0 of the derivative $f'(x)$ is an extreme point must be determined using the first derivative test. Specifically, we must first check if the function $f(x)$ itself is continuous at the point x_0. If it is discontinuous (that is, $f(x_0)=\frac{0}{0}$, $0\cdot\infty$, $\frac{\infty}{\infty}$ or

$f(x_0)=\infty$), x_0 is not an extreme point. If the function $f(x)$ is continuous at x_0, at the same time, the derivative $f'(x)$ exists on the left and right neighbourhood of the point x_0, we then check for a sign change in the derivative. If the signs of $f'(x)$ are opposite on either side of x_0, then $(x_0, f(x_0))$ is an extreme point. If the signs are the same, it is not an extreme point.

(2) When $f'(x_0)=f''(x_0)=0$, Theorem 5.4 is invalid. For example, the function $f(x)=x^3$ has $f'(0)=f''(0)=0$, yet the point $x=0$ is not an extreme point, as shown in Figure 5.3.3. Conversely, the function $f(x)=x^4$ has $f'(0)=f''(0)=0$; in this case, the point $x=0$ is a local minimum.

Example 3. Find the extreme values of the function $f(x)=2x^3-x+1$.

Solution.

$$f'(x)=6x^2-1$$

$$f''(x)=12x$$

To find the x-value of the stationary point, set the derivative to zero: $f'(x)=0$.

$$6x^2-1=0$$

$$x=\pm\sqrt{\frac{1}{6}}=\pm\frac{\sqrt{6}}{6}$$

Since $$f''(\frac{\sqrt{6}}{6})=12\cdot\frac{\sqrt{6}}{6}=2\sqrt{6}>0$$

thus $$f(\frac{\sqrt{6}}{6})=2(\frac{\sqrt{6}}{6})^3-\frac{\sqrt{6}}{6}+1=\frac{9-\sqrt{6}}{9}$$ is a local minimum.

Since $$f''(-\frac{\sqrt{6}}{6})=12\cdot(-\frac{\sqrt{6}}{6})=-2\sqrt{6}<0$$

thus $f(-\frac{\sqrt{6}}{6}) = 2(-\frac{\sqrt{6}}{6})^3 - (-\frac{\sqrt{6}}{6}) + 1 = \frac{9+\sqrt{6}}{9}$ is a local maximum.

as shown in Figure 5.3.10.

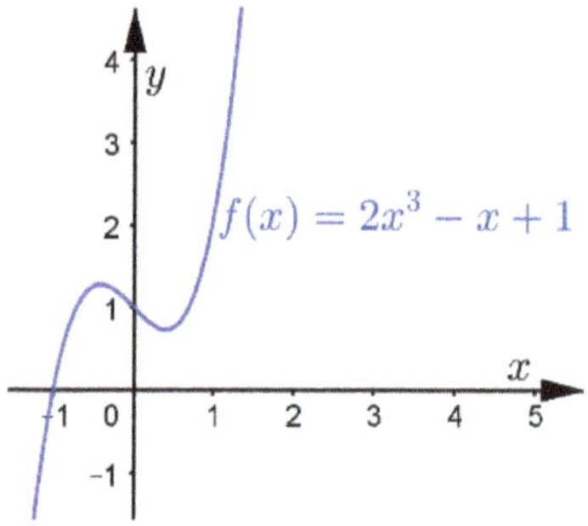

Figure 5.3.10

Exercise 5.3

1. Find the extreme values of the following functions.

a. $f(x) = x^3 - 3x^2 - 5$ **b.** $f(x) = (x-3)^2(x-2)$

c. $f(x) = (x+1)^{\frac{1}{3}}(x-5)^2$ **d.** $f(x) = x^4 + 12x^3 + 20x^2$

e. $f(x) = (3x+5)^{-\frac{2}{3}} + x$ **f.** $f(x) = x^2e^{-x}$

g. $f(x) = \frac{2x}{1+x^2}$ **h.** $f(x) = \sqrt{2+x-x^2}$

i. $f(x) = x^2 - \frac{3}{x}$ **j.** $f(x) = 2x + \frac{1}{x}$

k. $f(x) = \sqrt[3]{x}(x-4)$ **l.** $f(x) = \sqrt{x} - \sqrt[4]{x}$

2. Find the coordinates of stationary points on the curve $f(x) = 3x^3 + 2x^2 - x + 4$ and determine their nature using the first or second derivative test.

3. Find the coordinates and nature of stationary points on the curve

$$f(x) = x^4 - 8x^2 + 16$$

4. Find the intervals of increase and decrease, and determine the extreme values of the function $f(x) = 1 + 3x - x^3$.

5. Find the value of the constant a for which the function

$$f(x) = a\sin x + \frac{1}{3}\sin(3x)$$

has an extreme value at $x = \frac{\pi}{3}$. Find all extreme values of the function on the interval $[0, 2\pi]$.

6. Find the extreme values of $f(x) = x\sin x + \cos x$ on the interval $[0, \pi]$.

7. Given that $y = f(x)$ satisfies the equation

$$y'' + y' - e^{\sin x} = 0$$

and $f'(x_0) = 0$, determine whether $f(x)$ has an extreme value at $x = x_0$. If it does, determine its nature.

8. $y = f(x)$ is a solution to the equation

$$y'' - 2y' + 4y = 0$$

If $f(x_0) > 0$ and $f'(x_0) = 0$, determine whether x_0 is an extreme value of $f(x)$. If it is, determine its nature.

5.4. Maximum and minimum applications of extreme values

5.4.1. Maximum and minimum

Extreme value is a **local** concept, while maximum (or minimum) is a **global** concept. A maximum (or minimum) value is the greatest (or least) value of all functional values of a function $f(x)$ on the interval under investigation. In contrast, an extreme value is the maximum or minimum of the function $f(x)$ within a specific neighbourhood of the extreme point.

In general, the absolute maximum and minimum of a continuous function $f(x)$ on $[a,b]$ can be found by comparing the values at the endpoints, $f(a)$ and $f(b)$, with the values at the points where $f'(x)=0$ or $f'(x)$ does not exist on the interval. The largest of these values is the maximum value of the function $f(x)$ on $[a,b]$ and the smallest is the minimum value of the function $f(x)$ on $[a,b]$.

Note: The following are two special cases.

(1) If the function $f(x)$ is **monotonically increasing** on $[a,b]$, then $f(a)$ is the minimum, and $f(b)$ is the maximum, as shown in Figure 5.4.1.

If the function $f(x)$ is **monotonically decreasing** on $[a,b]$, then $f(a)$ is the maximum, and $f(b)$ is the minimum, as shown in Figure 5.4.2.

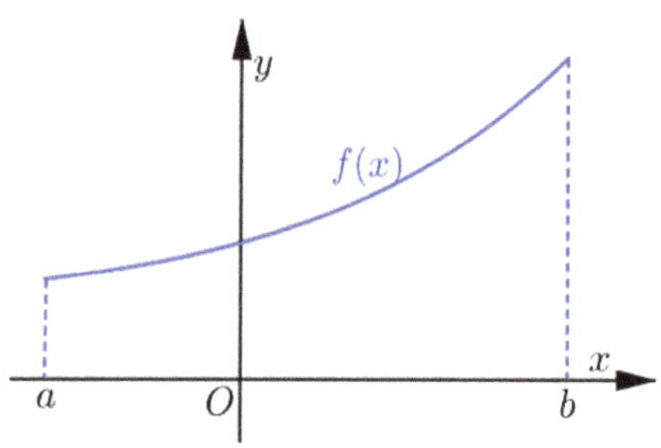

Figure 5.4.1

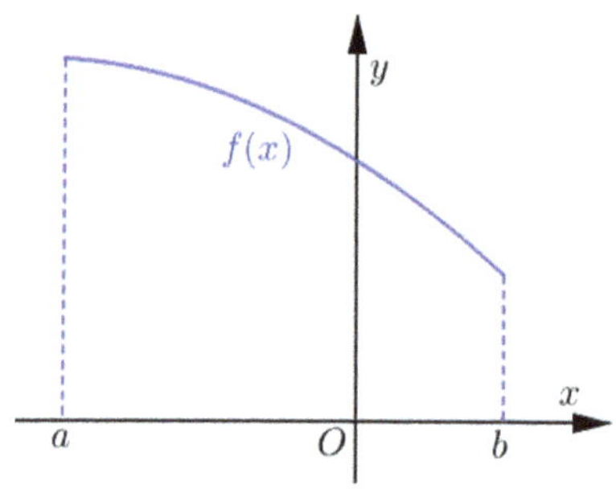

Figure 5.4.2

(2) If the continuous function $f(x)$ has only one local maximum on $[a,b]$ and no other extreme value, then this local maximum is also the global maximum, as shown in Figure 5.4.3.

If the continuous function $f(x)$ has only one local minimum on $[a,b]$ and no other extreme value, then this local minimum is also the global minimum, as shown in Figure 5.4.4.

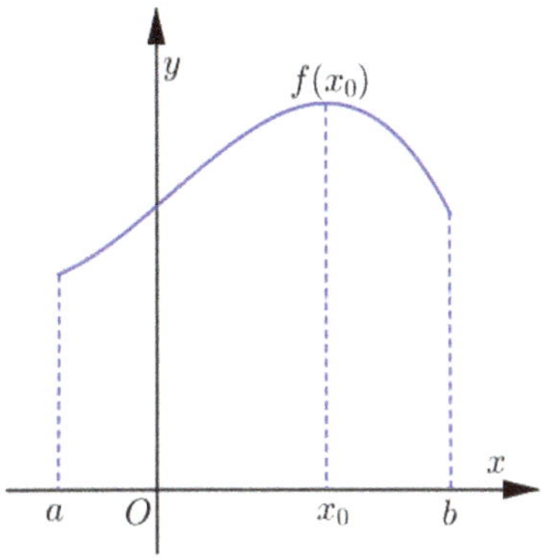

Figure 5.4.3

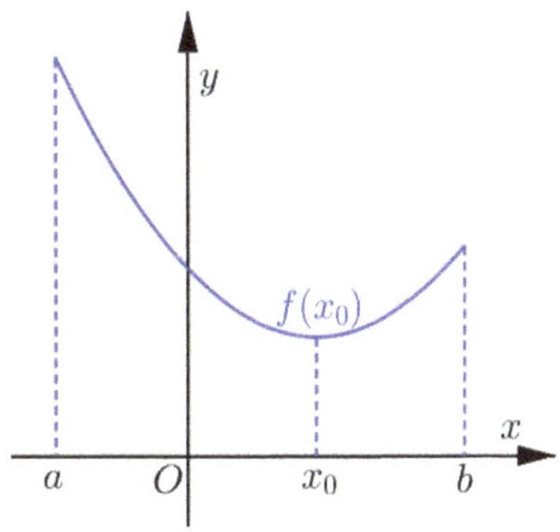

Figure 5.4.4

Many applied problems involving the search for a maximum or minimum are of this type and can be solved simply by finding the local extremum.

5.4.2. Application problems of extreme values

Example 1. A 10-metre wire is cut into two pieces. One piece is bent into a square and the other into a circle. Find the length of each piece that minimises the sum of the areas of the square and circle.

Solution. The wire is cut into two pieces. Let the length of one piece be x metres; then the length of the other piece will be $(10 - x)$ metres.

Bend the x-metre piece into a square; the side of the square is then $\frac{x}{4}$.

Bend the (10 – x)-metre piece into a circle; the perimeter of the circle is then

$2\pi r = 10 - x$, and the radius of the circle is $r = \dfrac{10-x}{2\pi}$.

The sum of the areas of the square and circle is

$$S = (\frac{x}{4})^2 + \pi(\frac{10-x}{2\pi})^2 = \frac{\pi x^2 + 400 - 80x + 4x^2}{16\pi}$$

When $S' = 0$, the sum of areas will be the smallest.

$$S' = (\frac{\pi x^2 + 400 - 80x + 4x^2}{16\pi})' = \frac{2\pi x - 80 + 8x}{16\pi} = 0$$

Solving gives $\quad x = \dfrac{40}{\pi + 4}$

Since $\quad S'' = (\dfrac{2\pi x - 80 + 8x}{16\pi})' = \dfrac{2\pi + 8}{16\pi} > 0$

the sum of the areas of the square and circle will be minimised when one piece of the wire has a length of $\dfrac{40}{\pi+4}$ metres, The length of the other piece of the wire is then

$10 - \dfrac{40}{\pi+4} = \dfrac{10\pi}{\pi+4}$ metres.

Example 2. The distance between stations A and B on a railway line is 200 kilometres. The distance between factory C and station B is 30 kilometres, BC ⊥ AB, as shown in Figure 5.4.5. It is known that the ratio of road freight costs to railway freight costs per kilometre is 5 : 3. If a road is built from some point D on the railway line (between station A and station B) to factory C, where should the point D be located to minimise the total freight cost from station A to factory C?

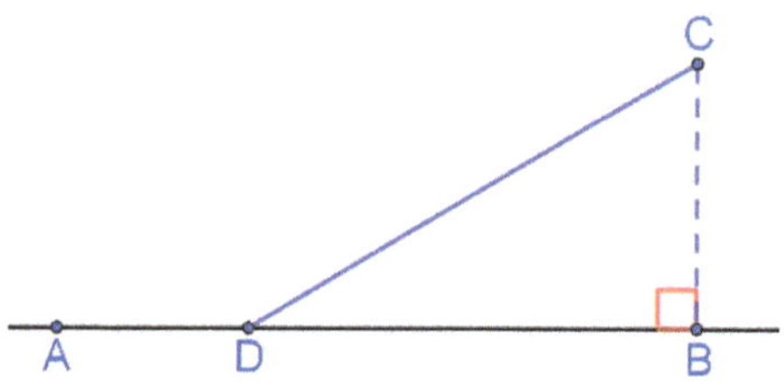

Figure 5.4.5

Solution. Let $\text{BD} = x$, then $\text{AD} = 200 - x$, $\text{CD} = \sqrt{30^2 + x^2} = \sqrt{900 + x^2}$.

Suppose that the road freight cost is $5k$ and the railway freight cost is $3k$.

Total freight cost from station A to factory C is

$$y = 5k \cdot \text{CD} + 3k \cdot \text{AD} = 5k\sqrt{900 + x^2} + 3k(200 - x)$$

To minimise the freight cost, set $y' = 0$.

$$y' = 5k \cdot \frac{d\sqrt{900 + x^2}}{d(900 + x^2)} \cdot \frac{d(900 + x^2)}{dx} + 3k(200 - x)'$$

$$= \frac{5kx}{\sqrt{900 + x^2}} - 3k = 0$$

Solving gives $x = 22.5$

Since

$$y'' = \left(\frac{5kx}{\sqrt{900 + x^2}} - 3k\right)' = \frac{5k\sqrt{900 + x^2} - 5kx \cdot \dfrac{x}{\sqrt{900 + x^2}}}{900 + x^2} = \frac{4500k}{(900 + x^2)^{\frac{3}{2}}} > 0$$

when the distance between the point D and station B is 22.5 kilometres, the freight cost from station A to factory C is minimised.

Example 3. A 2-metre-high oil painting hangs on the wall. Its lower edge is 2 metres above the ground level. The viewer's eyes are 1.7 metres above the ground level. How far from the wall should the viewer stand to maximise their view of the painting?

Solution. As shown in Figure 5.4.6, let the upper edge of the oil painting be A and the lower edge be B. The viewer's eyes are at F and the distance between the viewer and the wall is $\text{ED} = x$. We have AB = 2, BC = 0.3 (since the eye level is at 1.7 m and the lower edge is at 2 m) and CD = 1.7. Let $\angle \text{BFC} = \alpha$ and $\angle \text{AFB} = \beta$.

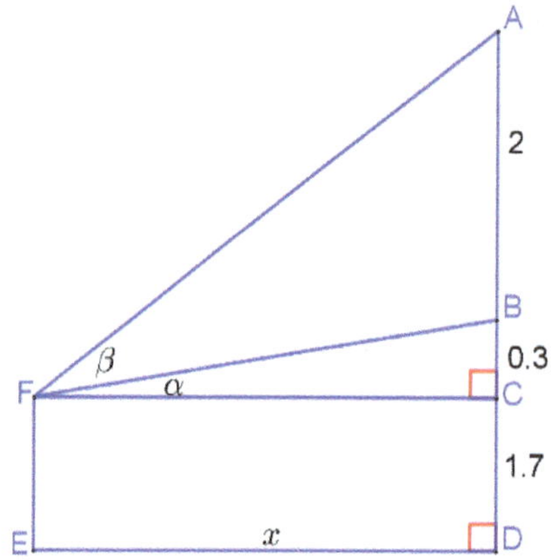

Figure 5.4.6

In the right-angled triangle ΔAFC

$$\tan(\alpha+\beta)=\frac{\text{AC}}{\text{FC}}=\frac{2.3}{x}$$

In the right-angled triangle ΔBFC

$$\tan\alpha=\frac{\text{BC}}{\text{FC}}=\frac{0.3}{x}$$

Using the formula, we have

$$\tan(\alpha+\beta)=\frac{\tan\alpha+\tan\beta}{1-\tan\alpha\tan\beta}=\frac{\frac{0.3}{x}+\tan\beta}{1-\frac{0.3}{x}\tan\beta}$$

Thus
$$\frac{\frac{0.3}{x}+\tan\beta}{1-\frac{0.3}{x}\tan\beta}=\frac{2.3}{x}$$

$$\frac{0.3+x\tan\beta}{x-0.3\tan\beta}=\frac{2.3}{x}$$

Rearranging gives

$$\tan\beta=\frac{2x}{x^2+0.69}$$

Since $\tan\beta$ is monotonically increasing on $(0,\frac{\pi}{2})$, when $\tan\beta$ achieves maximum, β achieves maximum.

$$(\tan\beta)' = \left(\frac{2x}{x^2+0.69}\right)' = \frac{2(x^2+0.69)-4x^2}{(x^2+0.69)^2} = \frac{2(0.69-x^2)}{(x^2+0.69)^2}$$

Set $(\tan\beta)' = \dfrac{2(0.69-x^2)}{(x^2+0.69)^2} = 0$ and it gives $x = \sqrt{0.69}$.

x	$(0, \sqrt{0.69})$	$\sqrt{0.69}$	$(\sqrt{0.69}, \infty)$
$(\tan\beta)'$	+	0	–
	↗	–	↘

Therefore, when $x = \sqrt{0.69}$ metres, the viewer's view of the painting is maximised.

Example 4. A sector with a central angle β (in radians) is cut from a circle of radius R and formed into a cone. For what value of β is the volume of the cone is maximised?

Solution. Let the radius of the base of the cone be r and the height be h, as shown in Figure 5.4.7.

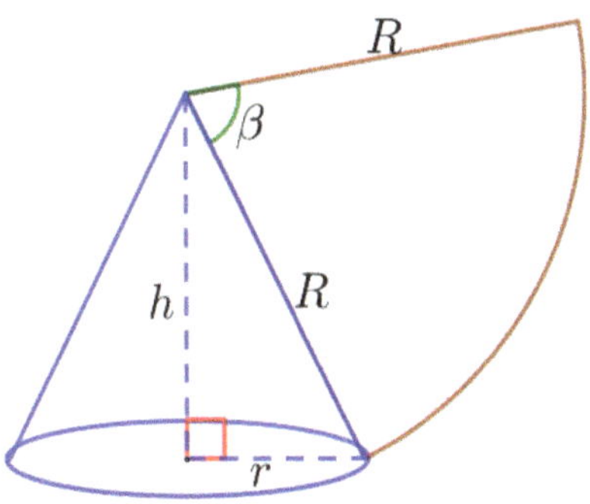

Figure 5.4.7

Then $\qquad h^2 = R^2 - r^2$

$$2\pi r = \beta R$$

The volume of the cone is

$$V = \frac{1}{3}\pi r^2 h = \frac{1}{3}\pi r^2\sqrt{R^2 - r^2}$$

When the volume of the cone is the biggest, $V' = 0$,

$$V'=(\frac{1}{3}\pi r^2\sqrt{R^2-r^2})'=\frac{1}{3}\pi\left[2r\sqrt{R^2-r^2}+r^2\cdot\frac{d(R^2-r^2)^{\frac{1}{2}}}{d(R^2-r^2)}\cdot\frac{d(R^2-r^2)}{dr}\right]$$

$$=\frac{1}{3}\pi\left[2r\sqrt{R^2-r^2}-\frac{r^3}{\sqrt{R^2-r^2}}\right]=\frac{1}{3}\pi\cdot\frac{2R^2r-3r^3}{\sqrt{R^2-r^2}}=0$$

which gives $\qquad 2R^2r-3r^3=0$

$$r(2R^2-3r^2)=0$$

$$r=0 \text{ (rejected)} \quad \text{or} \quad r=\frac{\sqrt{6}}{3}R$$

r	$(0,\frac{\sqrt{6}}{3}R)$	$\frac{\sqrt{6}}{3}R$	$(\frac{\sqrt{6}}{3}R,R)$
V'	+	0	−
	↗	−	↘

Therefore, when $r=\frac{\sqrt{6}}{3}R$, the volume of the cone is the biggest. From $2\pi r=\beta R$, we obtain

$$\beta=\frac{2\pi r}{R}=\frac{2\pi}{R}\cdot\frac{\sqrt{6}}{3}R=\frac{2\sqrt{6}}{3}\pi$$

Hence, when $\beta=\frac{2\sqrt{6}}{3}\pi$, the volume of the cone is the biggest.

Example 5. A sphere of radius r circumscribes a right cone. For what value of the height of the right cone is the volume of the cone, V, minimised? Find this minimum value.

Solution. As shown in Figure 5.4.8, O is the centre of the sphere, the radius of the sphere is $\text{OC}=\text{OA}=r$. Let A be the centre of the cone's base circle, let the height of the cone be $\mathrm{AD} = h$, and let the radius of the cone's base circle be $\mathrm{AB} = R$.

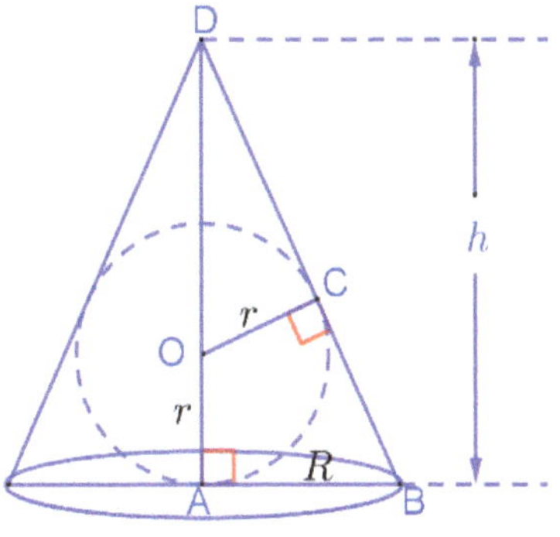

Figure 5.4.8

Since AD is the height of the cone, thus

$$\mathrm{AD} \perp \mathrm{AB} \qquad \Rightarrow \qquad \angle\mathrm{DAB} = 90^\circ$$

Because the sphere circumscribes the right cone, thus

$$\mathrm{OC} \perp \mathrm{CD} \qquad \Rightarrow \qquad \angle\mathrm{OCD} = 90^\circ$$

And $\angle$ODC is the common angle for triangle ΔBDA and triangle ΔODC, hence,

$$\Delta\mathrm{BDA} \sim \Delta\mathrm{ODC}$$

Since corresponding sides of similar triangles are proportional, we have

$$\frac{\mathrm{AB}}{\mathrm{OC}} = \frac{\mathrm{AD}}{\mathrm{CD}}$$

$$\frac{R}{r} = \frac{h}{\sqrt{(h-r)^2 - r^2}}$$

$$R = \frac{hr}{\sqrt{h^2 - 2hr}}$$

Therefore, the volume of the cone is

$$V = \frac{\pi}{3} R^2 h = \frac{\pi}{3} \cdot \frac{h^2 r^2}{h^2 - 2hr} \cdot h = \frac{\pi r^2}{3} \cdot \frac{h^2}{h-2r}$$

When the volume of the cone V is the smallest, $V' = 0$,

$$V' = \left(\frac{\pi r^2}{3} \cdot \frac{h^2}{h-2r}\right)' = \frac{\pi r^2}{3} \cdot \frac{(h^2)'(h-2r) - h^2(h-2r)'}{(h-2r)^2} = \frac{\pi r^2}{3} \cdot \frac{h^2 - 4hr}{(h-2r)^2} = 0$$

which gives

$$h^2 - 4hr = 0$$

$$h = 0 \text{ (rejected)} \quad \text{or} \quad h = 4r$$

When $h = 4r$, the volume of the cone is a minimum.

$$V_{\min} = \frac{\pi r^2}{3} \cdot \frac{h^2}{h-2r} = \frac{\pi r^2}{3} \cdot \frac{(4r)^2}{4r-2r} = \frac{8\pi r^3}{3}$$

Exercise 5.4

1. A ball is thrown vertically upwards. After t seconds, its height (h) in metres above the ground is given by

$$h = 14.5t - 4t^2$$

Find the maximum height the ball will reach and the time it is at its maximum height.

2. The profit, y euros, from the sale of x thousand items of a product is given by the equation

$$y = 180x + 15x^2 - x^3$$

Find the number of items that should be sold to maximise profit, and calculate the value of this maximum profit.

3. A piece of string of length 80 cm is cut into two pieces. One piece is bent to form a square of side x cm, and the other piece is bent to form a rectangle, which is three times as long as it is wide. Find, in terms of x, an expression for

a. the width of the rectangle.

b. the length of the rectangle.

c. the area of the rectangle.

Given that the sum of the two areas is a minimum

d. find the value of x.

4. A rectangular plot of land is enclosed using 400 metres of fencing, with a long straight river forming one side, as shown in Diagram 5.4.1 below.

Diagram 5.4.1

For the maximum area enclosed, find the dimensions of the plot.

5. A rectangular sheet of paper measures 36 cm by 24 cm. Four equal squares are cut from each corner, as shown in Diagram 5.4.2, and the remaining paper is folded to form an open box. Find the length of the side of the square that maximises the volume of the box. Calculate the value of this maximum volume.

Diagram 5.4.2

6. The perimeter of a rectangle is 24 cm. Find the maximum area of the rectangle.

7. Find the point on the curve $y = x^3$ that is the closest to the point (5, 0).

8. A rectangle is bounded by the parabola $y = 16 - x^2$ and the x-axis. One side of the rectangle is on the x-axis, and two other vertices are on the parabola, as shown in Diagram 5.4.3. Find the dimensions of the rectangle that give the maximum area.

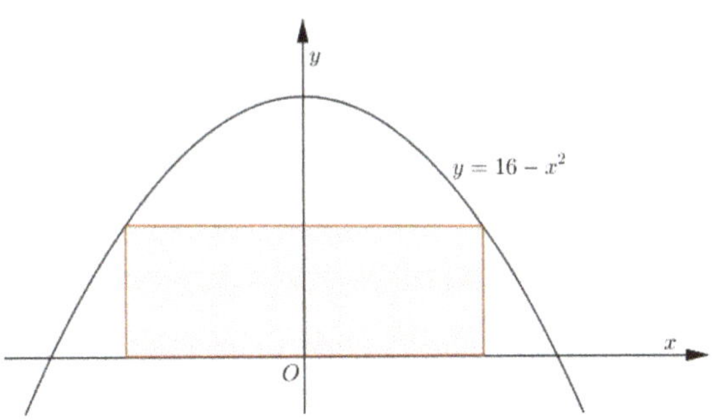

Diagram 5.4.3

9. Find the coordinates of the point P on the curve $y = \frac{1}{2}x^2 + 1$ that has the shortest distance from the point (3, $-\frac{1}{2}$).

10. An isosceles triangle is inscribed in a circle of radius R. Find the length of the base of the triangle such that the sum of the base and the height on the base is maximised.

11. In a right-angled triangle ABC, AC is the hypotenuse. Given that the sum of the lengths of BC and AC is 24 cm, find the maximum area of the triangle ABC.

12. The cost (in euros) of producing x thousand units of a product is given by the formula C = 100 + 5x + $\frac{1}{10}x^2$. The sale price per thousand units is P euros, where $P = a + \frac{x}{b}$ (a and b are constants). Assuming all units are sold, the profit is maximised when the production quantity is 150 thousand units, and the sale price is 40 euros for 150 thousand units. Find the values of a and b.

13. An isosceles triangle is inscribed in a circle of radius R. Find the height of the triangle such that its area is the largest.

14. An open rectangular box with a square base has a fixed volume of 108 m^3. Calculate the minimum total surface area required to construct the box.

15. If $x + y = 10$, find the minimum value of $N = x^3 + y^2$.

16. An open cylindrical tank is to be constructed to hold 300 m^3 of water. Given that the cost to construct the base is twice that of the curved surface of the cylinder, find the dimensions of the tank that minimise the total construction cost.

17. A window is to be constructed in the shape of a semicircle on top of a rectangle. The diameter of the semicircle is equal to width of the rectangle. Given that the perimeter of the window is 20 metres, find the radius of the semicircle and the dimensions of the rectangle that will allow the maximum amount of light through.

18. The cross-section ABCD of a water channel is an isosceles trapezium, as shown in Diagram 5.4.4. The area of the cross-section is a constant S, and its height is a constant h. To make the total length of AB + BC + CD the shortest, what is the size of angle β?

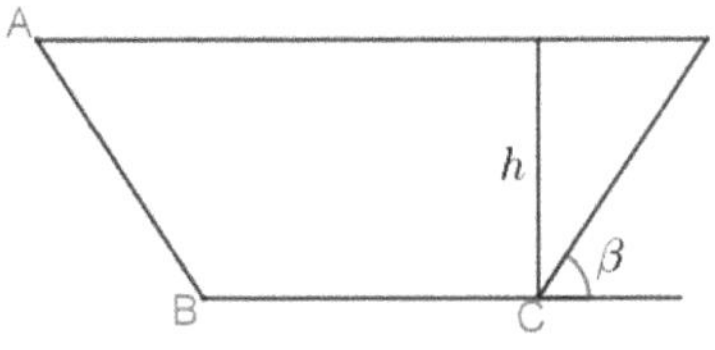

Diagram 5.4.4

19. A tourist wishes to take a photo of a statue, 6 metres tall, which is situated on a 10-metre-high marble stand. At what distance from the base of the marble stand should the tourist be positioned so that the statue subtends the largest possible angle at the camera lens? Assume the camera is held at 1.60 metres above the horizontal ground.

20. Two factories are located on the same side of a long straight road, as shown in Diagram 5.4.5. AB is on the road. AB = 20 kilometres. Factory C is 3 kilometres from the road and factory D is 5 kilometres from the road. The two factories want to build a storehouse E between A and B. Two new roads will then be constructed to connect the factories to the storehouse E. Where should the location of E be such that the total length of the roads CE and DE is the shortest?

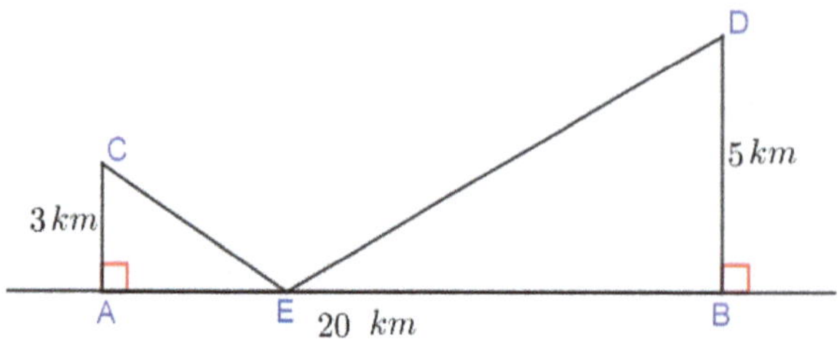

Diagram 5.4.5

21. An object consists of a vertical cylinder and a hemisphere. Given that the volume of the object is a constant, V, find the dimensions of the object such that its total surface area is a minimum.

22. The area of a rectangle is A. Find the dimensions of the rectangle such that its perimeter is a minimum.

23. Find the height of a cylinder with the largest volume that can be inscribed in a sphere of radius R. Also, calculate the value of this largest volume.

24. A rectangle ADEF is inscribed in the right-angled triangle ABC, as shown in Diagram 5.4.6. Given that AC = 60 cm and AB = 80 cm, find the dimensions of the rectangle ADEF that maximises its area.

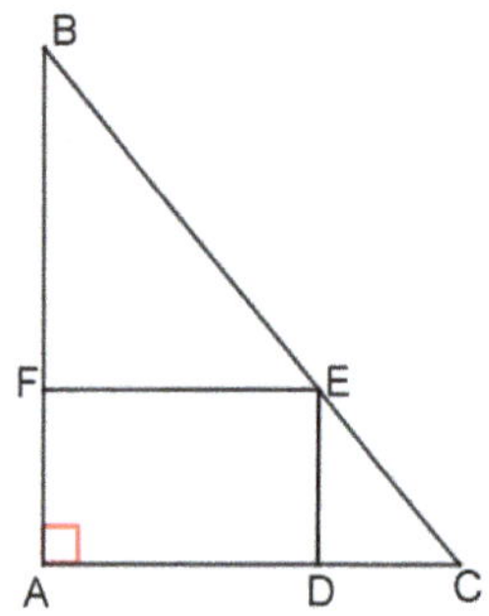

Diagram 5.4.6

25. A triangle ABC is inscribed in a circle with centre O and radius r. AC is a diameter of the circle. Find the length of the sides AB and BC such that the area of the triangle ABC is a maximum.

26. A pyramid has a square base with a side length of x and height of h. Find the value of x such that the volume of the pyramid is 2000 m^3 and its surface area is a minimum.

27. Triangle ABC is a right-angled triangle, $\angle ACB = x$ and $BC = h$. The radius of the inscribed circle is r, as shown in Diagram 5.4.7.

a. Express r in terms of the angle x and the length of the hypotenuse h.

b. Assuming h is constant and x varies, find the value of x for which r is a maximum.

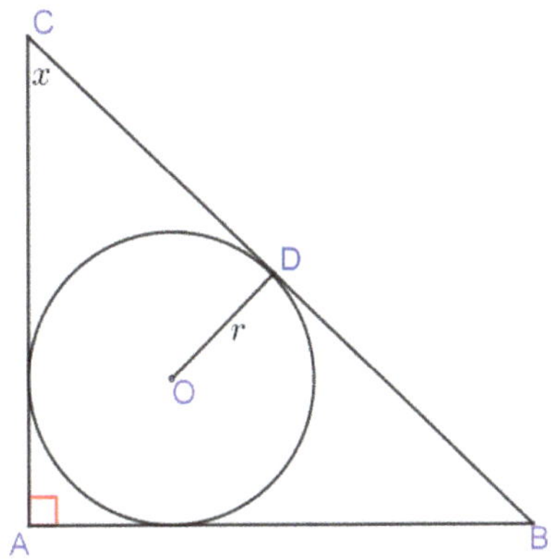

Diagram 5.4.7

28. A tangent is drawn to the curve $\frac{x^2}{a^2}+\frac{y^2}{b^2}=1$ (where a and b are constants) at a point $P(x_0, y_0)$ on the curve. This tangent intersects the x-axis at A and the y-axis at B. Find the coordinates of P such that the area of triangle OAB is a minimum, calculate this minimum area.

29. OAB is a triangle. The coordinates of A are (20, 0) and B are (10, 20), as shown in Diagram 5.4.8. A rectangle is inscribed in the triangle such that one of its sides lies on OA. Find the dimensions of the rectangle that maximise its area.

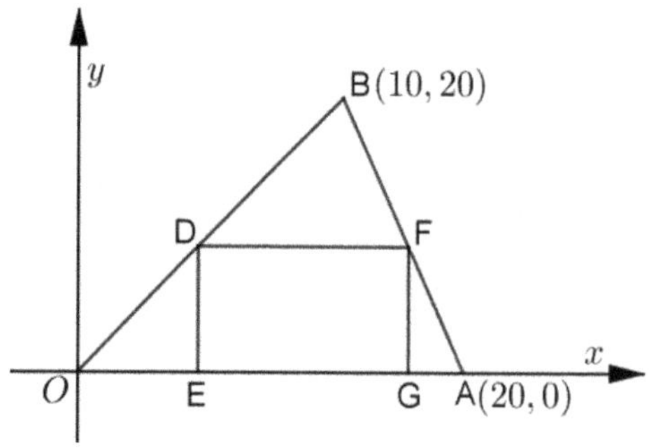

Diagram 5.4.8

30. A corridor of width a meets another corridor perpendicularly, as shown in Diagram 5.4.9. If a thin rod of length $8a$ can pass horizontally around a corner, what is the minimum width of the second corridor?

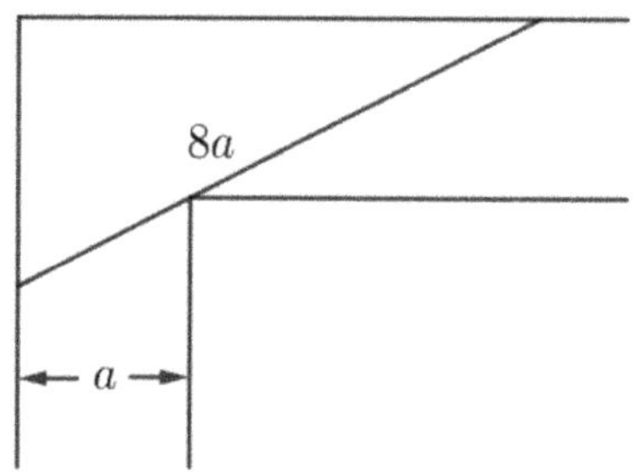

Diagram 5.4.9

31. A circular table has a radius a. A lamp is to be installed at a height directly above the centre of the table. The illumination intensity, I, at a point on the table is given by the formula

$$I = k\frac{\sin\theta}{x^2}$$

where x is the distance between the lamp and the point being illuminated.

θ is the angle between the light and the table surface of the point being illuminated.

k is a constant.

Diagram 5.4.10 illustrates the situation. The lamp is at point A, at a height h directly above the centre of the table. The centre of the table is at point O. $\angle AEO = \theta$.

Determine the height at which the lamp should be installed to maximise the illumination at the point E at the edge of the table.

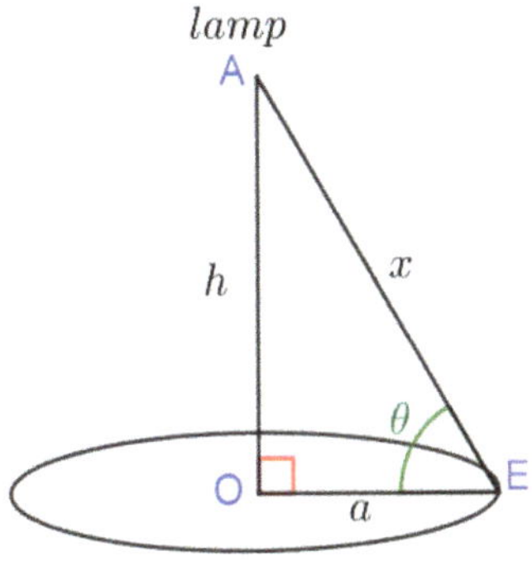

Diagram 5.4.10

5.5. Concavity and points of inflection of curves

When studying the shape of a function's graph, understanding the patterns of **increase and decrease** is helpful, but these do not fully describe its behaviour. As shown in Figure 5.5.1, on the interval (a,b), even though the function $f(x)$ is strictly increasing, its **curvature** varies. From left to right, the curve first **bends downwards**. After passing the point P, it changes its direction of curvature and **bends upwards**. Therefore, to fully understand a function, it is necessary to examine its **concavity** and the points where its curvature changes.

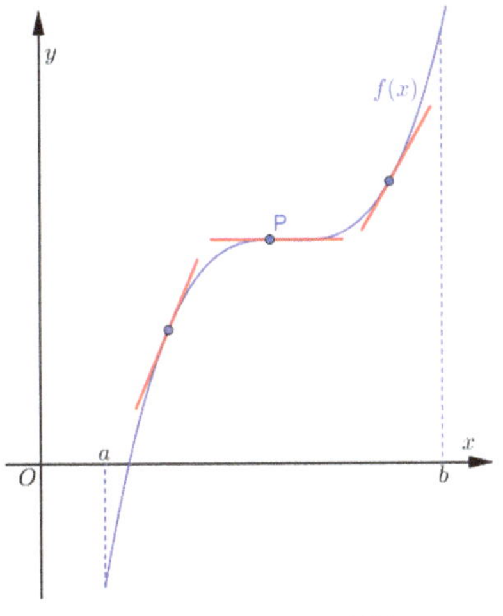

Figure 5.5.1

From Figure 5.5.1, it is evident that the curve bends downwards to the left of the point P, where the curve **lies below** the tangent at any point. To the right of the point P, the curve bends upwards and **lies above** the tangent. Based on this observation, we can establish the following definition:

Definition 5.2

If the curve lies above the tangent at each point in a given interval, then it is **concave up** in this interval, as shown in Figures 5.5.2 and 5.5.3.

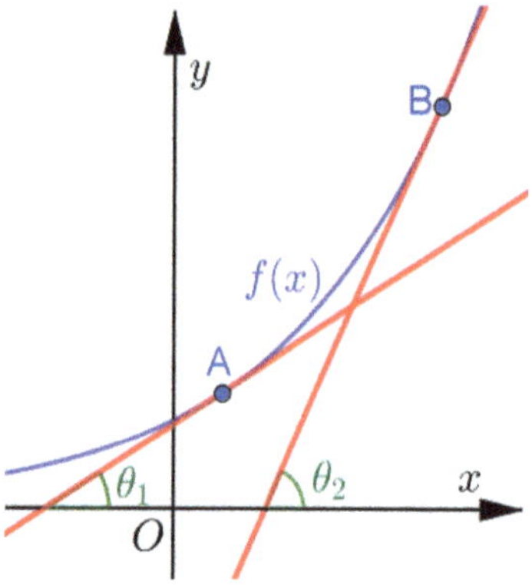

Figure 5.5.2 $\tan\theta$ increases as x increases (becomes more positive)

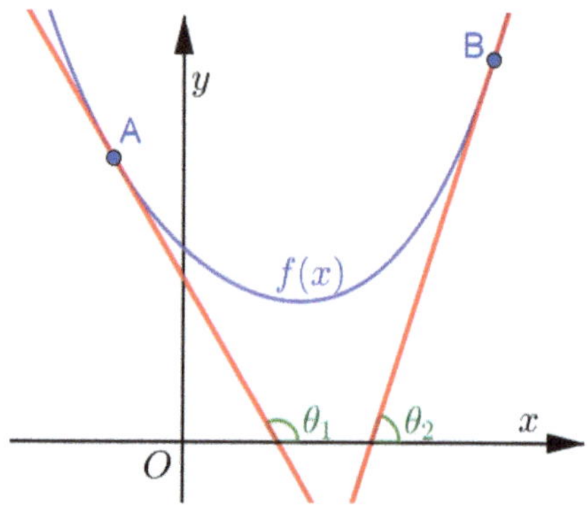

Figure 5.5.3 $\tan\theta$ increases as x increases (changing from negative to positive)

If the curve lies below the tangent at each point in a given interval, then the curve is **concave down** in this interval, as shown in Figures 5.5.4 and 5.5.5.

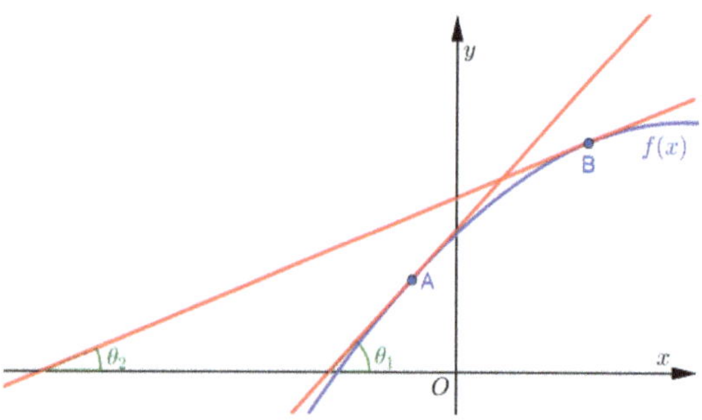

Figure 5.5.4 $\tan\theta$ decreases as x increases

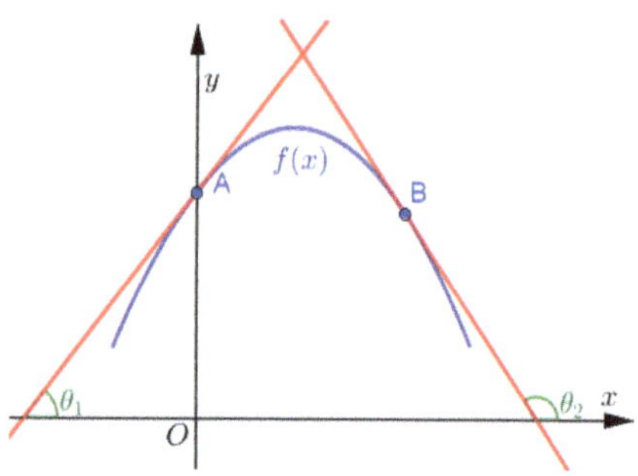

Figure 5.5.5 $\tan\theta$ decreases as x increases (changing from positive to negative)

Theorem 5.5

Let the function $f(x)$ have a second derivative on the interval (a,b).

(1) If $f''(x) > 0$ for all $x \in (a,b)$, then $f'(x) = \tan\theta$ is monotonically increasing on the interval (a,b). This implies that $\tan\theta$ increases as x increases. As shown in Figures 5.5.2 and 5.5.3, the curve $y = f(x)$ is concave up on the interval (a,b).

(2) If $f''(x) < 0$ for all $x \in (a,b)$, then $f'(x) = \tan\theta$ is monotonically decreasing on the interval (a,b). This implies that $\tan\theta$ decreases as x increases. As shown in Figures 5.5.4 and 5.5.5, the curve $y = f(x)$ is concave down on the interval (a,b).

Definition 5.3

The point at which the curve changes from being concave up to concave down or vice versa is called a **point of inflection**. Since a point of inflection is the dividing point between concave up and concave down, $f''(x)$ must have different signs on either side of the point. Therefore, at a point of inflection, either $f''(x) = 0$ or $f''(x)$ does not exist ($f''(x) = \infty$).

If the curve $f(x)$ has a point of inflection at the point x_0, then either $f''(x_0) = 0$ or $f''(x)$ does not exist ($f''(x_0) = \infty$).

I. When $f'(x_0) = 0$, the inflection point of the curve $f(x)$ at the point x_0 is a **stationary point of inflection** (or a **horizontal inflection point**), such as point P in Figure 5.5.1.

II. When $f'(x_0) \neq 0$, the inflection point of the curve $f(x)$ at the point x_0 is a **non-stationary point of inflection**, such as point A in Figure 5.5.6.

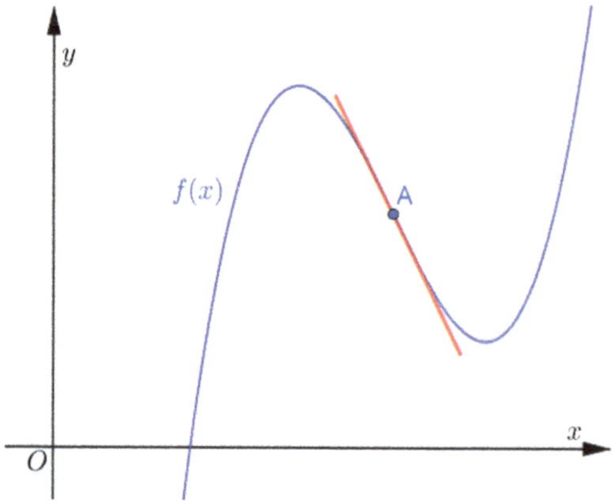

Figure 5.5.6

Note:

The second derivative $f''(x_0)$ does not exist ($f''(x_0) = \infty$). However, if the function $f(x)$ is continuous at the discontinuity point x_0 of $f''(x)$, and the second derivative $f''(x)$ exists on either side of x_0 with opposite signs, then $(x_0, f(x_0))$ is a point of inflection. If the signs are the same, it is not a point of inflection.

Example 1. Find the values of x such that $y = x^4 - 2x^3 + 1$ is concave up.

Solution. When $y'' > 0$, y is concave up.

$$y' = 4x^3 - 6x^2$$

$$y'' = 12x^2 - 12x = 12x(x - 1)$$

Therefore, $\quad 12x(x-1) > 0$

$$x > 1 \quad \text{or} \quad x < 0$$

Example 2.

(a) Find the coordinates of the stationary points of the curve

$$f(x) = x^3 + x^2 - 5x - 24$$

(b) Determine the nature of the stationary points.

(c) Find the coordinates of all points of inflection.

(d) Sketch the graph of the curve.

Solution. (a) $f(x) = x^3 + x^2 - 5x - 24$

$$f'(x) = 3x^2 + 2x - 5$$

Set $f'(x) = 0$

that is, $3x^2 + 2x - 5 = 0$

which gives the stationary points at

$$x_1 = 1 \qquad x_2 = -\frac{5}{3}$$

When $x_1 = 1$, $f(1) = 1^3 + 1^2 - 5 \cdot 1 - 24 = -27$

When $x_2 = -\frac{5}{3}$, $f(-\frac{5}{3}) = (-\frac{5}{3})^3 + (-\frac{5}{3})^2 - 5 \cdot (-\frac{5}{3}) - 24 = -\frac{473}{27}$

Therefore, the coordinates of the stationary points are $(1, -27)$, $(-\frac{5}{3}, -\frac{473}{27})$.

(b) Since $f'(x) = 3x^2 + 2x - 5$

$$f''(x) = 6x + 2$$

$$f''(1) = 6 \cdot 1 + 2 = 8 > 0$$

$$f''(-\frac{5}{3}) = 6 \cdot (-\frac{5}{3}) + 2 = -8 < 0$$

$f(1) = -27$ is a local minimum of $f(x)$. $f(-\frac{5}{3}) = -\frac{473}{27}$ is a local maximum of $f(x)$.

(c) As $f''(x) = 6x + 2$

set $f''(x) = 6x + 2 = 0$

Solving gives the point of inflection at

$$x = -\frac{1}{3}$$

Since $\qquad f'(x) = 3x^2 + 2x - 5$

$$f'(-\frac{1}{3}) = 3 \cdot (-\frac{1}{3})^2 + 2 \cdot (-\frac{1}{3}) - 5 = -\frac{16}{3} \neq 0$$

there is no stationary point of inflection; there is only a non-stationary point of inflection. Since $f''(x)$ exists on either side of the point $x = -\frac{1}{3}$ and has opposite signs

$$f(-\frac{1}{3}) = (-\frac{1}{3})^3 + (-\frac{1}{3})^2 - 5 \cdot (-\frac{1}{3}) - 24 = -\frac{601}{27}$$

thus, the non-stationary point of inflection is $(-\frac{1}{3}, -\frac{601}{27})$.

(d)

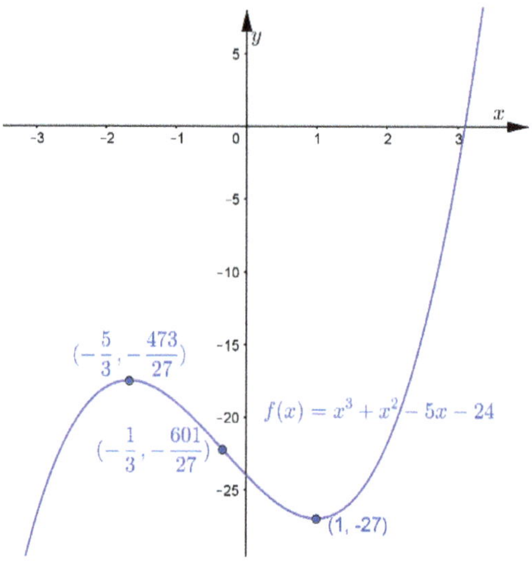

Figure 5.5.7

Example 3. On the curve $f(x) = ax^3 + bx^2 + cx + d$, the curve has a local minimum at the point (-3, -15) and a local maximum at $x = \frac{1}{3}$. The curve also passes through the point (0, 3).

(a) Find the values of the constants a, b, c and d.

(b) Find the coordinates of all points of inflection.

(c) Sketch the graph of the curve.

Solution. (a) Since the curve passes through the point (0, 3), substituting gives

$$3 = a\cdot 0^3 + b\cdot 0^2 + c\cdot 0 + d$$

$$d = 3$$

As there is a local minimum at (-3, -15), the point (-3, -15) lies on the curve.

$$-15 = a\cdot(-3)^3 + b\cdot(-3)^2 + c\cdot(-3) + 3$$

$$-18 = -27a + 9b - 3c$$

$$9a - 3b + c = 6 \qquad (1)$$

Moreover,

$$f'(x) = (ax^3 + bx^2 + cx + d)' = 3ax^2 + 2bx + c$$

There is a local minimum at (-3, -15).

$$f'(-3) = 3a\cdot(-3)^2 + 2b\cdot(-3) + c = 0$$

$$27a - 6b + c = 0 \qquad (2)$$

There is a local maximum at $x = \frac{1}{3}$.

$$f'(\frac{1}{3}) = 3a\cdot(\frac{1}{3})^2 + 2b\cdot(\frac{1}{3}) + c = 0$$

$$\frac{1}{3}a + \frac{2}{3}b + c = 0$$

$$a + 2b + 3c = 0 \qquad (3)$$

Solving equations (1), (2) and (3) simultaneously gives

$$a = -1, \quad b = -4, \quad c = 3$$

Therefore, the curve is

$$f(x) = -x^3 - 4x^2 + 3x + 3$$

(b) $$f(x) = -x^3 - 4x^2 + 3x + 3$$

$$f'(x) = (-x^3 - 4x^2 + 3x + 3)' = -3x^2 - 8x + 3$$

$$f''(x) = (-3x^2 - 8x + 3)' = -6x - 8 = 0$$

$$x = -\frac{4}{3}$$

Because $f'(-\frac{4}{3}) \neq 0$, there is no stationary point of inflection; there is only a non-stationary point of inflection. Since $f''(x)$ exists on either side of the point $x = -\frac{4}{3}$ and has opposite signs

$$f(-\frac{4}{3}) = -(-\frac{4}{3})^3 - 4 \cdot (-\frac{4}{3})^2 + 3 \cdot (-\frac{4}{3}) + 3 = -\frac{155}{27}$$

thus, $(-\frac{4}{3}, -\frac{155}{27})$ is the non-stationary point of inflection.

(c)

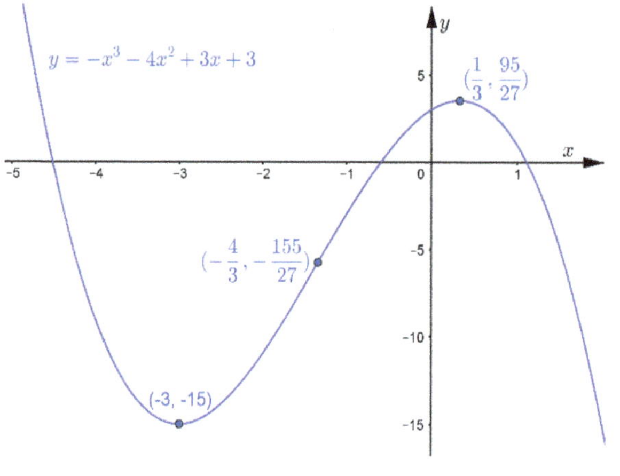

Figure 5.5.8

Example 4. Figure 5.5.9 shows the graph of $y = f'(x)$. On the graph

(a) mark the point corresponding to the local maximum of $f(x)$.

(b) mark the point corresponding to the local minimum of $f(x)$.

(c) mark the point corresponding to the point of inflection of $f(x)$.

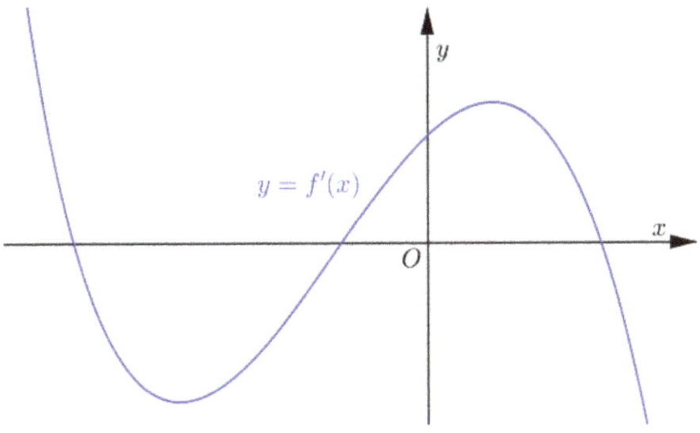

Figure 5.5.9

Solution. First, find the stationary points where $f'(x)=0$, then determine their nature by using either the first or second derivative test.

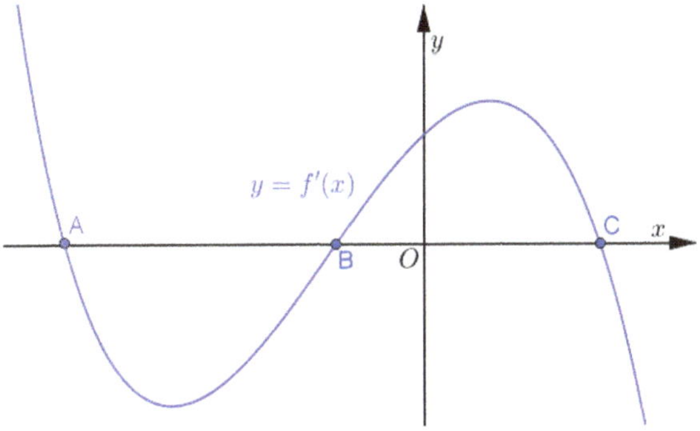

Figure 5.5.10

x	to the left of point A	point A	to the right of point A
y'	positive	0	negative
	↗	–	↘

Therefore, point A corresponds to the local maximum of $f(x)$.

x	to the left of point B	point B	to the right of point B
y'	negative	0	positive
	↘	–	↗

Therefore, point B corresponds to the local minimum of $f(x)$.

Point C is similar to point A, and it also corresponds to the local maximum of $f(x)$.

On the graph of $y = f'(x)$, find the points where the gradient of the tangent is zero, namely, points J and K, as shown in Figure 5.5.11.

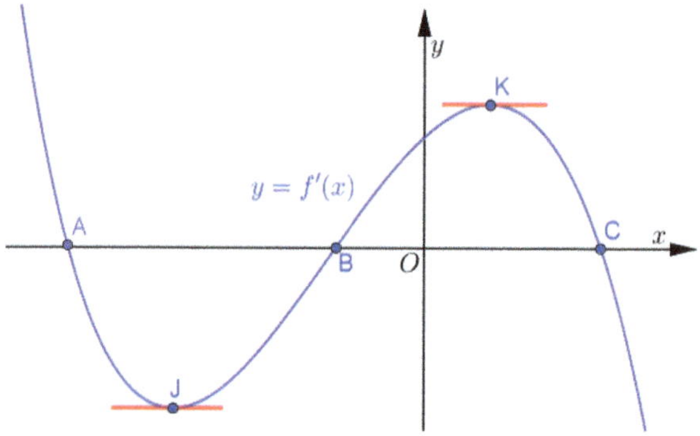

Figure 5.5.11

x	to the left of point J	point J	to the right of point J
y''	negative	0	positive
	↘	–	↗

Therefore, point J corresponds to the point of inflection of $f(x)$.

x	to the left of point K	point K	to the right of point K
y''	positive	0	negative
	↗	–	↘

Therefore, point K corresponds to the point of inflection of $f(x)$.

Example 5. $f(x) = x^4$. Does the curve have a point of inflection at $x = 0$?

Solution. $f'(x) = 4x^3$ $f''(x) = 12x^2$

Set $f'(x) = 4x^3 = 0$

Solving gives a stationary point at $x = 0$.

Since $f(x) = x^4 \geq 0$

thus $f(0) = 0$ is the minimum of $f(x)$.

Set $f''(x) = 12x^2 = 0$

Solving gives $\quad x = 0$

Since $\quad f''(x) = 12x^2 \geq 0$

for $x < 0$ or $x > 0$, there is always $f''(x) > 0$.

Because the signs of $f''(x)$ does not change on either side of $x = 0$, there is no point of inflection at $x = 0$, as shown in Figure 5.5.12.

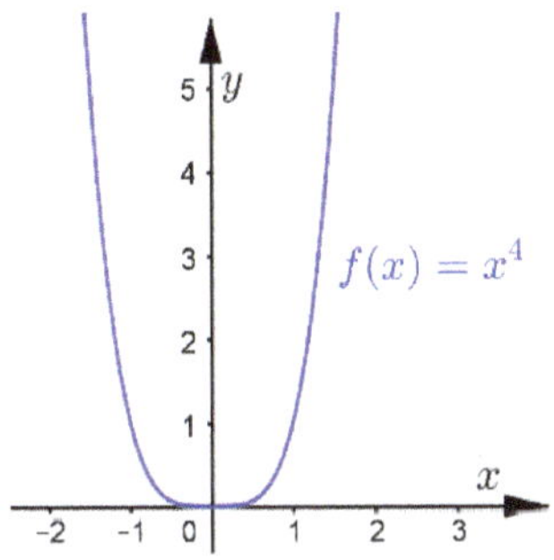

Figure 5.5.12

Exercise 5.5

1. Find the intervals on which the function $y = 3x^4 - 6x^2$ is concave up.

2. Find the intervals on which the function $y = \ln(x^2 + 1)$ is concave up. Also, find the coordinates of any points of inflection.

3. $f(x) = 2x^5 - 3x^3$

a. Find the coordinates of the stationary points.

b. Determine the nature of the stationary points.

c. Find the coordinates of any points of inflection.

d. Determine the intervals on which the function is concave up.

e. Sketch the graph of the curve.

4. Find and justify the occurrence for all maximum points, minimum points and points of inflections for the following curves. For each curve, determine the intervals on which it is concave down.

a. $y = \dfrac{2x}{1+x^2}$

b. $y = \ln(1+x^2)$

c. $y = \sqrt{x} + \dfrac{2}{\sqrt{x}}$

d. $y = (x+1)^2(3-x)$

e. $y = x^2e^{-2x}$

f. $y = \sqrt{1+x^2}$

g. $y = 2x + x^{\frac{5}{3}}$

h. $y = e^{-x^2}$

i. $y = x - e^x$

j. $y = x - \ln(1+x)$

5. The function $y = \dfrac{a}{x^2 + bx + c}$ has the value 0.5 when $x = 0$ and has a stationary point at $x = 2.5$. When $x = 0$, the graph has a gradient $-\dfrac{5}{12}$.

a. Find the values of a, b, and c.

b. Find the intervals on which the function is concave up.

c. Find the coordinates of any points of inflection.

6. The function $f(x)$ is defined by $f(x) = e^{-x}(x^2 - 4)$.

a. Find all stationary points and determine their nature.

b. Find the coordinates of any points of inflection.

c. Find the intervals on which the function is concave down.

7. The function is defined by $y = 2\cos x + \cos(2x)$.

a. Find the coordinates of all stationary points on the interval $[0, 2\pi]$.

b. Find the coordinates of any points of inflection.

c. Find the intervals on which the function is concave up.

8. The function $f(x)$ is defined by $f(x) = x\ln(\frac{1}{x^2})$.

a. Find all stationary points and determine their nature.

b. Find the coordinates of any points of inflection.

c. Find the intervals on which the function is concave up.

9. Diagram 5.5.1 shows the graph of $y = f'(x)$. On the diagram

a. mark the point corresponding to the local maximum of $f(x)$.

b. mark the point corresponding to the local minimum of $f(x)$.

c. mark the point corresponding to the point of inflection of $f(x)$.

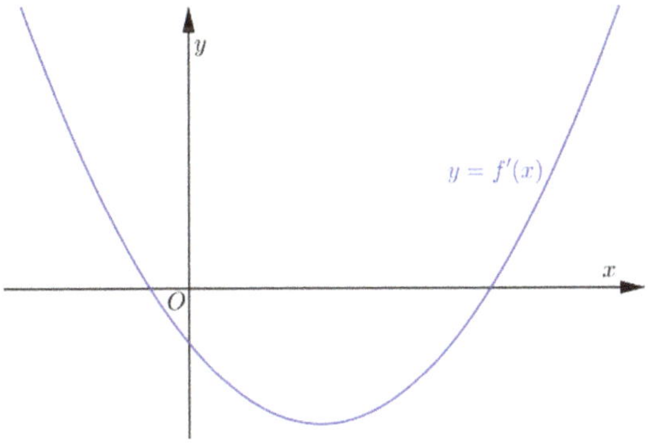

Diagram 5.5.1

10. Diagram 5.5.2 shows the graph of $y = f'(x)$. On the diagram

a. mark the point corresponding to the local maximum of $f(x)$.

b. mark the point corresponding to the local minimum of $f(x)$.

c. mark the point corresponding to the point of inflection of $f(x)$.

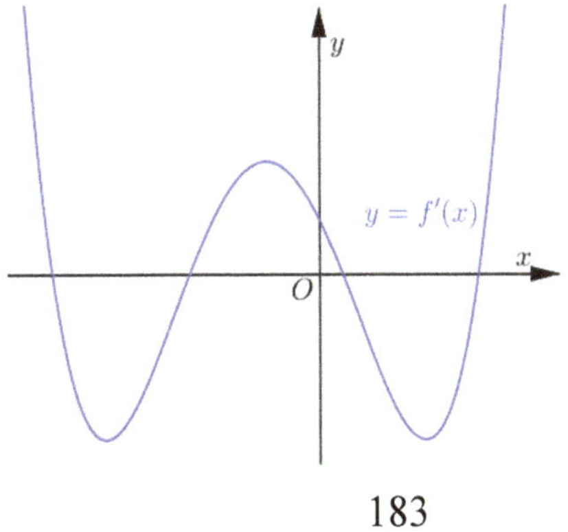

Diagram 5.5.2

11. Diagram 5.5.3 shows the graph of $y = f'(x)$. On the diagram

a. mark the point corresponding to the local maximum of $f(x)$.

b. mark the point corresponding to the local minimum of $f(x)$.

c. mark the point corresponding to the point of inflection of $f(x)$.

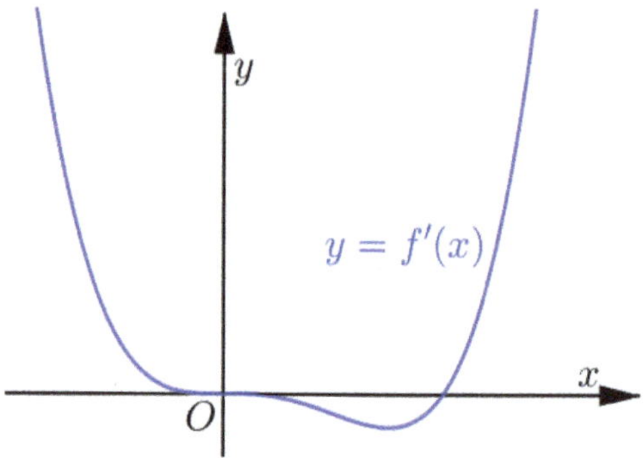

Diagram 5.5.3

5.6. Curve sketching

The previous sections discussed the relationship between a function's first and second derivatives and the behaviour of its graph. These discussions can now be applied to sketching the curves of various functions. We also introduce asymptotes, as they are essential for sketching certain types of graphs.

5.6.1. Asymptotes of a curve

For some curves, the domain and range are both finite, meaning the graph is contained within a limited region (for example, a circle and an ellipse). For other functions, the domain and range may be infinite, and the graphs extend indefinitely, as seen with hyperbolas and parabolas. Certain curves that extend to infinity approach a straight line indefinitely without ever reaching it. This straight line is called an **asymptote** of the curve.

Definition 5.4

If the distance between a point on a curve and a fixed straight line approaches zero as the point moves indefinitely along the curve, then the straight line is called an **asymptote** of the curve.

Given the equation of a curve $y = f(x)$, how do we determine whether it has an asymptote? Furthermore, if an asymptote exists, how can it be found? We shall consider these questions by examining three specific cases.

(1) Horizontal Asymptote

If the domain of the curve $y = f(x)$ is an infinite interval, and

$$\lim_{x \to -\infty} f(x) = b \qquad \text{or} \qquad \lim_{x \to +\infty} f(x) = b$$

then the line $y = b$ is a **horizontal asymptote** of the curve $y = f(x)$, as shown in Figures 5.6.1 and 5.6.2.

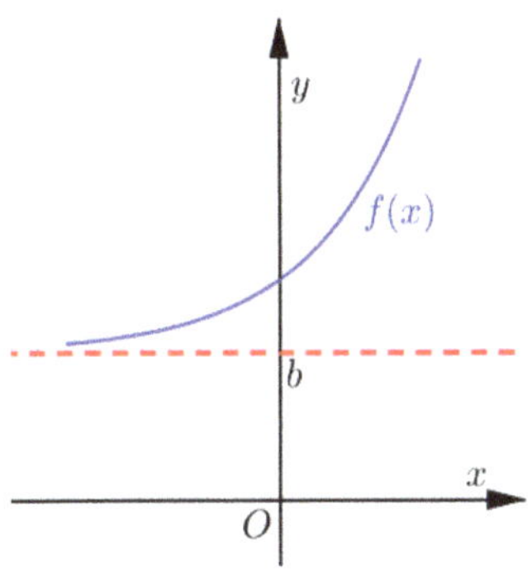

Figure 5.6.1

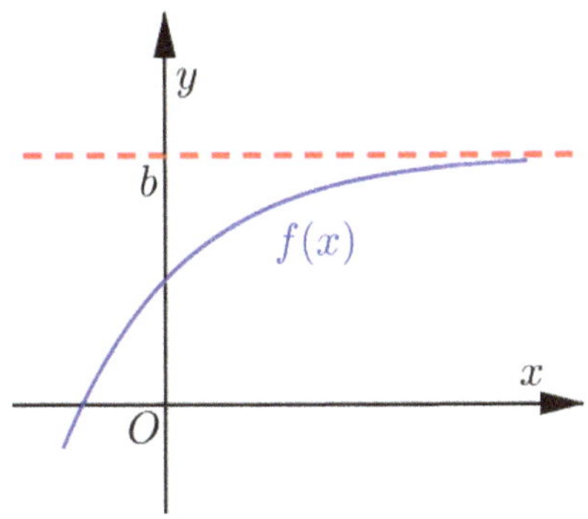

Figure 5.6.2

(2) Vertical Asymptote

If the curve $y = f(x)$ has a discontinuity at the point c, and

$$\lim_{x \to c^-} f(x) = \infty \qquad \text{(or } -\infty\text{)}$$

and

$$\lim_{x \to c^+} f(x) = -\infty \qquad \text{(or } \infty\text{)}$$

then, the line $x = c$ is a **vertical asymptote** of the curve $y = f(x)$, as shown in Figures 5.6.3 and 5.6.4.

Specifically, as $\frac{a}{0} = \pm\infty$, $a \neq 0$, it follows that when the denominator of a rational function $f(x)$ is zero, the value $x = c$, which is obtained from equating the denominator to zero, is a discontinuity. If substituting $x = c$ into the function results in a non-zero numerator over a zero denominator (i.e. $f(c) = \frac{a}{0} \neq \frac{0}{0}$), then the line $x = c$ is a vertical asymptote.

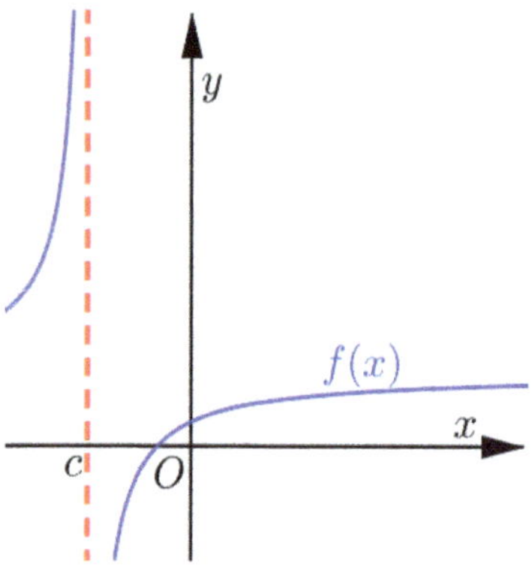

Figure 5.6.3

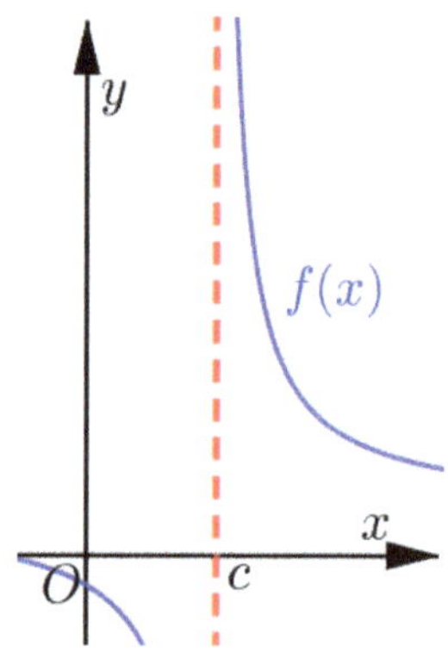

Figure 5.6.4

(3) Oblique Asymptote

If $\lim_{x \to \pm\infty} [f(x) - (ax + b)] = 0$, then the line $y = ax + b$ is an oblique asymptote of the curve $y = f(x)$, as shown in Figure 5.6.5.

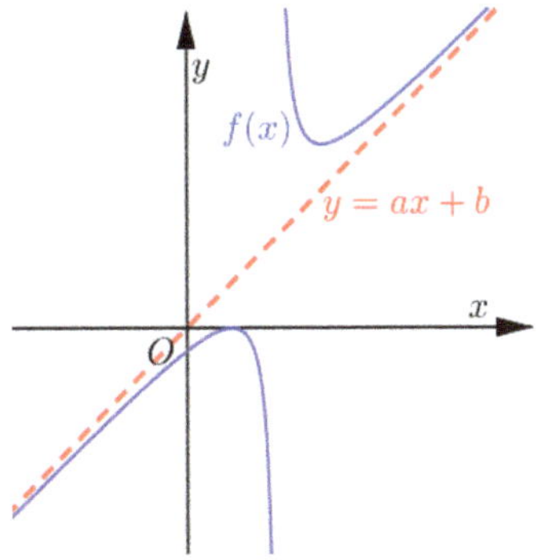

Figure 5.6.5

In the following, find the formula to calculate a and b.

From $\lim_{x\to\pm\infty}[f(x)-(ax+b)]=0$ we have

$$\lim_{x\to\pm\infty} x\left[\frac{f(x)}{x}-a-\frac{b}{x}\right]=0$$

Since x is an infinite quantity, we have

$$\lim_{x\to\pm\infty}\left(\frac{f(x)}{x}-a-\frac{b}{x}\right)=\lim_{x\to\pm\infty}\left(\frac{f(x)}{x}-a\right)=0$$

That is, $$a=\lim_{x\to\pm\infty}\frac{f(x)}{x}$$

Once the value of a has been found, substitute it into the original equation to determine b.

$$b=\lim_{x\to\pm\infty}[f(x)-ax]$$

Note: If a curve $y=f(x)$ has a horizontal asymptote $y=b$, it cannot have an oblique asymptote. Conversely, if a curve has an oblique asymptote, it cannot have a horizontal asymptote.

Proof. If $f(x)$ has a horizontal asymptote $y=b_1$, then

$$b_1=\lim_{x\to\pm\infty}f(x) \qquad b_1\in R$$

If we also examine the oblique asymptote, then we have

$$b_2 = \lim_{x\to\pm\infty}\left[f(x)-ax\right] = \lim_{x\to\pm\infty} f(x) - \lim_{x\to\pm\infty} ax = b_1 - a\cdot(\pm\infty) = \pm\infty$$

For the opposite case, if $f(x)$ has an oblique asymptote $y = ax + b_2$, then

$$b_2 = \lim_{x\to\pm\infty}\left[f(x)-ax\right] \qquad b_2 \in R$$

Hence,
$$b_1 = \lim_{x\to\pm\infty} f(x) = b_2 + \lim_{x\to\pm\infty} ax = b_2 + a\cdot(\pm\infty) = \pm\infty$$

Therefore, horizontal and oblique asymptotes cannot exist simultaneously. □

Example 1. Find the equations of all the asymptotes of the curve $f(x) = \dfrac{x-1}{x+3} + 1$.

Solution. (1) Horizontal asymptote

Since $\displaystyle\lim_{x\to\infty}\left(\frac{x-1}{x+3}+1\right) = 1+1 = 2, \qquad \lim_{x\to-\infty}\left(\frac{x-1}{x+3}+1\right) = 1+1 = 2$

the line $y = 2$ is a horizontal asymptote of the curve $f(x)$.

(2) Vertical asymptote

Setting the denominator of the curve $f(x) = \dfrac{x-1}{x+3} + 1$ equal to zero $(x + 3 = 0)$ gives a discontinuity at $x = -3$. Since $f(-3) \neq \dfrac{0}{0}$, the line $x = -3$ is a vertical asymptote of the curve $f(x)$.

(3) Oblique asymptote

Since $\displaystyle a = \lim_{x\to\pm\infty}\frac{f(x)}{x} = \lim_{x\to\pm\infty}\frac{\frac{x-1}{x+3}+1}{x} = \lim_{x\to\pm\infty}\left(\frac{x-1}{x^2+3x} + \frac{1}{x}\right) = \lim_{x\to\pm\infty}\frac{x-1}{x^2+3x} = 0$

$$b = \lim_{x\to\pm\infty}\left[f(x)-ax\right] = \lim_{x\to\pm\infty}\left(\frac{x-1}{x+3}+1\right) = 2$$

however, $y = 2$ is a horizontal asymptote, the curve $f(x)$ cannot have an oblique asymptote, as shown in Figure 5.6.6.

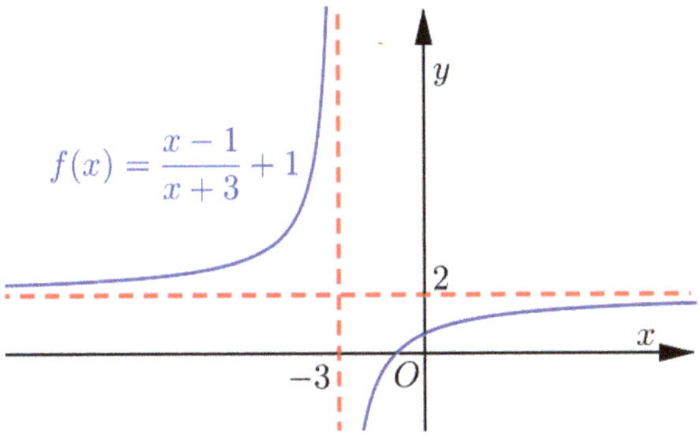

Figure 5.6.6

Example 2. Find the equations of all the asymptotes of the curve $f(x) = \dfrac{x^2 - 2}{x - 2}$.

Solution. (1) Horizontal asymptote

Since $\quad \lim\limits_{x\to\infty}(\dfrac{x^2-2}{x-2}) = \infty, \qquad \lim\limits_{x\to-\infty}(\dfrac{x^2-2}{x-2}) = -\infty$

the curve $f(x)$ does not have a horizontal asymptote.

(2) Vertical asymptote

Setting the denominator of the curve $f(x) = \dfrac{x^2-2}{x-2}$ to zero ($x - 2 = 0$) gives a discontinuity at $x = 2$. Since $f(2) \neq \dfrac{0}{0}$, the line $x = 2$ is a vertical asymptote of the curve $f(x)$.

(3) Oblique asymptote

Since $a = \lim\limits_{x\to\pm\infty} \dfrac{f(x)}{x} = \lim\limits_{x\to\pm\infty} \dfrac{\frac{x^2-2}{x-2}}{x} = \lim\limits_{x\to\pm\infty} \dfrac{x^2-2}{x^2-2x} = 1$

$$b = \lim_{x\to\pm\infty}[f(x)-ax] = \lim_{x\to\pm\infty}(\frac{x^2-2}{x-2}-x) = \lim_{x\to\pm\infty}\frac{x^2-2-x^2+2x}{x-2} = \lim_{x\to\pm\infty}\frac{2x-2}{x-2} = 2$$

thus, the line $y = x+2$ is the oblique asymptote of the curve $f(x)$, as shown in Figure 5.6.7.

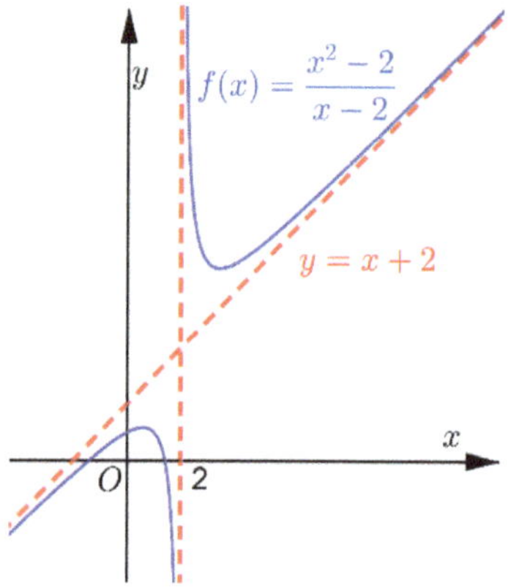

Figure 5.6.7

Example 3. Find the equations of all the asymptotes of the curve $f(x) = e^{\frac{1}{x}}$.

Solution. (1) Horizontal asymptote

Since $\lim_{x\to\infty} e^{\frac{1}{x}} = e^{\lim_{x\to\infty}\frac{1}{x}} = e^0 = 1$, $\lim_{x\to-\infty} e^{\frac{1}{x}} = e^{\lim_{x\to-\infty}\frac{1}{x}} = e^0 = 1$

the line $y = 1$ is a horizontal asymptote of the curve $f(x)$.

(2) Vertical asymptote

Since when $x = 0$, $e^{\frac{1}{0}} = \infty$, thus $x = 0$ is the discontinuity point of the curve $f(x)$.

But as $x \to 0^-$, that is, $x = 0 - a$, $e^{\frac{1}{x}} = e^{\frac{1}{-a}} = \frac{1}{e^{\frac{1}{a}}} \to 0$

as $x \to 0^+$, that is, $x = 0 + a$, $e^{\frac{1}{x}} = e^{\frac{1}{a}} \to \infty$

where $a \to 0$, $\frac{1}{a} \to \infty$.

Therefore, $\lim_{x\to 0^-} e^{\frac{1}{x}} = 0$, $\lim_{x\to 0^+} e^{\frac{1}{x}} = \infty$

the curve $f(x)$ does not have a vertical asymptote.

(3) Oblique asymptote

Since $a = \lim_{x\to\pm\infty} \frac{f(x)}{x} = \lim_{x\to\pm\infty} \frac{e^{\frac{1}{x}}}{x} = 0$

the curve $f(x)$ does not have an oblique asymptote, as shown in Figure 5.6.8.

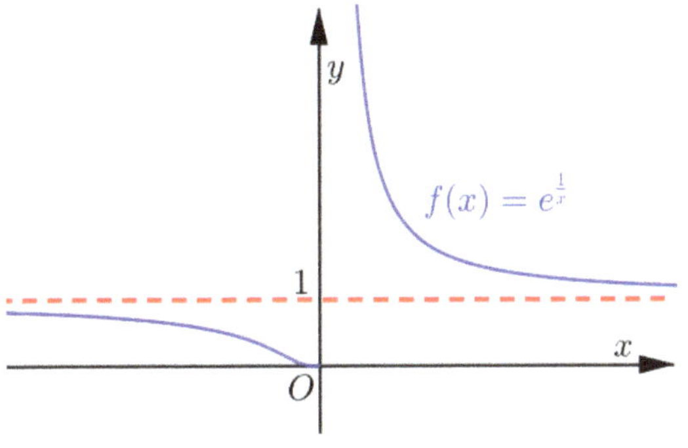

Figure 5.6.8

5.6.2. Sketching graphs of functions

In the previous few sections, we have discussed various properties of functions, which can be applied to sketching graphs. Generally, the following steps should be considered when sketching the curve of a function:

(1) Determine the domain and range of the function.

(2) Check for symmetry and periodicity.

(3) Find the intervals of monotonicity and turning points of the function (local maxima and minima).

(4) Determine the concavity and find any points of inflection.

(5) Identify any asymptotes of the function.

(6) Calculate the coordinates of specific points, especially the x- and y-intercepts.

Example 1. $f(x)=\dfrac{4(x+1)}{x^2}$.

(a) Find

(i) the coordinates of the x-intercepts of $f(x)$.

(ii) the equations of all the asymptotes.

(iii) the coordinates of the local minimum and justify that it is a local minimum.

(iv) the coordinates of the point of inflection and justify that it is a point of inflection.

(v) the intervals where $f(x)$ is concave up.

(b) Use your results from part (a) to sketch the graph of the curve $f(x)$ without the use of a calculator.

Solution. (a) (i) To find the x-intercepts of $f(x)$, set $f(x)=0$.

$$\frac{4(x+1)}{x^2}=0$$

$$4(x+1)=0$$

$$x=-1$$

The x-intercept of $f(x)$ is (-1, 0).

(ii) To find the vertical asymptote, set the denominator of the fraction equal to zero; that is, $x^2=0$. This gives a point of discontinuity at $x=0$. Since $f(0)\neq\frac{0}{0}$, it follows that the line $x=0$ is a vertical asymptote.

To find the horizontal asymptote, calculate the limit of $f(x)$ as $x\to\pm\infty$.

Since $\lim\limits_{x\to\infty}\dfrac{4(x+1)}{x^2}=0$, $\lim\limits_{x\to-\infty}\dfrac{4(x+1)}{x^2}=0$

thus, the line $y=0$ is a horizontal asymptote of the curve.

(iii) A local minimum exists where $f'(x) = 0$ and $f''(x) > 0$.

$$f'(x) = \frac{(4x+4)'x^2 - (4x+4)(x^2)'}{x^4} = \frac{4x^2 - 2x(4x+4)}{x^4} = \frac{-4x^2 - 8x}{x^4} = \frac{-4x-8}{x^3}$$

$$f''(x) = \frac{(-4x-8)'x^3 - (-4x-8)(x^3)'}{x^6} = \frac{-4x^3 + (4x+8)3x^2}{x^6}$$

$$= \frac{-4x + 3(4x+8)}{x^4} = \frac{-4x+12x+24}{x^4} = \frac{8x+24}{x^4}$$

Set $$f'(x) = \frac{-4x-8}{x^3} = 0$$

$$-4x - 8 = 0$$

which gives a stationary point at

$$x = -2$$

Since $$f''(-2) = \frac{8(-2)+24}{(-2)^4} = \frac{8}{16} = \frac{1}{2} > 0$$

and $$f(-2) = \frac{4(-2+1)}{(-2)^2} = -1$$

thus, the point (-2, -1) is a local minimum of the function $f(x)$.

(iv) A point of inflection occurs where $f''(x) = 0$ or where $f''(x)$ does not exist ($f''(x) = \infty$).

Set $$f''(x) = \frac{8x+24}{x^4} = 0, \qquad f''(x) = \frac{8x+24}{x^4} = \infty$$

Solving $8x + 24 = 0$ gives a point of inflection at $x = -3$. Note that $f''(x)$ is undefined at $x = 0$, which is a point of discontinuity of $f''(x)$.

$$f(-3) = \frac{4(x+1)}{x^2} = \frac{4(-3+1)}{(-3)^2} = -\frac{8}{9}$$

$$f(0) = \frac{4(x+1)}{x^2} = \frac{4(0+1)}{0^2} = \infty, \ f(0) \text{ does not exist.}$$

Since $f''(x)$ exists on either side of $x = -3$ and has opposite signs, the point $(-3, -\frac{8}{9})$ is a point of inflection.

(v) When $x < -3$, multiplying both sides of the inequality $x < -3$ by 8 and adding 24 gives

$$8x + 24 < 8 \cdot (-3) + 24 = 0, \qquad x^4 > 0$$

Hence, $$f''(x) = \frac{8x+24}{x^4} < 0$$

When $x > -3$, $8x + 24 > 0$, $x^4 > 0$

then $$f''(x) = \frac{8x+24}{x^4} > 0$$

Therefore, when $x < -3$, $f''(x) < 0$ and $f(x)$ is concave down. The interval where $f(x)$ is concave down is $(-\infty, -3)$. When $x > -3$, $f''(x) > 0$ and $f(x)$ is concave up. The interval where $f(x)$ is concave up is $(-3, \infty)$.

(b)

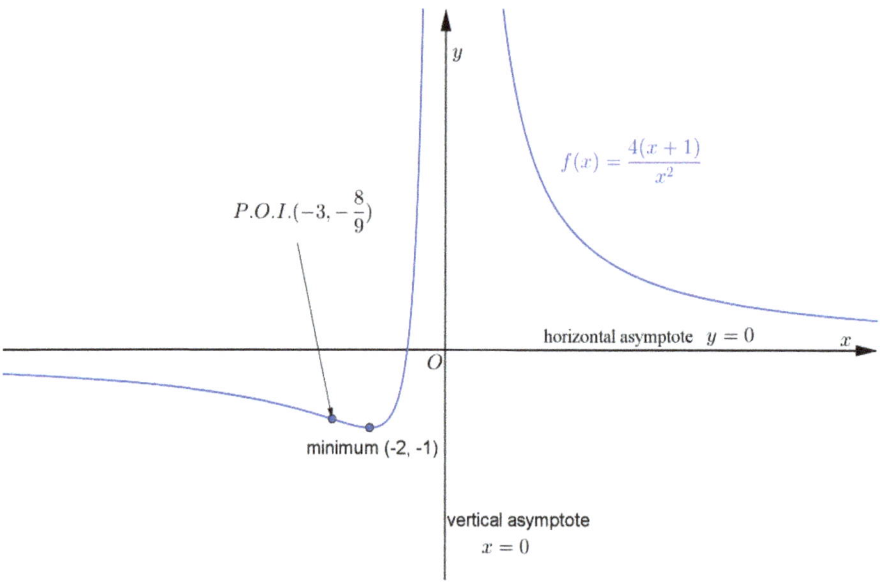

Figure 5.6.9

Example 2. $f(x) = \dfrac{x-1}{x+2}$.

(a) Find

(i) the coordinates of the intercepts of the curve $f(x)$ with the axes.

(ii) the stationary points and the point of inflection.

(iii) the equation of the vertical asymptote.

(iv) the equation of the horizontal asymptote.

(b) Sketch the graph of the curve $f(x)$.

Solution. (a) (i) To find the x-intercepts of the curve $f(x)$, set $f(x) = 0$.

$$f(x) = \frac{x-1}{x+2} = 0$$

$$x - 1 = 0$$

$$x = 1$$

Therefore, the x-intercept of the curve $f(x)$ is (1, 0).

To find the y-intercept of the curve $f(x)$, set $x = 0$.

$$f(0) = \frac{0-1}{0+2} = -\frac{1}{2}$$

Therefore, the y-intercept of the curve $f(x)$ is $(0,\ -\frac{1}{2})$.

(ii) At a stationary point, $f'(x) = 0$.

$$f'(x) = (\frac{x-1}{x+2})' = \frac{(x-1)'(x+2)-(x-1)(x+2)'}{(x+2)^2} = \frac{x+2-x+1}{(x+2)^2} = \frac{3}{(x+2)^2} \neq 0$$

In this case, there are no stationary points.

If a point of inflection exists, then either $f''(x) = 0$ or $f''(x) = \infty$.

Since $$f''(x) = [\frac{3}{(x+2)^2}]' = \frac{3d(x+2)^{-2}}{d(x+2)} \cdot \frac{d(x+2)}{dx} = \frac{-6}{(x+2)^3} \neq 0$$

$$f''(x) = -\frac{6}{(x+2)^3} = \infty$$

which gives the discontinuity point of $f''(x)$ is at $x = -2$.

However, $f(-2) = \dfrac{x-1}{x+2} = \dfrac{-2-1}{-2+2} = -\infty$ $f(-2)$ does not exist.

Therefore, the function $f(x)$ has no point of inflection.

(iii) To find the vertical asymptote, set the denominator of the fraction to zero ($x + 2 = 0$); this gives a discontinuity at $x = -2$. Since $f(-2) \neq \dfrac{0}{0}$, the line $x = -2$ is a vertical asymptote of the curve.

(iv) To find the horizontal asymptote, calculate the limit of $f(x)$ as $x \to \pm\infty$.

Since $\lim\limits_{x\to\infty} \dfrac{x-1}{x+2} = 1$, $\lim\limits_{x\to-\infty} \dfrac{x-1}{x+2} = 1$

the line $y = 1$ is a horizontal asymptote of the curve.

(b)

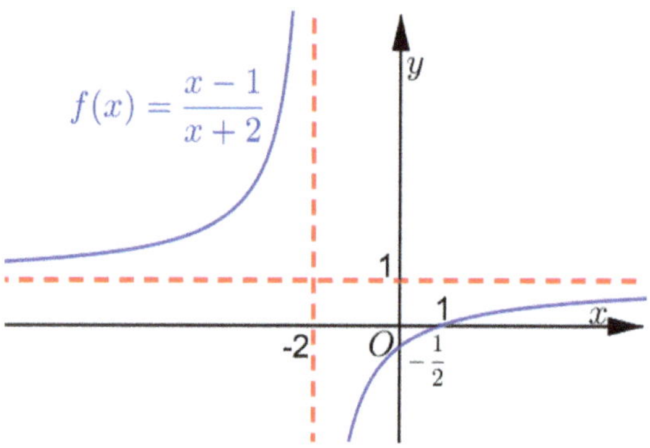

Figure 5.6.10

Example 3. Sketch the graph of function $\varphi(x) = \dfrac{1}{\sqrt{2\pi}} e^{-\frac{x^2}{2}}$.

Solution. The domain of the function is $(-\infty, \infty)$.

Since $\varphi(-x)=\varphi(x)$, $\varphi(x)$ is an even function; its graph is therefore symmetrical about the y-axis.

$$\varphi'(x) = -\frac{x}{\sqrt{2\pi}} e^{-\frac{x^2}{2}}, \qquad \varphi''(x) = \frac{(x+1)(x-1)}{\sqrt{2\pi}} e^{-\frac{x^2}{2}}$$

Setting $\varphi'(x)=0$ gives $x=0$.

Setting $\varphi''(x)=0$ gives $x=1$ and $x=-1$.

List in a table.

x	$(-\infty,-1)$	-1	$(-1,0)$	0	$(0,1)$	1	$(1,\infty)$
y'	+		+	0	−		−
y''	+	0	−		−	0	+
y	↗ concave up	$\frac{1}{\sqrt{2\pi e}}$ point of inflection	↗ concave down	$\frac{1}{\sqrt{2\pi}}$ local maximum	↘ concave down	$\frac{1}{\sqrt{2\pi e}}$ point of inflection	↘ concave up

where $\frac{1}{\sqrt{2\pi e}} \approx 0.24$ $\qquad \frac{1}{\sqrt{2\pi}} \approx 0.4$

To find the horizontal asymptote, calculate the limit of $f(x)$ as $x \to \pm\infty$.

$$\lim_{x\to\pm\infty} \frac{1}{\sqrt{2\pi}} e^{-\frac{x^2}{2}} = 0$$

Therefore, the line $y=0$ is a horizontal asymptote of the curve.

First sketch the graph for the interval $[0,\infty[$, then use symmetry about the y-axis to sketch the graph for the interval $]-\infty,0]$, as shown in Figure 5.6.11.

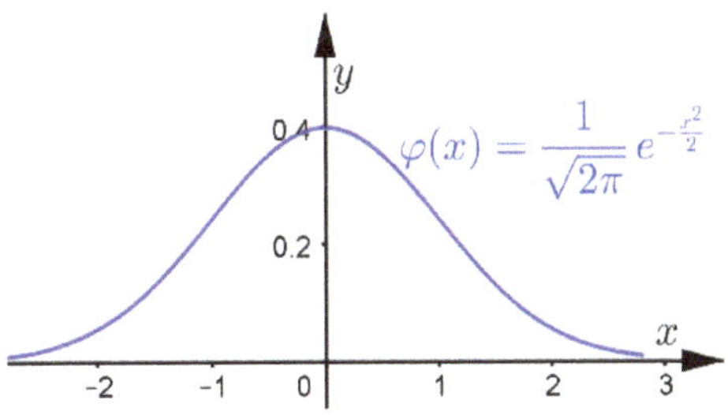

Figure 5.6.11

The curve of this function is a very important curve in probability theory and mathematical statistics.

Example 4. $f(x) = 1 - \dfrac{x^2}{x-2}$.

(a) Find

(i) the coordinates of the intercepts of the curve $f(x)$ with the axes.

(ii) the equations of all the asymptotes.

(iii) the coordinates of the stationary points and determine their nature.

(iv) the coordinates of all points of inflection.

(b) Sketch the graph of the curve $f(x)$.

Solution. (a) (i) To find the x-intercepts of the curve $f(x)$, set $f(x) = 0$.

$$f(x) = 1 - \frac{x^2}{x-2} = 0$$

$$1 = \frac{x^2}{x-2}$$

$$x^2 - x + 2 = 0$$

The equation has no real roots; therefore, the curve $f(x)$ has no x-intercept.

To find the y-intercept of the curve $f(x)$, set $x = 0$.

$$f(0) = 1 - \frac{0^2}{0-2} = 1$$

The y-intercept is (0, 1).

(ii)

- To find the vertical asymptote, set the denominator of the fraction to zero

($x-2=0$); this gives a discontinuity point at $x=2$. Since $f(2)\neq\frac{0}{0}$, the line $x=2$ is a vertical asymptote of the curve.

- To find the horizontal asymptote, calculate the limit of $f(x)$ as $x\to\pm\infty$.

  Since $\lim\limits_{x\to\infty}(1-\frac{x^2}{x-2})=-\infty$, $\lim\limits_{x\to-\infty}(1-\frac{x^2}{x-2})=\infty$

  it follows that the curve $f(x)$ has no horizontal asymptote.

- Now consider the oblique asymptote.

Since $$a=\lim_{x\to\pm\infty}\frac{f(x)}{x}=\lim_{x\to\pm\infty}\frac{1-\frac{x^2}{x-2}}{x}=\lim_{x\to\pm\infty}(\frac{1}{x}-\frac{x}{x-2})=-1$$

$$b=\lim_{x\to\pm\infty}\left[f(x)-ax\right]=\lim_{x\to\pm\infty}(1-\frac{x^2}{x-2}+x)=\lim_{x\to\pm\infty}\frac{x-2-x^2+x^2-2x}{x-2}=\lim_{x\to\pm\infty}\frac{-x-2}{x-2}=-1$$

thus, the line $y=-x-1$ is an oblique asymptote of the curve $f(x)$.

(iii) $$f(x)=1-\frac{x^2}{x-2}$$

$$f'(x)=(1-\frac{x^2}{x-2})'=-\frac{(x^2)'(x-2)-x^2(x-2)'}{(x-2)^2}=-\frac{2x(x-2)-x^2}{(x-2)^2}=-\frac{x^2-4x}{(x-2)^2}$$

Set $$f'(x)=0$$

That is, $$-\frac{x^2-4x}{(x-2)^2}=0$$

$$x^2-4x=0$$

Solving gives the stationary points are at

$$x=0 \quad \text{or} \quad x=4$$

Since $$f'(x)=-\frac{x^2-4x}{(x-2)^2}$$

$$f''(x)=-[\frac{x^2-4x}{(x-2)^2}]'=-\frac{(x^2-4x)'(x-2)^2-(x^2-4x)[(x-2)^2]'}{(x-2)^4}$$

$$= -\frac{(2x-4)(x-2)^2-(x^2-4x)\cdot 2(x-2)}{(x-2)^4}$$

$$= -\frac{2(x-2)(x^2-4x+4-x^2+4x)}{(x-2)^4} = -\frac{8}{(x-2)^3}$$

$$f''(0) = -\frac{8}{(0-2)^3} = 1 > 0$$

$$f''(4) = -\frac{8}{(4-2)^3} = -1 < 0$$

Thus $f(0) = 1 - \frac{0^2}{0-2} = 1$ is a local minimum.

$f(4) = 1 - \frac{4^2}{4-2} = -7$ is a local maximum.

Hence, the local minimum point is at (0, 1) and the local maximum is at (4, -7).

(iv) A point of inflection occurs where $f''(x) = 0$ or $f''(x) = \infty$.

Since $f''(x) = -\frac{8}{(x-2)^3} \neq 0$

from $f''(x) = -\frac{8}{(x-2)^3} = \infty$

we find that $x = 2$ is a discontinuity point of $f''(x)$.

However, $f(2) = 1 - \frac{x^2}{x-2} = 1 - \frac{2^2}{2-2} = -\infty$ $f(2)$ does not exist.

Therefore, the curve $f(x)$ has no point of inflection.

(b)

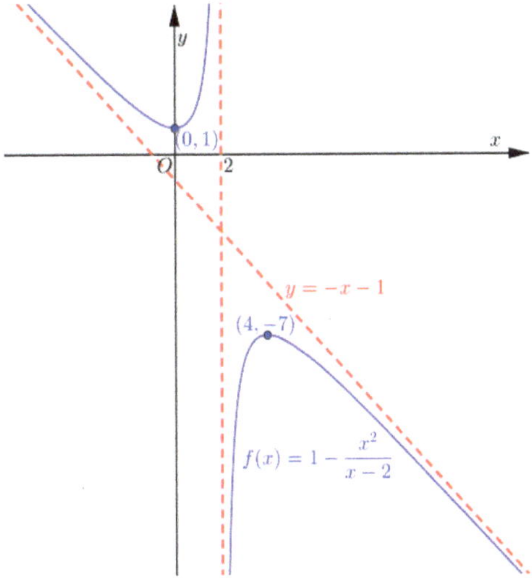

Figure 5.6.12

5.6.3. Sketching the gradient function

If the graph of a function is given but the algebraic expression is not, how can we sketch the graph of its derivative?

General steps to sketch the gradient function

1) On the original curve, first identify the stationary points (where the gradient of the tangent is equal to zero).
2) Use these stationary points to divide the x-axis into several intervals.
3) Analyse the behaviour of the gradient on each interval; for example, determine where the gradient is positive or negative, and where it is increasing or decreasing.
4) Using this analysis, sketch the gradient function, $y = f'(x)$.

Example. Figure 5.6.13 shows the graph of the function $f(x)$. Sketch the graph of the gradient function, $y = f'(x)$.

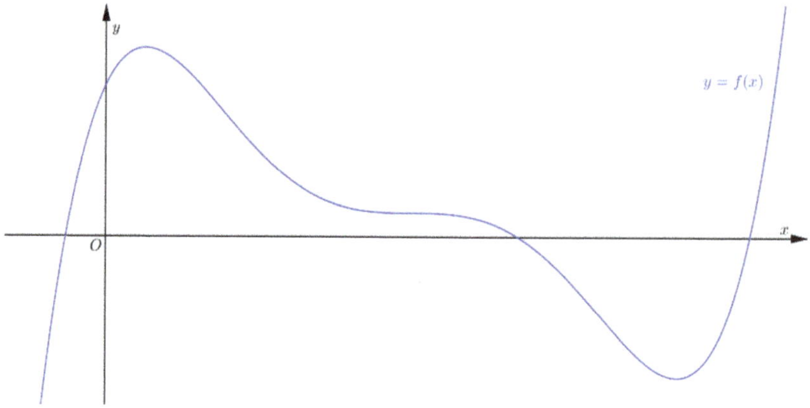

Figure 5.6.13

Solution.

Step 1. First, identify the stationary points on the curve where the gradient is zero. As shown in Figure 5.6.14, at the points A, B, and C, the gradient of the tangent is zero.

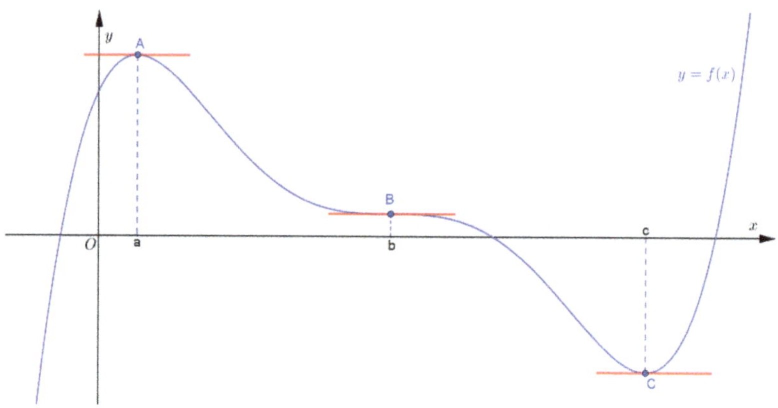

Figure 5.6.14

Step 2. Use these stationary points to divide the x-axis into several intervals. As shown in Figure 5.6.14, we can consider four intervals: $x \leq$ a, a $\leq x \leq$ b, b $\leq x \leq$ c and $x \geq$ c.

Step 3. Analyse the behaviour of the gradient on each interval.

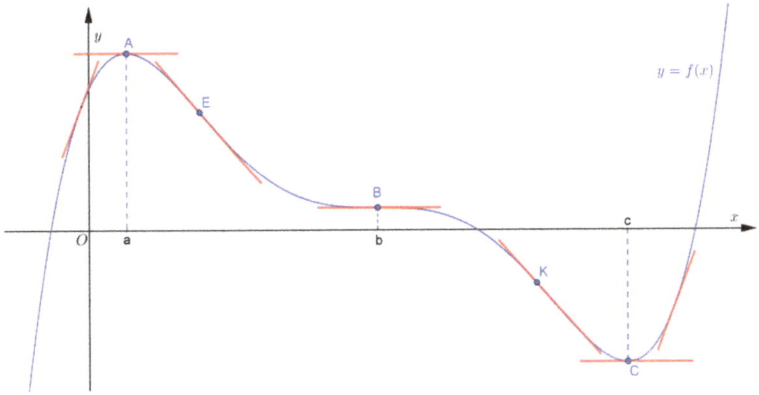

Figure 5.6.15

As shown in Figure 5.6.15

- on the interval $x < \text{a}$, the gradient is positive ($f'(x) > 0$). As $x \to -\infty$, the curve is very steep, and the gradient is a very large positive number. As $x \to \text{a}$, the gradient approaches zero ($f'(x) \to 0$). That is, on this interval, as we move along the curve from the left towards the point A, the gradient decreases gradually from a very large positive number to zero.
- on the interval $\text{a} < x < \text{b}$, the gradient is negative ($f'(x) < 0$). As we move along the curve from the point A towards the point B, the gradient first decreases from zero gradually to become increasingly negative until it reaches a local minimum at the point E; it then increases gradually from this minimum to zero at the point B.
- on the interval $\text{b} < x < \text{c}$, the gradient is negative ($f'(x) < 0$). As we move along the curve from the point B towards the point C, the gradient first decreases from zero to become increasingly negative until it reaches a local minimum at the point K. It then increases gradually from this minimum to zero at the point C.
- on the interval $x > \text{c}$, the gradient is positive ($f'(x) > 0$). As we move along the curve to the right of the point C, the gradient increase from zero and tends to infinity.

Step 4. Based on this analysis, we can sketch the gradient function, $y = f'(x)$, as shown in Figure 5.6.16. Note that since the algebraic expression for the function is not given, we cannot determine the exact value of the gradient at each point. Instead, we can sketch the graph by referring to the behaviour of the graph of a tangent function.

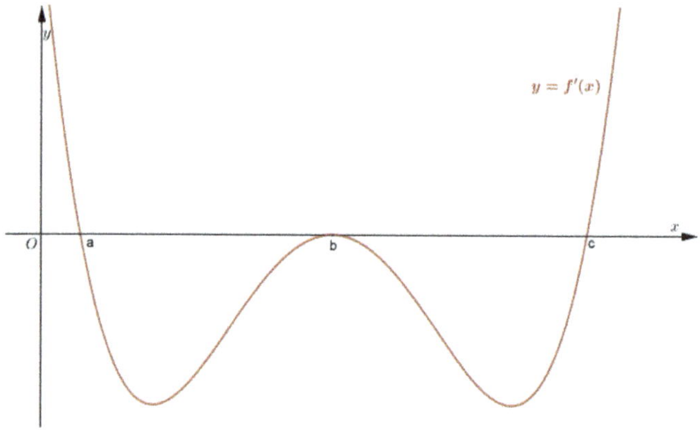

Figure 5.6.16

Similarly, we can sketch the graph of $f''(x)$.

Exercise 5.6

1. Determine the equations of the asymptotes of each of the following.

a. $f(x) = e^x$ **b.** $f(x) = -x^2$

c. $f(x) = \ln(x+1)$ **d.** $f(x) = e^{-\frac{1}{x}}$

e. $f(x) = \dfrac{x+3}{x-4}$ **f.** $f(x) = \dfrac{3x+2}{5x+1}$

g. $f(x) = 2x - \dfrac{1}{2x}$ **h.** $f(x) = 6 + \dfrac{1}{x+3}$

i. $f(x) = \dfrac{5 - x^3}{x^2}$

j. $f(x) = \dfrac{x^2}{x^2 - 3x + 2}$

2. Sketch the curves of the functions in question 1.

3. Sketch the following curves, clearly labelling all intercepts, local maxima, local minima, points of inflection and the equations of any asymptotes.

a. $f(x) = \dfrac{4}{x+3}$

b. $f(x) = \dfrac{1}{(x+1)(x-3)}$

c. $f(x) = \dfrac{x}{x+2}$

d. $f(x) = \dfrac{2x-5}{x+4}$

e. $f(x) = \dfrac{x}{x^2-4}$

f. $f(x) = \dfrac{x^2}{x^2-4}$

g. $f(x) = \dfrac{x^2+2x+3}{x^2-2x-3}$

h. $f(x) = \dfrac{(x-3)^2}{4(x-1)}$

i. $f(x) = \dfrac{2x^3+3x}{x^2}$

j. $f(x) = \dfrac{3x^2-2x-3}{x^2}$

4. Diagram 5.6.1 shows part of the graph of

$$f(x) = \frac{ax-3}{x+k}$$

Find the values of a and k.

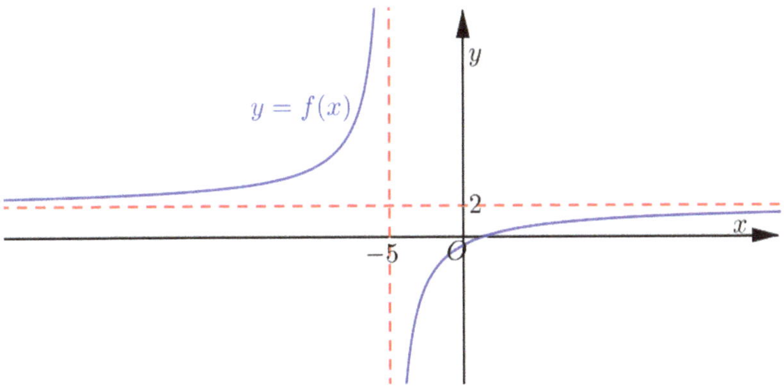

Diagram 5.6.1

5. $f(x)=\dfrac{x}{(1+x)(1-x)^2}$

a. Find

(i) the coordinates of the intercepts of $f(x)$ with the axes.

(ii) the equations of the asymptotes of the curve.

(iii) the coordinates of the point of inflection and justify that it is a point of inflection.

(iv) the intervals on which $f(x)$ is concave up.

b. Use the results from part (a) to sketch the graph of the curve $f(x)$.

6. $f(x)=\dfrac{x^3}{(x-1)^2}$

a. Find

(i) the coordinates of the intercepts of $f(x)$ with the axes.

(ii) the equations of the asymptotes of the curve.

(iii) the coordinates of the stationary points and determine their nature.

(iv) the coordinates of any points of inflection.

b. Sketch the graph of the curve $f(x)$.

7. $f(x)=(x+1)(x-2)^2$

a. Find

(i) the coordinates of the intercepts of the curve $f(x)$ with the axes.

(ii) the coordinates of the stationary points and determine their nature.

(iii) the coordinates of any points of inflection.

b. Sketch the graph of the curve $f(x)$.

8. $f(x)=\dfrac{x^4}{(1+x)^3}$

a. Find

(i) the coordinates of the intercepts of the curve $f(x)$ with the axes.

(ii) the equations of the asymptotes of the curve.

(iii) the coordinates of the stationary points and determine their nature.

(iv) the coordinates of any points of inflection.

b. Sketch the graph of the curve $f(x)$.

9. $f(x)=\dfrac{x^2-9}{x^2-16}$

a. Find

(i) the coordinates of the intercepts of the curve $f(x)$ with the axes.

(ii) the equations of the asymptotes of the curve.

(iii) the coordinates of the stationary points and determine their nature.

(iv) the coordinates of any points of inflection.

b. Sketch the graph of the curve $f(x)$.

10. $f(x)=\dfrac{x^2-5x+4}{x-2}$

a. Find

(i) the coordinates of the intercepts of the curve with the axes.

(ii) the equations of the asymptotes of the curve.

(iii) the coordinates of the stationary points and determine their nature.

(iv) the coordinates of any points of inflection.

b. Sketch the graph of the curve $f(x)$.

11. $f(x)=2+\dfrac{3}{x}-\dfrac{1}{x^3}$

a. Find

(i) the coordinates of the intercepts of the curve with the axes.

(ii) the equations of the asymptotes of the curve.

(iii) the coordinates of the stationary points and determine their nature.

(iv) the coordinates of any points of inflection.

b. Sketch the graph of the curve $f(x)$.

12. The following diagrams show the graphs of functions $y = f(x)$. Sketch the gradient functions $y = f'(x)$ and $y = f''(x)$ for each of the given functions.

a.

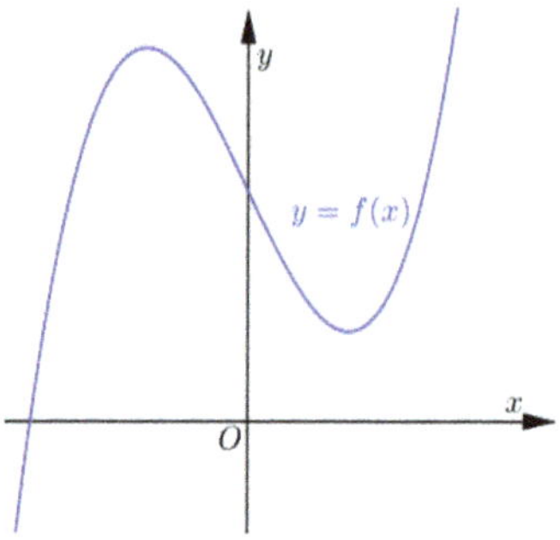

b.

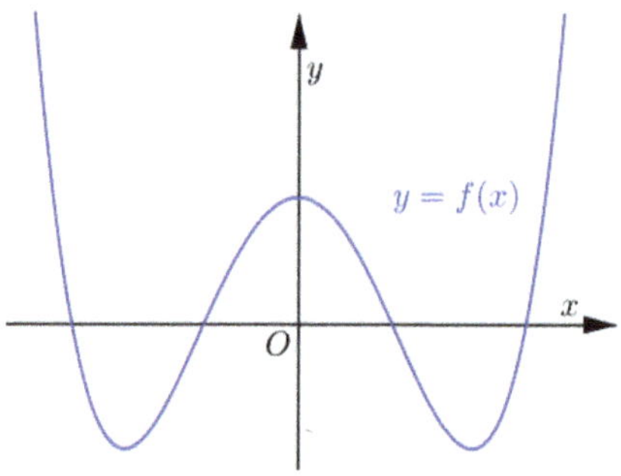

c.

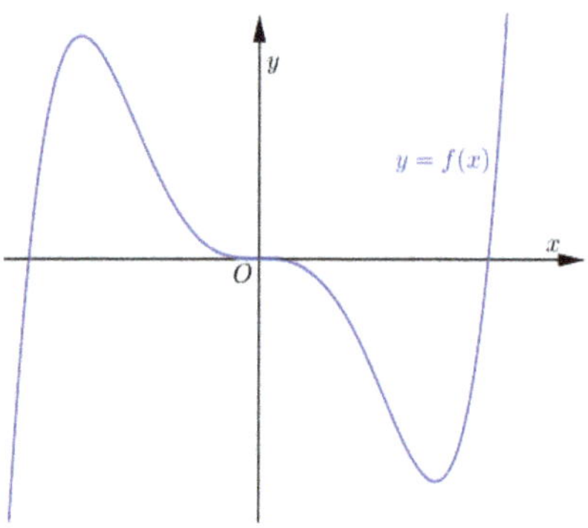

d.

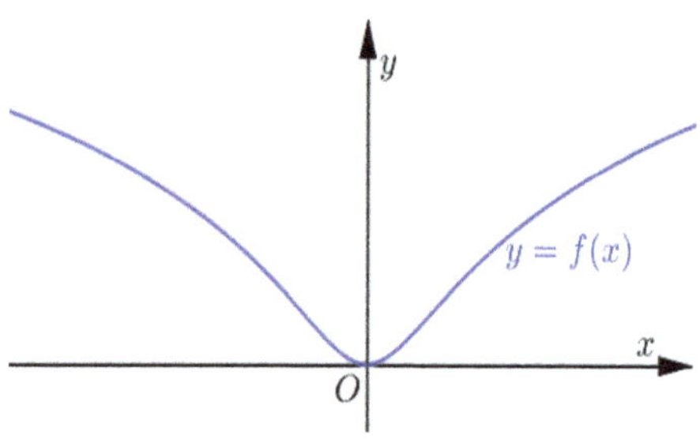

e.

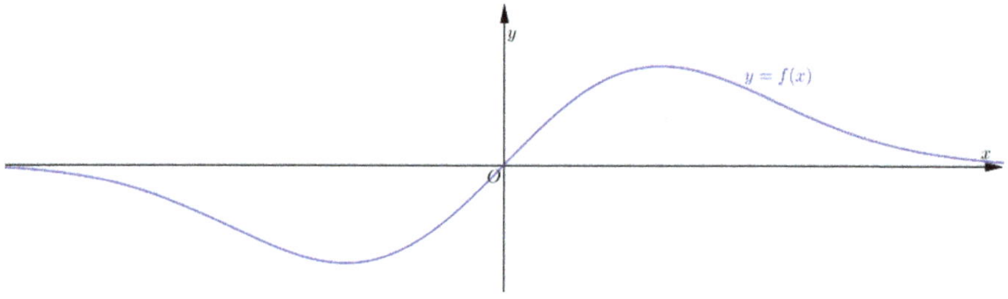

f.

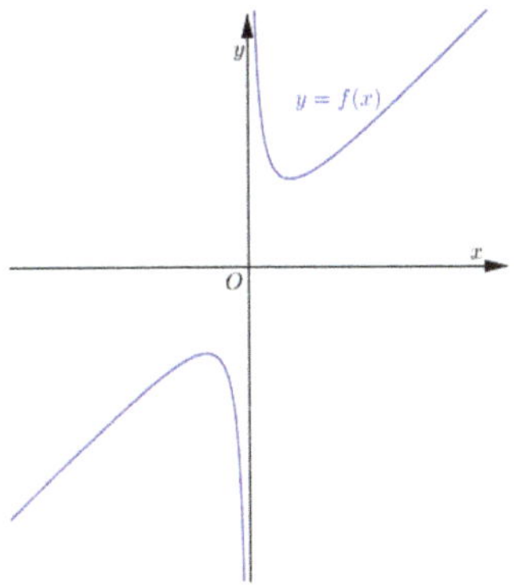

g.

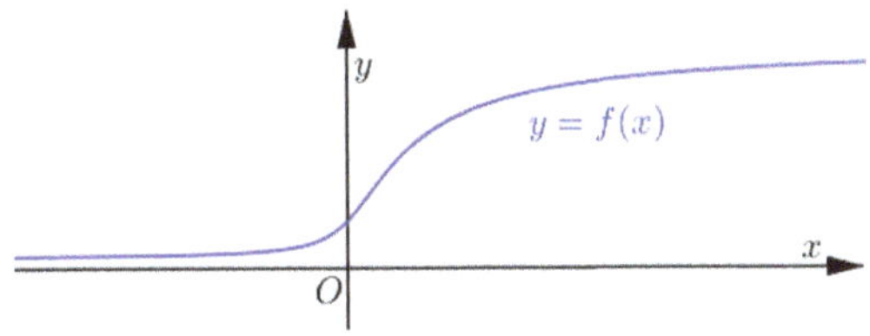

5.7. Related rates

In the previous study, we know that if a relationship exists between y and x, then $\frac{dy}{dx}$ tells us how y changes with respect to x. If x also changes with a third variable – such as time (t) – and $\frac{dx}{dt}$ exists, then y will change alongside this third variable as well. This rate of change of y with respect to this third variable is linked to $\frac{dx}{dt}$.

Finding the change rate of y with respect to a third variable is known as finding **related rates**.

General steps for solving related rates problems:

(1) **Establish the relationship** between the variables, such as y and x.

(2) **Differentiate both sides** of the relationship with respect to the third variable (frequently time, t), applying the chain rule where necessary.

(3) **Substitute the known values** into the resulting derivative equation to calculate the required **related rates**.

Example 1. The side of a square is x and is increasing at a rate of 2 cm/min. When the side length is 6 cm, what is the rate of change of the area of the square?

Solution. Given that $\frac{dx}{dt}=2$, find $\left.\frac{dA}{dt}\right|_{x=6}=?$

The relationship between the area A of the square and its side x is

$$A=x^2$$

Differentiating both sides with respect to time gives

$$\frac{dA}{dt}=\frac{dx^2}{dt}$$

$$=\frac{dx^2}{dx}\cdot\frac{dx}{dt}=2x\cdot\frac{dx}{dt}$$

Substituting $x=6$ and $\frac{dx}{dt}=2$ gives

$$\frac{dA}{dt}=2\cdot6\cdot2=24\,\text{cm}^2/\text{min}$$

When the side is equal to 6 cm, the rate of change of the area of the square is 24 cm^2/min.

Example 2. Air is pumped into a spherical balloon at a rate of 20 cm^3/min. When the volume of the balloon is 10 cm^3, what is the rate of change of its radius?

Solution. Given that $\frac{dV}{dt}=20$, find $\left.\frac{dr}{dt}\right|_{V=10}=?$

The relationship between the volume of a sphere and its radius is

$$V=\frac{4}{3}\pi r^3$$

Differentiating both sides with respect to t gives

$$\frac{dV}{dt} = \frac{4}{3}\pi\frac{dr^3}{dt}$$

$$= \frac{4}{3}\pi\frac{dr^3}{dr}\cdot\frac{dr}{dt} = 4\pi r^2\frac{dr}{dt}$$

When $V = \frac{4}{3}\pi r^3 = 10 \quad \Rightarrow \quad r = \sqrt[3]{\frac{15}{2\pi}}$

Substituting $\frac{dV}{dt} = 20$ and $r = \sqrt[3]{\frac{15}{2\pi}}$ gives

$$20 = 4\pi\cdot\left(\sqrt[3]{\frac{15}{2\pi}}\right)^2\frac{dr}{dt}$$

$$\frac{dr}{dt} = \frac{5}{\pi}\left(\frac{15}{2\pi}\right)^{-\frac{2}{3}} \approx 0.891 \text{ cm/min}$$

When the volume of the spherical balloon is 10 cm^3, the rate of change of its radius is approximately 0.891 cm/min.

Example 3. A 15-metre ladder is leaning against a vertical wall. Its lower end on the ground level slides away from the wall at a rate of 2 m/s. When the lower end of the ladder is 10 metres from the wall, at what rate is the top of the ladder sliding down?

Solution. As shown in Figure 5.7.1, AC represents the ladder. Let AC = 15, AB = y, and BC = x.

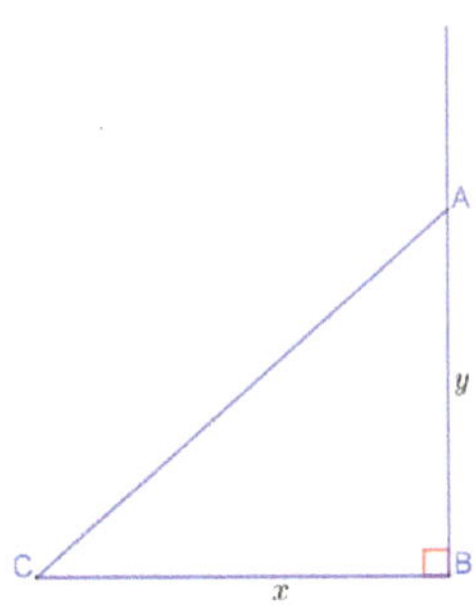

Figure 5.7.1

Given that $\frac{dx}{dt}=2$, find $\left.\frac{dy}{dt}\right|_{x=10}=?$

The relationship between x and y is

$$x^2+y^2=15^2$$

Differentiate both sides of the equation with respect to t.

$$\frac{dx^2}{dt}+\frac{dy^2}{dt}=\frac{d15^2}{dt}$$

$$\frac{dx^2}{dx}\cdot\frac{dx}{dt}+\frac{dy^2}{dy}\cdot\frac{dy}{dt}=0$$

$$2x\frac{dx}{dt}+2y\frac{dy}{dt}=0$$

$$x\frac{dx}{dt}+y\frac{dy}{dt}=0$$

When $x=10$, $y=\sqrt{15^2-10^2}=5\sqrt{5}$.

Substituting $\frac{dx}{dt}=2$ and $y=5\sqrt{5}$ gives

$$10\cdot 2+5\sqrt{5}\frac{dy}{dt}=0$$

$$\frac{dy}{dt}=-\frac{4\sqrt{5}}{5}\approx -1.79 \text{ m/s}$$

When the lower end of the ladder is 10 metres from the wall, its top is sliding down at a rate of 1.79 m/s. The negative sign in the calculation of $\frac{dy}{dt}$ indicates that the height y is decreasing.

Example 4. A hot air balloon rises vertically from the point A on the ground at a rate of 20 m/min. A camera is positioned on the ground at the point C, 50 metres away from point A, and tracks the balloon's ascent. When the balloon is 50 metres above the ground, what is the rate of change of the angle of elevation from the camera to the balloon?

Solution. As shown in Figure 5.7.2, B represents the hot air balloon and C represents the camera. Let AC = 50 metres and AB = x.

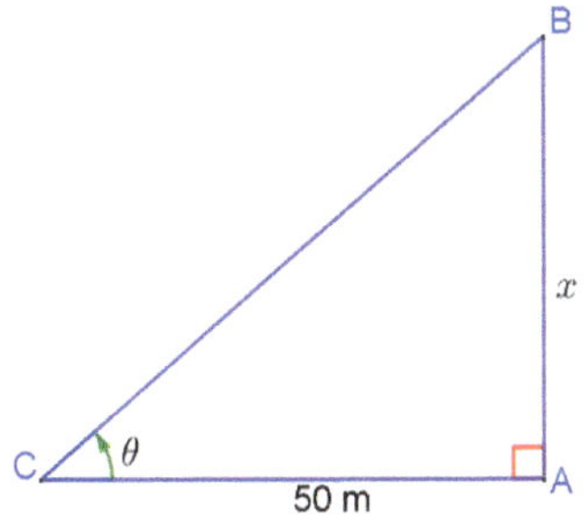

Figure 5.7.2

Given that $\frac{dx}{dt}=20$, find $\left.\frac{d\theta}{dt}\right|_{x=50}=?$

The relationship between x and θ is

$$\tan\theta=\frac{x}{50}$$

$$50\tan\theta=x$$

Differentiate both sides of the equation with respect to t.

$$50\frac{d\tan\theta}{dt}=\frac{dx}{dt}$$

$$50\frac{d\tan\theta}{d\theta}\cdot\frac{d\theta}{dt}=\frac{dx}{dt}$$

$$50\sec^2\theta\cdot\frac{d\theta}{dt}=\frac{dx}{dt}$$

When $x=50$, $\tan\theta=\frac{50}{50}=1$.

Hence, $\sec^2\theta=1+\tan^2\theta=1+1^2=2$

Substituting $\frac{dx}{dt}=20$ and $\sec^2\theta=2$ gives

$$50\cdot 2\cdot\frac{d\theta}{dt}=20$$

$$\frac{d\theta}{dt} = 0.2 \text{ rad/min}$$

When the hot air balloon rises to 50 metres above the ground, the rate of change of the angle of elevation from the camera to the balloon is 0.2 rad/min.

Example 5. An inverted right circular cone container is filled with sand. The height of the container is 10 metres, and the radius of its circular top is 5 metres. Sand leaks from the container at a rate of 10 m^3/min. When the surface area of the remaining sand is 20π m^2, what is the rate of change of the depth of sand?

Solution. As shown in Figure 5.7.3, AB = 10 metres, BC = 5 metres, AD = h metres, and DE = r metres. Given that $\frac{dV}{dt} = -10$ m^3/min, find $\left.\frac{dh}{dt}\right|_{\pi r^2 = 20\pi} = ?$

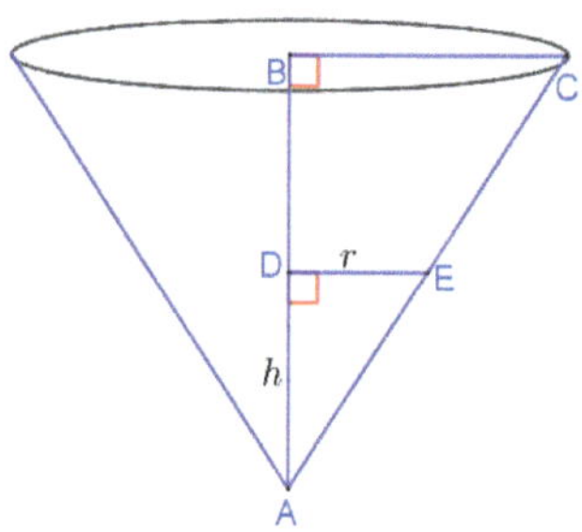

Figure 5.7.3

Using similar triangles gives

$$\frac{r}{h} = \frac{\text{BC}}{\text{AB}} = \frac{5}{10} = \frac{1}{2}$$

Hence, $$r = \frac{1}{2}h$$

When $\pi r^2 = 20\pi$, $r = \sqrt{20}$.

The relationship between the volume of sand V and the depth of sand h is

$$V = \frac{1}{3}\pi r^2 h$$

Substituting $r = \frac{1}{2}h$ gives

$$V = \frac{1}{12}\pi h^3$$

Differentiate both sides of the equation with respect to t.

$$\frac{dV}{dt} = \frac{1}{12}\pi \frac{dh^3}{dt}$$

$$= \frac{1}{12}\pi \frac{dh^3}{dh} \cdot \frac{dh}{dt} = \frac{1}{4}\pi h^2 \frac{dh}{dt}$$

Substituting $\frac{dV}{dt} = -10$ and $h = 2r = 2\sqrt{20}$ gives

$$-10 = \frac{1}{4}\pi (2\sqrt{20})^2 \frac{dh}{dt}$$

$$\frac{dh}{dt} = -\frac{1}{2\pi} \approx -0.159 \text{ metres/min}$$

The depth of sand changes at a rate of -0.159 metres/min. The negative sign indicates that the depth of sand is decreasing.

Exercise 5.7

1. The radius of a circle is increasing at a rate of 5 cm/s.

- **a.** Find the rate at which the circumference of the circle is increasing when the radius is 3 cm.
- **b.** Find the rate at which the area of the circle is increasing when the circumference is 16π cm.

2. The side length of a cube is decreasing at a rate of 2 cm/s.

- **a.** Find the rate at which the surface area is decreasing when the side length is 50 cm.
- **b.** Find the rate at which the volume of the cube is changing when the side

length of the cube is 8 cm.

c. Find the rate at which the diagonal of the cube is changing when the side of the cube is 10 cm.

3. In an isosceles triangle ABC, AB = BC = 20 cm. Angle ABC is increasing at a rate of $\frac{\pi}{24}$ radian/s. When angle ABC = $\frac{\pi}{3}$, what is the rate of change of the side AC?

4. A closed right-circular cylinder's radius is decreasing at a rate of 2 cm/s, and its height is increasing at 4 cm/s. Find the rate of change at which the cylinder's surface area and volume change when its height is 40 cm and radius is 20 cm.

5. A conical water tank has a vertex down. Water is poured into the tank at a rate of 2 m^3 per minute. Find the rate at which the water level in the tank rises when the depth is 5 metres and the radius of the water surface is 2 metres.

6. A rocket, rising vertically, is tracked by a radar station on the ground, 10 km from the launchpad. Find the rate of change in the angle of elevation of the tracking radar of the rocket when the rocket is rising at a rate of 800 km per hour and is 20 km directly above the launchpad.

7. An airplane is flying horizontally at a constant altitude of 4 kilometres above a fixed observer on the ground at a constant speed, as shown in Diagram 5.7.1. At a particular instant, the angle of elevation $\theta = \frac{\pi}{3}$ and is decreasing at a rate of $\frac{1}{80}$ radian per second. How fast is the distance between the airplane and the observer changing at this instant?

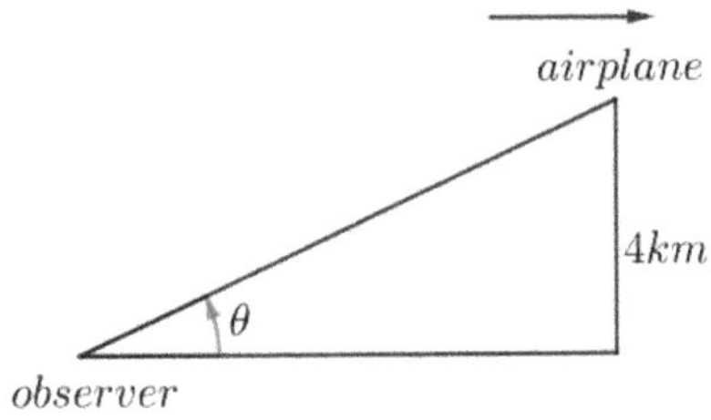

Diagram 5.7.1

8. Sand pouring from the conveyor belt on a freighter down to the harbour forms a conical pile whose height is always equal to its diameter. Given that the height is increasing at a constant rate of 3 metres per minute, find the rate of change of the volume of sand when the pile is 10 metres high.

9. A streetlight is 4 metres high. A girl, who is 1.6 metres tall, is walking away from the streetlight in a straight line at a rate of 80 metres per minute, as shown in Diagram 5.7.2.

a. At what rate is the length of her shadow changing?

b. How fast is the tip of her shadow moving?

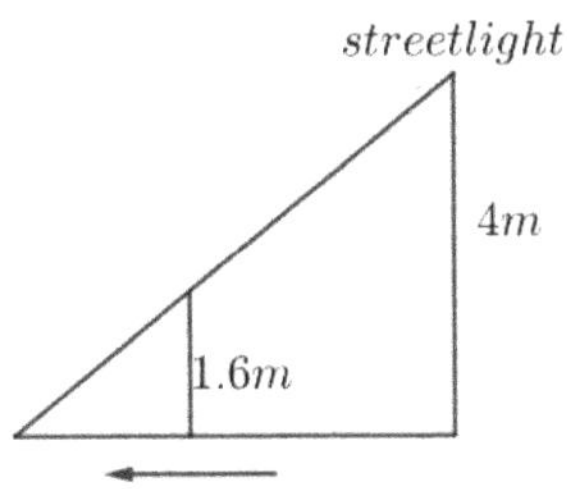

Diagram 5.7.2

10. Each side of an equilateral triangle is *a* cm in length. Given that the side length is increasing at a rate of 5 cm per minute and the area of the triangle is increasing at a rate of 5 cm^2 per minute, find the value of *a*.

11. A boat is pulled towards a dock by a rope attached to a pulley on the dock. The pulley is at a height of 4 metres above the boat. If the rope is pulled through the pulley at a rate of 2 metres per second, at what rate does the boat approach the dock at the instant when it is 8 metres from the dock?

12. A point P is moving along the curve whose equation is $y^4 + 2x^2y + 5 = 0$. Given that $x = 2$ and $\frac{dx}{dt} = 4$ when t = 5. Find the value of $\frac{dy}{dt}$ at this instant.

13. The base of a triangle is increasing at a rate of 2 cm per minute, and its height is increasing at a rate of 6 cm per minute. Find the rate of change of the area of the triangle when the base is 10 cm and the height is 12 cm.

14. A lighthouse is situated on an island 10 kilometres from the nearest point P on a straight shoreline. The beam of light rotates at a constant rate of 8 revolutions per minute. How fast is the beam of light moving along the shore at the instant when it is 3 kilometres from P?

15. The length and width of a rectangle are x and y, respectively. Given that the length x is decreasing at a rate of 0.01 m/s and the width y is increasing at a rate of 0.02 m/s. Find the rate of change of the area and the rectangle's diagonal at the instant when $x = 20$ and $y = 15$.

16. Two boats leave the same harbour at the same time. Boat A sails north at a speed of 30 km/h. Boat B sails east at a speed of 40 km/h. Find the rate of change of distance between the two boats after 2 hours.

17. At 12 o'clock, Boat A sails east at a speed of 6 km/h. At the same time, Boat B is 16 km north of Boat A and sails south at a speed of 8 km/h. Find the rate of change of the distance between two boats at precisely 13 : 00.

Chapter 6 Indefinite integrals

6.1. Concept and properties of indefinite integrals and basic integration formulae

In differential calculus, we have discussed the problems of finding the derivatives (or differentials) of given functions. However, in many situations in science, technology, and economics, there is a need to solve the inverse problem: to determine a function from its given derivative (or differential). This type of problem – finding an antiderivative from a given derivative (or differential) – is one of the fundamental tasks in integral calculus: finding the indefinite integral.

Antiderivative

Definition

Let $f(x)$ be a given function defined on an interval. If there exists a function $F(x)$ such that

$$F'(x) = f(x) \qquad \text{or} \qquad dF(x) = f(x)dx$$

for all x in that interval, then the function $F(x)$ is called an **antiderivative** of the given function $f(x)$ on that interval.

For example, on the interval $(-\infty, \infty)$, consider the function $f(x) = 2x$. Since the function $F(x) = x^2$ satisfies $F'(x) = 2x$, $F(x) = x^2$ is an antiderivative of $f(x) = 2x$. Similarly, $x^2 - 1$, $x^2 + 2$ are also antiderivatives of $2x$.

Indefinite integral

From the example above, we can observe that there is more than one antiderivative for a given function. In fact, there is an infinite number of antiderivatives for any given function. This is because if $F(x)$ is an antiderivative of the function $f(x)$, then the function $F(x) + C$ also satisfies

$$[F(x) + C]' = F'(x) = f(x)$$

Therefore, functions of the form $F(x) + C$ are all antiderivatives of $f(x)$.

Definition

The set of all antiderivatives of a function $f(x)$ is called the **indefinite integral** of $f(x)$, denoted by $\int f(x)dx$.

If $F(x)$ is an antiderivative of $f(x)$, then from the definition, we have

$$\int f(x)dx = F(x) + C \qquad \text{(C is an arbitrary constant.)}$$

where $\int$ is called the **integral sign**;

x is called the **variable of integration**;

$f(x)$ is called the **integrand**;

$f(x)dx$ is called the **integrand expression**;

C is called the **constant of integration**.

Hence, finding the indefinite integral can be summarised as finding one of its antiderivatives and then adding an arbitrary constant C.

Example 1. Find the indefinite integral of $f(x) = 3x^2$.

Solution. Since $(x^3)' = 3x^2$

thus $\int 3x^2 dx = x^3 + C$

Example 2. Find the indefinite integral of $f(x) = \dfrac{1}{x}$.

Solution. Since $(\ln x)' = \frac{1}{x}$

thus $\int \frac{1}{x} dx = \ln|x| + C \qquad (x \neq 0)$

If an initial condition is given, a specific value of C can be determined, and a particular antiderivative can be found.

Example 3. Find the equation of the curve that passes through the point (1, 3), given that the gradient of its tangent is $2x$.

Solution. Given that $m_t = \frac{dy}{dx} = 2x$, we have

$$y = \int dy = \int 2x dx = x^2 + C$$

That is, we obtain a family of curves.

$$y = x^2 + C$$

Substituting $x = 1$ and $y = 3$ into the equation, we obtain

$$C = 2$$

Hence, the equation of the required curve is

$$y = x^2 + 2$$

Properties of the indefinite integral

Finding the indefinite integral and finding the derivative or differential are inverse operations.

(1) $[\int f(x)dx]' = f(x)$

Alternatively, $d\int f(x)dx = f(x)\,dx$

(2) $\int F'(x)dx = F(x) + C$

Alternatively, $\int dF(x) = F(x) + C$

A non-zero constant factor may be moved outside the integral sign.

$$\int af(x)dx = a\int f(x)dx$$

The indefinite integral of the algebraic sum of two functions is equal to the algebraic sum of the indefinite integrals of those functions.

$$\int [f(x) \pm g(x)]dx = \int f(x)dx \pm \int g(x)dx$$

Basic Integration Formulae

Since finding the indefinite integral is the inverse operation of finding the derivative, the basic integration formulae can be obtained directly from the corresponding basic derivative formulae.

Basic derivative formulae	Corresponding basic integration formulae
$(kx)' = k$	$\int kdx = kx + C$
$(x^{n+1})' = (n+1)x^n$	$\int x^n dx = \frac{x^{n+1}}{n+1} + C \qquad n \neq -1$
$(\ln x)' = \frac{1}{x}$	$\int \frac{1}{x} dx = \ln\|x\| + C$
$(e^x)' = e^x$	$\int e^x dx = e^x + C$
$(a^x)' = a^x \ln a$	$\int a^x dx = \frac{1}{\ln a} a^x + C$
$(\sin x)' = \cos x$	$\int \cos x dx = \sin x + C$
$(\cos x)' = -\sin x$	$\int \sin x dx = -\cos x + C$
$(\tan x)' = \sec^2 x$	$\int \sec^2 x dx = \tan x + C$
$(\arcsin x)' = \frac{1}{\sqrt{1-x^2}}$	$\int \frac{1}{\sqrt{1-x^2}} dx = \arcsin x + C$
$(\arctan x)' = \frac{1}{1+x^2}$	$\int \frac{1}{1+x^2} dx = \arctan x + C$

Example 4. Find (a) $\int 4x^2 dx$ (b) $\int (9+5\sqrt[3]{x^2}-e^x)dx$

(c) $\int (x+1)(x-2)dx$ (d) $\int \frac{x^5+4x^3-2x}{x^4}dx$

Solution. (a) $\int 4x^2 dx = 4\int x^2 dx = 4\times\frac{x^{2+1}}{2+1} = \frac{4}{3}x^3 + C$

(b) $\int (9+5\sqrt[3]{x^2}-e^x)dx = \int 9dx + 5\int x^{\frac{2}{3}}dx - \int e^x dx$

$$= 9x + 5\times\frac{x^{\frac{2}{3}+1}}{\frac{2}{3}+1} - e^x = 9x + 3x^{\frac{5}{3}} - e^x + C$$

(c) $\int (x+1)(x-2)dx = \int (x^2-x-2)dx = \int x^2 dx - \int xdx - \int 2dx$

$$= \frac{x^{2+1}}{2+1} - \frac{x^{1+1}}{1+1} - 2x = \frac{x^3}{3} - \frac{x^2}{2} - 2x + C$$

(d) $\int \frac{x^5+4x^3-2x}{x^4}dx = \int \left(\frac{x^5}{x^4}+\frac{4x^3}{x^4}-\frac{2x}{x^4}\right)dx$

$$= \int \left(x+\frac{4}{x}-2x^{-3}\right)dx = \frac{x^2}{2} + 4\ln|x| + \frac{1}{x^2} + C$$

Example 5. Given that $f''(x) = 3x+7$, $f(0) = 5$, $f(1) = 12$, find $f(x)$.

Solution. Since

$$\frac{df'(x)}{dx} = f''(x) \qquad \Rightarrow \qquad df'(x) = f''(x)dx$$

$$\frac{df(x)}{dx} = f'(x) \qquad \Rightarrow \qquad df(x) = f'(x)dx$$

thus $f'(x) = \int df'(x) = \int f''(x)dx = \int (3x+7)dx = \frac{3x^2}{2} + 7x + C$

$$f(x)=\int df(x)=\int f'(x)dx=\int(\frac{3x^2}{2}+7x+C)dx$$

$$=\frac{3}{2}\times\frac{x^3}{3}+7\frac{x^2}{2}+Cx+D=\frac{x^3}{2}+\frac{7}{2}x^2+Cx+D$$

To find *C* and *D*, substituting $f(0)=5$ and $f(1)=12$ into $f(x)$, we obtain

$f(0)=0+D=5$ thus $D=5$

$f(1)=\frac{1}{2}+\frac{7}{2}+C+5=12$ thus $C=3$

Therefore, $f(x)=\frac{x^3}{2}+\frac{7}{2}x^2+3x+5$

Exercise 6.1

1. Find the following indefinite integrals.

a. $\int(x^3+x^2-2)dx$

b. $\int(5x+\frac{1}{x}-\frac{1}{x^3})dx$

c. $\int(x-1)(x-2)dx$

d. $\int\frac{4}{\sqrt[3]{x}}dx$

e. $\int(3-x^2)^2dx$

f. $\int(\sin x+\sqrt{3}\cos x+\sqrt[3]{x})dx$

g. $\int(3e^x-5\sec^2 x+\frac{2}{\sqrt{1-x^2}})dx$

h. $\int(2-\frac{1}{x})^3dx$

i. $\int 4x\sqrt{x}dx$

j. $\int\frac{1}{x^2\sqrt{x}}dx$

k. $\int x^2\sqrt[4]{x}dx$

l. $\int\frac{3x^4+2x^2+5}{x^3}dx$

2. Find the following indefinite integrals.

a. $\int \frac{2+3x}{x^6} dx$

b. $\int \sqrt{x}(3x - \frac{1}{x^2})dx$

c. $\int \sqrt{\frac{5}{x^5}} dx$

d. $\int \frac{x^2 + \sqrt{x}}{x^3} dx$

e. $\int \frac{(2x+1)(x-1)}{x} dx$

f. $\int x^{\frac{2}{3}}(x^{\frac{1}{2}} + x^{-\frac{1}{2}})dx$

g. $\int \left(\frac{x^2 - 1}{x} \right)^2 dx$

h. $\int \frac{\sqrt{x} - 2\sqrt[5]{x^3} + 1}{\sqrt[5]{x}} dx$

i. $\int \frac{(1-x)^3}{x\sqrt{x}} dx$

j. $\int (4e^x + \sec^2 x + \frac{1}{1+x^2})dx$

k. $\int (7 - 3^x)dx$

l. $\int (6 \sec x \tan x - \frac{2}{x})dx$

3. Given that $f'(x) = x - \frac{1}{x^2}$ and $f(1) = 1$, find $f(x)$.

4. Given that $f'(x) = 5x^2 - \sqrt[3]{x}$ and $f(0) = 5$, find $f(x)$.

5. Find the equation of the curve with gradient $\frac{dy}{dx} = 3x - 2$, given that it passes through the point (6, -2).

6. Given that the gradient of the function $y = f(x)$ is $f'(x) = x + 2$. When $x = 0$, $y = 5$. Find the function $y = f(x)$.

7. A curve has the gradient function $\frac{dy}{dx} = 2e^x - 3x$ and passes through (2, e^2). Find the equation of this curve.

8. A curve has the gradient function $g'(x) = bx - 2$, where b is a constant. Find $g(x)$ given that $g(0) = 5$ and $g(-2) = -7$.

9. The gradient of a function $y = h(x)$ is given by $h'(x) = ax^2 + bx + c$, where a, b, and c are constants. Find $y = h(x)$ given that $h(0) = 3$, $h(-1) = 5$, $h(2) = 14$, and $h(1) = 7$.

10. Find the function $f(x)$ using the given information.

a. $f''(x) = x - 1,\ f'(-1) = 1,\ f(1) = \frac{1}{3}$

b. $f''(x) = x + \cos x,\ f'(0) = 2,\ f(0) = 1$

c. $f''(x) = \frac{1}{2}e^x - \sin x,\ f'(0) = \frac{3}{2},\ f(0) = 3$

d. $f''(x) = 3\cos x + \sin x,\ f'(\frac{3\pi}{2}) = 2,\ f(\frac{\pi}{2}) = \frac{3\pi}{2}$

6.2. Integration by substitution

We can use basic integral formulae and properties to obtain indefinite integrals for some functions. However, we cannot use the basic integral formulae (e.g., $\int \cos x dx = \sin x + C$) to directly integrate certain functions (e.g., $\int \cos 3x dx$). Therefore, we need other methods to find this type of indefinite integral. Here we introduce the most fundamental and effective method: **integration by substitution**.

The purpose of the substitution method is to perform a suitable change of variable on the integrand, thereby transforming the indefinite integral into the form in the basic integral formula or a simpler expression that is easier to evaluate.

Steps for finding an indefinite integral by the substitution method:

$$\int f(x)dx$$

Let $u = u(x)$ then $du = u'(x)dx$

Substituting gives

$$\int f(x)dx = \int f(u)du = F(u) + C = F(x) + C$$

Example 1. Find $\int (3x+7)^4 \, dx$.

Solution. Since the integrand $(3x + 7)^4$ is of the same form as the integrand in the basic integral formula $\int x^n dx$,

let $u = 3x + 7$.

Then $\dfrac{du}{dx} = 3$, thus $dx = \dfrac{1}{3} du$.

$$\int (3x+7)^4 \, dx = \int u^4 \frac{du}{3} = \frac{1}{3}\int u^4 du = \frac{1}{3}\cdot\frac{u^{4+1}}{4+1} = \frac{1}{15}u^5 = \frac{1}{15}(3x+7)^5 + C$$

Example 2. Find $\int \cos 5x dx$.

Solution. Since the integrand $\cos 5x$ is of the same form as the integrand in the basic integral formula $\int \cos x dx$,

let $u = 5x$.

Then $\dfrac{du}{dx} = 5$, thus $dx = \dfrac{du}{5}$.

$$\int \cos 5x dx = \int \cos u \frac{du}{5} = \frac{1}{5}\int \cos u du = \frac{1}{5}\sin u = \frac{1}{5}\sin 5x + C$$

Example 3. Find $\int \frac{7}{2x+5} dx$.

Solution. Since the integrand $\frac{7}{2x+5}$ is of the same form as the integrand in the basic integral formula $\int \frac{1}{x} dx$,

let $u = 2x + 5$.

Then $\frac{du}{dx} = 2$, thus $dx = \frac{du}{2}$.

$$\int \frac{7}{2x+5} dx = \int \frac{7}{u} \cdot \frac{du}{2} = \frac{7}{2} \int \frac{1}{u} du = \frac{7}{2} \ln|u| = \frac{7}{2} \ln|2x+5| + C$$

Example 4. Find $\int e^{-4x} dx$.

Solution. Since the integrand e^{-4x} is of the same form as the integrand in the basic integral formula $\int e^{x} dx$,

let $u = -4x$.

Then $\frac{du}{dx} = -4$, thus $dx = \frac{du}{-4}$.

$$\int e^{-4x} dx = \int e^{u} \cdot \frac{du}{-4} = -\frac{1}{4} \int e^{u} du = -\frac{1}{4} e^{u} = -\frac{1}{4} e^{-4x} + C$$

Example 5. Find $\int 2x(x^2+7)^3 dx$.

Solution. Since the derivative of x^2+7 is $2x$,

let $u = x^2 + 7$.

Then $\frac{du}{dx} = 2x$, thus $dx = \frac{du}{2x}$.

$$\int 2x(x^2+7)^3 dx = \int 2x \cdot u^3 \cdot \frac{du}{2x} = \int u^3 du = \frac{u^4}{4} = \frac{(x^2+7)^4}{4} + C$$

Example 6. Find $\int x\cos x^2 dx$.

Solution. Since the derivative of x^2 is $2x$,

let $\quad u = x^2$.

Then $\quad \dfrac{du}{dx} = 2x$, thus $\quad dx = \dfrac{du}{2x}$.

$$\int x\cos x^2 dx = \int x\cos u\frac{du}{2x} = \frac{1}{2}\int \cos u du = \frac{1}{2}\sin u = \frac{1}{2}\sin x^2 + C$$

Example 7. Find $\int \sec^2 x e^{\tan x} dx$.

Solution. Since the derivative of $\tan x$ is $\sec^2 x$,

let $\quad u = \tan x$.

Then $\quad \dfrac{du}{dx} = \sec^2 x$, thus $\quad dx = \dfrac{du}{\sec^2 x}$.

$$\int \sec^2 x e^{\tan x} dx = \int \sec^2 x e^u \cdot \frac{du}{\sec^2 x} = \int e^u du = e^u = e^{\tan x} + C$$

Example 8. Find $\int \dfrac{4x^2}{\left(x^3+5\right)^2} dx$.

Solution. Let $\quad u = x^3 + 5$.

Then $\quad \dfrac{du}{dx} = 3x^2$, thus $\quad dx = \dfrac{du}{3x^2}$.

$$\int \frac{4x^2}{\left(x^3+5\right)^2} dx = \int \frac{4x^2}{u^2} \cdot \frac{du}{3x^2} = \frac{4}{3}\int u^{-2} du = \frac{4}{3} \cdot \frac{u^{-2+1}}{-2+1} = -\frac{4}{3u} = -\frac{4}{3\left(x^3+5\right)} + C$$

Example 9. Find $\int \dfrac{1}{x}\ln x dx$.

Solution. Let $u = \ln x$.

Then $\frac{du}{dx} = \frac{1}{x}$, thus $du = \frac{1}{x}dx$.

$$\int \frac{1}{x}\ln x dx = \int u du = \frac{u^2}{2} = \frac{(\ln x)^2}{2} + C$$

Example 10. Find $\int \sin^2 x \cos x dx$.

Solution. Let $u = \sin x$.

Then $\frac{du}{dx} = \cos x$, thus $dx = \frac{du}{\cos x}$.

$$\int \sin^2 x \cos x dx = \int u^2 \cos x \cdot \frac{du}{\cos x} = \int u^2 du = \frac{u^3}{3} = \frac{\sin^3 x}{3} + C$$

Example 11. Find $\int \frac{e^x}{1+e^x}dx$.

Solution. Let $u = 1 + e^x$.

Then $\frac{du}{dx} = e^x$, thus $dx = \frac{du}{e^x}$.

$$\int \frac{e^x}{1+e^x}dx = \int \frac{e^x}{u}\cdot\frac{du}{e^x} = \int \frac{1}{u}du = \ln|u| = \ln\left|1+e^x\right| + C$$

Example 12. Find $\int \frac{e^x - e^{-x}}{e^x + e^{-x}}dx$.

Solution. Let $u = e^x + e^{-x}$.

Then $\frac{du}{dx} = e^x - e^{-x}$, thus $dx = \frac{du}{e^x - e^{-x}}$.

$$\int \frac{e^x - e^{-x}}{e^x + e^{-x}} dx = \int \frac{e^x - e^{-x}}{u} \cdot \frac{du}{e^x - e^{-x}} = \int \frac{1}{u} du = \ln|u| = \ln\left|e^x + e^{-x}\right| + C$$

Example 13. Find $\int e^x \sin\left(2e^x\right) dx$.

Solution. Let $u = 2e^x$.

Then $\frac{du}{dx} = 2e^x$, thus $dx = \frac{du}{2e^x}$.

$$\int e^x \sin\left(2e^x\right) dx = \int e^x \sin u \cdot \frac{du}{2e^x} = \frac{1}{2} \int \sin u du = -\frac{1}{2} \cos u = -\frac{1}{2} \cos\left(2e^x\right) + C$$

Example 14. Find $\int 3 \sin\left(\frac{x}{2}\right) e^{\cos\left(\frac{x}{2}\right)} dx$.

Solution. Let $u = \cos\left(\frac{x}{2}\right)$.

Then $\frac{du}{dx} = \frac{d \cos\left(\frac{x}{2}\right)}{d(\frac{x}{2})} \cdot \frac{d(\frac{x}{2})}{dx} = -\sin\left(\frac{x}{2}\right) \cdot \frac{1}{2}$, thus $dx = \frac{-2du}{\sin\left(\frac{x}{2}\right)}$.

$$\int 3 \sin\left(\frac{x}{2}\right) e^{\cos\left(\frac{x}{2}\right)} dx = \int 3 \sin\left(\frac{x}{2}\right) e^u \cdot \frac{-2du}{\sin\left(\frac{x}{2}\right)} = -6 \int e^u du = -6e^u = -6e^{\cos\left(\frac{x}{2}\right)} + C$$

Example 15. Find $\int \frac{2}{x} \cos(\ln x) dx$.

Solution. Let $u = \ln x$.

Then $\frac{du}{dx} = \frac{1}{x}$, thus $dx = x du$.

$$\int \frac{2}{x}\cos(\ln x)\,dx = \int \frac{2}{x}\cos u \cdot x\,du = 2\int \cos u\,du = 2\sin u = 2\sin(\ln x) + C$$

Example 16. Find $\int \frac{4\sec^2 3x}{1+\tan 3x}dx$.

Solution. Let $u = 1 + \tan 3x$.

Then $\frac{du}{dx} = \frac{d(1+\tan 3x)}{d(3x)} \cdot \frac{d(3x)}{dx} = 3\sec^2 3x$, thus $dx = \frac{du}{3\sec^2 3x}$.

$$\int \frac{4\sec^2 3x}{1+\tan 3x}dx = \int \frac{4\sec^2 3x}{u} \cdot \frac{du}{3\sec^2 3x} = \frac{4}{3}\int \frac{1}{u}du = \frac{4}{3}\ln|u| = \frac{4}{3}\ln|1+\tan 3x| + C$$

Example 17. Find $\int x\sqrt{2x-1}\,dx$.

Solution. Let $u = 2x - 1$, then $x = \frac{u+1}{2}$.

Since $\frac{du}{dx} = 2$, thus $dx = \frac{du}{2}$.

$$\int x\sqrt{2x-1}\,dx = \int \frac{u+1}{2} \cdot \frac{u^{\frac{1}{2}}}{2}du = \frac{1}{4}\int \left(u^{\frac{3}{2}} + u^{\frac{1}{2}}\right)du = \frac{1}{4}\left(\frac{u^{\frac{3}{2}+1}}{\frac{3}{2}+1} + \frac{u^{\frac{1}{2}+1}}{\frac{1}{2}+1}\right)$$

$$= \frac{1}{4}\left(\frac{2}{5}u^{\frac{5}{2}} + \frac{2}{3}u^{\frac{3}{2}}\right) = \frac{1}{4}\left[\frac{2}{5}(2x-1)^{\frac{5}{2}} + \frac{2}{3}(2x-1)^{\frac{3}{2}}\right] + C$$

Example 18. Find $\int \frac{e^{2x}}{1+e^x}dx$.

Solution. Let $u = 1 + e^x$, then $e^x = u - 1$.

Since $\dfrac{du}{dx} = e^x$, thus $dx = \dfrac{du}{e^x}$.

$$\int \frac{e^{2x}}{1+e^x}dx = \int \frac{e^{2x}}{u} \cdot \frac{du}{e^x} = \int \frac{e^x}{u}du = \int \frac{u-1}{u}du = \int \left(1 - \frac{1}{u}\right)du$$

$$= u - \ln|u| = 1 + e^x - \ln\left|1+e^x\right| + C$$

Example 19. Find $\displaystyle\int \frac{x}{\sqrt{x-1}}dx$.

Solution. Let $u = x - 1$ then $x = u + 1$, $du = dx$.

$$\int \frac{x}{\sqrt{x-1}}dx = \int \frac{u+1}{u^{\frac{1}{2}}}du = \int \left(u^{\frac{1}{2}} + u^{-\frac{1}{2}}\right)du = \frac{u^{\frac{1}{2}+1}}{\frac{1}{2}+1} + \frac{u^{-\frac{1}{2}+1}}{-\frac{1}{2}+1}$$

$$= \frac{2}{3}u^{\frac{3}{2}} + 2u^{\frac{1}{2}} = \frac{2}{3}(x-1)^{\frac{3}{2}} + 2(x-1)^{\frac{1}{2}} + C$$

Example 20. Find $\displaystyle\int \frac{x-2}{x+1}dx$.

Solution. Let $u = x + 1$ then $x = u - 1$, $du = dx$.

$$\int \frac{x-2}{x+1}dx = \int \frac{u-3}{u}du = \int \left(1 - \frac{3}{u}\right)du = u - 3\ln|u| = x + 1 - 3\ln|x+1| + C$$

Example 21. Find $\displaystyle\int \frac{1}{9+16x^2}dx$.

Solution. **Method 1**

$$\int \frac{1}{9+16x^2}dx = \int \frac{1}{16\left(\frac{9}{16}+x^2\right)}dx = \frac{1}{16}\int \frac{1}{\left(\frac{3}{4}\right)^2+x^2}dx$$

$$= \frac{1}{16}\cdot\frac{4}{3}\arctan\left(\frac{4x}{3}\right) = \frac{1}{12}\arctan\left(\frac{4x}{3}\right)+C$$

Method 2

Let $u = 4x$ then $du = 4dx$.

$$\int \frac{1}{9+16x^2}dx = \int \frac{1}{3^2+u^2}\cdot\frac{du}{4} = \frac{1}{4}\cdot\frac{1}{3}\arctan\left(\frac{u}{3}\right) = \frac{1}{12}\arctan\left(\frac{4x}{3}\right)+C$$

- In the following, the integrand of an indefinite integral is a trigonometric function.

Example 22. Find $\int \sin^2 x dx$.

Solution. From the double angle theorem $\cos 2x = 1 - 2\sin^2 x$ we obtain

$$\sin^2 x = \frac{1}{2}(1-\cos 2x)$$

$\int \cos 2x dx$	
Let	$u = 2x$
then	$\frac{du}{dx} = 2$

Hence, $\int \sin^2 x dx = \int \frac{1}{2}(1-\cos 2x)dx = \frac{1}{2}\int dx - \frac{1}{2}\int \cos 2x dx$

$$= \frac{1}{2}x - \frac{1}{2}\int \cos u \cdot \frac{du}{2} = \frac{1}{2}x - \frac{1}{4}\sin u = \frac{1}{2}x - \frac{1}{4}\sin 2x + C$$

Example 23. Find $\int \cos^2 x dx$.

Solution. From the double angle theorem $\cos 2x = 2\cos^2 x - 1$, we obtain

$$\cos^2 x = \frac{\cos 2x + 1}{2}$$

$$\int \cos^2 x dx = \frac{1}{2}\int \cos 2x dx + \frac{1}{2}\int dx = \frac{1}{4}\sin 2x + \frac{1}{2}x + C$$

Example 24. Find $\int \sin^3 x dx$.

Solution. $\int \sin^3 x dx = \int \sin^2 x \sin x dx = \int \left(1-\cos^2 x\right)\sin x dx$

$$= \int \sin x dx - \int \cos^2 x \sin x dx$$

For $\int \cos^2 x \sin x dx$

let $u = \cos x$.

Then $\frac{du}{dx} = -\sin x$, thus $dx = \frac{du}{-\sin x}$.

$$\int \cos^2 x \sin x dx = \int u^2 \sin x \cdot \frac{du}{-\sin x} = -\frac{u^3}{3} = -\frac{\cos^3 x}{3}$$

Therefore, $\int \sin^3 x dx = \int \sin x dx - \int \cos^2 x \sin x dx = -\cos x + \frac{\cos^3 x}{3} + C$

Example 25. Find $\int \cos^3 x dx$.

Solution. $\int \cos^3 x dx = \int \cos x \cos^2 x dx = \int \cos x\left(1-\sin^2 x\right)dx$

$$= \int \cos x dx - \int \cos x \sin^2 x dx$$

For $\int \cos x \sin^2 x dx$

let $u = \sin x$.

Then $\frac{du}{dx} = \cos x$, thus $dx = \frac{du}{\cos x}$.

$$\int \cos x \sin^2 x dx = \int \cos x u^2 \cdot \frac{du}{\cos x} = \frac{u^3}{3} = \frac{\sin^3 x}{3}$$

Therefore, $\int \cos^3 x dx = \int \cos x dx - \int \cos x \sin^2 x dx = \sin x - \frac{\sin^3 x}{3} + C$

Example 26. Find $\int \sin^4 x dx$.

Solution. Since $\sin^2 x = \frac{1}{2}(1-\cos 2x)$

thus $\sin^4 x = \frac{1}{4}(1-\cos 2x)^2 = \frac{1}{4}(1-2\cos 2x + \cos^2 2x)$

$$\int \sin^4 x dx = \frac{1}{4}\int (1-2\cos 2x + \cos^2 2x)dx = \frac{1}{4}(\int dx - \int 2\cos 2x dx + \int \cos^2 2x dx)$$

From the double angle theorem $\cos 4x = 2\cos^2 2x - 1$, we obtain

$$\cos^2 2x = \frac{\cos 4x + 1}{2}$$

$$\int \cos^2 2x dx = \frac{1}{2}\int dx + \frac{1}{2}\int \cos 4x dx = \frac{1}{2}x + \frac{1}{8}\sin 4x$$

Hence, $\int \sin^4 x dx = \frac{1}{4}\int (1-2\cos 2x + \cos^2 2x)dx = \frac{1}{4}(x - \sin 2x + \frac{1}{2}x + \frac{1}{8}\sin 4x) + C$

Example 27. Find $\int \sin x \cos x dx$.

Solution. Method 1

Let $u = \sin x$.

Then $\frac{du}{dx} = \cos x$, thus $dx = \frac{du}{\cos x}$.

$$\int \sin x \cos x dx = \int u \cos x \frac{du}{\cos x} = \int u du = \frac{u^2}{2} = \frac{\sin^2 x}{2} + C$$

Method 2 From the double angle theorem $\sin 2x = 2\sin x \cos x$, we obtain

$$\sin x \cos x = \frac{\sin 2x}{2}.$$

$$\int \sin x \cos x dx = \int \frac{\sin 2x}{2} dx = -\frac{1}{4}\cos 2x + C$$

Example 28. Find $\int \sin^3 x \cos^2 x dx$.

Solution. Let $u = \cos x$.

Then $\frac{du}{dx} = -\sin x$, thus $dx = \frac{du}{-\sin x}$.

$$\int \sin^3 x \cos^2 x dx = \int \sin^3 x u^2 \cdot \frac{du}{-\sin x} = -\int \sin^2 x u^2 du = -\int (1-\cos^2 x) u^2 du$$

$$= -\int (1-u^2) u^2 du = \int (u^4 - u^2) du = \frac{u^5}{5} - \frac{u^3}{3}$$

$$= \frac{\cos^5 x}{5} - \frac{\cos^3 x}{3} + C$$

Example 29. Find $\int \tan x dx$.

Solution. $\int \tan x dx = \int \frac{\sin x}{\cos x} dx$

Let $u = \cos x$.

Then $\frac{du}{dx} = -\sin x$, thus $dx = \frac{du}{-\sin x}$.

$$\int \tan x dx = \int \frac{\sin x}{\cos x} dx = \int \frac{\sin x}{u} \cdot \frac{du}{-\sin x} = -\int \frac{1}{u} du$$

$$= -\ln|u| = -\ln|\cos x| = \ln|\sec x| + C$$

Example 30. Find $\int \cot x dx$.

Solution. $\int \cot x dx = \int \frac{\cos x}{\sin x} dx$

Let $u = \sin x$.

Then $\frac{du}{dx} = \cos x$, thus $dx = \frac{du}{\cos x}$.

$$\int \cot x dx = \int \frac{\cos x}{\sin x} dx = \int \frac{\cos x}{u} \cdot \frac{du}{\cos x} = \ln|u| = \ln|\sin x| + C$$

Example 31. Find $\int \tan^2 x dx$.

Solution. $\int \tan^2 x dx = \int (\sec^2 x - 1) dx = \tan x - x + C$

Example 32. Find $\int \tan^4 x dx$.

Solution. $\int \tan^4 x dx = \int \tan^2 x \tan^2 x dx = \int (\sec^2 x - 1) \tan^2 x dx$

$$= \int \sec^2 x \tan^2 x dx - \int \tan^2 x dx$$

For $\int \sec^2 x \tan^2 x dx$

let $u = \tan x$.

Then $\frac{du}{dx} = \sec^2 x$, thus $dx = \frac{du}{\sec^2 x}$.

Therefore, $\int \sec^2 x \tan^2 x dx = \int \sec^2 x u^2 \cdot \frac{du}{\sec^2 x} = \frac{u^3}{3} = \frac{\tan^3 x}{3}$

Hence, $\int \tan^4 x dx = \frac{\tan^3 x}{3} - \tan x + x + C$

In addition, some integrals contain radicals. In such cases, trigonometric substitution can be used to simplify the integrand and facilitate evaluation. Common substitutions include:

(1) if the integrand contains $\sqrt{x}$, let $\sqrt{x} = u$;

(2) if the integrand contains $\sqrt{a^2 - x^2}$, a is a constant, let $x = a \sin u$ or

$x = a\cos u$;

(3) if the integrand contains $\sqrt{a^2 + x^2}$, a is a constant, let $x = a\tan u$;

(4) if the integrand contains $\sqrt{x^2 - a^2}$, a is a constant, let $x = a\sec u$.

Example 33. Find $\int\left(1+\sqrt{x}\right)dx$.

Solution. Let $\sqrt{x} = u$ then $x = u^2$.

Hence, $\frac{dx}{du} = 2u$, $dx = 2udu$.

Therefore, $\int\left(1+\sqrt{x}\right)dx = \int(1+u)2udu = 2\int\left(u+u^2\right)du = 2\left(\frac{u^2}{2}+\frac{u^3}{3}\right)$

$$= 2\left(\frac{\left(\sqrt{x}\right)^2}{2}+\frac{\left(\sqrt{x}\right)^3}{3}\right) = 2\left(\frac{x}{2}+\frac{x^{\frac{3}{2}}}{3}\right)+C$$

Example 34. Find $\int\sqrt{a^2 - x^2}dx$.

Solution. Let $x = a\cos u$.

Then $\frac{dx}{du} = -a\sin u$, thus $dx = -a\sin udu$.

Use the trigonometric identity $\sin^2 u + \cos^2 u = 1$, that is, $\sin^2 u = 1 - \cos^2 u$, we obtain

$$\int\sqrt{a^2 - x^2}dx = \int\sqrt{a^2 - a^2\cos^2 u}\times -a\sin udu = -a^2\int\sin^2 udu$$

Using the result from Example 22 $\int\sin^2 xdx = \frac{1}{2}x - \frac{1}{4}\sin 2x + C$, we obtain

$$\int\sqrt{a^2 - x^2}dx = -a^2\int\sin^2 udu = -a^2\left(\frac{1}{2}u - \frac{1}{4}\sin 2u\right)$$

$$= a^2(\frac{1}{2}\sin u\cos u - \frac{1}{2}u) = \frac{a^2}{2}(\sin u\cos u - u)$$

Finally, we must express the result in terms of x.

From $x = a\cos u$ we obtain $u = \arccos\frac{x}{a}$.

To express $\sin u$ in terms of x, we can use $\cos u = \frac{x}{a}$ to construct a right-angled triangle, as shown in Figure 6.2.1.

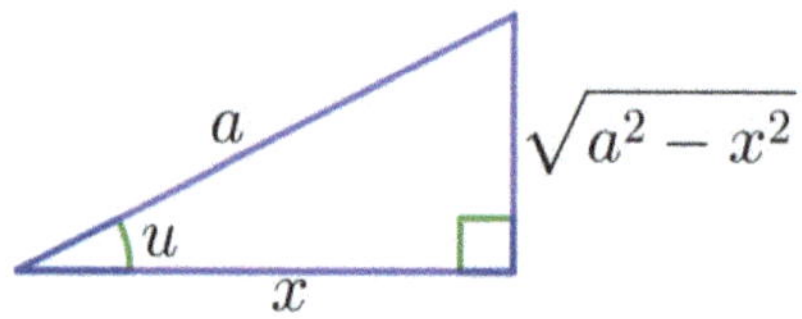

Figure 6.2.1

We obtain $\sin u = \frac{\sqrt{a^2 - x^2}}{a}$.

Hence, $\int\sqrt{a^2 - x^2}dx = \frac{a^2}{2}(\sin u\cos u - u) = \frac{a^2}{2}\left(\frac{\sqrt{a^2 - x^2}}{a}\times\frac{x}{a} - \arccos\frac{x}{a}\right)$

$$= \frac{a^2}{2}\left(\frac{x\sqrt{a^2 - x^2}}{a^2} - \arccos\frac{x}{a}\right) + C$$

Example 35. Find $\int\frac{dx}{\sqrt{a^2 + x^2}}$.

Solution. Let $x = a\tan u$.

Then $\frac{dx}{du} = a\sec^2 u$, thus $dx = a\sec^2 u du$.

Using the trigonometric identity $1 + \tan^2 x = \sec^2 x$, we obtain

$$\int\frac{dx}{\sqrt{a^2 + x^2}} = \int\frac{a\sec^2 u du}{\sqrt{a^2 + a^2\tan^2 u}} = \int\frac{a\sec^2 u du}{a\sec u} = \int\sec u du = \int\frac{\sec u(\sec u + \tan u)}{\sec u + \tan u}du$$

Let $t = \sec u + \tan u$.

Then $\frac{dt}{du} = \sec u \tan u + \sec^2 u$, thus $\frac{dt}{\sec u \tan u + \sec^2 u} = du$.

Hence,

$$\int \frac{dx}{\sqrt{a^2 + x^2}} = \int \frac{\sec u(\sec u + \tan u)}{\sec u + \tan u} du = \int \frac{\sec u(\sec u + \tan u)}{t} \times \frac{dt}{\sec u \tan u + \sec^2 u}$$

$$= \int \frac{1}{t} dt = \ln|\sec u + \tan u| + C$$

Finally, we must express the result in terms of x.

From $x = a \tan u$ we obtain $\tan u = \frac{x}{a}$.

To express $\sec u$ in terms of x, we can use $\tan u = \frac{x}{a}$ to construct a right-angled triangle, as shown in Figure 6.2.2.

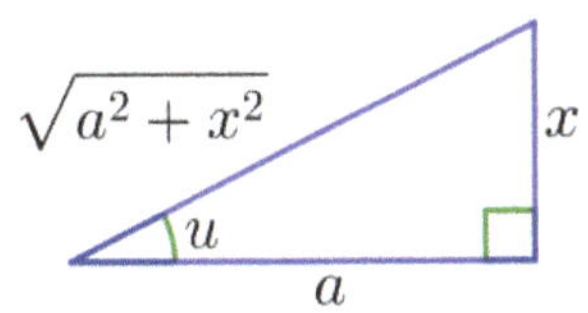

Figure 6.2.2

We obtain $\sec u = \frac{\sqrt{a^2 + x^2}}{a}$.

Therefore, $\int \frac{dx}{\sqrt{a^2 + x^2}} = \ln\left|\frac{\sqrt{a^2 + x^2}}{a} + \frac{x}{a}\right| + C$

Example 36. Find $\int \frac{dx}{\sqrt{x^2 - a^2}}$.

Solution. Let $x = a \sec u$.

Then $\frac{dx}{du} = a \sec u \tan u$, thus $dx = a \sec u \tan u du$.

Using the result from Example 35, we obtain

$$\int \frac{dx}{\sqrt{x^2-a^2}} = \int \frac{1}{\sqrt{a^2\sec^2 u - a^2}} \times a\sec u\tan u du = \int \sec u du = \ln\left|\sec u + \tan u\right| + C$$

Finally, we must express the result in terms of x.

From $\quad x = a\sec u \quad$ we obtain $\quad \sec u = \dfrac{x}{a}$.

To express $\tan u$ in terms of x, we can use $\sec u = \dfrac{x}{a}$ to construct a right-angled triangle, as shown in Figure 6.2.3.

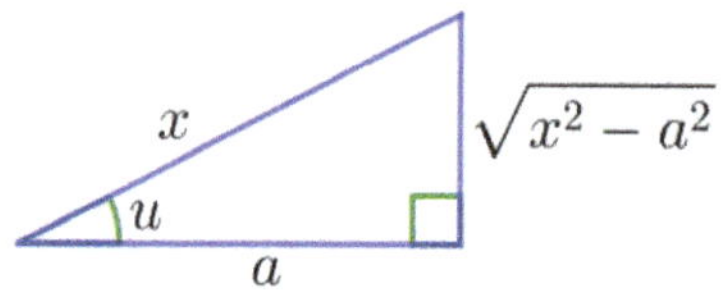

Figure 6.2.3

We obtain $\quad \tan u = \dfrac{\sqrt{x^2-a^2}}{a}$.

Therefore, $$\int \frac{dx}{\sqrt{x^2-a^2}} = \ln\left|\frac{x}{a} + \frac{\sqrt{x^2-a^2}}{a}\right| + C$$

Exercise 6.2

1. Using the given substitution for u, find each of the following indefinite integrals.

a. $\int (2x+5)^{10} dx$, $u = 2x+5$

b. $\int \sqrt[3]{1-4x} dx$, $u = 1-4x$

c. $\int \frac{4}{\sqrt{3-6x}} dx$, $u = 3-6x$

d. $\int \frac{1}{(5x+3)^3} dx$, $u = 5x+3$

e. $\int \cos(7x+1) dx$, $u = 7x+1$

f. $\int 3\sin(2-3x) dx$, $u = 2-3x$

g. $\int \frac{2x}{1+x^2} dx$, $u = 1+x^2$

h. $\int \frac{3}{2x+7} dx$, $u = 2x+7$

i. $\int x^2\sqrt[3]{1+x^3} dx$, $u = 1+x^3$

j. $\int xe^{-x^2} dx$, $u = -x^2$

k. $\int \frac{2x^4}{(x^5+3)^6}dx$, $u=x^5+3$

l. $\int \sin x e^{\cos x}dx$, $u=\cos x$

m. $\int \cos x \sin^7 x dx$, $u=\sin x$

n. $\int 2x^2 e^{x^3+6}dx$, $u=x^3+6$

o. $\int x\sqrt[4]{6x+7}dx$, $u=6x+7$

p. $\int (1+\ln x)^3 \frac{1}{x}dx$, $u=1+\ln x$

q. $\int \frac{1-\sin x}{x+\cos x}dx$, $u=x+\cos x$

r. $\int \frac{\arctan x}{1+x^2}dx$, $u=\arctan x$

2. Find the following indefinite integrals.

a. $\int \frac{x}{\sqrt{x-1}}dx$

b. $\int \frac{2x-4}{x^2-4x+3}dx$

c. $\int x^2 e^{-x^3}dx$

d. $\int \frac{(\ln x)^3}{x}dx$

e. $\int x\sqrt{x^2+3}dx$

f. $\int \frac{3x}{5-2x^2}dx$

g. $\int (e^{-3x}+e^{-x})dx$

h. $\int \frac{1}{(\arcsin x)^2\sqrt{1-x^2}}dx$

i. $\int 2\cos(5x)e^{\sin(5x)}dx$

j. $\int \frac{x-1}{x^2+1}dx$

k. $\int \frac{1}{x\ln x}dx$

l. $\int \frac{\cos(2x+\frac{\pi}{3})}{\sin^3(2x+\frac{\pi}{3})}dx$

m. $\int \frac{3}{\sqrt{9-4x^2}}dx$

n. $\int \frac{5}{64+16x^2}dx$

3. Find the following indefinite integrals.

a. $\int \sqrt[7]{5-3x}dx$

b. $\int \frac{2}{(4x+6)^{\frac{5}{3}}}dx$

c. $\int \frac{x}{\sqrt{1-x^2}}dx$

d. $\int \frac{1}{x^2}\sin\frac{1}{x}dx$

e. $\int \frac{2e^x}{3+e^x}dx$

f. $\int \frac{\sin x}{\sqrt{\cos^5 x}}dx$

g. $\int \sec^2(2x)e^{\tan(2x)}dx$

h. $\int \frac{\sin x}{1+\cos x}dx$

i. $\int \frac{x}{\sqrt[5]{1-2x}}dx$

j. $\int (\cos\frac{x}{3})^3 \sin(\frac{x}{3})dx$

k. $\int (2x+3)(5-x)^5 dx$

l. $\int (x+4)\sqrt{x+3}dx$

m. $\int (e^{2x-3}-2)^2 dx$

n. $\int \frac{x^3}{\sqrt{9+x^4}}dx$

o. $\int \frac{x^3}{\sqrt{x^2+16}}dx$

p. $\int \tan^2(3x)\sec^4(3x)dx$

4. Find the following indefinite integrals.

a. $\int \frac{2x+3}{(3x-1)^5}dx$

b. $\int \frac{1}{1+9x^2}dx$

c. $\int \frac{7}{25+16x^2}dx$

d. $\int \frac{1}{\sqrt{16-x^2}}dx$

e. $\int \frac{x^3+1}{x^4+4x}dx$

f. $\int (1-e^{-2x})^2 dx$

g. $\int \frac{e^{4x}}{\sqrt{e^{4x}-2}}dx$

h. $\int \frac{(e^{-x}+5)^2}{e^x}dx$

i. $\int \frac{2x-1}{x^2-x}dx$

j. $\int 5x\cos(x^2-8)dx$

k. $\int \frac{x-\cos x}{\sqrt{x^2-2\sin x}}dx$

l. $\int 4\cos x(5+\sin x)^6 dx$

m. $\int \sin^2 x\cos^2 x dx$

n. $\int \sin^3 x\cos^5 x dx$

o. $\int \sec^4(\frac{x}{2})\tan(\frac{x}{2})dx$

p. $\int \left(\frac{\sec(2x+5)}{1+\tan(2x+5)}\right)^2 dx$

5. Find the following indefinite integrals.

a. $\int x^3\sec^2(5x^4)dx$

b. $\int \frac{1}{\sqrt{x}}\sin(\sqrt{x})dx$

c. $\int 2\sin^2(\frac{x}{2})dx$

d. $\int \tan(4x)dx$

e. $\int \cot(3x)dx$

f. $\int \cos x \cos ecxdx$

g. $\int \cos(2x)\sin^5(2x)dx$

h. $\int \sin^3(2x)dx$

i. $\int \sin(3x)\cos^4(3x)dx$

j. $\int \cos^3 x\tan xdx$

k. $\int \cos^2(3x)dx$

l. $\int \cos^4 xdx$

m. $\int \sin(2x)\cos^3 xdx$

n. $\int \sin x\cos(\frac{x}{2})dx$

o. $\int \sec^2(5x)\tan(5x)dx$

p. $\int 2\cos^2(4x)dx$

6. Find the following indefinite integrals.

a. $\int 4\cos^2(2x-\frac{\pi}{2})dx$

b. $\int \tan^5(x+\frac{\pi}{3})\sec^2(x+\frac{\pi}{3})dx$

c. $\int \frac{\sec^2 x}{\sqrt{1-9\tan^2 x}}dx$

d. $\int \frac{5\cos^3(3x)}{\sqrt{\sin(3x)}}dx$

e. $\int \sin^3 x\cos^4 xdx$

f. $\int \sqrt[3]{\sin x}\cos^3 xdx$

g. $\int \sin^3(\frac{2x}{5})dx$

h. $\int 6\sin^2(3x+1)\cos^3(3x+1)dx$

i. $\int 3\tan x\sec^3 xdx$

j. $\int 2\sin^2 x\tan xdx$

k. $\int 6\cos^2 x\tan^2 xdx$

l. $\int 4\sin^5(\frac{x}{2})\cot^2(\frac{x}{2})dx$

7. Use a suitable trigonometric substitution, such as $x = a\sin u$, $x = a\cos u$, $x = a\tan u$, $x = a\sec u$, to find the following indefinite integrals.

a. $\int \sqrt{9 - x^2}\, dx$ **b.** $\int \frac{x}{\sqrt{16 + x^2}} dx$

c. $\int \frac{\sqrt{x^2 - 16}}{x} dx$ **d.** $\int \frac{1}{(x^2 + 1)^2} dx$

e. $\int \frac{x}{\sqrt{25 - x^2}} dx$ **f.** $\int \frac{x}{\sqrt{x^2 - 4}} dx$

g. $\int \frac{dx}{x^4 \sqrt{1 + x^2}}$ **h.** $\int \frac{\sqrt{x^2 - 9}}{x} dx$

i. $\int \frac{dx}{\sqrt{(1 - x^2)^3}}$ **j.** $\int \frac{dx}{\sqrt{(x^2 + a^2)^3}}$

k. $\int \frac{x^2 dx}{\sqrt{a^2 - x^2}}$ **l.** $\int \frac{\sqrt{a^2 - x^2}}{x^4} dx$

6.3. Integration by parts

The integration by substitution method above has indeed solved many integration problems and is a vital integration method. However, the substitution method cannot be used for integrals, such as $\int x \cos x dx$ or $\int e^x \sin x dx$. Next, we will introduce another important method of indefinite integration: **integration by parts**.

If both $u = u(x)$ and $v = v(x)$ are functions differentiable with respect to x, then the derivative formula for the product of the two functions is

$$\frac{d(uv)}{dx} = v\frac{du}{dx} + u\frac{dv}{dx}$$

Multiplying both sides by dx and then integrating both sides gives

$$d(uv) = vdu + udv$$

$$\int d(uv) = \int vdu + \int udv$$

$$uv = \int vdu + \int udv$$

Rearranging gives

$$\int udv = uv - \int vdu$$

This result is known as the **integration by parts formula**.

Generally, when the integrand is a product of two different types of functions, the method of integration by parts may be applied.

When using integration by parts, first divide the integrand into two parts: one part be u and the remaining part be dv. The choice of u and dv should satisfy the following conditions:

- from dv, it is easy to get v;
- it is simpler to integrate $\int vdu$ than $\int udv$.

Example 1. Find $\int x\cos xdx$.

Solution. Let $u = x$ $\quad dv = \cos xdx$

then $du = dx$ $\quad v = \int \cos xdx = \sin x$

Therefore, $\int x\cos xdx = x\sin x - \int \sin xdx = x\sin x + \cos x + C$

Example 2. Find $\int (2x-1)\sin 2xdx$.

Solution. Let $u = 2x - 1$ $dv = \sin 2x dx$

then $du = 2dx$ $v = \int \sin 2x dx = -\frac{1}{2}\cos 2x$

Therefore, $$\int (2x-1)\sin 2x dx = -\frac{1}{2}(2x-1)\cos 2x - \int -\frac{1}{2}\cos 2x \times 2dx$$

$$= -\frac{1}{2}(2x-1)\cos 2x + \frac{1}{2}\sin 2x + C$$

Example 3. Find $\int \ln x dx$.

Solution. Let $u = \ln x$ $dv = dx$

then $du = \frac{1}{x}dx$ $v = x$

Therefore, $$\int \ln x dx = x\ln x - \int x \cdot \frac{1}{x}dx = x\ln x - x + C$$

Example 4. Find $\int (3x-1)e^{-2x}dx$.

Solution. Let $u = 3x - 1$ $dv = e^{-2x}dx$

then $du = 3dx$ $v = \int e^{-2x}dx = -\frac{1}{2}e^{-2x}$

Therefore, $$\int (3x-1)e^{-2x}dx = -\frac{1}{2}(3x-1)e^{-2x} - \int -\frac{1}{2}e^{-2x} \times 3dx$$

$$= -\frac{1}{2}(3x-1)e^{-2x} + \frac{3}{2}\int e^{-2x}dx$$

$$= -\frac{1}{2}(3x-1)e^{-2x} - \frac{3}{4}e^{-2x} + C$$

Example 5. Find $\int x^2 \ln x dx$.

Solution. Let $u = \ln x$ $dv = x^2 dx$

then $du = \frac{1}{x}dx$ $v = \int x^2 dx = \frac{x^3}{3}$

Therefore, $\int x^2 \ln x dx = \frac{x^3}{3}\ln x - \int \frac{x^3}{3}\cdot\frac{1}{x}dx = \frac{x^3}{3}\ln x - \int \frac{x^2}{3}dx = \frac{x^3}{3}\ln x - \frac{x^3}{9} + C$

Example 6. Find $\int \arccos x dx$.

Solution. Let $u = \arccos x$ $dv = dx$

then $du = -\frac{1}{\sqrt{1-x^2}}dx$ $v = x$

Therefore, $\int \arccos x dx = x\arccos x + \int \frac{x}{\sqrt{1-x^2}}dx$

For $\int \frac{x}{\sqrt{1-x^2}}dx$

Let $T = 1 - x^2$ then $dT = -2xdx$

Hence, $\int \frac{x}{\sqrt{1-x^2}}dx = \int \frac{x}{T^{\frac{1}{2}}}\cdot\frac{dT}{-2x} = -\frac{1}{2}\cdot\frac{T^{-\frac{1}{2}+1}}{-\frac{1}{2}+1} = -T^{\frac{1}{2}} = -(1-x^2)^{\frac{1}{2}}$

Therefore, $\int \arccos x dx = x\arccos x - (1-x^2)^{\frac{1}{2}} + C$

Example 7. Find $\int e^{\sqrt{x}}dx$.

Solution. In some cases, it is necessary to use both integration by substitution and integration by parts.

Let $t = \sqrt{x}$.

Then $x = t^2$, $dx = 2tdt$

Therefore, $\int e^{\sqrt{x}}dx = 2\int e^{t}tdt$

Let $\quad u = t \qquad dv = e^{t}dt$

then $\quad du = dt \qquad v = \int e^{t}dt = e^{t}$

Hence, $\quad 2\int e^{t}tdt = 2\left(te^{t} - \int e^{t}dt\right) = 2\left(te^{t} - e^{t}\right) = 2e^{t}(t-1)$

Therefore, $\quad \int e^{\sqrt{x}}dx = 2e^{\sqrt{x}}\left(\sqrt{x}-1\right)+C$

Example 8. Find $\int x^{2}e^{x}dx$.

Solution. Let $\quad u = x^{2} \qquad dv = e^{x}dx$

then $\quad du = 2xdx \qquad v = \int e^{x}dx = e^{x}$

Therefore, $\quad \int x^{2}e^{x}dx = x^{2}e^{x} - 2\int xe^{x}dx$

For $\int xe^{x}dx$, it is necessary to apply integration by parts a second time.

Let $\quad u = x \qquad dv = e^{x}dx$

then $\quad du = dx \qquad v = \int e^{x}dx = e^{x}$

Hence, $\quad \int xe^{x}dx = xe^{x} - \int e^{x}dx = xe^{x} - e^{x}$

Therefore, $\quad \int x^{2}e^{x}dx = x^{2}e^{x} - 2\left(xe^{x} - e^{x}\right)+C$

In some cases, it is necessary to apply integration by parts multiple times to evaluate an integral. It should be noted that for each application of this method, the choice of *u* (or *dv*) must remain consistent in terms of function type. For example, in Example 8, u is chosen as a polynomial for both integrations, while dv is an exponential function. Failing to maintain this consistency may result in an integral that is more complex than the original, or one that returns to the starting expression.

Example 9. Find $\int x^2 \cos 2x dx$.

Solution. Let $u = x^2$ $dv = \cos 2x dx$

then $du = 2x dx$ $v = \int \cos 2x dx = \frac{1}{2}\sin 2x$

Hence, $\int x^2 \cos 2x dx = \frac{x^2}{2}\sin 2x - \int 2x \times \frac{1}{2}\sin 2x dx = \frac{x^2}{2}\sin 2x - \int x \sin 2x dx$

For $\int x \sin 2x dx$, it is necessary to apply integration by parts a second time.

Let $u = x$ $dv = \sin 2x dx$

then $du = dx$ $v = \int \sin 2x dx = -\frac{1}{2}\cos 2x$

Thus $\int x \sin 2x dx = -\frac{x}{2}\cos 2x + \frac{1}{2}\int \cos 2x dx = -\frac{x}{2}\cos 2x + \frac{1}{4}\sin 2x$

Therefore, $\int x^2 \cos 2x dx = \frac{x^2}{2}\sin 2x - \left(-\frac{x}{2}\cos 2x + \frac{1}{4}\sin 2x\right)$

$$= \frac{x^2}{2}\sin 2x + \frac{x}{2}\cos 2x - \frac{1}{4}\sin 2x + C$$

Example 10. Find $\int e^x \sin x dx$.

Solution. Let $u = e^x$ $dv = \sin x dx$

then $du = e^x dx$ $v = \int \sin x dx = -\cos x$

Hence, $\int e^x \sin x dx = -e^x \cos x + \int e^x \cos x dx$

For $\int e^x \cos x dx$, it is necessary to apply integration by parts a second time.

Let $u = e^x$ $dv = \cos x dx$

then $du = e^x dx$ $v = \int \cos x dx = \sin x$

Thus $\int e^x \cos x dx = e^x \sin x - \int e^x \sin x dx$

Therefore, $\int e^x \sin x dx = -e^x \cos x + e^x \sin x - \int e^x \sin x dx$

$$2\int e^x \sin x dx = -e^x \cos x + e^x \sin x$$

$$\int e^x \sin x dx = \frac{-e^x \cos x + e^x \sin x}{2} + C$$

Example 11. Find $\int x \arccos x dx$.

Solution. We know the derivative of $\arccos x$ is $(\arccos x)' = -\frac{1}{\sqrt{1-x^2}}$. However, we do not know the result of $\int \arccos x dx$, hence,

let $u = \arccos x$ $dv = x dx$

then $du = -\frac{1}{\sqrt{1-x^2}} dx$ $v = \int x dx = \frac{x^2}{2}$

Therefore, $\int x \arccos x dx = \frac{x^2}{2} \arccos x + \frac{1}{2}\int \frac{x^2}{\sqrt{1-x^2}} dx$

For $\int \frac{x^2}{\sqrt{1-x^2}} dx$

let $x = \cos\theta$.

Then $\frac{dx}{d\theta} = -\sin\theta$, thus $dx = -\sin\theta d\theta$

Hence, $\int \frac{x^2}{\sqrt{1-x^2}} dx = \int \frac{\cos^2\theta}{\sqrt{1-\cos^2\theta}} \times -\sin\theta d\theta = -\int \cos^2\theta d\theta$

From the double angle identity $\cos 2\theta = 2\cos^2\theta - 1$ we obtain

$$\cos^2\theta = \frac{\cos 2\theta + 1}{2}$$

$$\int \cos^2\theta d\theta = \frac{1}{2}\int \cos 2\theta d\theta + \frac{1}{2}\int d\theta = \frac{1}{4}\sin 2\theta + \frac{1}{2}\theta$$

Therefore, $\frac{1}{2}\int \frac{x^2}{\sqrt{1-x^2}}dx = -\frac{1}{2}\left(\frac{1}{4}\sin 2\theta + \frac{1}{2}\theta\right)$

Now, we need to express the result as a function of x.

From $x = \cos\theta$ we obtain $\theta = \arccos x$

To express $\sin 2\theta = 2\sin\theta\cos\theta$ as a function of x, we can use $x = \cos\theta$ to construct a right-angled triangle, as shown in Figure 6.3.1.

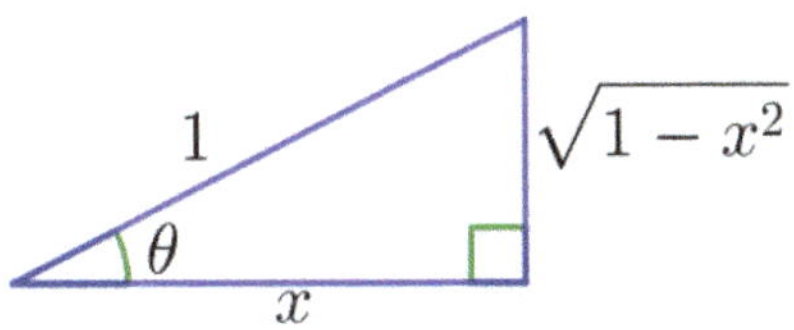

Figure 6.3.1

We obtain $\sin 2\theta = 2\sin\theta\cos\theta = 2x\sqrt{1-x^2}$

Therefore, $\int x\arccos x dx = \frac{x^2}{2}\arccos x - \frac{1}{2}\left(\frac{1}{4}\times 2x\sqrt{1-x^2} + \frac{1}{2}\arccos x\right)$

$$= \frac{x^2}{2}\arccos x - \frac{x}{4}\sqrt{1-x^2} - \frac{1}{4}\arccos x + C$$

Example 12. Find $\int x\arctan x dx$.

Solution. We know the derivative of $\arctan x$ is $(\arctan x)' = \frac{1}{1+x^2}$. However, we do not know the result of $\int \arctan x dx$, hence,

let $u = \arctan x$ $dv = xdx$

then $du = \frac{1}{1+x^2}dx$ $v = \int xdx = \frac{x^2}{2}$

Hence, $\int x\arctan x dx = \frac{x^2}{2}\arctan x - \frac{1}{2}\int \frac{x^2}{1+x^2}dx$

For $\int \frac{x^2}{1+x^2}dx$, we can integrate with two methods.

Method 1

$$\int \frac{x^2}{1+x^2}dx = \int \frac{x^2+1-1}{1+x^2}dx = \int \frac{x^2+1}{1+x^2}dx - \int \frac{1}{1+x^2}dx = x - \arctan x$$

Method 2

Let $x = \tan\theta.$

Then $\frac{dx}{d\theta} = \sec^2\theta$

Hence, $$\int \frac{x^2}{1+x^2}dx = \int \frac{\tan^2\theta}{1+\tan^2\theta} \times \sec^2\theta d\theta = \int \frac{\tan^2\theta}{\sec^2\theta} \times \sec^2\theta d\theta = \int \tan^2\theta d\theta$$

$$= \int \left(\sec^2\theta - 1\right)d\theta \ = \tan\theta - \theta = x - \arctan x$$

Therefore, $\int x\arctan x dx = \frac{x^2}{2}\arctan x - \frac{1}{2}\left(x - \arctan x\right) + C$

Example 13. Find $\int \arctan x dx$.

Solution. Let $u = \text{arctan}x$ $dv = dx$

then $du = \frac{1}{1+x^2}dx$ $v = x$

Hence, $\int \arctan x dx = x\arctan x - \int \frac{x}{1+x^2}dx$

For $\int \frac{x}{1+x^2}dx$

let $T = 1 + x^2.$

Then $\frac{dT}{dx} = 2x$, thus $dx = \frac{dT}{2x}$

Hence, $\int \frac{x}{1+x^2}dx = \int \frac{x}{T} \times \frac{dT}{2x} = \frac{1}{2}\ln|T| = \frac{1}{2}\ln\left|1+x^2\right|$

Therefore, $\int \arctan x dx = x\arctan x - \frac{1}{2}\ln\left|1+x^2\right| + C$

Exercise 6.3

1. Find the following indefinite integrals.

a. $\int xe^x dx$ **b.** $\int x\cos x dx$

c. $\int x^{-2}\ln x dx$ **d.** $\int x\ln(1+x)dx$

e. $\int x^3 e^{-x^2} dx$ **f.** $\int x\sin(2x)dx$

g. $\int e^{3x}\cos(5x)dx$ **h.** $\int (4x+9)\sin(3x)dx$

i. $\int \sqrt{x}\ln^2 x dx$ **j.** $\int (2x-1)e^{-3x}dx$

k. $\int (6x+7)\cos(2x)dx$ **l.** $\int \ln(4x)dx$

2. Find the following indefinite integrals.

a. $\int x\cos(3x)dx$ **b.** $\int 4x\sin(\frac{x}{6})dx$

c. $\int x\ln(3x)dx$ **d.** $\int x^2 e^{-x}dx$

e. $\int x^2\sin(3x)dx$ **f.** $\int x^3\ln^2 x dx$

g. $\int \ln^3 x dx$ **h.** $\int x\sec^2 x dx$

i. $\int e^{-4x}\sin(4x)dx$ **j.** $\int x^3\cos(2x)dx$

k. $\int \cos x\ln(\sin x)dx$ **l.** $\int x^2\arctan(\frac{x}{2})dx$

3. Find the following indefinite integrals.

a. $\int \sin x \cos(2x)dx$

b. $\int e^{2x} \cos(e^x)dx$

c. $\int x^2 \sin(5x)dx$

d. $\int \frac{x}{\sin^2 x}dx$

e. $\int x^2 \arcsin x dx$

f. $\int x^4 \sin(2x)dx$

g. $\int e^{2x} \arctan(e^x)dx$

h. $\int (\arcsin x)^2 dx$

i. $\int \frac{\arctan\sqrt{x}}{\sqrt{x}}dx$

j. $\int \sin\sqrt{x+3}dx$

k. $\int \cos\sqrt{3x+5}dx$

l. $\int xe^{\sqrt{x}}dx$

m. $\int \arccos(2x)dx$

n. $\int \cos(\ln x)dx$

4. Find the following indefinite integrals.

a. $\int x^2 \arctan x dx$

b. $\int \sin(\ln x)dx$

c. $\int e^x \sin^2 x dx$

d. $\int x \cos^2 x dx$

e. $\int \frac{\arcsin x}{\sqrt{1-x}}dx$

f. $\int \frac{x^2}{\sqrt{16-x^2}}dx$

g. $\int \frac{x}{\cos^2 x}dx$

h. $\int x \sin x \cos x dx$

6.4. Integration of rational functions

A rational function can always be written as the ratio of two polynomials.

$$f(x) = \frac{P(x)}{Q(x)} = \frac{a_0x^n + a_1x^{n-1} + \ldots + a_{n-1}x + a_n}{b_0x^m + b_1x^{m-1} + \ldots + b_{m-1}x + b_m}$$

where n, m are positive integers, $a_0 \neq 0$, $b_0 \neq 0$.

Assume that there are no common factors between the numerator and the denominator. When $m > n$, the expression is called a proper fraction. When $n \geq m$, it is called an improper fraction.

An improper fraction can be expressed as the sum of a polynomial and a proper fraction using algebraic long division. Since polynomials can be easily integrated term by term, we need only focus on the integration of proper fractions.

6.4.1. Expressing proper rational fractions in partial fractions

From algebraic theory, the following conclusions hold.

(1) A polynomial with real coefficients of degree n

$$Q(x) = x^n + a_1 x^{n-1} + \ldots + a_{n-1}x + a_n$$

can always be factorised into a product of linear and quadratic factors with real coefficients. That is, there is always

$$Q(x) = (x-a)^k \cdots (x^2 + px + q)^t$$

where $a, \ldots, p, q$ are constants. $k, \ldots, t$ are positive integers.

(2) Once $Q(x)$ has been factorised into the form above, the proper fraction $\dfrac{P(x)}{Q(x)}$ can be uniquely expressed as a sum of partial fractions in the following way. That is, we have:

$$\frac{P(x)}{Q(x)} = \frac{A_1}{(x-a)} + \frac{A_2}{(x-a)^2} + \ldots + \frac{A_k}{(x-a)^k} + \frac{B_1 x + C_1}{x^2 + px + q}$$

$$+ \frac{B_2 x + C_2}{(x^2 + px + q)^2} + \ldots + \frac{B_t x + C_t}{(x^2 + px + q)^t}$$

where $A_1, A_2, \ldots, A_k, B_1, B_2, \ldots, B_t, C_1, C_2, \ldots, C_t$ are constants.

For example,

- $$\frac{2x-1}{x^2-5x+6}=\frac{2x-1}{(x-3)(x-2)}=\frac{A}{x-3}+\frac{B}{x-2}$$
- $$\frac{x^2+1}{x(x-1)^2}=\frac{A}{x}+\frac{B}{x-1}+\frac{C}{(x-1)^2}$$
- $$\frac{x^2+2x-1}{(x-1)(x^2-x+1)}=\frac{A}{x-1}+\frac{Bx+C}{x^2-x+1}$$

A, B, C in the equations above are the coefficients to be determined. Generally, there are two methods to determine these coefficients.

Example 1. Express $\frac{2x-1}{x^2-5x+6}$ in partial fractions.

Solution. **Method 1**

Let $$\frac{2x-1}{x^2-5x+6}=\frac{2x-1}{(x-3)(x-2)}=\frac{A}{x-3}+\frac{B}{x-2}$$

Multiplying both sides of the equation by $(x-3)(x-2)$ to clear the denominators

gives $$2x-1=A(x-2)+B(x-3)$$

That is, $$2x-1=(A+B)x-(2A+3B)$$

Comparing coefficients of like terms on both sides gives

$$A+B=2$$

$$2A+3B=1$$

Solving simultaneously gives

$$A=5 \qquad B=-3$$

Therefore, $$\frac{2x-1}{x^2-5x+6}=\frac{5}{x-3}-\frac{3}{x-2}$$

Method 2

$$2x-1=A(x-2)+B(x-3)$$

Let $x=2$

$$2\cdot 2-1=A(2-2)+B(2-3)$$

then $\quad B=-3$

Let $x=3$

$$2\cdot 3-1=A(3-2)+B(3-3)$$

then $\quad A=5$

As shown above, both methods yield the same result.

Example 2. Express $\dfrac{x^2+2x-1}{(x-1)(x^2-x+1)}$ in partial fractions.

Solution. Let $\quad \dfrac{x^2+2x-1}{(x-1)(x^2-x+1)}=\dfrac{A}{x-1}+\dfrac{Bx+C}{x^2-x+1}$

Multiplying both sides of the equation by $(x-1)(x^2-x+1)$ to clear the denominators gives

$$x^2+2x-1=A(x^2-x+1)+(Bx+C)(x-1)$$

Let $x=1$: we have $A=2$

Let $x=0$: we have $-1=A-C$ thus $C=3$

Let $x=2$: we have $7=3A+2B+C$ thus $B=-1$

Therefore, $\quad \dfrac{x^2+2x-1}{(x-1)(x^2-x+1)}=\dfrac{2}{x-1}-\dfrac{x-3}{x^2-x+1}$

Example 3. Express $\dfrac{x^2+1}{x(x-1)^2}$ in partial fractions.

Solution. Let $\quad \dfrac{x^2+1}{x(x-1)^2}=\dfrac{A}{x}+\dfrac{B}{x-1}+\dfrac{C}{(x-1)^2}$

Multiplying both sides of the equation by $x(x-1)^2$ to clear the denominators gives

$$x^2+1 = A(x-1)^2 + Bx(x-1) + Cx$$

Let $x = 0$: we have $A = 1$

Let $x = 1$: we have $C = 2$

Let $x = 2$: we have $5 = A + 2B + 2C$ thus $B = 0$

Therefore, $$\frac{x^2+1}{x(x-1)^2} = \frac{1}{x} + \frac{2}{(x-1)^2}$$

6.4.2. Expressing improper rational fractions ($n \geq m$) in partial fractions

Example 1. Express $\dfrac{3x^2+16}{(x-5)(x+2)}$ in partial fractions.

Solution. Since $n = m$

let $$\frac{3x^2+16}{(x-5)(x+2)} = A + \frac{B}{x-5} + \frac{C}{x+2}$$

Multiplying both sides of the equation by $(x-5)(x+2)$ to clear the denominators gives

$$3x^2+16 = A(x-5)(x+2) + B(x+2) + C(x-5)$$

Let $x = -2$: we have $C = -4$

Let $x = 5$: we have $B = 13$

Let $x = 0$: we have $A = 3$

Therefore, $$\frac{3x^2+16}{(x-5)(x+2)} = 3 + \frac{13}{x-5} - \frac{4}{x+2}$$

Example 2. Express $\dfrac{x^4-x^3+7x+6}{(x-2)(x^2+3)}$ in partial fractions.

Solution. Since $n > m$

let $$\frac{x^4 - x^3 + 7x + 6}{(x-2)(x^2+3)} = Ax + B + \frac{C}{x-2} + \frac{Dx+E}{x^2+3}$$

Multiplying both sides of the equation by $(x-2)(x^2+3)$ to clear the denominators gives

$$x^4 - x^3 + 7x + 6 = (Ax+B)(x-2)(x^2+3) + C(x^2+3) + (Dx+E)(x-2)$$

Let $x = 2$: we have $C = 4$

Let $x = 0$: we have $3B + E = 3$ (1)

Let $x = 1$: we have $4A + 4B + D + E = 3$ (2)

Let $x = -1$: we have $4A - 4B - D + E = 5$ (3)

Let $x = 3$: we have $36A + 12B + 3D + E = 33$ (4)

Solving equations (1) to (4) simultaneously gives

$$A = 1, \quad B = 1, \quad D = -5, \quad E = 0$$

Therefore, $$\frac{x^4 - x^3 + 7x + 6}{(x-2)(x^2+3)} = x + 1 + \frac{4}{x-2} - \frac{5x}{x^2+3}$$

6.4.3. Integration of proper rational fractions

From the partial fraction decomposition of proper rational fractions above, it can be seen that the integral of any proper rational fraction can be transformed into integral of the following four types.

(1) $$\int \frac{A}{x-a} dx$$

(2) $$\int \frac{A}{(x-a)^n} dx$$

(3) $$\int \frac{Mx+N}{x^2+px+q} dx$$

(4) $\int \frac{Mx+N}{(x^2+px+q)^n}dx$

The results of the integrals for (1) and (2) are:

$$\int \frac{A}{x-a}dx = A\ln|x-a|+C$$

$$\int \frac{A}{(x-a)^n}dx = \frac{A}{(1-n)(x-a)^{n-1}}+C$$

The results of the integrals for (3) and (4) are illustrated in Examples 1 – 3.

Example 1. Find $\int \frac{x+1}{x^2-2x+5}dx$.

Solution. Let $u = x^2-2x+5$.

Then $du = (2x-2)dx$

$$x+1 = \frac{1}{2}(2x-2+4) = \frac{1}{2}(2x-2)+2$$

Therefore, $\int \frac{x+1}{x^2-2x+5}dx = \frac{1}{2}\int \frac{2x-2}{x^2-2x+5}dx + 2\int \frac{dx}{x^2-2x+5}$

$$= \frac{1}{2}\int \frac{du}{u} + 2\int \frac{d(x-1)}{(x-1)^2+2^2}$$

$$= \frac{1}{2}\ln|u| + \arctan\frac{x-1}{2}$$

That is, $\int \frac{x+1}{x^2-2x+5}dx = \frac{1}{2}\ln|x^2-2x+5| + \arctan\frac{x-1}{2}+C$

Example 2. Find $I_n = \int \frac{dx}{(x^2+a^2)^n}$, where n is a positive integer.

Solution. Let $u = x^2+a^2$ then $du = 2xdx$

$$\frac{d\frac{1}{u^{n-1}}}{du} = \frac{du^{-(n-1)}}{du} = -(n-1)u^{-n}$$

That is, $$d\frac{1}{u^{n-1}} = -(n-1)\frac{du}{u^n}$$

$$1 = \frac{a^2}{a^2} = \frac{1}{a^2}(x^2 + a^2 - x^2)$$

Therefore, $$I_n = \int \frac{dx}{(x^2+a^2)^n} = \frac{1}{a^2}\int \frac{x^2+a^2-x^2}{(x^2+a^2)^n}dx$$

$$= \frac{1}{a^2}\int \frac{dx}{(x^2+a^2)^{n-1}} - \frac{1}{a^2}\int \frac{x^2}{(x^2+a^2)^n}dx$$

$$= \frac{1}{a^2}I_{n-1} - \frac{1}{2a^2}\int \frac{xdu}{u^n}$$

$$= \frac{1}{a^2}I_{n-1} - \frac{1}{2a^2}\int x \cdot \frac{1}{-(n-1)} d\frac{1}{u^{n-1}}$$

The second integral on the right-hand side is

$$-\frac{1}{2a^2}\int x \cdot \frac{1}{-(n-1)} d\frac{1}{u^{n-1}} = \frac{1}{2(n-1)a^2}\int xd\frac{1}{u^{n-1}}$$

Integrating by parts gives

$$\int xd\frac{1}{u^{n-1}} = x \cdot \frac{1}{u^{n-1}} - \int \frac{1}{u^{n-1}}dx$$

$$= x \cdot \frac{1}{(x^2+a^2)^{n-1}} - \int \frac{1}{(x^2+a^2)^{n-1}}dx$$

$$= \frac{x}{(x^2+a^2)^{n-1}} - I_{n-1}$$

Thus $$I_n = \int \frac{dx}{(x^2+a^2)^n} = \frac{1}{a^2}I_{n-1} + \frac{1}{2(n-1)a^2}\left[\frac{x}{(x^2+a^2)^{n-1}} - I_{n-1}\right]$$

Rearranging gives

$$I_n = \frac{x}{2(n-1)a^2(x^2+a^2)^{n-1}} + \frac{2(n-1)-1}{2(n-1)a^2}I_{n-1} \qquad (n \geq 2)$$

where $$I_{n-1} = \int \frac{dx}{(x^2+a^2)^{n-1}} \qquad (n-1 \geq 1)$$

The result of the integral I_n is a recursive formula.

Since $$I_1 = \int \frac{dx}{x^2+a^2} = \frac{1}{a}\arctan\frac{x}{a} + C$$

using the formula successively gives

$$I_2 = \frac{x}{2a^2(x^2+a^2)} + \frac{1}{2a^3}\arctan\frac{x}{a} + C$$

$$I_3 = \frac{x}{4a^2(x^2+a^2)^2} + \frac{3x}{8a^4(x^2+a^2)} + \frac{3}{8a^5}\arctan\frac{x}{a} + C$$

$$\cdots$$

Example 3. Find $\int \frac{x-1}{(x^2+2x+5)^2}dx$.

Solution. Let $u = x^2+2x+5 = (x+1)^2 + 2^2$.

Then $du = (2x+2)dx$

Therefore, $$\int \frac{x-1}{(x^2+2x+5)^2}dx = \frac{1}{2}\int \frac{2x+2-4}{(x^2+2x+5)^2}dx$$

$$= \frac{1}{2}\int \frac{2x+2}{(x^2+2x+5)^2}dx - 2\int \frac{dx}{(x^2+2x+5)^2}$$

$$= \frac{1}{2}\int \frac{du}{u^2} - 2\int \frac{d(x+1)}{\left[(x+1)^2+2^2\right]^2}$$

$$= -\frac{1}{2}\cdot\frac{1}{u} - 2I_2$$

$$= -\frac{1}{2}\cdot\frac{1}{(x^2+2x+5)} - 2\left[\frac{x+1}{2\cdot 2^2\left[(x+1)^2+2^2\right]} + \frac{1}{2\cdot 2^3}\arctan\frac{x+1}{2}\right]$$

$$= -\frac{x+3}{4(x^2+2x+5)} - \frac{1}{8}\arctan\frac{x+1}{2} + C$$

That is, $\int \frac{x-1}{(x^2+2x+5)^2}dx = -\frac{x+3}{4(x^2+2x+5)} - \frac{1}{8}\arctan\frac{x+1}{2} + C$

Example 4. Find $\int \frac{2x}{x^2-5x+6}dx$.

Solution. First, express $\frac{2x}{x^2-5x+6}$ in partial fractions.

$$\frac{2x}{x^2-5x+6} = \frac{2x}{(x-3)(x-2)} = \frac{A}{x-3} + \frac{B}{x-2}$$

Multiplying both sides of the equation by $(x-3)(x-2)$ gives

$$2x = A(x-2) + B(x-3)$$

Let $x = 2$: we have $B = -4$

Let $x = 3$: we have $A = 6$

Therefore, $\int \frac{2x}{x^2-5x+6}dx = \int\left(\frac{6}{x-3} - \frac{4}{x-2}\right)dx = \int \frac{6}{x-3}dx - \int \frac{4}{x-2}dx$

$$= 6\ln|x-3| - 4\ln|x-2| + C$$

Example 5. Find $\int \frac{3x^2-x+2}{x-1}dx$.

Solution. **Method 1** First, express $\frac{3x^2-x+2}{x-1}$ in partial fractions.

$$\frac{3x^2-x+2}{x-1} = Ax + B + \frac{C}{x-1}$$

Multiplying both sides of the equation by $(x-1)$ gives

$$3x^2 - x + 2 = Ax(x-1) + B(x-1) + C$$

Let $x = 1$: we have $C = 4$

Let $x = 0$: we have $B = 2$

Let $x = 2$: we have $A = 3$

$$\int \frac{3x^2 - x + 2}{x-1} dx = \int \left(3x + 2 + \frac{4}{x-1} \right) dx = \frac{3}{2}x^2 + 2x + 4\ln|x-1| + C$$

Method 2

Let $u = x - 1$ then $x = u + 1$

$du = dx$

$$\int \frac{3x^2 - x + 2}{x-1} dx = \int \frac{3(u+1)^2 - (u+1) + 2}{u} du = \int \frac{3u^2 + 6u + 3 - u + 1}{u} du$$

$$= \int \left(3u + 5 + \frac{4}{u} \right) du = \frac{3}{2}u^2 + 5u + 4\ln|u|$$

$$= \frac{3}{2}(x-1)^2 + 5(x-1) + 4\ln|x-1| + C$$

Exercise 6.4

1. Find

a. $\int \frac{2x+5}{(x-2)(x+5)} dx$

b. $\int \frac{3x-1}{(x+3)(x-2)} dx$

c. $\int \frac{3x}{x^2-16} dx$

d. $\int \frac{5x-1}{x^2-7x+10} dx$

e. $\int \frac{x+3}{6x^2-5x+1} dx$

f. $\int \frac{2x+3}{5x^2+13x-6} dx$

g. $\int \frac{2}{(x+2)(x+3)(x-1)} dx$

h. $\int \frac{2x^2-7x+1}{(2x+1)(x-3)(x+4)} dx$

i. $\int \frac{4x^2-3x+2}{(x^2+3)(x-2)}dx$

j. $\int \frac{2x+3}{(x+3)^2}dx$

k. $\int \frac{5x+3}{x^2-8x+16}dx$

l. $\int \frac{x^2+5x-2}{(x+4)(x-2)}dx$

2. Find

a. $\int \frac{x+10}{(3x-1)(x+4)}dx$

b. $\int \frac{3}{4x^2+5x+1}dx$

c. $\int \frac{7}{8x^2-14x+3}dx$

d. $\int \frac{x-6}{x^3-x^2-9x+9}dx$

e. $\int \frac{2x+7}{x^3-2x^2-5x+6}dx$

f. $\int \frac{2x^2-x-21}{(x^2+5)(x+7)}dx$

g. $\int \frac{3x^3+2x^2-7x+2}{(x-2)(x+7)}dx$

h. $\int \frac{x^2-3}{(x+1)^2(x+2)}dx$

i. $\int \frac{5x+2}{x^3-6x^2+9x}dx$

j. $\int \frac{x^2+4x+7}{(x^2+5x+2)(x-3)}dx$

k. $\int \frac{2x^3+x+1}{(x^2+1)(x^2+4)}dx$

l. $\int \frac{3x^2-5x+4}{(x-1)(x^2+1)}dx$

3. Use an appropriate method to find each of the following indefinite integrals.

a. $\int \frac{2}{1+x}dx$

b. $\int \frac{x}{1+x}dx$

c. $\int \frac{5x}{1+x^2}dx$

d. $\int \frac{4}{1+x^2}dx$

e. $\int \frac{3x}{(1+x)^2}dx$

f. $\int \frac{x}{\sqrt{1+x}}dx$

g. $\int \frac{4}{(1-x)^2}dx$

h. $\int \frac{5}{1-x^2}dx$

i. $\int \frac{7x}{1-x^2}dx$

j. $\int \frac{3}{\sqrt{1-x^2}}dx$

k. $\int \frac{2x}{\sqrt{1-x^2}}dx$

l. $\int \frac{7}{\sqrt{1+x}}dx$

4. Find the following indefinite integrals.

a. $\int \frac{x^3}{(2x-3)(x+4)}dx$

b. $\int \frac{x^5+3}{x^2(x+2)^2}dx$

c. $\int \frac{x^2+2x+5}{(x-3)^2(x+1)}dx$

d. $\int \frac{x+2}{x(x+5)}dx$

e. $\int \frac{3x+1}{x^2-9}dx$

f. $\int \frac{5x-9}{(x-2)(x+7)}dx$

g. $\int \frac{15}{x^2+3x-10}dx$

h. $\int \frac{8}{(x-3)^2(x+1)}dx$

i. $\int \frac{x^2+14x+39}{(x+2)(x+3)(x+5)}dx$

j. $\int \frac{x^2+13x+8}{x^3+5x^2+7x+35}dx$

k. $\int \frac{5x^3+5x^2-x-2}{(x^2+x)^2}dx$

l. $\int \frac{3x^2+8x+6}{(x-1)^2(x-2)}dx$

Chapter 7 Definite integrals

7.1. Definition and calculation formula of the definite integral

Two examples of the concept of the definite integral

Example 1. The area under a curve

In our previous study, we have already learned how to calculate the areas of circles and polygons. However, we do not know how to find the area of a plane shape bounded by curves. Therefore, we need to study definite integrals to calculate the areas of such plane shapes.

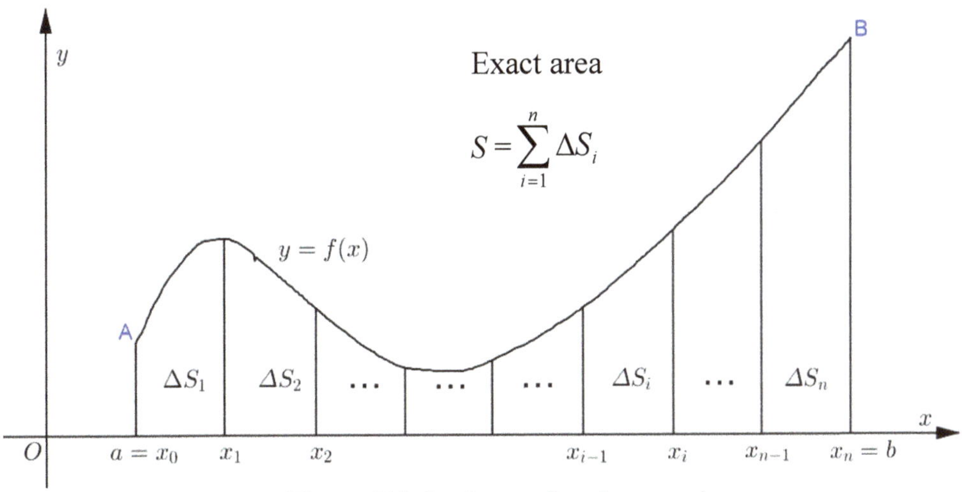

Figure 7.1.1 Area of a plane region

In Figure 7.1.1, the plane region A*ab*B is bounded by the curve $f(x)$, the lines $x = a, x = b$ and the *x*–axis. We need to calculate the area of the plane region A*ab*B.

As seen in Figure 7.1.1

(1) use points $a = x_0 < x_1 < x_2 < \ldots < x_{n-1} < x_n = b$ to divide the interval [*a*, *b*] into *n*

subintervals.

$$[x_0, x_1], [x_1, x_2], \ldots, [x_{n-1}, x_n]$$

The lengths of these subintervals are

$$\Delta x_1 = x_1 - x_0, \quad \Delta x_2 = x_2 - x_1, \ldots, \quad \Delta x_n = x_n - x_{n-1}$$

respectively. Through each point x_i ($i = 1, 2, \ldots, n-1$), draw a line perpendicular to the x-axis, dividing the plane region AabB into several small regions. Let S denote the exact area of plane region AabB, and let ΔS_i denote the exact area of the ith small region. Then we have

$$S = \Delta S_1 + \Delta S_2 + \ldots + \Delta S_n = \sum_{i=1}^{n} \Delta S_i$$

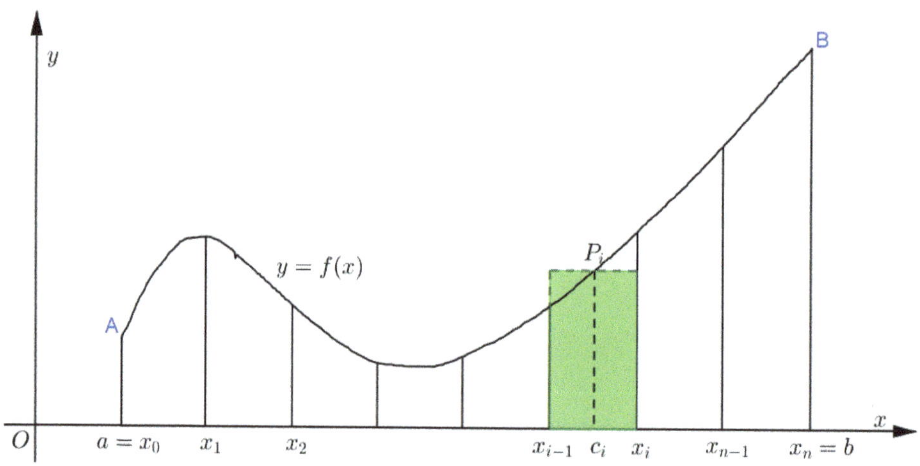

Figure 7.1.2 Area of a plane region

As seen in Figure 7.1.2

(2) randomly choose a point, c_i ($x_{i-1} \le c_i \le x_i$), from each subinterval $[x_{i-1}, x_i]$ ($i = 1, 2, \ldots, n$). Draw a line that passes through c_i, perpendicular to the x-axis and intersects the curve at $P_i(c_i, f(c_i))$. Use Δx_i as the base and $f(c_i)$ as the height to construct a rectangle. Use the area of this rectangle, $f(c_i)\Delta x_i$, to be the approximate value of the exact value, ΔS_i; that is,

$$\Delta S_i \approx f(c_i)\Delta x_i \qquad (i = 1, 2, \ldots, n)$$

The sum of approximate values is

$$S_n = f(c_1)\Delta x_1 + f(c_2)\Delta x_2 + \ldots + f(c_n)\Delta x_n = \sum_{i=1}^{n} f(c_i)\Delta x_i$$

Then S_n is an approximate value of the exact value, S.

(3) Use $\Delta x = \max\{\Delta x_i\}$ to represent the length of the longest subinterval among all subintervals. As the number of points $n \to \infty$, $\Delta x \to 0$. The limit of the sum of these approximate values, S_n, is defined as the exact area, S, of region AabB. That is,

$$S = \sum_{i=1}^{n} \Delta S_i = \lim_{\substack{n\to\infty \\ \Delta x\to 0}} S_n = \lim_{\substack{n\to\infty \\ \Delta x\to 0}} \sum_{i=1}^{n} f(c_i)\Delta x_i$$

Therefore, $\quad S = \lim_{\substack{n\to\infty \\ \Delta x\to 0}} \sum_{i=1}^{n} f(c_i)\Delta x_i$

Example 2. Distance of motion in a straight line with varying speed

When an object moves in a straight line with constant speed (V = constant), the distance it travels is equal to the product of speed and time. That is, $S = Vt$. If the speed, V, of the object changes with time, then V is a function of time t, $V = V(t)$. Speed V is a variable. Calculate the distance, S, travelled by the object over the time interval $[a, b]$.

(1) Use points

$$a = t_0 < t_1 < t_2 < \ldots < t_{n-1} < t_n = b$$

to divide the time interval $[a, b]$ into n subintervals.

$$[t_0, t_1], [t_1, t_2], \ldots, [t_{n-1}, t_n]$$

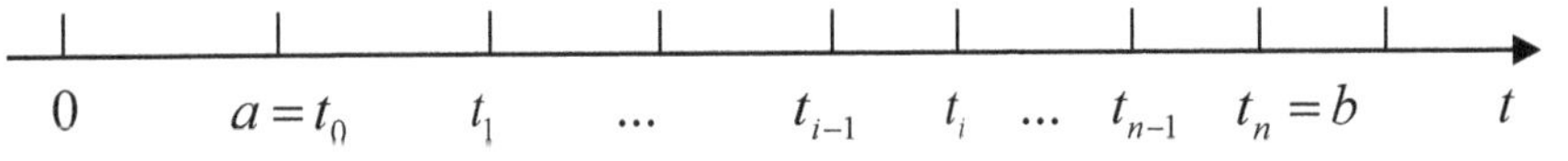

The length of each subinterval is respectively

$$\Delta t_1 = t_1 - t_0 \qquad \Delta t_2 = t_2 - t_1 \quad \dots \quad \Delta t_n = t_n - t_{n-1}$$

(2) Randomly choose a point, c_i $(t_{i-1} \le c_i \le t_i)$, from each subinterval $[t_{i-1}, t_i]$ ($i = 1, 2, \dots, n$). Use $V(c_i)\Delta t_i$ as the approximate value of the exact distance, ΔS_i, travelled by the object over the subinterval $[t_{i-1}, t_i]$; that is, $\Delta S_i \approx V(c_i)\Delta t_i$ ($i = 1, 2, \dots, n$).

On the time interval $[a, b]$, the exact distance travelled by the object is S, and its approximate value is S_n.

$$S_n = V(c_1)\Delta t_1 + V(c_2)\Delta t_2 + V(c_3)\Delta t_3 + \dots + V(c_n)\Delta t_n = \sum_{i=1}^{n} V(c_i)\Delta t_i$$

(3) As the number of points $n \to \infty$, $\Delta t \to 0$ ($\Delta t = \{\max\{\Delta t_i\}\}$), the limit of the sum of these approximate values, S_n, is the exact distance, S, that the object travels with varying speed $V(t)$ from moment a to moment b. That is,

$$S = \sum_{i=1}^{n} \Delta S_i = \lim_{\substack{n\to\infty \\ \Delta t\to 0}} S_n = \lim_{\substack{n\to\infty \\ \Delta t\to 0}} \sum_{i=1}^{n} V(c_i)\Delta t_i$$

Therefore, $$S = \lim_{\substack{n\to\infty \\ \Delta t\to 0}} \sum_{i=1}^{n} V(c_i)\Delta t_i$$

From the two examples above, we can observe that although the problems are different, solution methods are the same – finding the limit of the sum of the approximate values of a similar structure. In practice, solving many problems requires finding such limits. Therefore, we need to study it in abstract form. Hence, we introduce the concept of definite integral in mathematics.

Definition of the definite integral

The function $f(x)$ is defined on the interval $[a, b]$. Use points $a = x_0 < x_1 < x_2 < \ldots < x_{n-1} < x_n = b$ to divide the interval $[a, b]$ into n subintervals $[x_{i-1}, x_i]$ $(i = 1, 2, \ldots, n)$, and its length is $\Delta x_i = x_i - x_{i-1}$. Randomly choose a point, $c_i\,(x_{i-1} \le c_i \le x_i)$, from each subinterval $[x_{i-1}, x_i]$. Then the product $f(c_i)\Delta x_i$ $(i = 1, 2, \ldots, n)$ is called the **integration element**. The sum of approximate values, $S_n = \sum_{i=1}^{n} f(c_i)\Delta x_i$, is called the **integral sum**. If as $n \to \infty$, we have $\Delta x \to 0$ ($\Delta x = \{\max\{\Delta x_i\}\}$), the limit of the sum of approximate values, S_n, exists, and this limit does not depend on the division method of $[a, b]$ or the choice of c_i, then we say that the function $f(x)$ is integrable. We call this limit the **definite integral** of the function $f(x)$ on the interval $[a, b]$, denoted by

$$\int_a^b f(x)dx$$

Hence, we have $\int_a^b f(x)dx = \lim\limits_{\substack{n\to\infty \\ \Delta x\to 0}} S_n = \lim\limits_{\substack{n\to\infty \\ \Delta x\to 0}} \sum_{i=1}^{n} f(c_i)\Delta x_i = \sum_{i=1}^{\infty} f(x_i) \lim\limits_{\Delta x_i\to 0} \Delta x_i$

That is,

$$\int_a^b f(x)dx = \lim_{\substack{n\to\infty \\ \Delta x\to 0}} \sum_{i=1}^{n} f(c_i)\Delta x_i = \sum_{i=1}^{\infty} f(x_i) \lim_{\Delta x_i\to 0} \Delta x_i$$

where $f(x_i) \lim\limits_{\Delta x_i\to 0} \Delta x_i$ is the area of the line segment. The exact area value, S, is the algebraic sum of areas of infinitely many line segments.

$f(x)$	is called the **integrand**;
$f(x)dx$	is called the **integrand expression**;
x	is called the **variable of integration**;
$[a, b]$	is called the **interval of integration**;
a	is called the **lower limit of integration**;
b	is called the **upper limit of integration**.

Based on the definition of the definite integral, the two examples at the beginning of this chapter can be expressed as follows:

1. The area of the plane region, S, is the definite integral of the function $y = f(x)$ on the interval $[a, b]$.

That is, the exact area is $$S = \int_a^b f(x)dx \qquad (f(x) \geq 0)$$

2. The distance, S, that an object travels in a straight line with varying speed, V, is the definite integral of the speed function, $V = V(t)$, over the time interval $[a, b]$.

That is, the exact distance is $$S = \int_a^b V(t)dt$$

Calculation formula of the definite integral (Newton-Leibniz formula)

Fundamental Theorem of Calculus

If the function $f(x)$ is continuous on the interval $[a, b]$ and $F(x)$ is an antiderivative of $f(x)$, then the calculation formula of the definite integral is

$$\int_a^b f(x)dx = [F(x)]_a^b = F(b) - F(a)$$

Proof. Given that $F(x)$ is an antiderivative of $f(x)$, by Lagrange's Mean Value Theorem, we have

$$f(c) = F'(c) = \frac{\Delta F(x)}{\Delta x}$$

That is, $$f(c)\,\Delta x = \Delta F(x)$$

Since $$\int_a^b f(x)dx = \lim_{\substack{n\to\infty \\ \Delta x\to 0}} \sum_{i=1}^n f(c_i)\Delta x_i = \lim_{\substack{n\to\infty \\ \Delta x\to 0}} \sum_{i=1}^n \Delta F(x_i)$$

$$= \lim_{\substack{n\to\infty \\ \Delta x\to 0}} \{[F(x_1) - F(a)] + [F(x_2) - F(x_1)] + [F(x_3) - F(x_2)] + ... + [F(b) - F(x_{n-1})]\}$$

$$= \lim_{\substack{n\to\infty \\ \Delta x\to 0}} [F(b) - F(a)] = F(b) - F(a)$$

thus $$\int_a^b f(x)dx = F(b) - F(a)$$

Generally, we use $[F(x)]_a^b$ to denote $F(b) - F(a)$.

That is, the calculation formula of the definite integral is

$$\int_a^b f(x)dx = [F(x)]_a^b = F(b) - F(a)$$ □

When using the calculation formula for the definite integral, we first find an antiderivative $F(x)$ of $f(x)$, and then calculate the difference $F(b) - F(a)$.

Note:

Since the length of integration interval is $\Delta x = b - a$, corresponding to Δx, the integration result P must be $P = \Delta F(x)$, which is also a difference between the two terms.

Example 3. $\int_0^1 xdx = \left[\frac{x^2}{2}\right]_0^1 = \frac{1}{2} - 0 = \frac{1}{2}$

Example 4. $\int_3^6 \frac{1}{x}dx = [\ln x]_3^6 = \ln 6 - \ln 3 = \ln 2$

Basic properties of definite integrals

Basic properties	proof or example
1. $\int_a^a f(x)dx = 0$	Since $\int_a^a f(x)dx = F(a) - F(a) = 0$
2. $\int_a^b f(x)dx = -\int_b^a f(x)dx$	Since $\int_a^b f(x)dx = F(b) - F(a)$ $= -[F(a) - F(b)] = -\int_b^a f(x)dx$
3. $\int_a^b kf(x)dx = k\int_a^b f(x)dx$ k is a constant.	$\int_1^2 2xdx = 2\int_1^2 xdx = 2\left[\frac{x^2}{2}\right]_1^2 = 2(2-\frac{1}{2}) = 3$
4. $\int_a^b [f(x) + g(x)]dx$ $= \int_a^b f(x)dx + \int_a^b g(x)dx$	$\int_0^1 (x + x^2)dx = \int_0^1 xdx + \int_0^1 x^2dx$ $= \left[\frac{x^2}{2}\right]_0^1 + \left[\frac{x^3}{3}\right]_0^1 = \frac{5}{6}$
5. $\int_a^b f(x)dx = \int_a^c f(x)dx + \int_c^b f(x)dx$	$\int_2^6 e^x dx = \left[e^x\right]_2^6 = e^6 - e^2$ $\int_2^6 e^x dx = \int_2^4 e^x dx + \int_4^6 e^x dx = \left[e^x\right]_2^4 + \left[e^x\right]_4^6$ $= (e^4 - e^2) + (e^6 - e^4) = e^6 - e^2$
6. $\int_a^b f(\bullet)d(\bullet) = F(b) - F(a)$	The integration result $P = F(b) - F(a)$ is only related to the integrand $f(\bullet)$ and the integration interval $[a, b]$. It has nothing to do with the letter used to represent the variable of integration. That is, $\int_a^b f(x)dx = \int_a^b f(t)dt \quad (x = t)$

Example 5. Find $\int_{-1}^{3}|2-x|dx$.

Solution. Since the integrand contains an absolute value function, to remove the absolute value sign, we must determine on which interval the function is positive and on which interval it is negative. $|2-x|=\begin{cases}2-x, & x\le 2\\ x-2, & x>2\end{cases}$

Hence, $\int_{-1}^{3}|2-x|dx=\int_{-1}^{2}(2-x)dx+\int_{2}^{3}(x-2)dx=\left[2x-\frac{x^2}{2}\right]_{-1}^{2}+\left[\frac{x^2}{2}-2x\right]_{2}^{3}$

$$=\left[\left(2\times 2-\frac{2^2}{2}\right)-\left(2\times(-1)-\frac{(-1)^2}{2}\right)\right]+\left[\left(\frac{3^2}{2}-2\times 3\right)-\left(\frac{2^2}{2}-2\times 2\right)\right]=5$$

Example 6. $f(x)=\begin{cases}2x+1, & |x|\le 2\\ 1+x^2, & 2<x\le 4\end{cases}$. Find the value of *k* such that $\int_{k}^{3}f(x)dx=\frac{40}{3}$.

Solution. $\int_{k}^{3}f(x)dx=\int_{k}^{2}(2x+1)dx+\int_{2}^{3}\left(1+x^2\right)dx=\left[x^2+x\right]_{k}^{2}+\left[x+\frac{x^3}{3}\right]_{2}^{3}$

$$=\left(2^2+2\right)-\left(k^2+k\right)+\left(3+\frac{3^3}{3}\right)-\left(2+\frac{2^3}{3}\right)=\frac{40}{3}-\left(k^2+k\right)$$

Hence, $\frac{40}{3}-\left(k^2+k\right)=\frac{40}{3}$

$$k^2+k=0$$

$$k(k+1)=0$$

$$k=0 \quad \text{or} \quad k=-1$$

Example 7. Find $\int_{-1}^{1}e^{|x|}dx$.

Solution. $\int_{-1}^{1} e^{|x|}dx = \int_{-1}^{0} e^{-x}dx + \int_{0}^{1} e^{x}dx = \left[-e^{-x}\right]_{-1}^{0} + \left[e^{x}\right]_{0}^{1} = \left[-1-(-e)\right]+(e-1)$

$$= 2(e-1)$$

Example 8. Find $\int_{-1}^{3} \left|3x^2-6x\right|dx$.

Solution.

$3x^2-6x \geq 0$	$3x^2-6x<0$
$x(x-2)\geq 0$	$x(x-2)<0$
$x\geq 2$ or $x\leq 0$	$0<x<2$

Therefore, removing absolute value sign gives

$$\int_{-1}^{3}\left|3x^2-6x\right|dx = \int_{-1}^{0}\left(3x^2-6x\right)dx + \int_{0}^{2}\left(6x-3x^2\right)dx + \int_{2}^{3}\left(3x^2-6x\right)dx$$

$$= \left[x^3-3x^2\right]_{-1}^{0} + \left[3x^2-x^3\right]_{0}^{2} + \left[x^3-3x^2\right]_{2}^{3}$$

$$= 0-(-1-3)+(12-8)-0+(27-27)-(8-12) = 12$$

Exercise 7.1

1. Evaluate the following.

a. $\int_{1}^{3}(2x+\frac{3}{x}+\sqrt{x})dx$ **b.** $\int_{-1}^{1}\frac{3}{3x^2+4}dx$

c. $\int_{3}^{4}\frac{2x+1}{2x^2+x}dx$ **d.** $\int_{0}^{2\pi}\left|\cos x\right|dx$

e. $\int_{-3}^{3}\left|5x-10\right|dx$ **f.** $\int_{-2}^{4}\left|3-x^2\right|dx$

g. $\int_{0}^{3} |x^2 - 3x + 2| dx$

h. $\int_{-\frac{\pi}{2}}^{\frac{3\pi}{2}} |\sin x| dx$

2. Given that $\int_{-2}^{2} 2f(x)dx = 16$, $\int_{-2}^{4} f(x)dx = 5$, $\int_{-2}^{4} g(x)dx = 6$, find the values of the following.

a. $\int_{-2}^{2} f(x)dx$

b. $\int_{2}^{4} f(x)dx$

c. $\int_{4}^{-2} g(x)dx$

d. $\int_{-2}^{4} \frac{2}{3}[4f(x) - 2g(x)] dx$

3. Given that $\int_{0}^{k} (2x - 3x^2) dx = 0$, find the value of k.

4. Given that $\int_{0}^{1} (2x + k) dx = 2$, find the value of k.

7.2. Integration by substitution and integration by parts of definite integrals

<u>Integration by substitution of definite integrals</u>

Now, we need to use integration by substitution to find $\int_{a}^{b} f(x)dx$.

Let the intermediate variable u replace the original variable x.

Let $u = u(x)$ then $du = u'(x)dx$

When $x = a$, $u_1 = u(a)$

When $x = b$, $u_2 = u(b)$

Substituting into the definite integral $\int_a^b f(x)dx$ gives

$$\int_a^b f(x)dx = \int_{u_1}^{u_2} f(u)du = [F(u)]_{u_1}^{u_2} = F(u_2) - F(u_1)$$

Alternatively, change u back to the original variable x. That is,

$$[F(u)]_{u_1}^{u_2} = [F(x)]_a^b = F(b) - F(a)$$

Example 1. Find $\int_2^4 \frac{1}{x+1}dx$.

Solution. Let $u = x + 1$ then $du = dx$

When $x = 2$, $u_1 = 2 + 1 = 3$

When $x = 4$, $u_2 = 4 + 1 = 5$

Hence, $\int_2^4 \frac{1}{x+1}dx = \int_3^5 \frac{1}{u}du = [\ln u]_3^5 = \ln 5 - \ln 3$

Alternatively, $\int_2^4 \frac{1}{x+1}dx = \int_{u_1}^{u_2} \frac{1}{u}du = [\ln u]_{u_1}^{u_2} = [\ln(x+1)]_2^4 = \ln 5 - \ln 3$

Example 2. Find $\int_1^{\frac{5}{4}} \frac{1}{(3-4x)^3}dx$.

Solution. Let $u = 3 - 4x$ then $du = -4dx$, $dx = \frac{du}{-4}$

When $x = 1$, $u_1 = 3 - 4 = -1$

When $x = \frac{5}{4}$, $u_2 = 3 - 4 \cdot \frac{5}{4} = -2$

Hence, $\int_{1}^{\frac{5}{4}} \frac{1}{(3-4x)^3}dx = \int_{-1}^{-2} u^{-3} \cdot \frac{du}{-4} = \left[\frac{1}{8u^2}\right]_{-1}^{-2} = \frac{1}{8(-2)^2} - \frac{1}{8(-1)^2} = -\frac{3}{32}$

Example 3. Find $\int_{0}^{3} \sqrt{1+x}dx$.

Solution. Let $u = 1 + x$ then $du = dx$

When $x = 0$, $u_1 = 1 + 0 = 1$

When $x = 3$, $u_2 = 1 + 3 = 4$

Hence, $\int_{0}^{3} \sqrt{1+x}dx = \int_{1}^{4} u^{\frac{1}{2}}du = \left[\frac{2}{3}u^{\frac{3}{2}}\right]_{1}^{4} = \frac{2}{3}\left(4^{\frac{3}{2}} - 1^{\frac{3}{2}}\right) = \frac{14}{3}$

Example 4. Find $\int_{0}^{\ln 3} e^x\sqrt{1+e^x}dx$.

Solution. Let $u = 1 + e^x$ then $du = e^x dx$

When $x = 0$, $u_1 = 1 + e^0 = 2$

When $x = \ln 3$, $u_2 = 1 + e^{\ln 3} = 4$

$$\int_{0}^{\ln 3} e^x\sqrt{1+e^x}dx = \int_{2}^{4} e^x \cdot u^{\frac{1}{2}} \cdot \frac{du}{e^x} = \left[\frac{2}{3}u^{\frac{3}{2}}\right]_{2}^{4} = \frac{2}{3}\left(4^{\frac{3}{2}} - 2^{\frac{3}{2}}\right) = \frac{2}{3}\left(8-2\sqrt{2}\right) = \frac{4}{3}\left(4-\sqrt{2}\right)$$

Example 5. By using the substitution $x = 2\cos\theta$, prove that $\int_{1}^{2} \frac{1}{x^2\sqrt{4-x^2}}dx = \frac{\sqrt{3}}{4}$.

Solution. Let $x = 2\cos\theta$ then $dx = -2\sin\theta d\theta$

When $x = 1$, $\theta_1 = \arccos\frac{x}{2} = \arccos\frac{1}{2} = \frac{\pi}{3}$

When $x = 2$, $\theta_2 = \arccos\frac{2}{2} = 0$

Therefore, $\int_1^2 \frac{1}{x^2\sqrt{4-x^2}}dx = \int_{\frac{\pi}{3}}^{0} \frac{1}{4\cos^2\theta\sqrt{4-4\cos^2\theta}}(-2\sin\theta)d\theta$

$$= -\frac{1}{4}\int_{\frac{\pi}{3}}^{0} \frac{1}{\cos^2\theta 2\sin\theta}\cdot 2\sin\theta d\theta$$

$$= -\frac{1}{4}\int_{\frac{\pi}{3}}^{0} \sec^2\theta d\theta = -\frac{1}{4}\left[\tan\theta\right]_{\frac{\pi}{3}}^{0} = -\frac{1}{4}\left(\tan 0 - \tan\frac{\pi}{3}\right)$$

$$= -\frac{1}{4}\left(0-\sqrt{3}\right) = \frac{\sqrt{3}}{4}$$

Example 6. By using the substitution $x = a\sec\theta$, find $\int_{\frac{2}{\sqrt{3}}a}^{\sqrt{2}a} \frac{\sqrt{x^2-a^2}}{x}dx$.

Solution. Let $x = a\sec\theta$ then $dx = a\sec\theta\tan\theta d\theta$

When $x = \frac{2}{\sqrt{3}}a$, $a\sec\theta_1 = \frac{2}{\sqrt{3}}a$

$$\cos\theta_1 = \frac{\sqrt{3}}{2} \quad \Rightarrow \quad \theta_1 = \frac{\pi}{6}$$

When $x = \sqrt{2}a$, $a\sec\theta_2 = \sqrt{2}a$

$$\cos\theta_2 = \frac{1}{\sqrt{2}} \quad \Rightarrow \quad \theta_2 = \frac{\pi}{4}$$

Hence, $\int_{\frac{2}{\sqrt{3}}a}^{\sqrt{2}a} \frac{\sqrt{x^2-a^2}}{x}dx = \int_{\frac{\pi}{6}}^{\frac{\pi}{4}} \frac{\sqrt{a^2\sec^2\theta - a^2}}{a\sec\theta}\cdot a\sec\theta\tan\theta d\theta = a\int_{\frac{\pi}{6}}^{\frac{\pi}{4}} \tan^2\theta d\theta$

$$= a\int_{\frac{\pi}{6}}^{\frac{\pi}{4}} (\sec^2\theta - 1)\,d\theta = a[\tan\theta - \theta]_{\frac{\pi}{6}}^{\frac{\pi}{4}}$$

$$= a\left[\left(\tan\frac{\pi}{4} - \frac{\pi}{4}\right) - \left(\tan\frac{\pi}{6} - \frac{\pi}{6}\right)\right]$$

$$= a(1 - \frac{\sqrt{3}}{3} - \frac{\pi}{12})$$

Integration by parts of definite integrals

Integration by parts of definite integrals is similar to that of indefinite integrals.

Example 7. Find $\int_0^1 xe^x dx$.

Solution. Let $u = x$ $\quad dv = e^x dx$

then $du = dx$ $\quad v = \int e^x dx = e^x$

Therefore, $\int_0^1 xe^x dx = \left[xe^x\right]_0^1 - \int_0^1 e^x dx = \left[xe^x\right]_0^1 - \left[e^x\right]_0^1 = (e-0)-(e-1) = 1$

Example 8. It is given that, for non-negative integers n,

$$I_n = \int_0^{\frac{\pi}{2}} x^n \sin x dx.$$

(a) Show that, for $n \geq 2$,

$$I_n = n\left(\frac{\pi}{2}\right)^{n-1} - n(n-1)I_{n-2}$$

(b) Find I_4 in terms of π.

Solution. (a) $I_n = \int_0^{\frac{\pi}{2}} x^n \sin x dx$

Let $u = x^n$ $dv = \sin x dx$

then $du = nx^{n-1}dx$ $v = \int \sin x dx = -\cos x$

Therefore, $I_n = \int_0^{\frac{\pi}{2}} x^n \sin x dx = \left[-x^n \cos x\right]_0^{\frac{\pi}{2}} + n\int_0^{\frac{\pi}{2}} x^{n-1} \cos x dx = n\int_0^{\frac{\pi}{2}} x^{n-1} \cos x dx$

For $\int_0^{\frac{\pi}{2}} x^{n-1} \cos x dx$

let $u = x^{n-1}$ $dv = \cos x dx$

then $du = (n-1)x^{n-2}dx$ $v = \int \cos x dx = \sin x$

$$\int_0^{\frac{\pi}{2}} x^{n-1} \cos x dx = \left[x^{n-1} \sin x\right]_0^{\frac{\pi}{2}} - (n-1)\int_0^{\frac{\pi}{2}} x^{n-2} \sin x dx$$

$$= \left(\frac{\pi}{2}\right)^{n-1} - (n-1)\int_0^{\frac{\pi}{2}} x^{n-2} \sin x dx$$

Hence, $I_n = \int_0^{\frac{\pi}{2}} x^n \sin x dx = n\int_0^{\frac{\pi}{2}} x^{n-1} \cos x dx = n\left[\left(\frac{\pi}{2}\right)^{n-1} - (n-1)\int_0^{\frac{\pi}{2}} x^{n-2} \sin x dx\right]$

$$= n\left(\frac{\pi}{2}\right)^{n-1} - n(n-1)\int_0^{\frac{\pi}{2}} x^{n-2} \sin x dx = n\left(\frac{\pi}{2}\right)^{n-1} - n(n-1)I_{n-2}$$

Therefore, $I_n = n\left(\frac{\pi}{2}\right)^{n-1} - n(n-1)I_{n-2}$

(b) Since $I_n = n\left(\frac{\pi}{2}\right)^{n-1} - n(n-1)I_{n-2}$

thus $$I_4 = 4\left(\frac{\pi}{2}\right)^{4-1} - 4(4-1)I_{4-2} = 4\left(\frac{\pi}{2}\right)^3 - 12I_2$$

$$I_2 = 2\left(\frac{\pi}{2}\right)^{2-1} - 2(2-1)I_0 = \pi - 2I_0$$

$$I_0 = \int_0^{\frac{\pi}{2}} \sin x dx = [-\cos x]_0^{\frac{\pi}{2}} = 1$$

Therefore, $$I_4 = 4\left(\frac{\pi}{2}\right)^3 - 12(\pi - 2) = \frac{\pi^3}{2} - 12(\pi - 2)$$

Example 9. It is given that, for non-negative integers n,

$$I_n = \int_0^{\frac{\pi}{2}} \cos^n x dx \qquad n \geq 2$$

(a) Show that for $n \geq 2$,

$$I_n = \frac{n-1}{n} I_{n-2}.$$

(b) Show that $I_{2n+1} < I_{2n-1}$.

(c) Find I_{10} in terms of π.

(d) It is given that $I_{2n+1} < I_{2n} < I_{2n-1}$. By taking $n = 5$, determine the interval within which the value of π lies. Give the answers correct to seven decimal places.

Solution.

(a) $$I_n = \int_0^{\frac{\pi}{2}} \cos^n x dx = \int_0^{\frac{\pi}{2}} \cos^{n-1} x \cos x dx$$

Let $u = \cos^{n-1} x$ $\qquad dv = \cos x dx$

then $du = (n-1)\cos^{n-2} x(-\sin x)dx$ $\qquad v = \int \cos x dx = \sin x$

Hence, $I_n = \int_0^{\frac{\pi}{2}} \cos^{n-1}x \cos x dx = \left[\sin x \cos^{n-1} x\right]_0^{\frac{\pi}{2}} + (n-1)\int_0^{\frac{\pi}{2}} \sin^2 x \cos^{n-2} x dx$

$$= (n-1)\int_0^{\frac{\pi}{2}} (1-\cos^2 x)\cos^{n-2} x dx$$

$$= (n-1)\int_0^{\frac{\pi}{2}} \cos^{n-2} x dx - (n-1)\int_0^{\frac{\pi}{2}} \cos^n x dx$$

$$= (n-1)I_{n-2} - (n-1)I_n$$

Therefore, $I_n = \dfrac{n-1}{n} I_{n-2}$

(b) Since $I_n = \dfrac{n-1}{n} I_{n-2}$

thus $I_{2n+1} = \dfrac{2n+1-1}{2n+1} I_{2n+1-2} = \dfrac{2n}{2n+1} I_{2n-1}$

Since $\dfrac{2n}{2n+1} < 1$

thus $I_{2n+1} < I_{2n-1}$

(c) Since $I_n = \dfrac{n-1}{n} I_{n-2}$

thus $I_{10} = \dfrac{9}{10} I_8$, $I_8 = \dfrac{7}{8} I_6$, $I_6 = \dfrac{5}{6} I_4$, $I_4 = \dfrac{3}{4} I_2$, $I_2 = \dfrac{1}{2} I_0$

$$I_0 = \int_0^{\frac{\pi}{2}} \cos^0 x dx = \frac{\pi}{2}$$

Therefore, $I_{10} = \dfrac{9}{10} \cdot \dfrac{7}{8} \cdot \dfrac{5}{6} \cdot \dfrac{3}{4} \cdot \dfrac{1}{2} \cdot \dfrac{\pi}{2} = \dfrac{63\pi}{512}$

(d) By taking $n = 5$ and using a calculator, the following values are obtained

$$I_{11} = \int_0^{\frac{\pi}{2}} \cos^{11} x dx \approx 0.3694084$$

$$I_9 = \int_0^{\frac{\pi}{2}} \cos^9 x dx = \frac{128}{315} \approx 0.4063492$$

Using $I_{2n+1} < I_{2n} < I_{2n-1}$ gives

$$0.3694084 < \frac{63\pi}{512} < \frac{128}{315}$$

Rearranging gives

$$3.0021762 < \pi < 3.30239355$$

Exercise 7.2

1. Evaluate the following definite integrals, giving your answers in exact values.

a. $\int_2^4 \frac{2}{(5-3x)^3} dx$

b. $\int_2^4 \frac{(\ln x)^2}{x} dx$

c. $\int_0^1 x(2-x^2)^4 dx$

d. $\int_4^8 \frac{x}{\sqrt{x-3}} dx$

e. $\int_0^2 x\sqrt{10+x^2}\, dx$

f. $\int_{\frac{1}{2}}^1 \frac{x+1}{(x^2+2x+3)^2} dx$

g. $\int_2^3 \frac{x+1}{2x^2+4x+5} dx$

h. $\int_0^{\frac{\pi}{3}} \sin^3 x \cos x dx$

i. $\int_{\frac{\pi}{4}}^{\frac{3\pi}{2}} x^3 \sin(x^4+3) dx$

j. $\int_9^{16} \frac{3}{\sqrt{x}-2} dx$

k. $\displaystyle\int_{\frac{\pi}{4}}^{\frac{\pi}{3}} (1+\tan(3x))^3 \sec^2(3x)dx$ **l.** $\displaystyle\int_{-1}^{1} (e^{2x}+e^{-2x})^2 dx$

m. $\displaystyle\int_{-2}^{2} e^{3x}\cos(e^{3x})dx$ **n.** $\displaystyle\int_{2}^{3} \frac{1}{9+4x^2}dx$

2. Evaluate the following definite integrals. Give your answers in exact values.

a. $\displaystyle\int_{1}^{2} x(\ln x)^2 dx$ **b.** $\displaystyle\int_{0}^{\ln 2} xe^{-x} dx$

c. $\displaystyle\int_{0}^{\pi} x\sin x dx$ **d.** $\displaystyle\int_{0}^{2\pi} x^2 \cos x dx$

e. $\displaystyle\int_{0}^{1} \arccos x dx$ **f.** $\displaystyle\int_{0}^{\sqrt{3}} x\arctan x dx$

g. $\displaystyle\int_{0}^{\frac{\pi}{2}} e^{2x}\cos x dx$ **h.** $\displaystyle\int_{1}^{4} \frac{\ln x}{\sqrt{x}}dx$

i. $\displaystyle\int_{0}^{\ln 3} \frac{e^x}{1+e^x}dx$ **j.** $\displaystyle\int_{1}^{e} \frac{1}{x}\cos(\ln x)dx$

k. $\displaystyle\int_{3}^{4} \frac{2}{x^2+x-2}dx$ **l.** $\displaystyle\int_{0}^{1} xe^{2x} dx$

m. $\displaystyle\int_{0}^{\frac{\pi}{4}} \sin^3 x dx$ **n.** $\displaystyle\int_{0}^{\frac{\pi}{2}} e^x \sin^2 x dx$

3. Use $x = a\sin\theta$, find the exact value of $\displaystyle\int_{0}^{a} x^2\sqrt{a^2-x^2}\,dx$.

4. Use $x = 2\sin\theta$, find the exact value of $\displaystyle\int_{1}^{\sqrt{2}} \frac{x^2}{(4-x^2)^{\frac{3}{2}}}dx$.

5. Use $x = 1 + \tan\theta$, find the exact value of $\int_0^2 \frac{xdx}{(x^2 - 2x + 2)^2}$.

6. Use $x = e^\theta$, find the exact value of $\int_1^e \sin(\ln x)dx$.

7. Use the substitution $x = a(1 + \sin\theta)$ to evaluate $\int_a^{2a} \sqrt{2ax - x^2}\,dx$.

8. Use the substitution $x = \sin^2\theta$ to calculate $\int_0^{\frac{1}{2}} \sqrt{\frac{x}{1-x}}\,dx$.

9. Use the substitution $x = \sec\theta$ to find the exact value of $\int_{\sqrt{2}}^{2} (x^2 - 1)^{-\frac{3}{2}} dx$.

10. Use the substitution $x = \sin\theta$, find the exact value of $\int_{\frac{1}{2}}^{\frac{\sqrt{3}}{2}} \sqrt{9 - 9x^2}\,dx$.

11. It is given that, for non-negative integers n,

$$I_n = \int_0^{\frac{\pi}{4}} \tan^{2n} x dx$$

a. Show that for $n \geq 2$,

$$I_n = \frac{1}{2n-1} - I_{n-1}.$$

b. Find the value of I_5 in terms of π.

12. It is given that, for non-negative integers n,

$$I_n = \int_0^{\frac{\pi}{2}} \sin^n x dx$$

a. Show that for $n \geq 2$,

$$I_n = \frac{n-1}{n} I_{n-2}.$$

b. Show that $I_{2n+1} < I_{2n-1}$.

c. It is given that

$$I_{2n+1} < I_{2n} < I_{2n-1}$$

By taking $n = 5$, find the interval within which the value of π lies. Give your answers correct to eight decimal places.

13. It is given for non-negative integers n,

$$I_n = \int_0^1 x^n e^{-x} dx$$

a. Show that for $n \geq 1$

$I_n - nI_{n-1} = p$ for $n \geq 1$, where p is a constant to be determined.

b. Find the exact value of I_5.

c. Find the exact value of $I_9 - 504I_6$.

7.3. Mean value theorem for definite integrals and formula for finding the derivative of a definite integral

<u>**Mean Value Theorem for definite integrals**</u>

Theorem. If the function $f(x)$ is continuous on the closed interval $[a, b]$, then on the open interval (a, b), there exists at least one point c such that

$$\int_a^b f(x)dx = f(c)(b-a)$$

This equality is known as the **Mean Value Theorem for Definite Integrals**.

Proof. By Lagrange's Mean Value Theorem, we have

$$\int_a^b f(x)dx = F(b) - F(a) = F'(c)(b-a)$$

where $\qquad F'(c) = f(c)$.

That is, the Mean Value Theorem for definite integrals is

$$\int_a^b f(x)dx = f(c)(b-a) \qquad c \in (a, b) \qquad \square$$

Moreover, $\frac{1}{b-a}\int_a^b f(x)dx$ is called the average value of the function $f(x)$ on the interval $[a, b]$. That is, the formula of the average value of a function is

$$\overline{f(x)} = \frac{1}{b-a}\int_a^b f(x)dx$$

Proof. It can be seen from Figure 7.1.1 that the limit of the length of the subinterval $\Delta x_i \to 0$ is

$$\lim_{\Delta x_i \to 0} \Delta x_i = \lim_{n\to\infty} \frac{b-a}{n} \qquad (i = 1, 2, \ldots, n)$$

hence, $\qquad \frac{1}{b-a}\int_a^b f(x)dx = \frac{1}{b-a}\sum_{i=1}^{\infty} f(x_i) \lim_{\Delta x_i \to 0} \Delta x_i = \frac{1}{b-a}\sum_{i=1}^{\infty} f(x_i) \cdot \lim_{n\to\infty} \frac{b-a}{n}$

$$= \lim_{n\to\infty} \frac{1}{n} \sum_{i=1}^{n} f(x_i) = \overline{f(x)}$$

Therefore, $$f(c) = \overline{f(x)} = \frac{1}{b-a} \int_a^b f(x)dx = \frac{\int_a^b f(x)dx}{\int_a^b dx}$$ □

Further average formulae are $$\overline{f(x)} = \frac{\int_a^b uf(x)dx}{\int_a^b udx} \qquad \bar{u} = \frac{\int_a^b uf(x)dx}{\int_a^b f(x)dx}$$

The average value of discrete values is $$\bar{x} = \frac{\sum_{i=1}^{n} a_i x_i}{\sum_{i=1}^{n} a_i}$$

When $a_i = 1, i = 1, 2, \ldots, n$, then $\sum_{i=1}^{n} a_i = n$.

$$\bar{x} = \frac{\sum_{i=1}^{n} a_i x_i}{\sum_{i=1}^{n} a_i} = \frac{1}{n} \sum_{i=1}^{n} x_i$$ □

Example 1. Find the maximum and minimum of $f(x) = \sqrt{x^3 + 2}$ on the interval $0 \le x \le 3$. Use these values to find the upper and lower bounds of the integral

$$\int_0^3 \sqrt{x^3 + 2}dx$$

Solution. Since $f(x) = \sqrt{x^3 + 2}$ is a continuous increasing function on $0 \le x \le 3$,

thus $$f_{\max}(x) = f(3) = \sqrt{3^3 + 2} = \sqrt{29}$$

$$f_{\min}(x) = f(0) = \sqrt{0^3 + 2} = \sqrt{2}$$

$$f_{\min}(x) \le f(x) \le f_{\max}(x)$$

By the Mean Value Theorem for definite integrals

$$\int_a^b f(x)dx = f(c)(b-a) \qquad a < c < b$$

hence, $\int_0^3 \sqrt{x^3 + 2}dx = f(c)(3-0) \qquad 0 < c < 3$

Since $c \neq 0, 3$

thus $f(0) < f(c) < f(3)$

Multiplying both sides of the inequality by $(b - a) = 3$ gives

$$3 \cdot f(0) < 3 \cdot f(c) < 3 \cdot f(3)$$

Therefore, $3\sqrt{2} < \int_0^3 \sqrt{x^3 + 2}dx < 3\sqrt{29}$

Note: In this example $f_{\min}(x),\ f_{\max}(x) = f(a),\ f(b)$

Thus $f_{\min}(x) = f(a) < f(c) < f(b) = f_{\max}(x)$ where $c \neq a, b$

If $f_{\min}(x),\ f_{\max}(x) \neq f(a),\ f(b)$

then $f_{\min}(x) \le f(c) \le f_{\max}(x)$ where $c \neq a, b$

The formula for finding the derivative of a definite integral

Example 2. Find the derivative of the definite integral $\int_{x^2}^{\sin x} e^{-t^2}\,dt$.

Solution. Because $\int_{x^2}^{\sin x} e^{-t^2}\,dt = F(\sin x) - F(x^2)$

thus $[\int_{x^2}^{\sin x} e^{-t^2}dt]' = \frac{d[\int_{x^2}^{\sin x} e^{-t^2}dt]}{dx} = \frac{d\left[F(\sin x) - F(x^2)\right]}{dx} = \frac{dF(\sin x)}{dx} - \frac{dF(x^2)}{dx}$

$$= \frac{dF(\sin x)}{d\sin x}\cdot\frac{d\sin x}{dx} - \frac{dF(x^2)}{dx^2}\cdot\frac{dx^2}{dx} = \cos x\cdot F'(\sin x) - 2x\cdot F'(x^2)$$

$$= \cos x\cdot f(\sin x) - 2x\cdot f(x^2)$$

That is, $[\int_{x^2}^{\sin x} e^{-t^2}dt]' = \cos x\cdot f(\sin x) - 2x\cdot f(x^2)$

Since $f(t) = e^{-t^2}$

thus $f(\sin x) = e^{-\sin^2 x}$, $f(x^2) = e^{-x^4}$

Therefore, $[\int_{x^2}^{\sin x} e^{-t^2}dt]' = \cos x\cdot e^{-\sin^2 x} - 2x\cdot e^{-x^4}$ □

From this example, we can provide the formula for finding the derivative of a definite integral.

Because any definite integral is

$$\int_{v(x)}^{u(x)} f(t)dt = F[u(x)] - F[v(x)]$$

thus $[\int_{v(x)}^{u(x)} f(t)dt]' = \frac{d[\int_{v(x)}^{u(x)} f(t)dt]}{dx} = \frac{d\{F[u(x)] - F[v(x)]\}}{dx} = \frac{dF[u(x)]}{dx} - \frac{dF[v(x)]}{dx}$

$$= \frac{dF[u(x)]}{du(x)}\cdot\frac{du(x)}{dx} - \frac{dF[v(x)]}{dv(x)}\cdot\frac{dv(x)}{dx} = u'(x)F'[u(x)] - v'(x)F'[v(x)]$$

$$= u'(x)f[u(x)] - v'(x)f[v(x)]$$

That is, the formula for finding the derivative of a definite integral is

$$[\int_{v(x)}^{u(x)} f(t)dt]' = u'(x)f[u(x)] - v'(x)f[v(x)]$$

Example 3. Find the derivative of the definite integral $\int_{\frac{x}{2}}^{x^2} \ln t dt$.

Solution. The formula is $[\int_{v(x)}^{u(x)} f(t)dt]' = u'(x)f[u(x)] - v'(x)f[v(x)]$

Since $u(x) = x^2$ $\quad v(x) = \frac{x}{2}$ $\quad f(t) = \ln t$

thus $u'(x) = 2x$ $\quad v'(x) = \frac{1}{2}$

$f(x^2) = \ln x^2 = 2\ln x$ $\quad f(\frac{x}{2}) = \ln\frac{x}{2}$

Therefore, $[\int_{\frac{x}{2}}^{x^2} \ln t dt]' = 2x \cdot 2\ln x - \frac{1}{2} \cdot \ln\frac{x}{2} = 4x\ln x - \frac{1}{2}\ln\frac{x}{2}$

Exercise 7.3

1. Find the values of m and M such that $m \le x\sin x \le M$ for $0 \le x \le \pi$, and use these values to find bounds for the value of the integral $\int_0^{\pi} x\sin x dx$.

2. Find the derivative of the following $F(x)$.

a. $F(x) = \int_0^x \sqrt{1+t}\,dt$ $\quad$ **b.** $F(x) = \int_x^{-1} te^{-t}dt$

c. $F(x) = \int_0^{x^2} \frac{1}{\sqrt{1+t^4}}dt$ $\quad$ **d.** $F(x) = \int_{x^3}^{x^2} e^t dt$

e. $F(x)=\int_{1}^{x}\sin\sqrt{t}\,dt$ **f.** $F(x)=\int_{(x+1)^3}^{\sin x^2}\frac{dt}{1+\sqrt{t}}$

3. Find the maximum and minimum values of the following function on the interval [-1, 5].

$$F(x)=\int_{0}^{x}t(t-4)dt$$

4. Find the average value of each of the following functions on the given interval.

a. $f(x)=x^2$ $0\le x\le 2\pi$

b. $f(x)=\sqrt{x}$ $0\le x\le 50$

c. $f(x)=6+3\sin x+\cos x$ $0\le x\le 2\pi$

7.4. Applications of definite integrals

We have learned the basic properties and calculation methods of definite integrals and now we use them to solve practical problems.

7.4.1. Area between two curves

1. Area between a curve and the x-axis

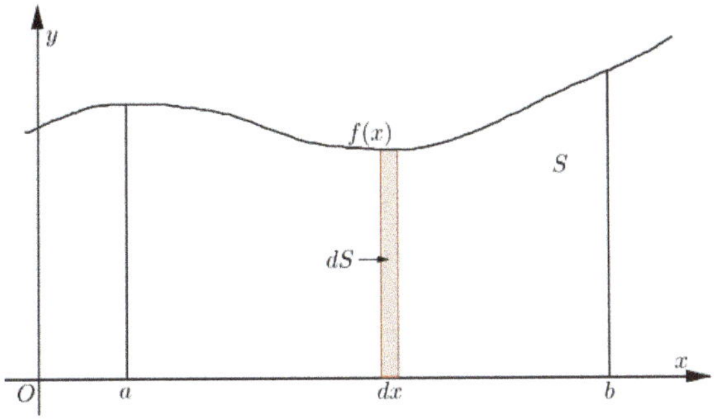

Figure 7.4.1.1

Now, we want to find the area S of the region bounded by the function $f(x)$, $x=a$, $x=b$ and the x-axis. As illustrated in Figure 7.4.1.1, on the infinitesimal interval dx, we obtain an infinitesimal rectangle.

the area of an infinitesimal rectangle (also known as the **area element**)

$$dS = f(x)dx$$

where height $\mathrm{h} = f(x)$, base $l = dx$

The area, S, is the algebraic sum of infinitely many infinitesimal rectangular areas, dS.

Therefore, $S = \int_0^S dS = \int_a^b f(x)dx$

That is, $S = \int_a^b f(x)dx$

If $y = f(x)$ ($f(x) \geq 0$) is continuous on the interval $[a, b]$, then the geometric significance of the definite integral $\int_a^b f(x)dx$ is the area enclosed by $y = f(x)$, the x-axis, the lines $x = a$ and $x = b$, as shown in Figure 7.4.1.2.

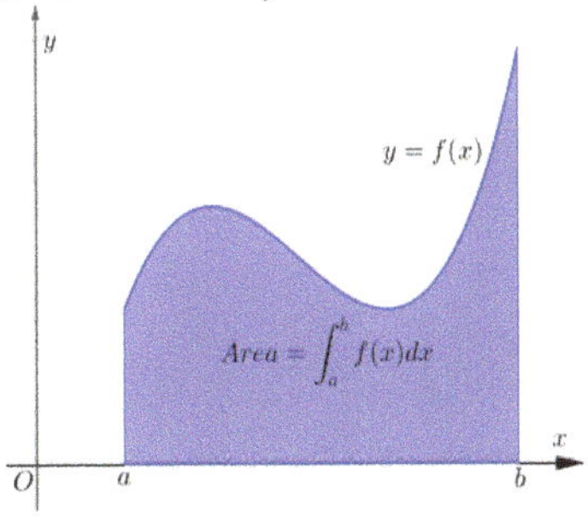

Figure 7.4.1.2

If $f(x) < 0$ on $[a, b]$, as in Figure 7.4.1.3, then

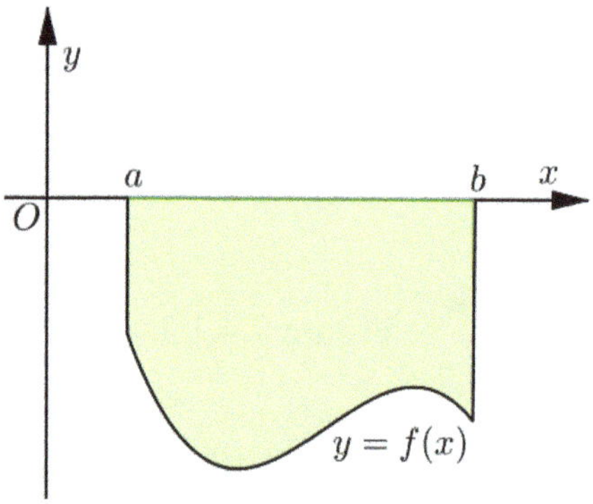

Figure 7.4.1.3

in Figure 7.4.1.3

the area enclosed by $y = f(x)$, the x-axis, the lines $x = a$ and $x = b$ is

$$S = \left| \int_a^b f(x)dx \right| = -\int_a^b f(x)dx$$

Note: Geometric area must be positive and cannot be negative. However, because the definite integral has other applications, its result can be negative when it does not represent a physical area.

If, on the interval $[a, b]$, $f(x)$ is positive on certain intervals and negative on the other intervals, as in Figure 7.4.1.4, then

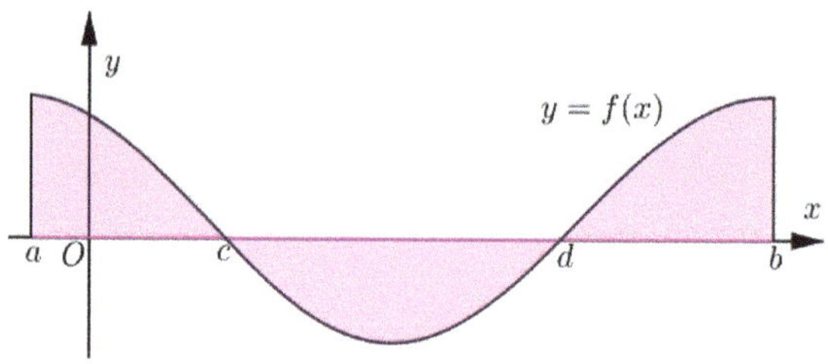

Figure 7.4.1.4

in Figure 7.4.1.4

the area enclosed by $y = f(x)$, the x-axis, the lines $x = a$ and $x = b$ is

$$S = \int_a^c f(x)dx + \left| \int_c^d f(x)dx \right| + \int_d^b f(x)dx$$

$$= \int_a^c f(x)dx - \int_c^d f(x)dx + \int_d^b f(x)dx$$

Example 1. Find the area enclosed by $y = \sin x$, the x-axis on the interval $[0, \frac{3\pi}{2}]$.

Solution. Sketch $y = \sin x$ on the interval $[0, \frac{3\pi}{2}]$ and shade the area we are looking for, as shown in Figure 7.4.1.5.

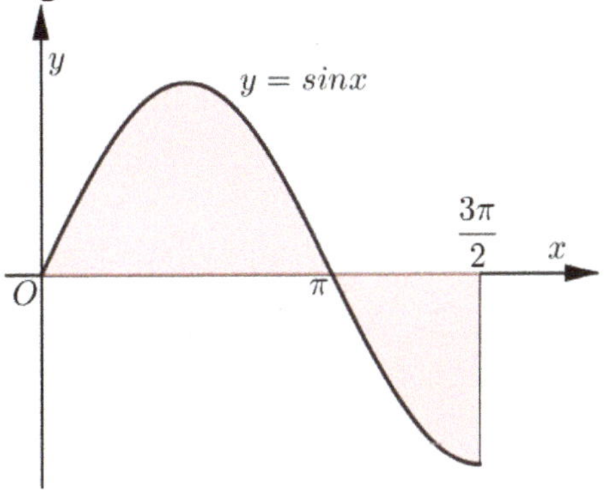

Figure 7.4.1.5

The area is $S = \int_0^{\pi} \sin x dx - \int_{\pi}^{\frac{3\pi}{2}} \sin x dx = [-\cos x]_0^{\pi} - [-\cos x]_{\pi}^{\frac{3\pi}{2}}$

$= [(-\cos\pi) - (-\cos 0)] - [(-\cos\frac{3\pi}{2}) - (-\cos\pi)]$

$= 2 - (-1) = 3$

Example 2. Find the area enclosed by $y = \sqrt{x+2}$, $y = -x^3$ and the x-axis.

Solution. Sketch $y = \sqrt{x+2}$ and $y = -x^3$ and shade the area we are looking for, as shown in Figure 7.4.1.6.

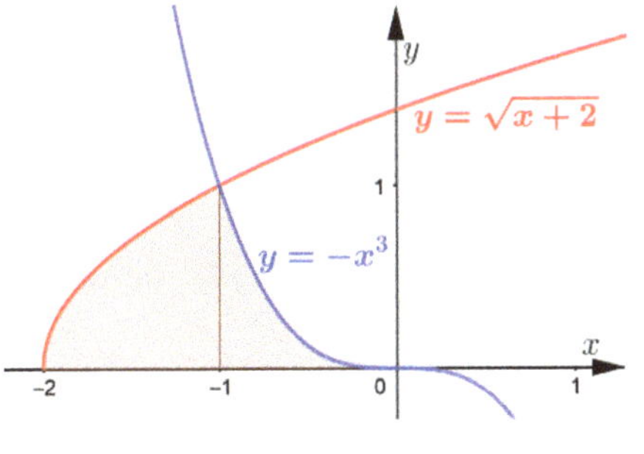

Figure 7.4.1.6

First, find the intersection points of the curves.

$$\sqrt{x+2} = -x^3$$

$$x+2 = x^6$$

$$x = -1$$

According to Figure 7.4.1.6, on the interval [-2, -1], $y=\sqrt{x+2}$ is the upper boundary of the area; on the interval [-1, 0], $y = -x^3$ is the upper boundary of the area. Therefore,

the area is $S=\int_{-2}^{-1}\sqrt{x+2}dx+\int_{-1}^{0}-x^3dx=\frac{2}{3}\left[(x+2)^{\frac{3}{2}}\right]_{-2}^{-1}-\left[\frac{x^4}{4}\right]_{-1}^{0}=\frac{2}{3}+\frac{1}{4}=\frac{11}{12}$

2. Area between two curves

If, on the interval $[a, b]$, $f(x)\geq g(x)$, as shown in Figure 7.4.1.7, then the area enclosed between the two curves $f(x)$ and $g(x)$, and the lines $x=a$ and $x=b$ is

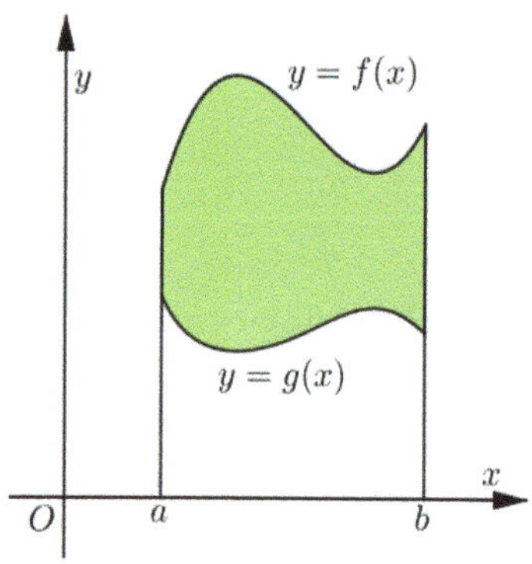

Figure 7.4.1.7

$$S=\int_a^b[f(x)-g(x)]dx$$

$$=\int_a^b f(x)dx-\int_a^b g(x)dx$$

Please note that:

$f(x)$ is the upper boundary of the enclosed area.

$g(x)$ is the lower boundary of the enclosed area.

This is because $\int_a^b f(x)dx$ denotes the area enclosed by the function $f(x)$, $x=a$, $x=b$ and the x-axis, as shown in Figure 7.4.1.8, $\int_a^b g(x)dx$ denotes the area enclosed by the function $g(x)$, $x=a$, $x=b$ and the x-axis, as shown in Figure 7.4.1.9.

Therefore, $\int_a^b f(x)dx - \int_a^b g(x)dx$ will give us the area bounded by the two curves $f(x)$ and $g(x)$, and the lines $x = a$ and $x = b$.

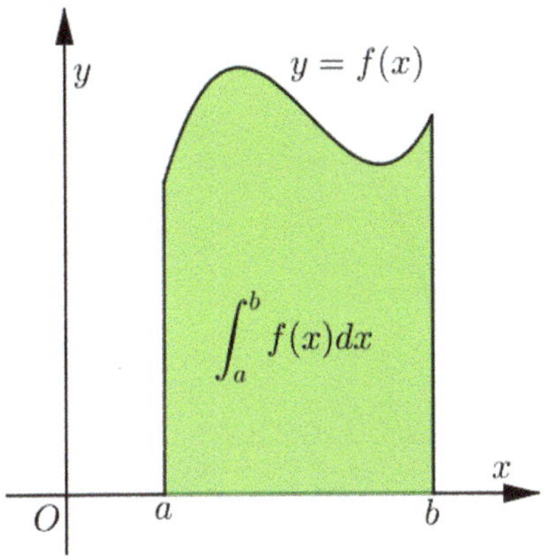

Figure 7.4.1.8

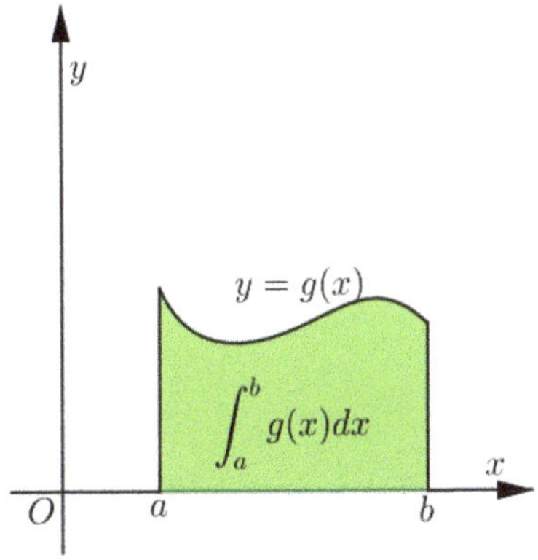

Figure 7.4.1.9

The formula also applies to the situation in Figure 7.4.1.10.

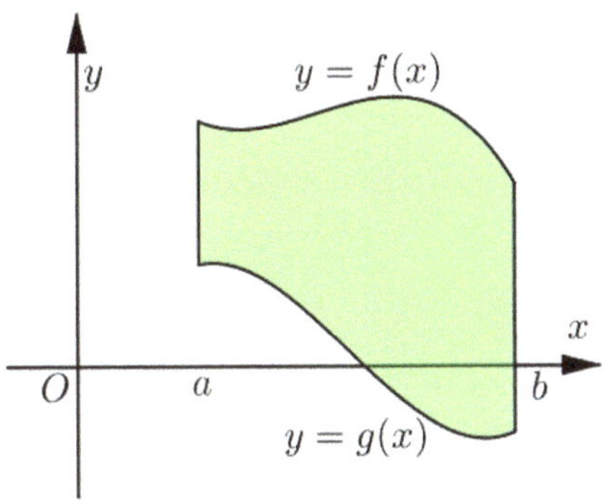

Figure 7.4.1.10

This is because if $f(x)$ and $g(x)$ are not both positive on the interval $[a, b]$, we can translate them by c units parallel to the y-axis. This ensures the curves $f(x)+c$ and $g(x)+c$ lie above the x-axis, creating a situation similar to that shown in Figure 7.4.1.7. In this case, their difference

$$[f(x)+c]-[g(x)+c]=f(x)-g(x)$$

remains unchanged. Therefore, the above formula also holds.

Example 3. Find the area enclosed by $y = 3x - 2$ and $y = x^2 - 2$.

Solution. Sketch $y = 3x - 2$ and $y = x^2 - 2$ and shade the area we are looking for, as shown in Figure 7.4.1.11.

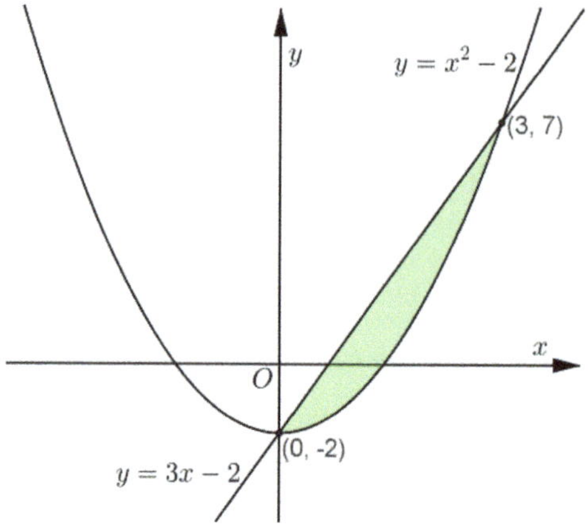

Figure 7.4.1.11

First, find the intersection points.

$$x^2 - 2 = 3x - 2$$

$$x^2 - 3x = 0$$

$$x(x - 3) = 0$$

$$x = 0 \quad \text{or} \quad x = 3$$

Hence, the area is $S = \int_0^3 \left[(3x-2)-(x^2-2)\right] dx$

$$= \int_0^3 \left[3x - x^2\right] dx = \left[\frac{3}{2}x^2 - \frac{1}{3}x^3\right]_0^3 = \left[\frac{3}{2}3^2 - \frac{1}{3}3^3\right] - 0 = 4\frac{1}{2}$$

Example 4. Find the area enclosed by $y = \cos x$, $y = \cos 2x$ and $x = \pi$ on the interval $[0, \pi]$.

Solution. Sketch $y = \cos x$, $y = \cos 2x$ and $x = \pi$ on the interval $[0, \pi]$ and shade the area we are looking for, as shown in Figure 7.4.1.12.

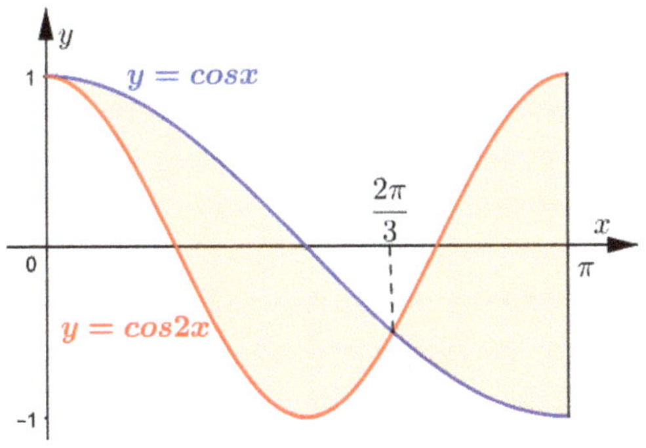

Figure 7.4.1.12

First, find the intersection points of the given curves.

$$\cos 2x = \cos x$$

$$2\cos^2 x - \cos x - 1 = 0$$

$$\cos x = 1 \quad \text{or} \quad \cos x = -0.5$$

$$x = 0 \quad \text{or} \quad x = \frac{2\pi}{3}$$

As seen in Figure 7.4.1.12, on the interval $[0, \frac{2\pi}{3}]$, $y = \cos x$ is the upper boundary and $y = \cos 2x$ is the lower boundary. On the interval $[\frac{2\pi}{3}, \pi]$, $y = \cos 2x$ is the upper boundary and $y = \cos x$ is the lower boundary.

Therefore, the area is

$$S = \int_0^{\frac{2\pi}{3}} (\cos x - \cos 2x)dx + \int_{\frac{2\pi}{3}}^{\pi} (\cos 2x - \cos x)dx$$

$$= \left[\sin x - \frac{1}{2}\sin 2x\right]_0^{\frac{2\pi}{3}} + \left[\frac{1}{2}\sin 2x - \sin x\right]_{\frac{2\pi}{3}}^{\pi}$$

$$= \left[\left(\sin\frac{2\pi}{3} - \frac{1}{2}\sin\frac{4\pi}{3}\right) - 0\right] + \left[\left(\frac{1}{2}\sin 2\pi - \sin\pi\right) - \left(\frac{1}{2}\sin\frac{4\pi}{3} - \sin\frac{2\pi}{3}\right)\right] = \frac{3\sqrt{3}}{2}$$

3. Area between a curve and the *y*-axis

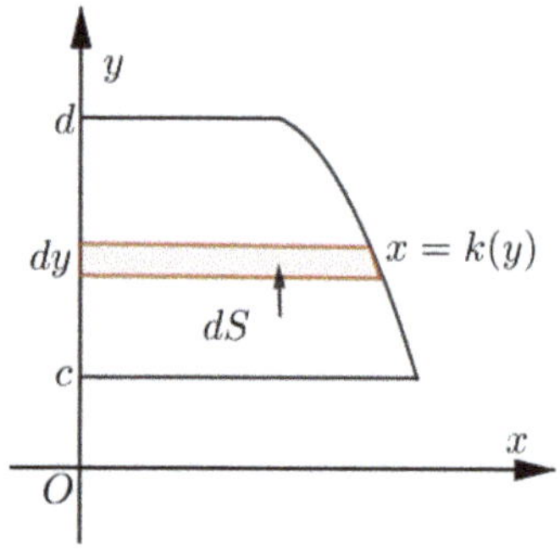

Figure 7.4.1.13

Now, we need to find the area bounded by $x = k(y)$, $y = c$, $y = d$ and the y-axis. If we view the area from a direction perpendicular to the y-axis, we can see from Figure 7.4.1.13 that

the infinitesimal area is $\quad dS = k(y)dy$

where $\quad$ height $\quad \text{h} = k(y)$, $\quad$ base $\quad l = dy$

The area, S, is the algebraic sum of infinitely many infinitesimal areas, dS.

Therefore, $\quad S = \int_0^S dS = \int_c^d k(y)dy$

That is, $\quad S = \int_c^d k(y)dy$

Example 5. Find the area enclosed by the curve $y = x^2$ between $y = 1$ and $y = 4$.

Solution. Sketch $y = x^2, y = 1, y = 4$ and shade the area we are looking for, as shown in Figure 7.4.1.14.

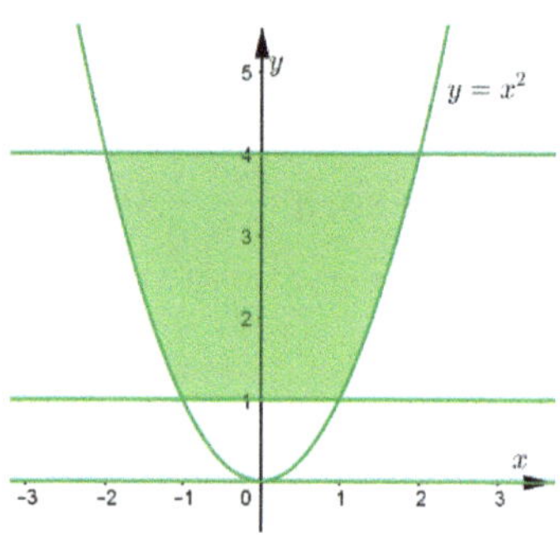

Figure 7.4.1.14

From $y = x^2$, we obtain $x = \pm\sqrt{y}$ and in the first quadrant, $x = \sqrt{y}$.

The area is $$S = 2\int_1^4 \sqrt{y}dy = 2\left[\frac{2}{3}y^{\frac{3}{2}}\right]_1^4 = \frac{4}{3}(4^{\frac{3}{2}} - 1^{\frac{3}{2}}) = \frac{28}{3}$$ □

And we can find the area bounded by $x = g(y)$, $x = h(y)$, $y = c$ and $y = d$.

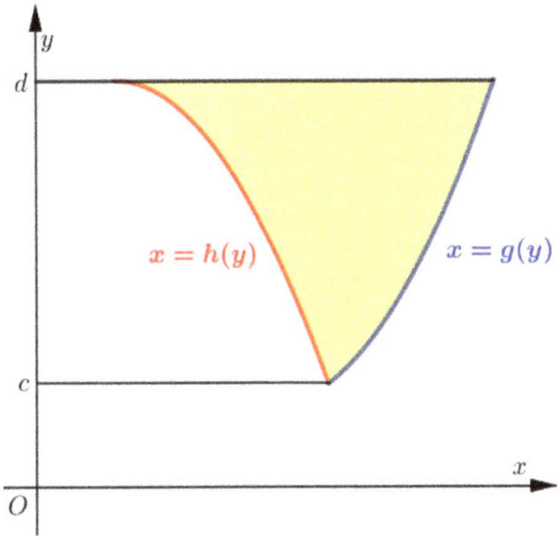

Figure 7.4.1.15

In Figure 7.4.1.15

$$S = \int_c^d [g(y) - h(y)]dy$$

Please note: when viewing the area from a direction perpendicular to the y-axis, $g(y)$ is the upper boundary of the enclosed area and $h(y)$ is the lower boundary of the enclosed area.

Example 6. Find the area enclosed by $y = e^{x-2}$, $y = \dfrac{1}{x-1}$, $y = 2$ and $y = 8$.

Solution. Sketch $y = e^{x-2}$, $y = \dfrac{1}{x-1}$, $y = 2$ and $y = 8$ and shade the area we are looking for, as shown in Figure 7.4.1.16.

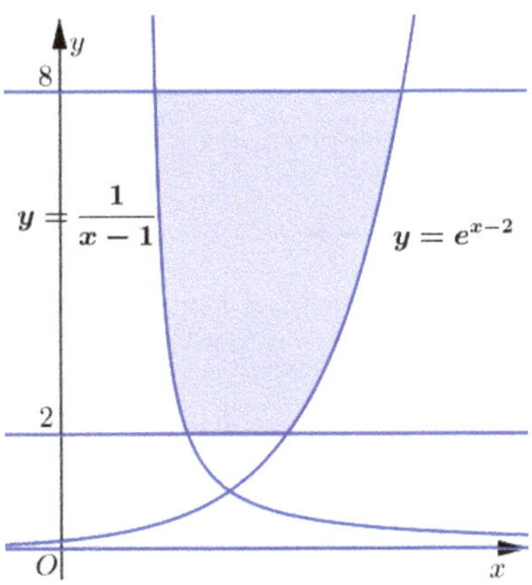

Figure 7.4.1.16

From $y = e^{x-2}$, we obtain

$$\ln y = x - 2$$

$$x = \ln y + 2$$

From $y = \dfrac{1}{x-1}$, we obtain

$$x - 1 = \frac{1}{y}$$

$$x = \frac{1}{y} + 1$$

Therefore, the area enclosed is $S = \displaystyle\int_2^8 \left[(\ln y + 2) - (\frac{1}{y} + 1) \right] dy \approx 13.9$

Sometimes, it is not easy to find the area using integration with respect to x. Then, we can use $x = f^{-1}(y)$ to find the area S by integrating with respect to y.

Example 7. Find the area enclosed between $y^2 = 2x$ and $y = x - 4$.

Solution. Sketch $y^2 = 2x$ and $y = x - 4$ and shade the area we are looking for, as shown in Figure 7.4.1.17.

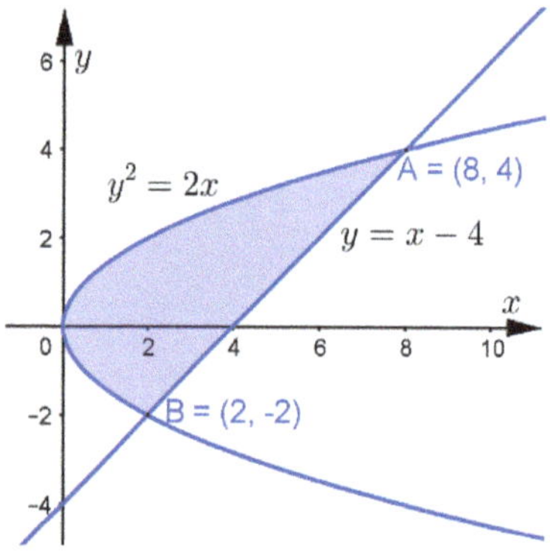

Figure 7.4.1.17

As seen in Figure 7.4.1.17, integrating with respect to x to find the area requires dividing the integration interval into two subintervals: [0, 2] and [2, 8]. On the interval [0, 2], $y=\sqrt{2x}$ is the upper boundary and $y=-\sqrt{2x}$ is the lower boundary. On the interval [2, 8], $y=\sqrt{2x}$ is the upper boundary and $y=x-4$ is the lower boundary.

Alternatively, integrating with respect to y over the interval [-2, 4] is simpler because $y=x-4$ is always the upper boundary and $y^2=2x$ is always the lower boundary. Therefore, we use $x=f^{-1}(y)$ and integrate with respect to y.

From $\quad y=x-4 \quad \Rightarrow \quad x=y+4$

From $\quad y^2=2x \quad \Rightarrow \quad x=\dfrac{y^2}{2}$

Find the intersection points of two curves. Substituting $y=x-4$ into $y^2=2x$ gives

$$(x-4)^2=2x$$

$$x^2-8x+16=2x$$

$$x^2-10x+16=0$$

$$(x-2)(x-8)=0$$

$$x_1=2 \qquad x_2=8$$

Thus $\quad y_1=-2 \qquad y_2=4$

Therefore, the intersection points are (2, -2) & (8, 4).

Hence, the area enclosed is $S=\int_{-2}^{4}[(y+4)-\frac{y^2}{2}]dy=\left[\frac{y^2}{2}+4y-\frac{y^3}{6}\right]_{-2}^{4}$

$$=\left(\frac{4^2}{2}+4\times4-\frac{4^3}{6}\right)-\left(\frac{(-2)^2}{2}+4\times(-2)-\frac{(-2)^3}{6}\right)=18$$

Note: If a calculator is allowed, we can find the area enclosed under a curve $f(x)$ by integrating $|f(x)|$ over the given interval. Thus, we do not need to manually determine where $f(x)$ is positive or negative.

The area is $\qquad S = \int_a^b |f(x)| dx$

For example,

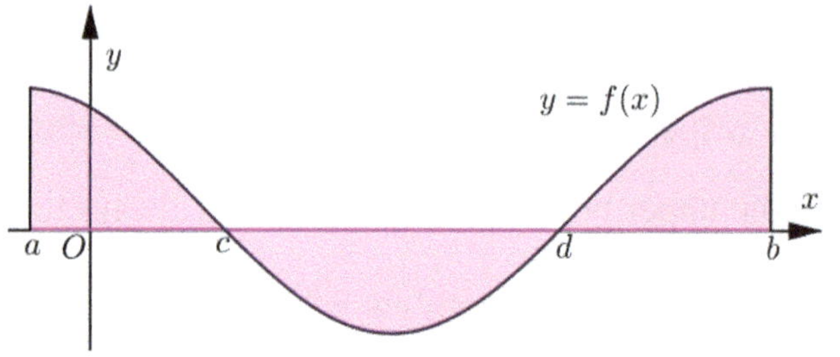

Figure 7.4.1.4

becomes

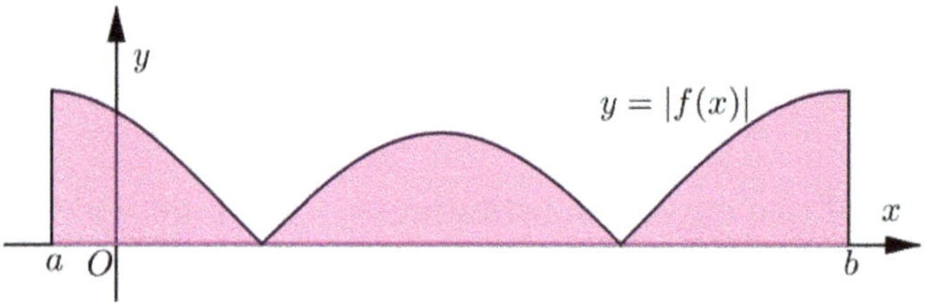

Figure 7.4.1.18

Exercise 7.4.1

1. Find the area of the region bounded by

a. $y = \frac{1}{x}$, $y = x$ and $x = 2$ **b.** $y = x^2$ and $x + y = 3$

c. $y = 3x - x^2$ and $x + y = 1$ **d.** $y = x^2 + 4$, $x = 0$, $x = 1$ and $y = 0$

e. $y = x^2$ and $y = 2 - x^2$ **f.** $y = e^x$, $y = e^{-x}$ and $x = 1$

g. $y = \sin x$, $x = 0$ and $y = 1$ on $[0, \frac{\pi}{2}]$

h. $y = x^2 - 8$, $2x + y + 8 = 0$ and $y = -4$

2. Find the area of the region bounded by

a. $y^2 = 2x$ and $y^2 = 4x - x^2$

b. $y = x^2$, $4y = x^2$ and $y = 1$

c. $y = x^3$ and $y = \sqrt[3]{x}$

d. $y = (x-1)(x-2)(x+1)$ and the x-axis

e. $y = \dfrac{5}{1+x^2}$, $y = 1$ and $y = 2$

f. $y = 6 - x^2$ and $y = 2|x-1|$

g. $y = 9a^2 - x^2$ and $y = 2ax$ $(a > 0)$

h. $y = x^2 - 5x + 4$, x-axis and $x = 5$

i. $y = \ln x$, $x = \dfrac{1}{e}$, $x = e$ and $y = 0$

j. $y = \sqrt{2x-4}$, x-axis, y-axis and $y = 2$

3. Find the value of k $(k > 0)$ such that the area bounded by $y = x^2$ and $y = kx^3$ is $\dfrac{2}{3}$.

4. Find the area bounded by $y = \sin(2x)$ and $y = 3\cos(2x)$ between $x = 0$ and $x = \pi$.

5. The area of the region enclosed by the curve $y^2 = 9kx$ and the line $x = k$ is pk^2 square units. Find the value of p.

6. The curve $y = \sqrt{2+x}$ meets the x-axis at A and the y-axis at B. The normal to the curve at B meets the x-axis at C.

a. Find the coordinates of A and B.

b. Find the equation of the normal to the curve at B.

c. Find the coordinates of C.

d. Find the area enclosed by the curve $y = \sqrt{2+x}$, the normal at B and the x-axis.

7. The area enclosed by the curve $y = \dfrac{x}{x^2+4}$, the x-axis and the lines $x = -3$ and $x = 3$ is $\ln a$. Find the value of a.

8. Two tangents are drawn from the point (2, 0) to the curve $y=x^3$. Find the area enclosed by these two tangents and the curve $y=x^3$.

9. Find the area of the region enclosed by the curves $y=\ln x$ and $y^2=x+2$ between $y=-1$ and $y=1$.

10. Find the area of the region enclosed by the curves $x=y^2-4y$ and $x=2y-y^2$.

11. Find the area of the region enclosed by the curves $x=-4y^2$ and $y-x=5$.

12. The parabola $y=6x-x^2$ meets the *x*-axis at the points O and A. The tangents to the parabola at O and A meet at the point T. Find the area bounded by OT, AT and the parabola.

13. Find the value of p such that the areas *A* and *B* are equal in Diagram 7.4.1.1.

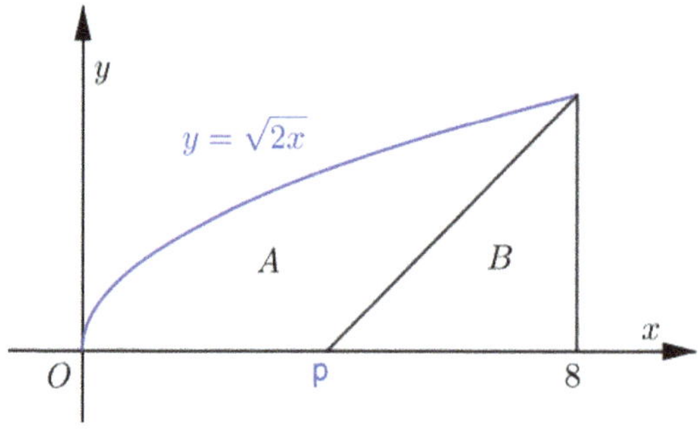

Diagram 7.4.1.1

14. Find the coordinates of the point A on the curve $y=x^2$ for $0 \le x \le 8$, such that the area of the region enclosed by the tangent at point A, $y=0$ and $x=8$ is a maximum.

15. Find the equation of a tangent to the curve $y=\ln x$ for $2 \le x \le 6$, such that the area of the region enclosed by the tangent, the lines $x=2, x=6$ and the curve $y=\ln x$ is the smallest.

16. Find the area of the region enclosed by the curves $f(x)=\sin\frac{x}{2}$ and $g(x)=\sin x$ over the interval $0\leq x\leq 2\pi$.

17. Find the area of the region bounded by the curve $f(x)=\frac{4}{(x-2)^2}$ and

a. the x-axis, the lines $x=4$ and $x=6$.

b. the lines $y=4$ and $y=6$.

18. **a**. Find the area of the region bounded by the curve $y=e^{x-3}$, the x-axis and the lines $x=0$ and $x=2$.

b. Hence, by using the result from a, evaluate $\int_{e^{-3}}^{e^{-1}}(\ln x+3)dx$.

19. In the first quadrant, the line $x+y=5$ is a tangent to the parabola

$$y=ax^2+bx \qquad (a<0, b>0)$$

The area bounded by the parabola and the x-axis is A.

a. Determine the values of a and b for which the area A is a maximum.

b. Calculate this maximum value of A.

20. The curves $y=a\sqrt{x}$ $(a>0)$ and $y=\ln\sqrt{x}$ have the same tangent at the point (x_0, y_0). Find

a. the value of constant a and the coordinates of (x_0, y_0).

b. the exact value of the area bounded by these two curves and the x-axis.

21. Let A be the area bounded by the curve $y = x^2$ and two tangents to the curve which are perpendicular to each other. One tangent touches the curve at the point B(a, a^2), where $a > 0$. Show that when $a = \frac{1}{2}$, the area A is the smallest.

7.4.2.* Area in polar coordinates

For certain plane figures, calculating area is simpler using polar coordinates. Consider a curved sector bounded by the curve $r = r(\theta)$ and the rays $\theta = \alpha$ and $\theta = \beta$. We are required to calculate its area, S (as illustrated in Figure 7.4.2.1).

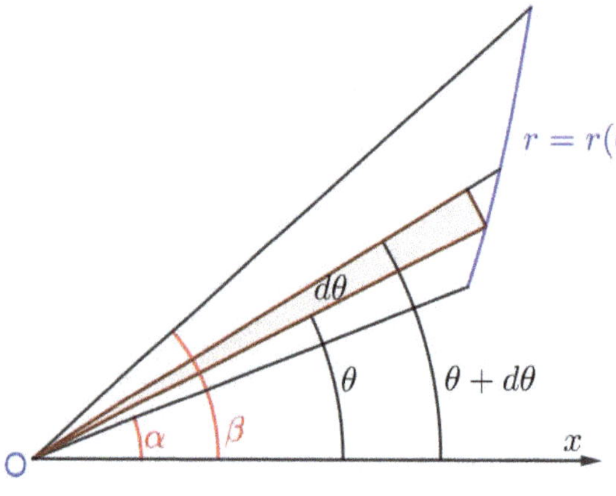

Figure 7.4.2.1

Here, $r(\theta)$ is continuous on $[\alpha, \beta]$, and $r(\theta) \geq 0$, $0 \leq \beta - \alpha \leq 2\pi$. Using the area formula of the sector

$$S = \frac{1}{2} R^2 \theta$$

we find the area of an infinitesimal sector.

$$dS = \frac{1}{2}[r(\theta)]^2 \, d\theta$$

where radius $R = r(\theta)$, central angle $\theta = d\theta$

The area S is the algebraic sum of the areas of infinitely many infinitesimal sectors.

Therefore, $$S = \int_0^S dS = \int_\alpha^\beta \frac{1}{2}[r(\theta)]^2 \, d\theta$$

That is, $$S = \int_\alpha^\beta \frac{1}{2}[r(\theta)]^2 \, d\theta$$

Example 1. Find the area of the region (as shown in Figure 7.4.2.2) enclosed by Archimedean spiral $r = a\theta$, $a > 0$, $0 \le \theta \le 2\pi$.

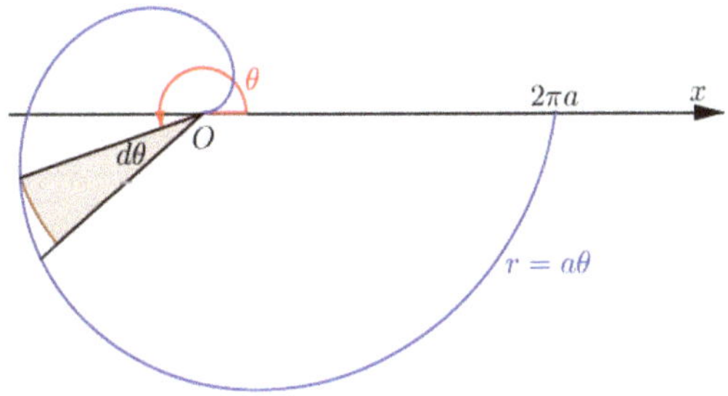

Figure 7.4.2.2

Solution. Since the area element is

$$dS = \frac{1}{2} r^2 d\theta = \frac{1}{2}(a\theta)^2 d\theta$$

thus, the area is $$S = \int_0^{2\pi} \frac{1}{2} a^2 \theta^2 d\theta = \frac{1}{2} a^2 \left[\frac{\theta^3}{3}\right]_0^{2\pi} = \frac{4}{3} a^2 \pi^3$$

Example 2. Find the area of the region (as shown in Figure 7.4.2.3) enclosed by cardioid $r = a(1 + \cos\theta)$, $a > 0$.

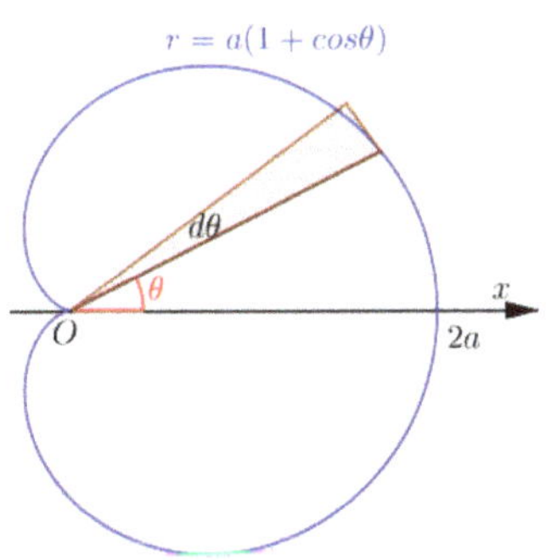

Figure 7.4.2.3

Solution. This graph is symmetrical about the polar axis; therefore, the required area S is twice the area S_1 of the region above the polar axis.

Since the area element is

$$dS = \frac{1}{2}r^2 d\theta = \frac{1}{2}a^2(1+\cos\theta)^2 d\theta$$

thus, the area is $S_1 = \int_0^\pi \frac{1}{2}a^2(1+\cos\theta)^2 d\theta = \frac{1}{2}a^2\int_0^\pi (1+2\cos\theta+\cos^2\theta)d\theta$

$$= \frac{1}{2}a^2\int_0^\pi (\frac{3}{2}+2\cos\theta+\frac{1}{2}\cos 2\theta)d\theta$$

$$= \frac{1}{2}a^2\left[\frac{3}{2}\theta+2\sin\theta+\frac{1}{4}\sin 2\theta\right]_0^\pi = \frac{3}{4}\pi a^2$$

The required area is

$$S = 2S_1 = \frac{3}{2}\pi a^2$$

Example 3. Find the area of the region (as shown in Figure 7.4.2.4) enclosed by a three-petal rose $r = a\sin(3\theta)$, $a > 0$.

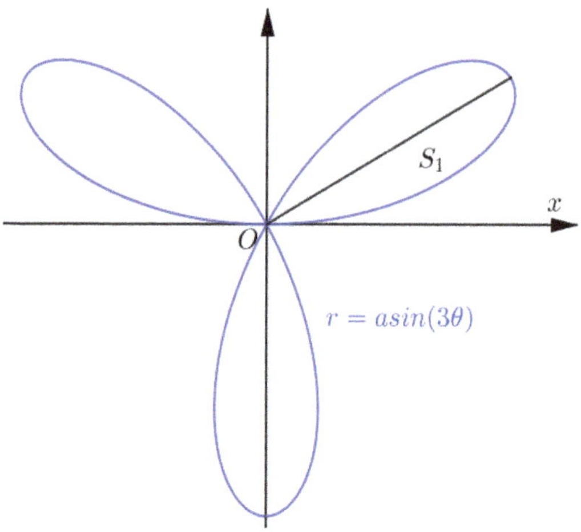

Figure 7.4.2.4

Solution. First, find the area of half a petal S_1 in the first quadrant, as shown in Figure 7.4.2.4.

Since $\quad 3\theta = \arcsin\frac{r}{a} \qquad |\sin(3\theta)| = \frac{r}{a} \le 1 \quad \Rightarrow \quad 0 \le r \le a$

thus $\quad$ when $r = 0$, $3\theta_1 = \arcsin\frac{0}{a} = 0$. $\quad$ That is, $\quad \theta_1 = 0$

$\quad$ when $r = a$, $3\theta_2 = \arcsin\frac{a}{a} = \frac{\pi}{2}$. $\quad$ That is, $\quad \theta_2 = \frac{\pi}{6}$

then $\qquad 0 \le \theta \le \frac{\pi}{6}$

Therefore,
$$S_1 = \int_{\alpha}^{\beta} \frac{1}{2}[r(\theta)]^2\, d\theta = \int_0^{\frac{\pi}{6}} \frac{1}{2} a^2 \sin^2(3\theta) d\theta = \frac{a^2}{4}\int_0^{\frac{\pi}{6}} [1 - \cos(6\theta)] d\theta$$

$$= \frac{a^2}{4}\left[\theta - \frac{1}{6}\sin(6\theta)\right]_0^{\frac{\pi}{6}} = \frac{\pi a^2}{24}$$

The required area is

$$S = 6S_1 = \frac{\pi a^2}{4}$$

Example 4. Calculate the area of the region (as shown in Figure 7.4.2.5) enclosed by the four-petal rose $r = a\cos(2\theta)$, $a > 0$.

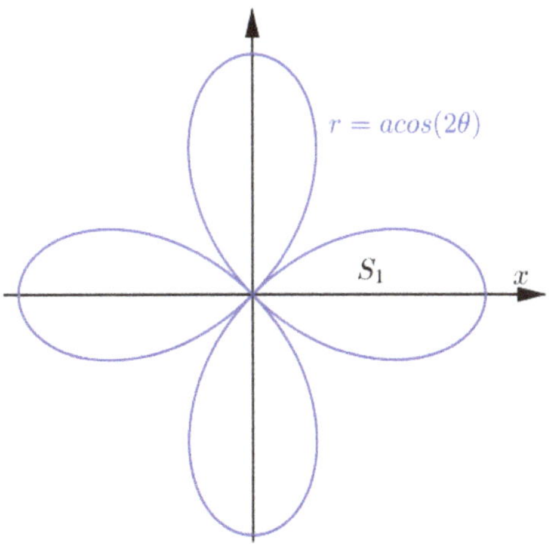

Figure 7.4.2.5

Solution. First, find the area of half a petal on the x-axis, as shown in Figure 7.4.2.5.

Since $\quad 2\theta = \arccos\frac{r}{a} \qquad |\cos(2\theta)| = \frac{r}{a} \le 1 \quad \Rightarrow \quad 0 \le r \le a$

thus $\quad$ when $r = a$, $2\theta_1 = \arccos\frac{a}{a} = 0$. That is, $\theta_1 = 0$

when $r = 0$, $2\theta_2 = \arccos\frac{0}{a} = \frac{\pi}{2}$. That is, $\theta_2 = \frac{\pi}{4}$

then $\quad 0 \le \theta \le \frac{\pi}{4}$

Hence, $S_1 = \int_\alpha^\beta \frac{1}{2}[r(\theta)]^2\, d\theta = \int_0^{\frac{\pi}{4}} \frac{1}{2}a^2\cos^2(2\theta)d\theta = \frac{a^2}{4}\int_0^{\frac{\pi}{4}}[1+\cos(4\theta)]d\theta$

$$= \frac{a^2}{4}\left[\theta + \frac{1}{4}\sin(4\theta)\right]_0^{\frac{\pi}{4}} = \frac{\pi a^2}{16}$$

The required area is

$$S = 8S_1 = \frac{\pi a^2}{2}$$

Exercise 7.4.2

1. Find the area of the region bounded by each of the following.

a. $r^2 = a^2\cos(2\theta)$ **b.** $r = ae^\theta$, $\theta = 0$, $\theta = 2\pi$

7.4.3. Volume of solids of revolution

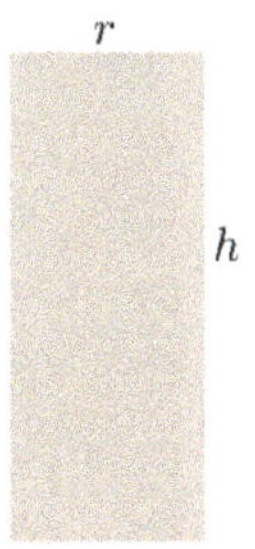

Figure 7.4.3.1

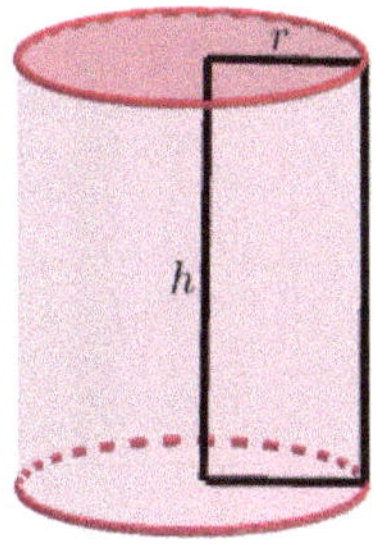

Figure 7.4.3.2

Consider a rectangle with width r and length h, as shown in Figure 7.4.3.1. If we use the length h of the rectangle as the axis of rotation and rotate the rectangle through 360° about h, we will obtain a cylinder with the base circle of radius r and height h, as shown in Figure 7.4.3.2.

The volume of the cylinder = Area of cross section × height = $\pi r^2 h$

This example illustrates that when a plane region is rotated through 360° about a specific axis of rotation, it generates a solid of revolution. We can use the definite integral to determine the volume of such a solid.

1. Volume of revolution about the *x*-axis

The region bounded by the function $f(x)$, the lines $x = a, x = b$ and the x-axis (as shown in Figure 7.4.3.3) is rotated through 360° about the x-axis.

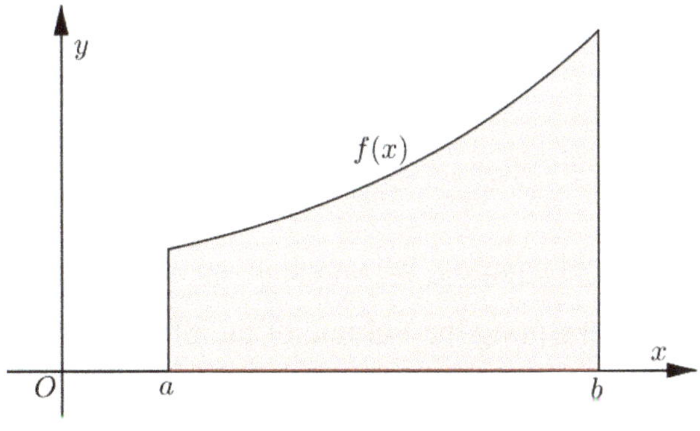

Figure 7.4.3.3

Figure 7.4.3.4 shows the obtained solid of revolution.

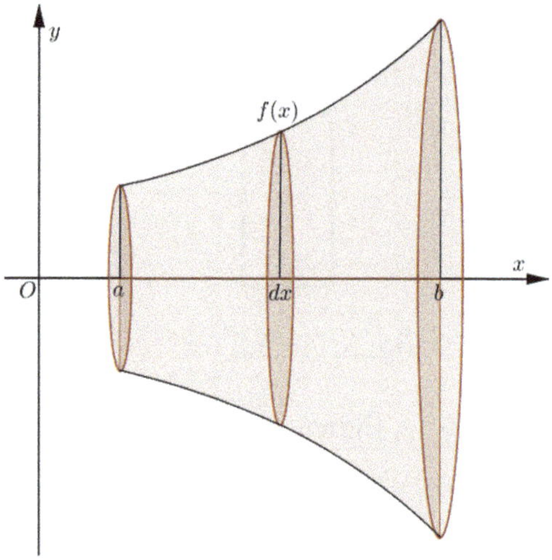

Figure 7.4.3.4

Figure 7.4.3.5

As seen in Figure 7.4.3.5, on the infinitesimal interval dx, we obtain an infinitesimal cylinder.

volume of an infinitesimal cylinder (also known as the **volume element**)

$$dV = Sdx = \pi[f(x)]^2 dx$$

where the area of the base of the cylinder $S = \pi r^2 = \pi[f(x)]^2$,

height of the cylinder $h = dx$.

The volume of revolution, V, is the algebraic sum of infinitely many infinitesimal cylindrical volumes, dV.

Therefore, $$V = \int_0^V dV = \int_a^b \pi[f(x)]^2 dx$$

That is, $$V = \int_a^b \pi[f(x)]^2 dx$$

Please note: regardless of the axis of rotation, we can analyse the volume of revolution in a similar way. In determining the volume of an infinitesimal cylinder, the radius of the base circle of the cylinder (that is, the radius of rotation r) is the distance between the outer boundary of the region being rotated and the axis of rotation.

Example 1. The region bounded by the curve $y = x^2 + 3$, the x-axis, the lines $x = 1$ and $x = 2$ is rotated through 360° about the x-axis (shown in Figure 7.4.3.6). Calculate the volume of the solid generated.

Solution.

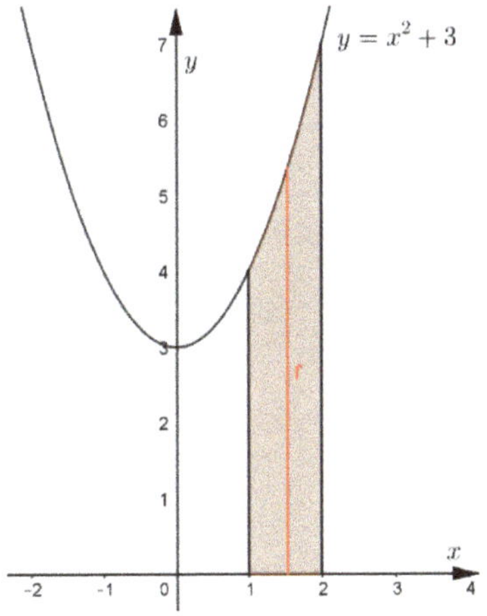

Figure 7.4.3.6

The radius of rotation is $r = y$.

$$\text{Volume } V = \pi\int_1^2 y^2 dx = \pi\int_1^2 \left(x^2+3\right)^2 dx$$

$$= \pi\int_1^2 \left(x^4 + 6x^2 + 9\right)dx$$

$$= \pi\left[\frac{x^5}{5} + 2x^3 + 9x\right]_1^2 = \frac{146\pi}{5}$$

Example 2. The region bounded by the curve $y = 9 - x^2$ and the line $y = 3x + 9$ is rotated through 360° about the x-axis. Calculate the volume of the solid generated.

Solution.

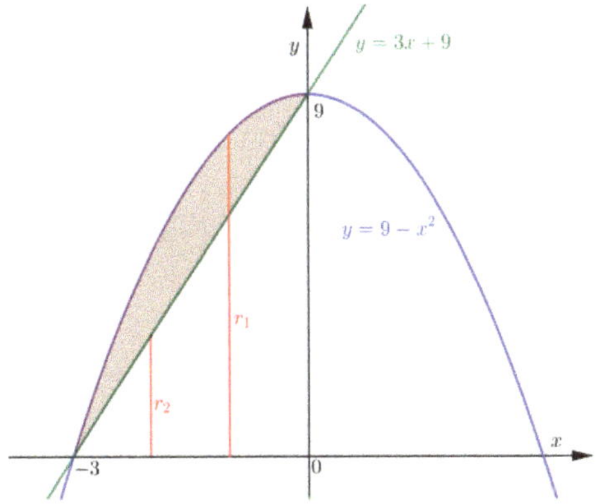

Figure 7.4.3.7

As illustrated in Figure 7.4.3.7, a gap exists between the region being rotated and the axis of rotation. The radius of rotation r should be the distance between the outermost boundary of the region being rotated and the axis of rotation.

Therefore, the volume of revolution is

$V = V_1 - V_2$ = Volume of the outer solid – Volume of the inner solid

V_1 = the volume of the solid formed by rotating the region bounded by the curve $y = 9 - x^2$ and the x-axis from $x = -3$ to $x = 0$ through 360° about the x-axis

$$= \int_{-3}^{0} \pi r_1^2 dx = \pi \int_{-3}^{0} (9 - x^2)^2 dx$$

V_2 = the volume of the solid formed by rotating the region bounded by the line $y = 3x + 9$ and the x-axis from $x = -3$ to $x = 0$ through 360° about the x-axis

$$= \int_{-3}^{0} \pi r_2^2 dx = \pi \int_{-3}^{0} (3x + 9)^2 dx$$

Hence,
$$V = V_1 - V_2 = \pi \int_{-3}^{0} (9 - x^2)^2 dx - \pi \int_{-3}^{0} (3x + 9)^2 dx$$

$$= \pi \int_{-3}^{0} (81 - 18x^2 + x^4) dx - \pi \int_{-3}^{0} (9x^2 + 54x + 81) dx$$

$$= \pi \left[81x - 6x^3 + \frac{x^5}{5} \right]_{-3}^{0} - \pi \left[3x^3 + 27x^2 + 81x \right]_{-3}^{0} = \frac{243\pi}{5}$$

Example 3. The region bounded by the curve $y = f(x)$ and the line $y = h$ from $x = a$ to $x = b$ (as shown in Figure 7.4.3.8) is rotated through 360° about the x-axis. Find the volume of the solid generated.

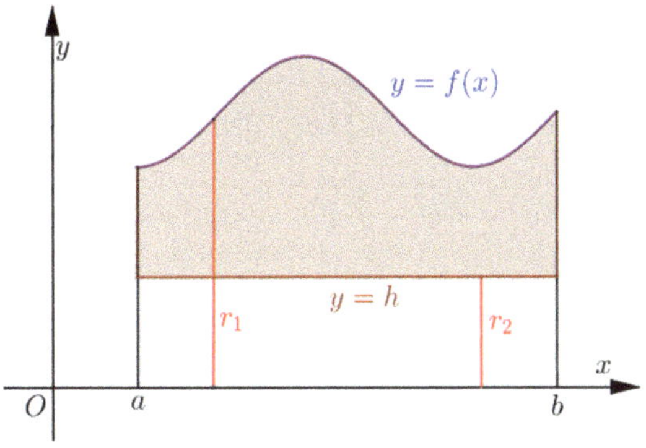

Figure 7.4.3.8

Solution. The volume of revolution is

$$V = V_1 - V_2 = \text{Volume of the outer solid} - \text{Volume of the inner solid}$$

$$= \pi\int_a^b r_1^2 dx - \pi\int_a^b r_2^2 dx = \pi\int_a^b [f(x)]^2 dx - \pi\int_a^b h^2 dx$$

$$= \pi\int_a^b [f(x)]^2 dx - \pi h^2 (b-a)$$

2. Volume of revolution about the *y*-axis

The region bounded by the curve $x = h(y)$, the lines $y = c, y = d$ and the *y*-axis (as shown in Figure 7.4.3.9) is rotated through 360° about the *y*-axis. Calculate the volume of the solid generated.

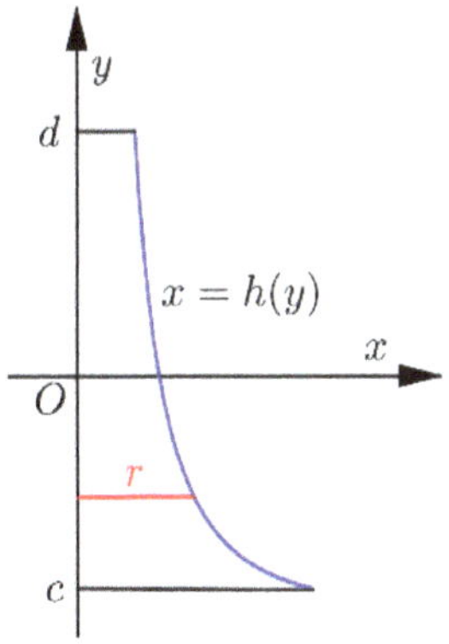

Figure 7.4.3.9

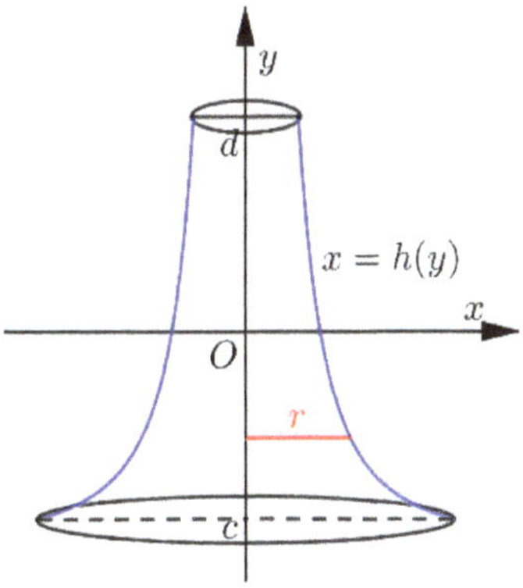

Figure 7.4.3.10

As shown in Figure 7.4.3.10, the radius of rotation is $r = x = h(y)$, therefore,

the volume of revolution is $$V = \pi\int_c^d [h(y)]^2 dy$$

Example 4. The region bounded by the curve $y = \dfrac{1}{x} - 3$ and the y-axis between $y = -2$ and $y = 7$ is rotated through 360° about the y-axis. Calculate the volume of the solid generated.

Solution.

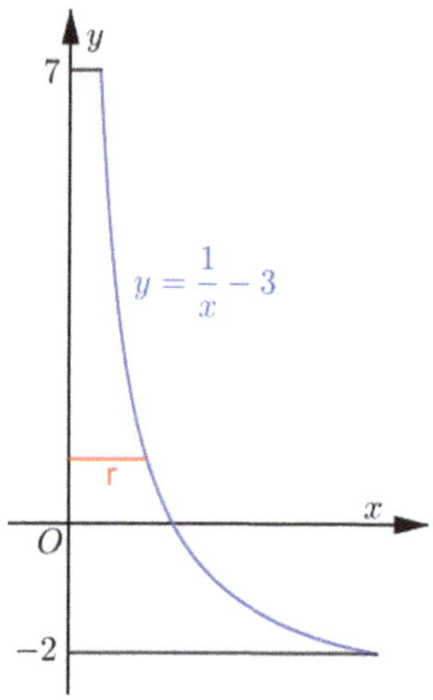

Figure 7.4.3.11

As shown in Figure 7.4.3.11, the radius of rotation is $r = x$, we need to express x in terms of y.

$$y = \frac{1}{x} - 3$$

Hence, $$x = \frac{1}{y+3}$$

Therefore, the volume of revolution is

$$V = \pi\int_{-2}^{7}\left(\frac{1}{y+3}\right)^2 dy = \frac{9\pi}{10}$$

Example 5. The region bounded by the curve $y = x^2 + 1$, the y-axis, the x-axis and the line $x = 2$ is rotated through 360° about the y-axis. Find the volume of the solid generated.

Solution.

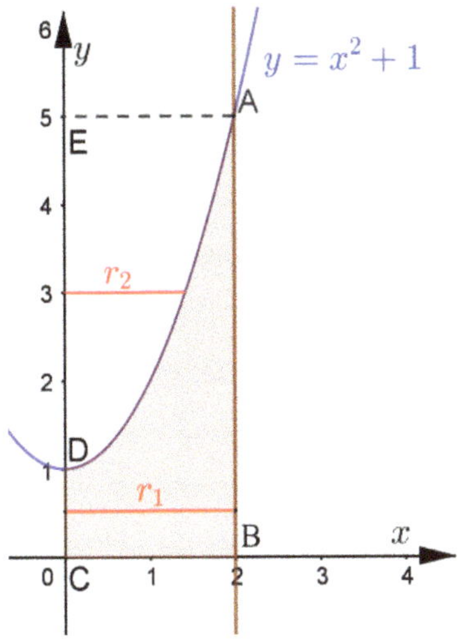

Figure 7.4.3.12

First, find the intersection point A of the curve $y = x^2 + 1$ and the line $x = 2$.

$$y = 2^2 + 1 = 5$$

Thus, the coordinates of A are (2, 5).

The intersection point of the curve $y = x^2 + 1$ and the line $x = 0$ is D (0, 1).

As shown in Figure 7.4.3.12

the volume of revolution is $V = V_1 - V_2$

V_1 = the volume of the solid formed by rotating the rectangle ABCE through 360° about the y-axis

$= \pi r_1^2 h = \pi \cdot 2^2 \cdot 5 = 20\pi$

V_2 = the volume of the solid formed by rotating the region bounded by $y = x^2 + 1$, the y-axis and $y = 5$ (that is, the region ADE) through 360° about the y-axis

$dV_2 = \pi r_2^2 dy$ $\quad r_2 = x$ $\quad$ From $y = x^2 + 1$ we obtain $x = \sqrt{y-1}$

Thus $$V_2 = \pi \int_1^5 \left(\sqrt{y-1}\right)^2 dy = \pi \int_1^5 (y-1)\, dy = \pi \left[\frac{y^2}{2} - y\right]_1^5 = 8\pi$$

Hence, the volume of revolution is $V = V_1 - V_2 = 20\pi - 8\pi = 12\pi$

Example 6. The region enclosed by the curve $y = \sqrt{x-1}$, the x-axis and the line $x = 5$ is rotated through 360° about the y-axis. Calculate the volume of the solid generated.

Solution.

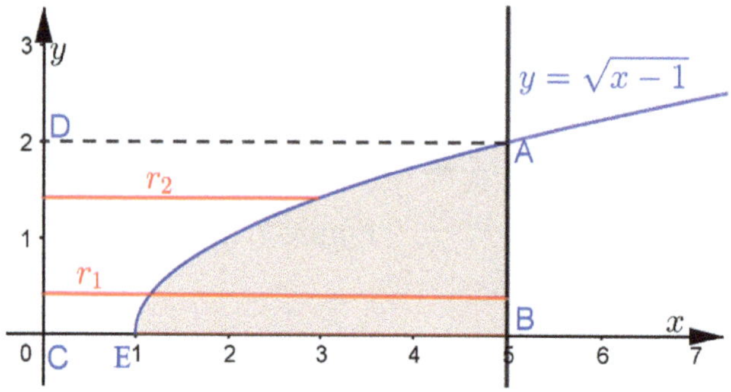

Figure 7.4.3.13

First, find the intersection point of the curve $y = \sqrt{x-1}$ and the line $x = 5$.

$$y = \sqrt{5-1} = 2$$

Thus, the coordinates of A are (5, 2).

As shown in Figure 7.4.3.13

the volume of revolution is $\quad V = V_1 - V_2$

V_1 = the volume of the solid formed by rotating the rectangle ABCD through 360° about the y-axis

$= \pi r_1^2 h = \pi \cdot 5^2 \cdot 2 = 50\pi$

V_2 = the volume of the solid formed by rotating the region bounded by $y = \sqrt{x-1}$, the x-axis, the y-axis and $y = 2$ (that is, the region AECD) 360° about the y-axis

$dV_2 = \pi r_2^2 dy \qquad r_2 = x \qquad$ From $y = \sqrt{x-1}$ we obtain $x = y^2 + 1$

Therefore, $V_2 = \pi \int_0^2 \left(y^2+1\right)^2 dy = \pi \int_0^2 \left(y^4 + 2y^2 + 1\right) dy = \pi \left[\frac{y^5}{5} + \frac{2y^3}{3} + y\right]_0^2 = \frac{206\pi}{15}$

Hence, $\qquad V = V_1 - V_2 = 50\pi - \frac{206\pi}{15} = \frac{544\pi}{15}$

3. Volume of revolution about the other axes

Example 7. The region enclosed by curve $y=\sqrt{x}$, the y-axis and the line $y=3$ is rotated through 360° about the line $y=3$. Calculate the volume of the solid generated.

Solution.

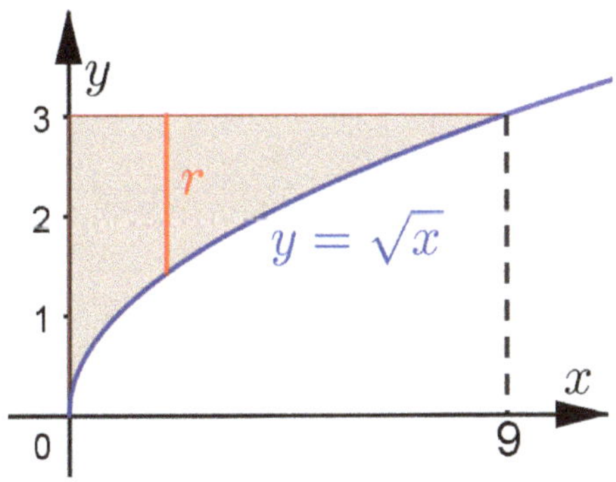

Figure 7.4.3.14

First, find the x-coordinate of the intersection of the curve $y=\sqrt{x}$ and the line $y=3$.

$$\sqrt{x}=3 \qquad \Rightarrow \qquad x=9$$

As illustrated in Figure 7.4.3.14

$$dV=\pi r^2 dx \qquad r=3-y=3-\sqrt{x}$$

Thus $$V=\pi\int_0^9\left(3-\sqrt{x}\right)^2 dx=\pi\int_0^9\left(9-6\sqrt{x}+x\right)dx=\pi\left[9x-4x^{\frac{3}{2}}+\frac{x^2}{2}\right]_0^9=\frac{27\pi}{2}$$

Example 8. The region enclosed by the curve $y=x^2-9$, the x-axis and the line $x=6$ is rotated through 360° about the line $x=7$. Calculate the volume of the solid generated.

Solution.

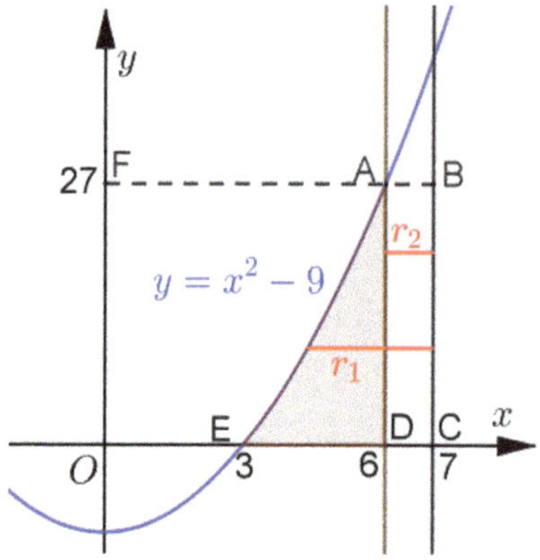

Figure 7.4.3.15

First, find the intersection point A of the curve $y = x^2 - 9$ and line $x = 6$.

$$y = 6^2 - 9 = 27$$

Thus, the coordinates of A are (6, 27).

As shown in Figure 7.4.3.15

the volume of revolution is $\quad V = V_1 - V_2$

V_1 = the volume of the solid formed by rotating the region bounded by the curve $y = x^2 - 9$, the x-axis and the line $x = 7$ between $y = 0$ and $y = 27$ (that is, the region ABCDE) through 360° about the line $x = 7$

$dV_1 = \pi r_1^2 dy$

$r_1 = 7 - x \quad$ From $y = x^2 - 9$ we obtain $x = \sqrt{y+9}$ thus $r_1 = 7 - \sqrt{y+9}$

Thus $\quad V_1 = \pi \int_0^{27} (7-x)^2 dy = \pi \int_0^{27} (7 - \sqrt{y+9})^2 dy = \dfrac{333\pi}{2}$

V_2 = the volume of the solid formed by rotating the rectangle ABCD 360° about $x = 7$

$= \pi r_2^2 h = \pi \cdot 1^2 \cdot 27 = 27\pi$

Therefore, $\quad V = V_1 - V_2 = \dfrac{333\pi}{2} - 27\pi = \dfrac{279\pi}{2}$

Example 9. The region enclosed by the curve $y = x^3$, the y-axis and the line $y = 1$ is rotated 360° about the line $y = 2$. Calculate the volume of the solid generated.

Solution.

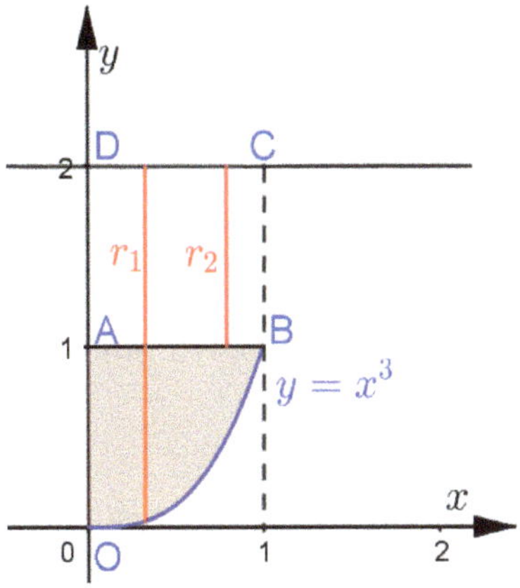

Figure 7.4.3.16

As shown in Figure 7.4.3.16

the volume of revolution is $V = V_1 - V_2$

V_1 = the volume of the solid formed by rotating the region bounded by $y = x^3$, $x = 1$ and the y-axis between $y = 0$ and $y = 2$ (that is, the region OADCB) 360° about $y = 2$

$$dV_1 = \pi r_1^2 dx \qquad r_1 = 2 - y = 2 - x^3$$

Thus $$V_1 = \pi\int_0^1 (2 - x^3)^2 dx = \pi\int_0^1 (4 - 4x^3 + x^6)dx = \pi\left[4x - x^4 + \frac{x^7}{7}\right]_0^1 = \frac{22\pi}{7}$$

V_2 = the volume of the solid formed by rotating the rectangle ABCD 360° about $y = 2$

$= \pi r_2^2 h = \pi \cdot 1^2 \cdot 1 = \pi$

Therefore, $$V = V_1 - V_2 = \frac{22\pi}{7} - \pi = \frac{15\pi}{7}$$

Exercise 7.4.3

1. For each of the following, find the volume of the solid generated when the region bounded by the given curves is rotated through

i. 360° about the x-axis;

ii. 180° about the y-axis.

a. $y = x$, $x = 2$ and the x-axis

b. $y = \sqrt{x}$, $x = 1$, $x = 4$ and $y = 0$

c. $y = \sin x$, $x = \dfrac{\pi}{2}$ and $y = 0$

d. $y = x^3$, $x = 2$ and $y = 0$

e. $y = \dfrac{1}{x}$, $x = 2$ and $y = 2$

f. $y = 2x$, $y = 3 - x^2$ and the y-axis in the first quadrant

2. Find the volume of the solid generated when the region bounded by $y = \sqrt{x+2}$, $y = \sqrt{3x}$ and $y = 0$ is rotated through 360° about the x-axis.

3. Find the volume of the solid generated when the region bounded by the ellipse

$$\frac{x^2}{a^2} + \frac{y^2}{b^2} = 1 \qquad (a > 0,\ b > 0)$$

is rotated through

a. 360° about the x-axis;

b. 360° about the y-axis.

4. Find the volume of the solid generated when the region bounded by $f(x) = \sqrt{64 - x^2}$ and the line $y = 5$ is rotated through 360° about

a. the x-axis;

b. the y-axis;

c. the line $y = 3$;

d. the line $x = 10$.

5. A tangent is drawn to the curve $y = x^2$ $(x \geq 0)$ at the point A. The area bounded by the curve $y = x^2$, this tangent and the x-axis is $\dfrac{1}{12}$.

a. Find the coordinates of the point A.

b. Find the equation of the tangent at the point A.

c. Find the volume of the solid generated when the area enclosed is rotated through 360° about the x-axis.

6. Find the volume of the solid generated when the region enclosed by the curve $y = \sec x$, the lines $y = 3$, $x = -\frac{\pi}{6}$ and $x = \frac{\pi}{6}$ is rotated through 360° about the x-axis.

7. The functions $y = f(x)$ and $y = g(x)$ are continuous on the interval $[a, b]$, where $g(x) < f(x) < k$ (k is a constant) for all x in $[a, b]$. Find the volume of the solid generated when the region bounded by the curves $y = f(x)$ and $y = g(x)$, and the lines $x = a$ and $x = b$ is rotated through 360° about the line $y = k$.

8. The curves $y = a\sqrt{x}$ $(a > 0)$ and $y = \ln\sqrt{x}$ have a common tangent at the point (x_0, y_0).

a. Find the value of constant a and the coordinates of (x_0, y_0).

b. Calculate the exact value of the volume of the solid generated when the region enclosed by the curves and the line $y = 0$ is rotated through 360° about the x-axis.

9. Find the volume of the solid generated when the region enclosed by the curve $y = 3 - |x^2 - 1|$ and the x-axis is rotated through 360° about the line $y = 3$.

10. The curves $y = ax^2$ $(a > 0, x \geq 0)$ and $y = 1 - x^2$ intersect at the point A. A line passes through the point A and the origin O.

a. Find the equation of the line OA.

The region bounded by the line OA and the curve $y = ax^2$ is rotated through 360° about the x-axis.

b. For what value of a, the volume generated is a maximum.

c. Show that this maximum volume is $\frac{32\sqrt{5}}{1875}\pi$.

11. A tangent is drawn to the curve $y = \ln x$ from the origin O. The tangent, the curve $y = \ln x$ and the x-axis bound a region S.

a. Find the area of the region S.

b. Calculate the exact value of the volume of the solid generated when the region S is rotated through 360° about the line $x = \mathrm{e}$.

12. Find the volume of the solid generated when the region bounded by the curve $y^2 = 2x$ and the line $x = \frac{1}{2}$ is rotated through 2π radians about the line $y = -1$.

13. Find the volume of the solid generated when the region enclosed by the curve $y = x^2$, the line $x = 1$ and the x-axis is rotated through 360° about the line $x = 1$.

14. The area of the trapezium bounded by the lines $y = ax + b$, $x = 0$, $x = 1$ and $y = 0$ is a constant A. Find the value of a and b such that the volume of the solid generated when the region is rotated through 360° about the x-axis is a minimum.

15. a. A region is bounded by the tangent at the point A(0, 4) to the curve $y = x^2 - 2x + 4$ and the curve $y^2 = 2(x-1)$. Find the area of the region.

b. Find the volume of the solid generated when the region is rotated through 360° about the y-axis.

16. A region is bounded by the curve $y = 4 - x^2$ and the line $y = -2x + 4$. Find the volume of the solid generated when the region is rotated through 360° about

a. the x-axis;

b. the y-axis;

c. the line $x = 4$;

d. the line $y = -2$.

17. The curve $a^2 y = x^2$ $(0 < a < 1)$ divides a unit square into two regions, part A and part B, as shown in Diagram 7.4.3.1.

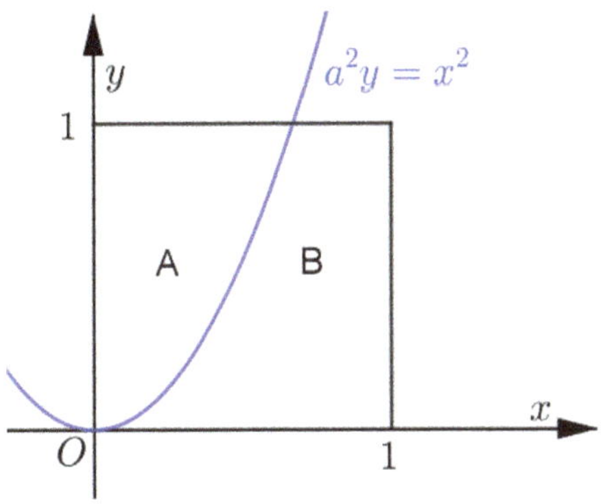

Diagram 7.4.3.1

Part A is rotated through 360° about the y-axis and part B is rotated through 360° about the x-axis. The volumes of the solids generated are V_A and V_B, respectively. Find

a. the value of a such that $V_A = V_B$.

b. the minimum value of $V_A + V_B$.

7.4.4.* Volume of a solid whose area of parallel cross-sections is known

If a solid is a solid of revolution, its cross-sectional area is circular; otherwise, its cross-sections may take other forms. Even if a solid is not generated by revolution, its volume can be calculated using definite integrals if the area of each cross-section perpendicular to a specific axis is known. As shown in Figure 7.4.4.1, let the fixed axis be the x-axis. If the cross-sectional area $S(x)$ perpendicular to the x-axis is a known continuous function, then the volume element is

$$dV = S(x)dx$$

By integrating over the closed interval $[a,b]$, we obtain the volume of the solid.

$$V = \int_a^b S(x)dx$$

Example. The base of a cylinder is a circle with radius R. A plane passing through the centre of the base circle intersects the cylinder at an angle α relative to the base (see Figure 7.4.4.1). Calculate the volume of the solid obtained by cutting the cylinder with the plane.

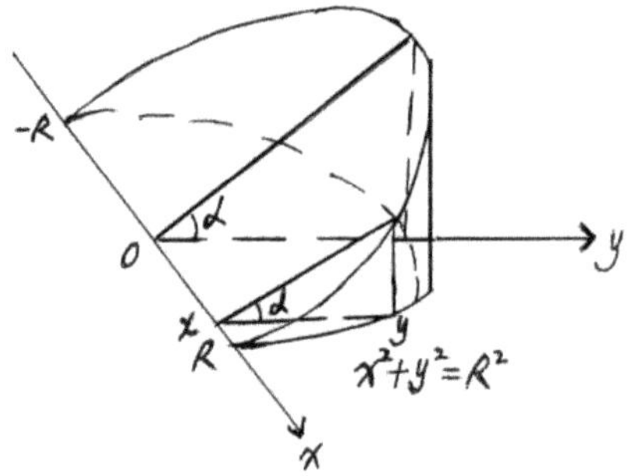

Figure 7.4.4.1

Solution. Set the x-axis along the intersection of the cutting plane and the cylinder's base. Let the y-axis be the line passing through the centre of the circle of the base perpendicular to the x-axis. The equation of the base circle is

$$x^2 + y^2 = R^2$$

A cross-section of the solid passing through point x on the x-axis and perpendicular to the x-axis is a right-angled triangle. The lengths of its two legs are y and $y \tan\alpha$, respectively; thus, the cross-sectional area is

$$S(x) = \frac{1}{2} y \cdot y \tan\alpha = \frac{1}{2}(R^2 - x^2)\tan\alpha$$

The volume of the required solid is

$$V = \int_{-R}^{R} \frac{1}{2}(R^2 - x^2)\tan\alpha dx = \int_{0}^{R}(R^2 - x^2)\tan\alpha dx$$

$$= \tan\alpha \left[R^2 x - \frac{1}{3}x^3 \right]_0^R = \frac{2}{3}R^3 \tan\alpha$$

Exercise 7.4.4

1. The base of a regular elliptical cylinder is an ellipse.

$$\frac{x^2}{a^2} + \frac{y^2}{b^2} = 1$$

The major axis of the ellipse is $2a$, and the minor axis is $2b$. A plane passes through the minor axis of the ellipse and intersects the ellipse at an angle α, which forms a wedge. Find the volume of this wedge.

7.4.5.* Arc length of a plane curve

We can use definite integrals to calculate the arc length of a curve.

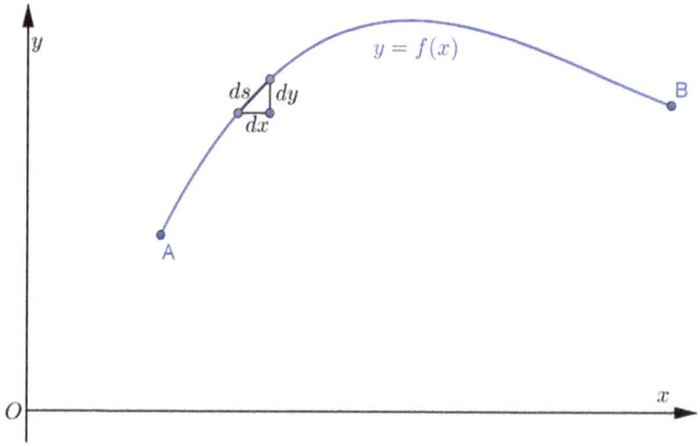

Figure 7.4.5.1

Let A and B be the two endpoints of the curve $y = f(x)$, as shown in Figure 7.4.5.1. Consider an infinitesimal arc length ds on the curve AB. By considering the right-angled triangle formed by ds, dx and dy, we can define the arc length element of the curve as

$$ds = \sqrt{(dx)^2 + (dy)^2} = \sqrt{(dx)^2 + (y'dx)^2} = \sqrt{1 + (y')^2}\,dx$$

The arc length, s, of the curve is the algebraic sum of infinitely infinitesimals, ds.

$$s = \int_0^s ds = \int_a^b \sqrt{1 + (y')^2}\,dx$$

That is, $$s = \int_a^b \sqrt{1 + (y')^2}\,dx$$

- When the arc length of the curve is given by the parametric equations,

$$\begin{cases} x = u(t) \\ y = v(t) \end{cases} \qquad \alpha \le t \le \beta$$

where $u(t)$, $v(t)$ have continuous derivatives on $\alpha \le t \le \beta$,

then, the arc length element of the curve is

$$ds = \sqrt{(dx)^2 + (dy)^2} = \sqrt{[u'(t)dt]^2 + [v'(t)dt]^2} = \sqrt{[u'(t)]^2 + [v'(t)]^2}\,dt$$

Therefore, the arc length of the curve is

$$s = \int_\alpha^\beta \sqrt{[u'(t)]^2 + [v'(t)]^2}\,dt$$

- When the arc length of the curve is given by the polar equation

$$r = r(\theta) \qquad \alpha \le \theta \le \beta$$

where $r(\theta)$ has continuous derivative on $\alpha \le \theta \le \beta$,

then, from the relationship between rectangular coordinates and polar coordinates, we can obtain

$$\begin{cases} x = r(\theta)\cos\theta \\ y = r(\theta)\sin\theta \end{cases}$$

Since

$$\frac{dx}{d\theta} = \cos\theta\frac{dr}{d\theta} - r\sin\theta \qquad \frac{dy}{d\theta} = \sin\theta\frac{dr}{d\theta} + r\cos\theta$$

$$(dx)^2 = (\cos\theta dr - r\sin\theta d\theta)^2 = \cos^2\theta(dr)^2 - 2r\sin\theta\cos\theta dr d\theta + (r\sin\theta)^2(d\theta)^2$$

$$(dy)^2 = (\sin\theta dr + r\cos\theta d\theta)^2 = \sin^2\theta(dr)^2 + 2r\sin\theta\cos\theta dr d\theta + (r\cos\theta)^2(d\theta)^2$$

Summing the left and right sides of the two equations gives

$$(dx)^2 + (dy)^2 = (dr)^2 + r^2(d\theta)^2 = (r'd\theta)^2 + r^2(d\theta)^2$$

Then, the arc length element of the curve is

$$ds = \sqrt{(dx)^2 + (dy)^2} = \sqrt{(r'd\theta)^2 + r^2(d\theta)^2} = \sqrt{r^2(\theta) + [r'(\theta)]^2}\,d\theta$$

Therefore, the arc length of the curve is

$$s = \int_{\alpha}^{\beta} \sqrt{r^2(\theta) + [r'(\theta)]^2}\,d\theta$$

Example 1. Calculate the arc length of the curve

$$y = \frac{2}{3}x^{\frac{3}{2}}$$

on the interval $a \le x \le b$ (as shown in Figure 7.4.5.2).

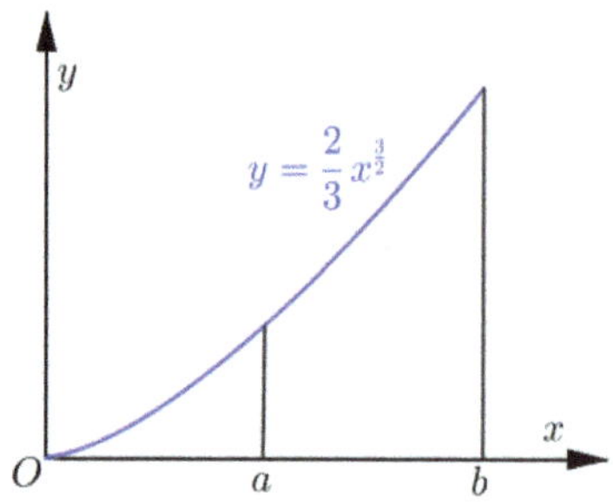

Figure 7.4.5.2

Solution. Since $y' = x^{\frac{1}{2}}$

the arc length element is given by

$$ds = \sqrt{1 + (x^{\frac{1}{2}})^2}\,dx = \sqrt{1+x}\,dx$$

Hence, the arc length of the curve is

$$s = \int_a^b \sqrt{1+x}\,dx = \left[\frac{2}{3}(1+x)^{\frac{3}{2}}\right]_a^b = \frac{2}{3}\left[(1+b)^{\frac{3}{2}} - (1+a)^{\frac{3}{2}}\right]$$

Example 2. Calculate the length of the cycloid (as shown in Figure 7.4.5.3).

$$\begin{cases} x = a(\theta - \sin\theta) \\ y = a(1-\cos\theta) \end{cases} \qquad 0 \le \theta \le 2\pi$$

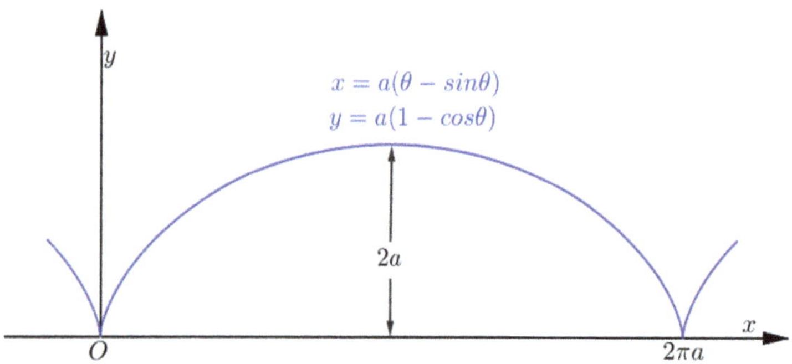

Figure 7.4.5.3

Solution. Since

$$\frac{dx}{d\theta} = a(1-\cos\theta) \qquad \frac{dy}{d\theta} = a\sin\theta$$

the arc length element is

$$ds = \sqrt{(dx)^2 + (dy)^2} = \sqrt{a^2(1-\cos\theta)^2 + a^2\sin^2\theta}d\theta$$

$$= a\sqrt{2(1-\cos\theta)}d\theta = 2a\sin\frac{\theta}{2}d\theta$$

Hence, the arc length of the curve is

$$s = \int_0^{2\pi} 2a\sin\frac{\theta}{2}d\theta = 2a\left[-2\cos\frac{\theta}{2}\right]_0^{2\pi} = 8a$$

Example 3. Calculate the arc length of the Archimedean spiral

$$r = a\theta \quad (a > 0)$$

on the interval $0 \le \theta \le 2\pi$ (as shown in Figure 7.4.5.4).

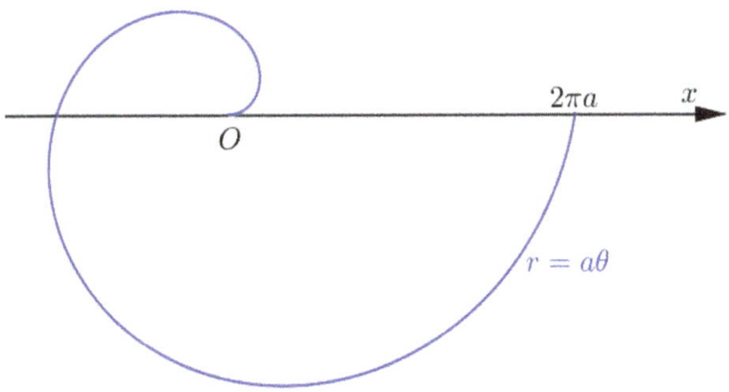

Figure 7.4.5.4

Solution. The arc length element is

$$ds = \sqrt{r^2 + (r')^2}d\theta = \sqrt{a^2\theta^2 + a^2}d\theta = a\sqrt{1+\theta^2}d\theta$$

Hence, the arc length of the curve is

$$s = a\int_0^{2\pi} \sqrt{1+\theta^2}d\theta = \frac{a}{2}\left[2\pi\sqrt{1+4\pi^2} + \ln(2\pi + \sqrt{1+4\pi^2})\right]$$

Exercise 7.4.5

Calculate the arc length of each of the following on the given interval.

1. $y = x^{\frac{3}{2}}$ on $0 \le x \le 4$

2. $r = e^{3\theta}$ from $\theta = 0$ to $\theta = 2$

3. $r = a\sin^3 \dfrac{\theta}{3}$ $\quad 0 \le \theta \le 3\pi$

4. $\begin{cases} x = a\cos^3\theta \\ y = a\sin^3\theta \end{cases}$ $\quad 0 \le \theta \le 2\pi$

7.4.6. Kinematics problems

I. Displacement

Consider an object as a particle. To describe the position change of the particle over a time period Δt, displacement is introduced, as shown in Figure 7.4.6.1.

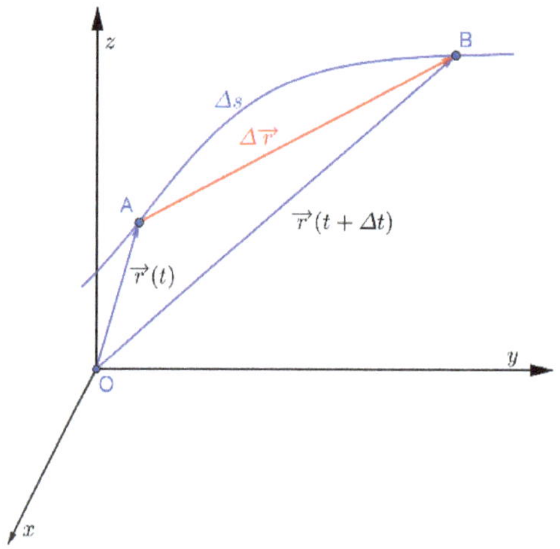

Figure 7.4.6.1

At time *t*, the particle is at the point A, and the position vector is $\vec{r}(t)$. After a time Δt, the particle is at the point B, and the position vector is $\vec{r}(t+\Delta t)$.

The vector $\Delta\vec{r} = \vec{r}(t+\Delta t) - \vec{r}(t) = \vec{r}_2 - \vec{r}_1$

is called the displacement of the particle over time Δt, and the displacement is a vector.

In the three-dimensional rectangular coordinate system

position vector is $\vec{r} = \vec{r}(t) = x\vec{i} + y\vec{j} + z\vec{k} = x(t)\vec{i} + y(t)\vec{j} + z(t)\vec{k}$

magnitude of position vector $\vec{r}$ is $|\vec{r}| = \sqrt{x^2 + y^2 + z^2}$ $\quad |\vec{r}| \geq 0$

displacement is $\Delta\vec{r} = \Delta x\vec{i} + \Delta y\vec{j} + \Delta z\vec{k}$

magnitude of displacement $\Delta\vec{r}$ is $|\Delta\vec{r}| = \sqrt{(\Delta x)^2 + (\Delta y)^2 + (\Delta z)^2}$ $\quad |\Delta\vec{r}| \geq 0$

$\Delta\vec{r} = \vec{r}_2 - \vec{r}_1 \neq \vec{r}_2 + \vec{r}_1$ Do not mix up.

$|\Delta\vec{r}|$ is the magnitude of displacement $\Delta\vec{r}$ and it represents the straight-line distance (i.e. length) between two points.

Note: The displacement, $\Delta\vec{r}$, is different from the change in the distance, Δs, travelled by the particle during this period. The change in the distance Δs of distance $s = s(t)$ is the length of the particle's motion path, which is a positive scalar, $\Delta s \geq 0$. In Figure 7.4.6.1, the arc length Δs and the length of $|\Delta\vec{r}|$ are not equal. For the general curved motion, displacement length $|\Delta\vec{r}| \neq \Delta s$ (arc length).

II. Velocity, acceleration

(1) The instantaneous velocity of the particle, referred to as velocity, is denoted by $\vec{V}$, that is,

$$\vec{V} = \lim_{\Delta t \to 0} \frac{\Delta\vec{r}}{\Delta t} = \frac{d\vec{r}}{dt}$$

Velocity, $\vec{V}$, is a vector that reflects the direction of motion and how fast the motion is.

(2) The instantaneous speed, referred to as speed, is also commonly used in physics. Speed is a scalar quantity that describes how fast a particle moves, i.e.

$$v = \lim_{\Delta t \to 0} \frac{\Delta s}{\Delta t} = \frac{ds}{dt}$$

where Δs is the distance travelled by the particle.

(3) The instantaneous acceleration of the particle, referred to as acceleration, is denoted by $\vec{a}$, that is,

$$\vec{a} = \lim_{\Delta t \to 0} \frac{\Delta \vec{V}}{\Delta t} = \frac{d\vec{V}}{dt} = \frac{d^2\vec{r}}{dt^2}$$

(4) Using the Mean Value Theorem of definite integrals, we can find the average velocity of $\vec{V}$ and the average acceleration of $\vec{a}$.

$$\Delta\vec{r} = \int_{\vec{r}_0}^{\vec{r}} d\vec{r} = \int_{t_0}^{t} \vec{V} dt = \overline{\vec{V}} \cdot \Delta t$$

That is, the average velocity is $\overline{\vec{V}} = \dfrac{\Delta \vec{r}}{\Delta t}$

Similarly, the average speed is $\bar{v} = \dfrac{\Delta s}{\Delta t}$

Similarly, $\Delta\vec{V} = \displaystyle\int_{\vec{V}_0}^{\vec{V}} d\vec{V} = \int_{t_0}^{t} \vec{a} dt = \overline{\vec{a}} \cdot \Delta t$

That is, the average acceleration is $\overline{\vec{a}} = \dfrac{\Delta \vec{V}}{\Delta t}$

- For the general curved motion, since $|\Delta\vec{r}| \neq \Delta s$, thus during the time interval

Δt, the magnitude of average velocity, $|\overline{\vec{V}}| = \left|\frac{\Delta \vec{r}}{\Delta t}\right|$, is obviously not equal to the average speed, $\bar{v} = \frac{\Delta s}{\Delta t}$. That is, $|\overline{\vec{V}}| \neq \bar{v}$. However, as $\Delta t \to 0$, $|d\vec{r}| = ds$. Hence, we have

$$|\vec{V}| = \left|\frac{d\vec{r}}{dt}\right| = \frac{|d\vec{r}|}{dt} = \frac{ds}{dt} = v$$

The expression above shows that at a certain moment, the magnitude of instantaneous velocity, $|\vec{V}|$, is equal to the instantaneous speed, v, at that moment.

(5) In the three-dimensional rectangular coordinate system

- The expression of the velocity vector in coordinates is

$$\vec{V} = v_x\vec{i} + v_y\vec{j} + v_z\vec{k}$$

or
$$\vec{V} = \frac{dx}{dt}\vec{i} + \frac{dy}{dt}\vec{j} + \frac{dz}{dt}\vec{k}$$

The magnitude of velocity $\vec{V}$, that is speed, is

$$v = |\vec{V}| = \sqrt{v_x^2 + v_y^2 + v_z^2}$$

- The expression of the acceleration vector in coordinates is

$$\vec{a} = a_x\vec{i} + a_y\vec{j} + a_z\vec{k}$$

or
$$\vec{a} = \frac{dv_x}{dt}\vec{i} + \frac{dv_y}{dt}\vec{j} + \frac{dv_z}{dt}\vec{k}$$

$$= \frac{d^2x}{dt^2}\vec{i} + \frac{d^2y}{dt^2}\vec{j} + \frac{d^2z}{dt^2}\vec{k}$$

The magnitude of acceleration $\vec{a}$ is

$$a = |\vec{a}| = \sqrt{a_x^2 + a_y^2 + a_z^2}$$

(6) Equations of motion

The position vector $\vec{r}=\vec{r}(t)$ and distance $s=s(t)$ are equations of motion, which reflect the law of change over time, through which the particle's velocity, speed and acceleration can be obtained. On the contrary, if the acceleration of the particle is known, the velocity of the particle and the equation of motion can be found.

Equations of motion $\vec{r}=\vec{r}(t)$ and $s=s(t)$ are different from equations of the path of motion $z=f(x,y)$ or $y=f(x)$, do not mix them up.

III. If the particle moves in a straight line, then in the rectangular coordinate system $y=0$, $z=0$, the equation of the path of linear motion is $f(x,0)=f(x)=0$, hence the coordinate expression is

$$\vec{r}=x\vec{i}$$

$$\vec{V}=v_x\vec{i}=\frac{dx}{dt}\vec{i}$$

$$\vec{a}=a_x\vec{i}=\frac{dv_x}{dt}\vec{i}=\frac{d^2x}{dt^2}\vec{i}$$

Let $x=s$ and it gives

$$\vec{r}=s\vec{i}$$

as shown in Figure 7.4.6.2.

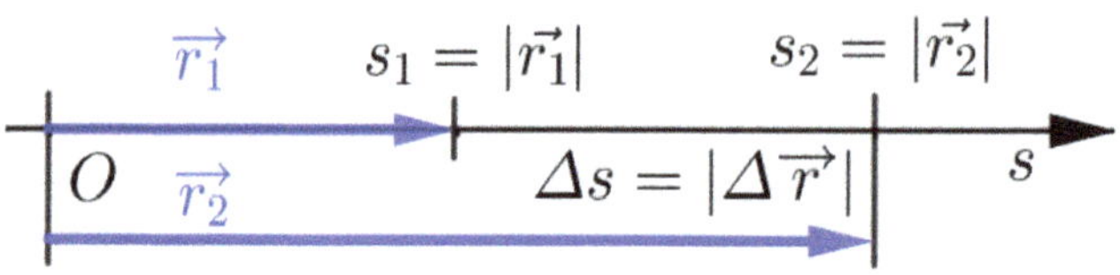

Figure 7.4.6.2

The scalar equation omits the subscript x is

length of displacement $|\Delta\vec{r}|=\Delta s$ (length of distance)

$$v=\frac{ds}{dt}$$

or $$v = \frac{ds}{dt} \cdot \frac{dv}{dv} = \frac{dv}{dt} \cdot \frac{ds}{dv} = a \cdot \frac{ds}{dv}$$

$$a = \frac{dv}{dt} = \frac{d^2 s}{dt^2}$$

or $$a = \frac{dv}{dt} \cdot \frac{ds}{ds} = \frac{ds}{dt} \cdot \frac{dv}{ds} = v \cdot \frac{dv}{ds}$$

If the particle moves in a straight line, scalars are $s = |\vec{r}|$, $v = |\vec{V}|$, $a = |\vec{a}|$.

Multiplying scalars s, v, a by unit vector $\vec{i}$, respectively will give vectors $\vec{r}$, $\vec{V}$, $\vec{a}$. Thus, we can use scalar equations to solve for the unknowns. In addition, because vectors can be positive or negative, scalars s, v, a, etc. can also be positive or negative. Positive means the movement direction is the same as the positive direction of the coordinate axis, and negative means the movement direction is opposite to the positive direction of the coordinate axis. Express the magnitude with an absolute value.

Example 1. A particle moves along a circle with radius R, and the distance is the arc length s. Its equation of motion is

$$s = v_0 t - \frac{1}{2} b t^2 \qquad v_0, b \text{ are positive constants.}$$

Find the speed v and motion path equation of the particle at any time moment t.

Solution. From the equation of motion of the particle, we can obtain the speed of the particle's motion as

$$v = \frac{ds}{dt} = \frac{d}{dt}(v_0 t - \frac{1}{2} b t^2) = v_0 - bt$$

That is, $$v = v_0 - bt$$

The path equation of circular motion is

$$x^2 + y^2 = R^2$$

Its equation of motion path is a circle.

Example 2. Suppose a particle undergoes projectile motion, as shown in Figure 7.4.6.3. Its equation of motion is

$$\vec{r} = x\vec{i} + y\vec{j} = 19.6t\vec{i} + (19.6t - 4.9t^2)\vec{j}$$

Find the velocity, acceleration and motion equation of the particle at the time $t = 1$.

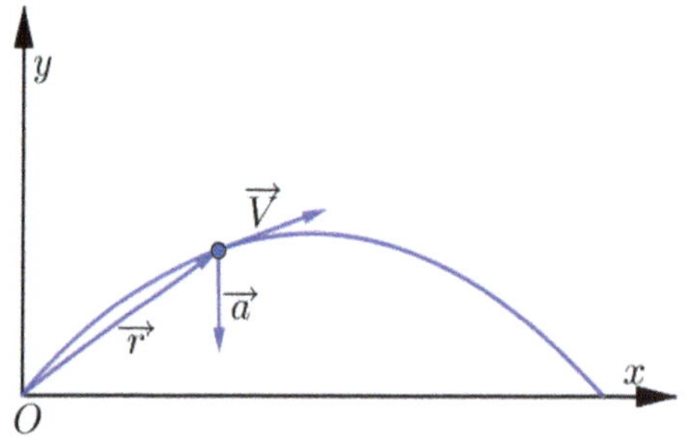

Figure 7.4.6.3

Solution. From the equation of motion of the particle, we can obtain the speed of the particle's motion as

$$\vec{V} = \frac{d\vec{r}}{dt} = \frac{d}{dt}[19.6t\vec{i} + (19.6t - 4.9t^2)\vec{j}] = 19.6\vec{i} + (19.6 - 9.8t)\vec{j}$$

Substituting $t = 1$ gives

$$\vec{V}(1) = 19.6\vec{i} + 9.8\vec{j}$$

The magnitude of the velocity of the particle is

$$\left|\vec{V}(1)\right| = \sqrt{v_x^2 + v_y^2} = \sqrt{19.6^2 + 9.8^2} \approx 21.9$$

The acceleration of the particle is

$$\vec{a} = \frac{d\vec{V}}{dt} = \frac{d}{dt}\left[19.6\vec{i} + (19.6 - 9.8t)\vec{j}\right] = -9.8\vec{j}$$

The magnitude of the acceleration of the particle is $|\vec{a}| = 9.8$ and the direction is vertically downward.

Since $x = 19.6t$, $y = 19.6t - 4.9t^2$

eliminating t from these two equations yields the motion path equation of the projectile.

$$y = x - \frac{5}{392}x^2$$

Its motion path is a parabola.

Example 3. An object moves in a straight line, and its equation of motion is

distance $\quad s = -t^3 + 3t^2 + 2t + 5 \quad t \geq 0$

where distance is measured in metres and time in seconds.

(a) What is the initial position of the object?

(b) What is the object's initial speed?

(c) When does the object come to rest?

(d) For $0 \leq t \leq 6$, what is the longest distance travelled by the object?

(e) For $0 \leq t \leq 6$, what is the maximum speed of the object?

(f) What is the acceleration of the object after 5 seconds? Explain the significance of this result.

Solution. (a) $\quad s = -t^3 + 3t^2 + 2t + 5$

When $t = 0$, $s = 5$.

Hence, the initial position is located 5 metres away from the origin in the positive direction.

(b) speed $\quad v = \frac{ds}{dt} = \frac{d}{dt}(-t^3 + 3t^2 + 2t + 5) = -3t^2 + 6t + 2$

When $t = 0$, $v = 2$.

Therefore, the initial speed is 2 m/s.

(c) When speed $v = 0$, the object is at rest.

$$-3t^2 + 6t + 2 = 0$$

$$t \approx -0.29 \text{ (rejected)} \quad \text{or} \quad t \approx 2.29$$

Hence, when $t \approx 2.29$, the object is at rest.

(d) When $\frac{ds}{dt}=0$, the object will travel the longest distance.

From (c), we get $t=2.29$.

Substituting $t=2.29$ into $s=-t^3+3t^2+2t+5$ gives

$$s=13.3$$

However, when $t=6$, $s=-91$.

Therefore, the longest distance the object travels is 91 metres.

(e) When $\frac{dv}{dt}=0$, the speed is the greatest.

$$\frac{dv}{dt}=\frac{d}{dt}(-3t^2+6t+2)=-6t+6=0$$

$$t=1$$

Substituting $t=1$ into $v=-3t^2+6t+2$ gives

$$v=5$$

However, when $t=6$, $v=-70$.

Hence, the maximum speed is 70 m/s.

(f) acceleration $a=\frac{dv}{dt}=-6t+6$

When $t=5$, $a=-24$.

Since $a=-24<0$ it follows that the speed of the object is decreasing.

Example 4. An object moves in a straight line and its speed is

$$v=\frac{2s-2}{5s+3}, \qquad s \text{ is distance.}$$

Find the acceleration of the object at the time when $s=2$.

Solution. $$a = \frac{dv}{dt} = \frac{dv}{dt} \cdot \frac{ds}{ds} = \frac{ds}{dt} \cdot \frac{dv}{ds} = v\frac{dv}{ds}$$

$$= \frac{2s-2}{5s+3} \cdot \frac{(2s-2)'(5s+3)-(2s-2)(5s+3)'}{(5s+3)^2} = \frac{16(2s-2)}{(5s+3)^3}$$

Therefore, $$a(2) = \frac{16(2 \cdot 2 - 2)}{(5 \cdot 2 + 3)^3} = \frac{32}{2197}$$

Example 5. An object moves in a straight line and its acceleration is

$$a = 10 - 6t$$

Given that when $t = 0$, speed $v = 0$.

(a) Find the speed of the object at any time.

(b) Find the distance travelled by the object in 7 seconds.

Solution. (a) $$v = \int a dt = \int (10 - 6t)dt = 10t - 3t^2 + C$$

When $t = 0$, $v = 0$, thus $C = 0$.

$$v = 10t - 3t^2$$

(b) $$s = \int_0^7 \left|10t - 3t^2\right| dt \approx 135$$

Example 6. The acceleration of an object is

$$a = \frac{1}{4 + 25s^2}, \quad s \text{ is distance.}$$

Given that when $s = 2$ metres, speed $v = 3$ m/s. Find the equation for speed as a function of s.

Solution. Since $$a = \frac{dv}{dt} = \frac{dv}{dt} \cdot \frac{ds}{ds} = \frac{ds}{dt} \cdot \frac{dv}{ds} = v\frac{dv}{ds}$$

thus $$a = v\frac{dv}{ds} = \frac{1}{4 + 25s^2}$$

$$vdv = \frac{1}{4+25s^2}ds$$

$$\int vdv = \int \frac{1}{4+25s^2}ds$$

$$\frac{v^2}{2} = \frac{1}{25}\int \frac{1}{(\frac{2}{5})^2 + s^2}ds$$

$$\frac{v^2}{2} = \frac{1}{25}\cdot\frac{5}{2}\arctan\frac{5s}{2} + C$$

That is, $$\frac{v^2}{2} = \frac{1}{10}\arctan\frac{5s}{2} + C$$

When $s = 2$, $v = 3$ and solving gives

$$C = \frac{9}{2} - \frac{1}{10}\arctan 5$$

Therefore, $$\frac{v^2}{2} = \frac{1}{10}\arctan\frac{5s}{2} + \frac{9}{2} - \frac{1}{10}\arctan 5$$

$$v^2 = \frac{1}{5}\arctan\frac{5s}{2} + 9 - \frac{1}{5}\arctan 5$$

Exercise 7.4.6

1. A particle moves in a straight line so that its displacement in metres, from a fixed origin O, at time t seconds, is given by the equation

$$s = t^3 - 15t^2 + 60t - 50 \qquad t \geq 0$$

a. Find the initial position of the particle.

b. Calculate the initial velocity and the initial acceleration.

c. When will the particle come to rest?

d. How many times will the particle pass through the origin?

e. How often does the particle change its direction of motion?

f. Find the intervals of time for which the particle's speed is increasing.

g. Find the intervals of time for which the particle's velocity is increasing.

h. What is the displacement of the particle for $0 \le t \le 10$?

i. What is the distance the particle travels for $0 \le t \le 10$?

2. A particle moves in a straight line so that its displacement in metres from a fixed origin O at time t seconds, is given by the equation

$$s = 4\sin(3t - \pi) - 1 \qquad t \ge 0$$

a. Find the initial position of the particle.

b. What is the maximum displacement from the origin?

c. What is the maximum displacement from the initial position?

d. What is the particle's maximum speed?

e. How long is it between consecutive times when the particle is at rest?

f. Find the particle's acceleration (i) in terms of t; (ii) in terms of displacement.

3. A particle moves with the given velocity. The particle is at the origin when $t = 0$.

(i) Find the displacement in terms of t.

(ii) Find the acceleration in terms of t.

a. $v = 3e^{-2t}$ **b.** $v = \dfrac{5}{t+4}$

c. $v = \dfrac{2t}{t^2+3}$ **d.** $v = \sqrt{t+2}$

4. A particle moves with the given velocity. Calculate

(i) the displacement between the given times.

(ii) the total distance travelled between the given times.

a. $v = 4e^{-t}$ between $t = 1$ and $t = 5$

b. $v = 4 - 6\sin t$ between $t = 0$ and $t = 2\pi$

c. $v = t^2 - 5t + 4$ between $t = 0$ and $t = 4$

5. An object moves in a straight line, and its velocity is given in terms of displacement. Find the acceleration of the object at the given value of the displacement or the velocity.

a. $v = 2e^{-3s}$ $\quad s = \ln 4$

b. $v = 3\cos(2s)$ $\quad s = \dfrac{\pi}{12}$

c. $v = \dfrac{s+3}{s-5}$ $\quad v = \dfrac{1}{4}$

d. $v = 4\ln(5s)$ $\quad v = 5$

6. An object moves in a straight line, and its acceleration and the initial velocity v_0 are given. Calculate

(i) the velocity of the object in terms of t.

(ii) the total distance travelled by the object between the given times.

a. $a = \dfrac{7}{(t+2)^2}$ $\quad v_0 = 3$ $\quad$ between $\mathrm{t} = 0$ and $\mathrm{t} = 4$

b. $a = -2\sin(3t)$ $\quad v_0 = 0$ $\quad$ between $\mathrm{t} = 0$ and $\mathrm{t} = \dfrac{3\pi}{2}$

c. $a = \dfrac{1}{\sqrt{4t+5}}$ $\quad v_0 = 2$ $\quad$ between $\mathrm{t} = 0$ and $\mathrm{t} = 3$

7. An object moves in a straight line and its acceleration, *a*, is given in terms of displacement, *s*.

$$a = \frac{s-1}{s+2}$$

Given that $v = 5$ when $s = 2$. Determine the expression for the velocity in terms of s.

7.5. Improper integral

In the previous sections, when we discuss definite integral, the integration interval $[a, b]$ for $\int_a^b f(x)dx$ is bounded and finite. Moreover, the integrand is a continuous function with a boundary. However, many applications require integrating over unbounded intervals or the integration of unbounded functions. These two types of integrals are called **improper integrals**.

Type 1: The integration interval is unbounded.

For example, $\int_1^\infty \frac{1}{x^2}dx \qquad \int_{-\infty}^\infty \frac{1}{x^2+1}dx \qquad \int_{-\infty}^{-2} \frac{1}{x}dx$

<u>Definition for integration over an unbounded interval</u>

Let the function $f(x)$ be continuous on the interval $[a, \infty[$. If the limit $\lim_{b\to\infty}\int_a^b f(x)dx$ $(a < b)$ exists, then this limit is called the improper integral of $f(x)$ on the interval $[a, \infty[$, denoted by

$$\int_a^\infty f(x)dx = \lim_{b\to\infty}\int_a^b f(x)dx$$

In such a case, we say that the improper integral $\int_a^\infty f(x)dx$ exists or converges. If $\lim_{b\to\infty}\int_a^b f(x)dx$ does not exist, then we say that the improper integral $\int_a^\infty f(x)dx$ does not exist or diverges.

Similarly, we can define the improper integral of $f(x)$ on]-∞, *b*] and]-∞, ∞[.

$$\int_{-\infty}^{b} f(x)dx = \lim_{a\to-\infty} \int_{a}^{b} f(x)dx$$

$$\int_{-\infty}^{\infty} f(x)dx = \int_{-\infty}^{a} f(x)dx + \int_{a}^{\infty} f(x)dx = \lim_{b\to-\infty} \int_{b}^{a} f(x)dx + \lim_{c\to\infty} \int_{a}^{c} f(x)dx$$

Example 1. Evaluate $\int_{1}^{\infty} \frac{1}{x^2} dx$.

Solution. $\int_{1}^{\infty} \frac{1}{x^2} dx = \lim_{a\to\infty} \int_{1}^{a} \frac{1}{x^2} dx = \lim_{a\to\infty} \left[-\frac{1}{x} \right]_1^a = \lim_{a\to\infty} \left(-\frac{1}{a} + 1 \right) = 1$

Example 2. Evaluate $\int_{0}^{\infty} xe^{-x^2} dx$.

Solution. Let $u = -x^2$ then $du = -2xdx$

$$\int_{0}^{a} xe^{-x^2} dx = \int_{u_1}^{u_2} xe^{u} \cdot \frac{du}{-2x} = -\frac{1}{2} \int_{u_1}^{u_2} e^{u} du = -\frac{1}{2} \left[e^{u} \right]_{u_1}^{u_2} = -\frac{1}{2} \left[e^{-x^2} \right]_0^a = -\frac{1}{2} \left(e^{-a^2} - 1 \right)$$

Hence, $\int_{0}^{\infty} xe^{-x^2} dx = \lim_{a\to\infty} \int_{0}^{a} xe^{-x^2} dx = \lim_{a\to\infty} -\frac{1}{2} \left(e^{-a^2} - 1 \right) = -\frac{1}{2}(0-1) = \frac{1}{2}$

Example 3. Evaluate $\int_{-\infty}^{\infty} \frac{1}{1+x^2} dx$.

Solution. $\int_{-\infty}^{\infty} \frac{1}{1+x^2} dx = \int_{-\infty}^{0} \frac{1}{1+x^2} dx + \int_{0}^{\infty} \frac{1}{1+x^2} dx$

$$\int_{-\infty}^{0} \frac{1}{1+x^2}dx = \lim_{a\to-\infty}\int_{a}^{0} \frac{1}{1+x^2}dx = \lim_{a\to-\infty}\left[\arctan x\right]_a^0$$

$$= \lim_{a\to-\infty}\left(\arctan 0 - \arctan a\right) = \frac{\pi}{2}$$

$$\int_{0}^{\infty} \frac{1}{1+x^2}dx = \lim_{b\to\infty}\int_{0}^{b} \frac{1}{1+x^2}dx = \lim_{b\to\infty}\left[\arctan x\right]_0^b$$

$$= \lim_{b\to\infty}\left(\arctan b - \arctan 0\right) = \frac{\pi}{2}$$

Therefore, $\displaystyle\int_{-\infty}^{\infty} \frac{1}{1+x^2}dx = \frac{\pi}{2}+\frac{\pi}{2} = \pi$

Example 4. Discuss whether $\displaystyle\int_{-\infty}^{\infty} \cos x dx$ converges.

Solution. $\displaystyle\int_{-\infty}^{\infty} \cos x dx = \int_{-\infty}^{0} \cos x dx + \int_{0}^{\infty} \cos x dx$

Since $\displaystyle\int_{-\infty}^{0} \cos x dx = \lim_{a\to-\infty}\int_{a}^{0} \cos x dx = \lim_{a\to-\infty}\left[\sin x\right]_a^0 = \lim_{a\to-\infty}\left[-\sin a\right]$

$\displaystyle\lim_{a\to-\infty}\left[-\sin a\right]$ does not exist, thus $\displaystyle\int_{-\infty}^{0} \cos x dx$ diverges.

Similarly, $\displaystyle\int_{0}^{\infty} \cos x dx$ diverges. Therefore, $\displaystyle\int_{-\infty}^{\infty} \cos x dx$ diverges.

Type 2: The integrand is unbounded on the integration interval.

For example, $\displaystyle\int_{0}^{1} \frac{1}{x}dx \qquad \int_{0}^{3} \ln x dx \qquad \int_{0}^{\frac{\pi}{2}} \tan x dx$

Definition for integration of unbounded functions

1) Let the function $f(x)$ be continuous on $[a, b[$. As $x \to b^-$, $f(x) \to \infty$ (as shown in Figure 7.5.1). If $\lim_{t \to b^-} \int_a^t f(x)dx$ exists, then this limit is called the improper integral of $f(x)$ on the interval $[a, b[$, denoted by

$$\int_a^b f(x)dx = \lim_{t \to b^-} \int_a^t f(x)dx$$

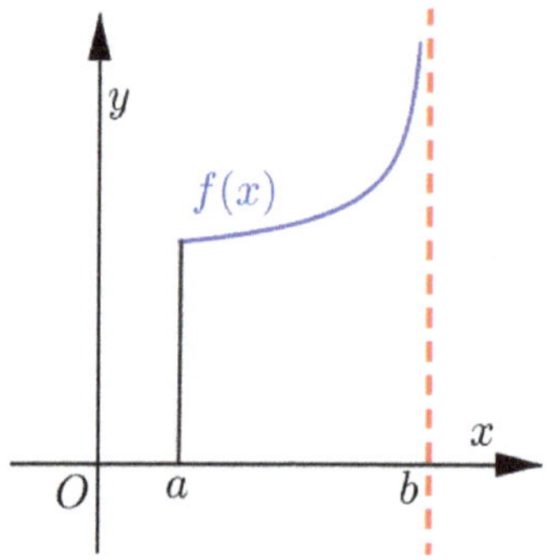

Figure 7.5.1

2) Let the function $f(x)$ be continuous on $]a, b]$. As $x \to a^+$, $f(x) \to \infty$ (as shown in Figure 7.5.2). If $\lim_{t \to a^+} \int_t^b f(x)dx$ exists, then this limit is called the improper integral of $f(x)$ on the interval $]a, b]$, denoted by

$$\int_a^b f(x)dx = \lim_{t \to a^+} \int_t^b f(x)dx$$

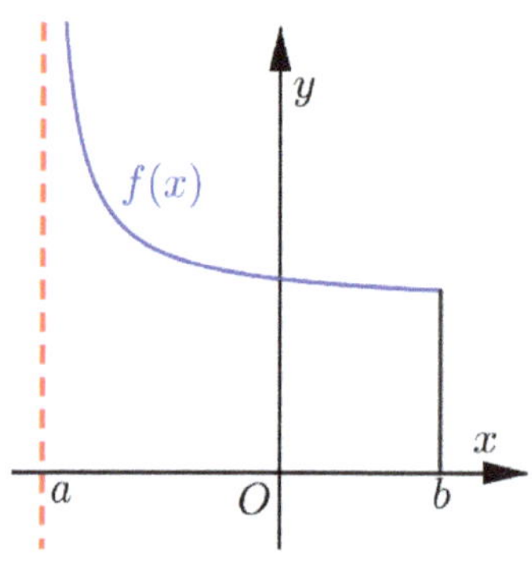

Figure 7.5.2

3) Let the function $f(x)$ be continuous on the interval $[a, b]$ except at the point c. As $x \to c, f(x) \to \infty$ (as shown in Figure 7.5.3). If both $\lim_{t \to c^-} \int_a^t f(x)dx$ and $\lim_{t \to c^+} \int_t^b f(x)dx$ exist, then the limit is called the improper integral of $f(x)$ on the interval $[a, b]$, denoted by

$$\int_a^b f(x)dx = \int_a^c f(x)dx + \int_c^b f(x)dx = \lim_{t \to c^-} \int_a^t f(x)dx + \lim_{t \to c^+} \int_t^b f(x)dx$$

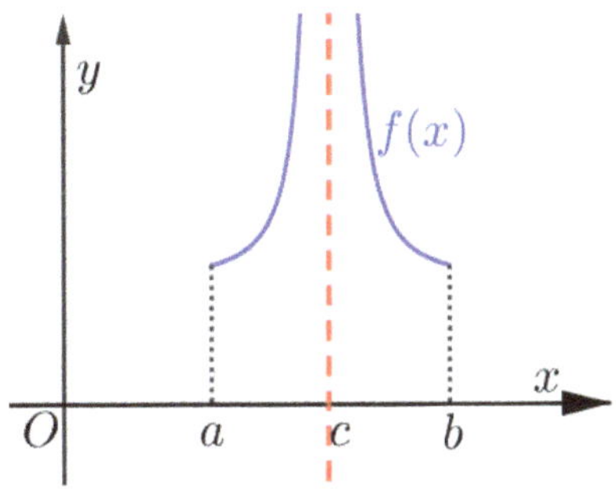

Figure 7.5.3

Example 5. Evaluate $\int_0^2 \frac{1}{\sqrt{2-x}}dx$.

Solution. The function $y = \frac{1}{\sqrt{2-x}}$ is undefined at the point $x = 2$. It follows that

$$\int_0^2 \frac{1}{\sqrt{2-x}}dx = \lim_{t \to 2^-} \int_0^t \frac{1}{\sqrt{2-x}}dx = \lim_{t \to 2^-} \left[-2\sqrt{2-x}\right]_0^t = \lim_{t \to 2^-} \left(-2\sqrt{2-t} + 2\sqrt{2}\right) = 2\sqrt{2}$$

Example 6. Evaluate $\int_{-3}^1 \frac{1}{x+3}dx$.

Solution. The function $y = \frac{1}{x+3}$ is undefined at the point $x = -3$. Hence,

$$\int_{-3}^{1}\frac{1}{x+3}dx=\lim_{t\to -3^+}\int_{t}^{1}\frac{1}{x+3}dx=\lim_{t\to -3^+}\left[\ln(x+3)\right]_t^1=\lim_{t\to -3^+}\left[\ln 4-\ln(x+3)\right]=\infty$$

Therefore, $\int_{-3}^{1}\frac{1}{x+3}dx$ diverges, and it does not exist.

Example 7. Evaluate $\int_{-1}^{1}\frac{1}{x^2}dx$.

Solution. The function is undefined at the point $x=0$. Hence,

$$\int_{-1}^{1}\frac{1}{x^2}dx=\int_{-1}^{0}\frac{1}{x^2}dx+\int_{0}^{1}\frac{1}{x^2}dx=\lim_{t\to 0^-}\int_{-1}^{t}\frac{1}{x^2}dx+\lim_{t\to 0^+}\int_{t}^{1}\frac{1}{x^2}dx$$

$$=\lim_{t\to 0^-}\left[-\frac{1}{x}\right]_{-1}^{t}+\lim_{t\to 0^+}\left[-\frac{1}{x}\right]_{t}^{1}=\lim_{t\to 0^-}\left[-\frac{1}{t}-1\right]+\lim_{t\to 0^+}\left[-1+\frac{1}{t}\right]$$

Both limits do not exist, therefore, $\int_{-1}^{1}\frac{1}{x^2}dx$ diverges and does not exist.

Exercise 7.5

1. Evaluate the following integrals.

a. $\int_{0}^{\infty}\frac{1}{4+x^2}dx$

b. $\int_{0}^{\infty}\frac{x^2}{1+x^3}dx$

c. $\int_{1}^{\infty}\frac{1}{x(x^2+1)}dx$

d. $\int_{0}^{1}\ln x dx$

e. $\int_{0}^{\infty}x\cos x^2 dx$

f. $\int_{0}^{\infty}\frac{\arctan x}{1+x^2}dx$

2. Evaluate the following integrals.

a. $\int_{-\infty}^{\infty} \frac{x}{(x^2+2)^3} dx$

b. $\int_{2}^{\infty} \frac{dx}{x^2+x-2}$

c. $\int_{-1}^{1} \frac{dx}{\sqrt{1-x^2}}$

d. $\int_{0}^{1} \frac{dx}{\sqrt{1-x}}$

e. $\int_{-\infty}^{\infty} \frac{dx}{e^x+e^{-x}}$

f. $\int_{1}^{\infty} \frac{\arctan x}{x^2} dx$

7.6.* Approximate calculation of definite integrals

Calculating a definite integral typically requires finding an antiderivative. However, in many cases, an antiderivative may be difficult to determine or impossible to express in terms of elementary functions. Furthermore, if only a set of observed or experimental data for the integrand are available, an analytical antiderivative cannot be obtained. Under these circumstances, the calculation formula of the definite integral cannot be used to find the exact value of the definite integral. Additionally, exact values are not always required in practical applications; often, an approximate value with a specific degree of accuracy is sufficient. Therefore, it is necessary to study the numerical methods for the approximation of definite integrals.

Geometrically, the definite integral represents the area enclosed by the curve. Hence, an approximate value for the definite integral can be obtained by estimating the area of the curved trapezoid with $[a, b]$ as the base and the curve $f(x)$ as the curved edge. In the following, we introduce three commonly used numerical methods for the approximation of definite integrals.

1. Rectangular method

Let the function $f(x)$ be continuous on the interval $[a, b]$. Use points $a = x_0 < x_1 < x_2 < ... < x_{n-1} < x_n = b$ to divide the interval $[a, b]$ into n subintervals of equal length. The length of each subinterval is $\Delta x = \dfrac{b-a}{n}$. Through each point, draw a straight line perpendicular to the x-axis and intersect $f(x)$ at points A_0, A_1, …, A_n. The y-coordinates of these points are y_0, y_1, y_2, …, y_n, respectively. Use Δx as the base and y_i as the height ($i = 1, 2, …, n$), we will obtain n rectangles, as shown in Figure 7.6.1.

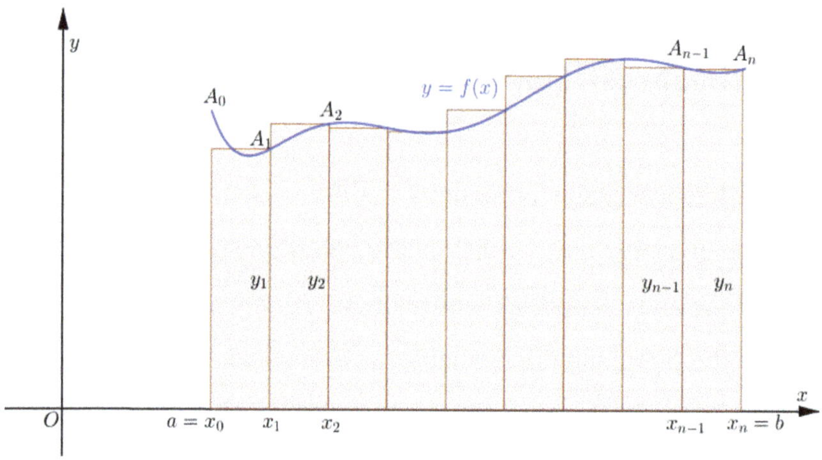

Figure 7.6.1

The areas of these rectangles are $\Delta x \cdot y_1$, $\Delta x \cdot y_2$, …, $\Delta x \cdot y_n$, respectively. The sum of the areas of these rectangles provides an approximation to the area of the plane figure aA_0A_nb; that is, it is an approximation to the definite integral $\int_a^b f(x)dx$.

Therefore,
$$\int_a^b f(x)dx = \Delta x(y_1 + y_2 + ... + y_n) + R \qquad (R \text{ is an error.})$$
$$= \frac{b-a}{n}\sum_{i=1}^{n} y_i + R$$

That is,
$$\int_a^b f(x)dx = \frac{b-a}{n}\sum_{i=1}^{n} y_i + R$$

Thus, we obtain the rectangular method formula for calculating an approximation to the definite integral.

$$\int_a^b f(x)dx \approx \frac{b-a}{n}\sum_{i=1}^{n} y_i$$

or $$\int_a^b f(x)dx \approx \Delta x\sum_{i=1}^{n} y_i$$

where $$\Delta x = \frac{b-a}{n}, \quad y_i = f(x_i) = f(a+i\Delta x), \quad i = 0, 1, 2, \ldots, n$$

Error $$R = -\frac{(b-a)^2}{2n} f'(c) \qquad c \in]a, b[$$

Absolute error $$e = |R| = \left|-\frac{(b-a)^2}{2n} f'(c)\right|$$

2. Trapezoidal method

The trapezoidal method is a refinement of the rectangular method. It approximates each curved subregion with a trapezoid, as illustrated in Figure 7.6.2. Under the same conditions as the rectangular method, we connect adjacent points with chords (i.e. straight line segments) A_0A_1, A_1A_2, …, $A_{n-1}A_n$. In this way, we obtain n trapezoids, as shown in Figure 7.6.2.

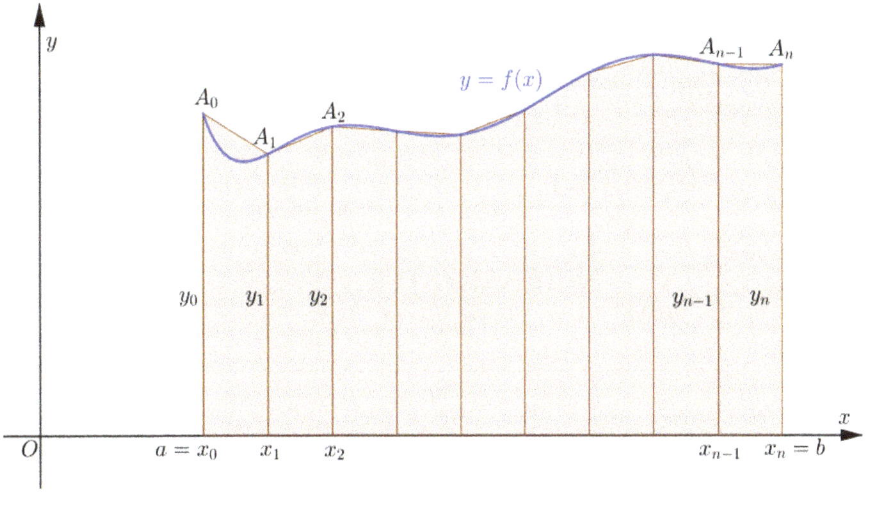

Figure 7.6.2

The areas of these trapezoids are $\frac{y_0 + y_1}{2}\Delta x$, $\frac{y_1 + y_2}{2}\Delta x$, …, $\frac{y_{n-1} + y_n}{2}\Delta x$, respectively. The sum of these areas provides an approximation to the area of the region aA_0A_nb; that is, it is an approximation to the definite integral $\int_a^b f(x)dx$.

Hence, $$\int_a^b f(x)dx = \Delta x\left(\frac{y_0 + y_1}{2} + \frac{y_1 + y_2}{2} + \ldots + \frac{y_{n-1} + y_n}{2}\right) + R$$ (R is an error.)

That is, $$\int_a^b f(x)dx = \Delta x\left(\frac{y_0 + y_n}{2} + y_1 + y_2 + \ldots + y_{n-1}\right) + R$$

or $$\int_a^b f(x)dx = \frac{\Delta x}{2}\left(y_0 + 2y_1 + 2y_2 + \ldots + 2y_{n-1} + y_n\right) + R$$

Thus, we obtain the trapezoidal method formula for calculating an approximation to the definite integral.

$$\int_a^b f(x)dx \approx \Delta x\left(\frac{y_0 + y_1}{2} + \frac{y_1 + y_2}{2} + \ldots + \frac{y_{n-1} + y_n}{2}\right)$$

That is, $$\int_a^b f(x)dx \approx \Delta x\left(\frac{y_0 + y_n}{2} + y_1 + y_2 + \ldots + y_{n-1}\right)$$

or $$\int_a^b f(x)dx \approx \frac{\Delta x}{2}\left(y_0 + 2y_1 + 2y_2 + \ldots + 2y_{n-1} + y_n\right)$$

where $\Delta x = \frac{b-a}{n}$, $y_i = f(x_i) = f(a + i\Delta x)$, $i = 0, 1, 2, \ldots, n$

Error $R = -\frac{(b-a)(\Delta x)^2}{12} f''(c)$ $\quad c \in \,]a, b[$

Absolute error $e = |R| = \left|-\frac{(b-a)(\Delta x)^2}{12} f''(c)\right|$

Example 1. Use the trapezoidal method with 6 intervals to approximate the value of

$$\int_1^7 e^x dx.$$

Solution. $f(x)=e^x$, $a=1$, $b=7$, $n=6$, $\Delta x=\frac{b-a}{n}=\frac{7-1}{6}=1$

$$y_0=f(a+0\cdot\Delta x)=f(a)=f(1)=e^1$$
$$y_1=f(a+1\cdot\Delta x)=f(1+1)=f(2)=e^2$$
$$y_2=f(a+2\cdot\Delta x)=f(1+2)=f(3)=e^3$$
$$y_3=f(a+3\cdot\Delta x)=f(1+3)=f(4)=e^4$$
$$y_4=f(a+4\cdot\Delta x)=f(1+4)=f(5)=e^5$$
$$y_5=f(a+5\cdot\Delta x)=f(1+5)=f(6)=e^6$$
$$y_6=f(a+6\cdot\Delta x)=f(1+6)=f(7)=e^7$$

Since

$$\int_a^b f(x)dx\approx\frac{\Delta x}{2}\left(y_0+2y_1+2y_2+\ldots+2y_{n-1}+y_n\right)$$

then $$\int_1^7 e^x dx\approx\frac{1}{2}\left[e+2(e^2+e^3+e^4+e^5+e^6)+e^7\right]\approx 1183.6$$

Example 2. (a) What is the error when using the trapezoidal method to find $\int_1^2 \frac{dx}{x}$?

(b) How many intervals are needed to make the error less than 10^{-3}?

Solution. (a) $f(x)=\frac{1}{x}$ $f''(x)=\frac{2}{x^3}$

When $1\le x\le 2$ $|f''(x)|=\frac{2}{x^3}\le 2$

Therefore, $e = |R| = \left|-\frac{(b-a)(\Delta x)^2}{12}f''(c)\right| < \left|-\frac{(2-1)(\Delta x)^2}{12}\cdot 2\right| = \frac{(\Delta x)^2}{6}$

(b) $$\frac{(\Delta x)^2}{6}<10^{-3}$$

$$\Delta x < \frac{\sqrt{15}}{50}$$

$$\Delta x = \frac{2-1}{n} < \frac{\sqrt{15}}{50}$$

$$n > 12.91$$

Therefore, $\quad n = 13$

3. Simpson's method (or parabolic method)

In the trapezoidal method, we replace the curve $f(x)$ in the small interval with a straight chord. Now, let us try using a parabola $y = ax^2 + bx + c$ to replace the curve on the corresponding small interval. The result obtained is generally more accurate.

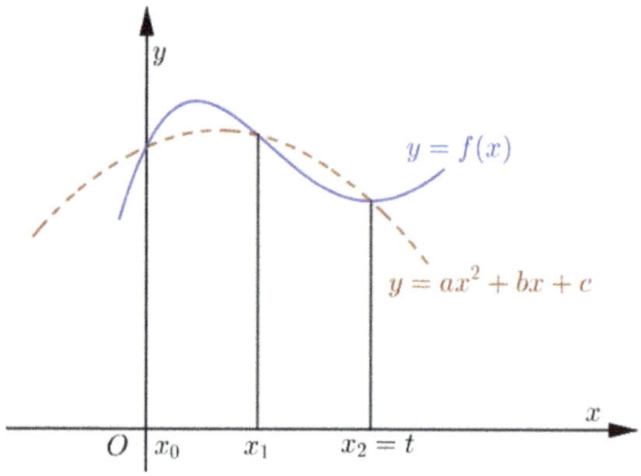

Figure 7.6.3

First, calculate the area A bounded by the parabola $y = ax^2 + bx + c$, the x-axis, the y-axis and the line $x = t$, as shown in Figure 7.6.3.

$$\mathrm{A} = \int_0^t \left(ax^2 + bx + c\right)dx = \left[\frac{ax^3}{3} + \frac{bx^2}{2} + cx\right]_0^t = \frac{t}{6}\left(2at^2 + 3bt + 6c\right)$$

Then, calculate the corresponding y-coordinates on the parabola when $x_0 = 0$, $x_1 = \frac{t}{2}$, $x_2 = t$.

$$y_0 = c$$

$$y_1 = \frac{at^2}{4} + \frac{bt}{2} + c$$

$$y_2 = at^2 + bt + c$$

Thus $\quad y_0 + 4y_1 + y_2 = 2at^2 + 3bt + 6c$

That is, $\quad \frac{t}{6}(y_0 + 4y_1 + y_2) = \frac{t}{6}(2at^2 + 3bt + 6c) = \mathrm{A}$

Hence, $\quad \mathrm{A} = \int_0^t (ax^2 + bx + c)dx = \frac{t}{6}(y_0 + 4y_1 + y_2)$

Therefore, we obtain formula (1):

$$\int_0^t f(x)dx = \int_{x_0}^{x_2} f(x)dx \approx \frac{x_2 - x_0}{6}[f(x_0) + 4f(x_1) + f(x_2)] \qquad (1)$$

where $x_1 = \frac{x_0 + x_2}{2}$ is the mid-point of the interval $[x_0, x_2]$.

Now we calculate the area bounded by the continuous curve $y = f(x)$ (where $y > 0$), the x-axis and the lines $x = a, x = b$, as shown in Figure 7.6.4.

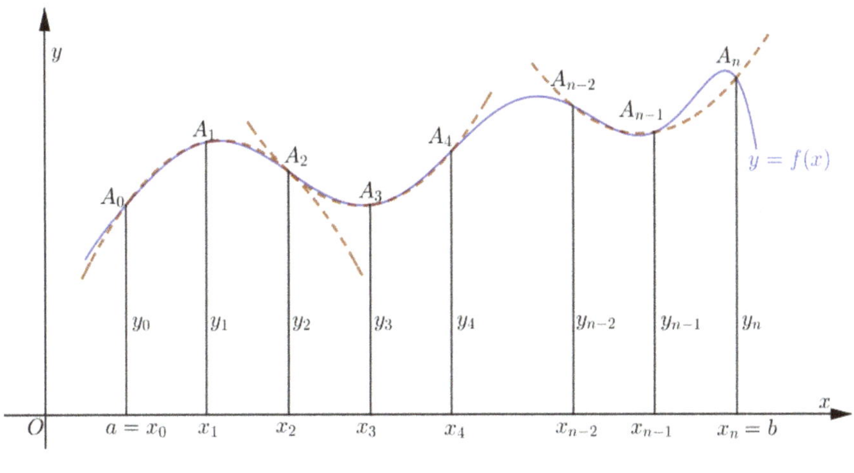

Figure 7.6.4

Use points $a = x_0 < x_1 < x_2 < ... < x_{n-1} < x_n = b$ to divide the interval $[a, b]$ into n subintervals of equal length. The length of each subinterval is $\Delta x = \frac{b-a}{n}$. The

y-coordinates of these points are $y_0, y_1, y_2, \ldots, y_{n-1}, y_n$, respectively. For each subinterval $[x_0, x_2], [x_2, x_4], \ldots, [x_{n-2}, x_n]$, using formula (1) gives

$$\int_a^b f(x)dx = \int_{x_0}^{x_2} f(x)dx + \int_{x_2}^{x_4} f(x)dx + \ldots + \int_{x_{n-2}}^{x_n} f(x)dx$$

$$= \frac{x_2 - x_0}{6}(y_0 + 4y_1 + y_2) + \frac{x_4 - x_2}{6}(y_2 + 4y_3 + y_4) + \ldots + \frac{x_n - x_{n-2}}{6}(y_{n-2} + 4y_{n-1} + y_n) + R$$

$$= \frac{b-a}{3n}(y_0 + 4y_1 + y_2) + \frac{b-a}{3n}(y_2 + 4y_3 + y_4) + \ldots + \frac{b-a}{3n}(y_{n-2} + 4y_{n-1} + y_n) + R$$

where $x_2 - x_0 = x_4 - x_2 = \ldots = x_n - x_{n-2} = 2\Delta x = \dfrac{2(b-a)}{n}$, R is an error.

Thus, we obtain Simpson's formula (or parabola formula) to calculate an approximation to the definite integral.

$$\int_a^b f(x)dx \approx \frac{\Delta x}{3}(y_0 + 4y_1 + 2y_2 + 4y_3 + 2y_4 + \ldots + 2y_{n-2} + 4y_{n-1} + y_n)$$

$$= \frac{\Delta x}{3}\left[(y_0 + y_n) + 4(y_1 + y_3 + \ldots + y_{n-1}) + 2(y_2 + y_4 + \ldots + y_{n-2})\right]$$

where $\Delta x = \dfrac{x_2 - x_0}{2} = \dfrac{x_4 - x_2}{2} = \ldots = \dfrac{x_n - x_{n-2}}{2} = \dfrac{b-a}{n}$

***n* can only be even numbers.**

$y_i = f(x_i) = f(a + i\Delta x)$, $\quad i = 0, 1, 2, \ldots, n$

Error $\quad R = -\dfrac{(b-a)(\Delta x)^4}{180} f^{(4)}(c) \qquad c \in\,]a, b[$

Absolute error $\quad e = |R| = \left| -\dfrac{(b-a)(\Delta x)^4}{180} f^{(4)}(c) \right|$

Example 3. Find the approximate value of $\displaystyle\int_1^4 \frac{dx}{x}$ using 6 intervals

a) with the trapezoidal method;

b) with Simpson's method.

Solution. a) Use the trapezoidal method. Divide the interval [1, 4] into 6 subintervals.

$$f(x)=\frac{1}{x},\quad a=1,\quad b=4,\quad n=6,\qquad \Delta x=\frac{b-a}{n}=\frac{4-1}{6}=0.5$$

$$y_0=f(a+0\cdot\Delta x)=f(a)=f(1)=\frac{1}{1}=1$$

$$y_1=f(a+1\cdot\Delta x)=f(1+0.5)=f(1.5)=\frac{1}{1.5}\approx 0.667$$

$$y_2=f(a+2\cdot\Delta x)=f(1+1)=f(2)=\frac{1}{2}=0.5$$

$$y_3=f(a+3\cdot\Delta x)=f(1+1.5)=f(2.5)=\frac{1}{2.5}=0.4$$

$$y_4=f(a+4\cdot\Delta x)=f(1+2)=f(3)=\frac{1}{3}\approx 0.333$$

$$y_5=f(a+5\cdot\Delta x)=f(1+2.5)=f(3.5)=\frac{1}{3.5}\approx 0.286$$

$$y_6=f(a+6\cdot\Delta x)=f(1+3)=f(4)=\frac{1}{4}=0.25$$

Since $$\int_a^b f(x)dx\approx\frac{\Delta x}{2}\left(y_0+2\sum_{i=1}^{n-1}y_i+y_n\right)$$

$$=\frac{\Delta x}{2}\left(y_0+2y_1+2y_2+\ldots+2y_{n-1}+y_n\right)$$

thus $$\int_1^4\frac{dx}{x}\approx\frac{0.5}{2}[1+2(0.667+0.5+0.4+0.333+0.286)+0.25]=1.4055$$

b) Use the Simpson's method

Since $$\int_a^b f(x)dx\approx\frac{\Delta x}{3}\left(y_0+4y_1+2y_2+4y_3+2y_4+\ldots+2y_{n-2}+4y_{n-1}+y_n\right)$$

$$=\frac{\Delta x}{3}\left[\left(y_0+y_n\right)+4\left(y_1+y_3+\ldots+y_{n-1}\right)+2\left(y_2+y_4+\ldots+y_{n-2}\right)\right]$$

thus $$\int_1^4\frac{dx}{x}\approx\frac{0.5}{3}[1+0.25+4(0.667+0.4+0.286)+2(0.5+0.333)]=1.388$$

In fact, $\int_1^4 \frac{dx}{x} = [\ln x]_1^4 = \ln 4 - \ln 1 \approx 1.386$

Therefore, the absolute error using the trapezoidal method is

$$e = |1.4055 - 1.386| = 0.0195$$

The absolute error using Simpson's method is

$$e = |1.388 - 1.386| = 0.002$$

Consequently, it is evident that using Simpson's method (i.e., the parabola method) to calculate the approximate value of the definite integral provides greater accuracy.

Example 4. (a) What is the error when using Simpson's method to approximate the value of $\int_1^2 \frac{dx}{x}$?

(b) How many intervals are needed to make the error less than 10^{-3}?

Solution. (a) $f(x) = \frac{1}{x}$ Its fourth derivative is $f^{(4)}(x) = \frac{24}{x^5}$

When $1 \le x \le 2$ $\quad |f^{(4)}(x)| = \frac{24}{x^5} \le 24$

Therefore, $|R| = \left| -\frac{(b-a)(\Delta x)^4}{180} f^{(4)}(c) \right| < \frac{(2-1)(\Delta x)^4}{180} \cdot 24 = \frac{2(\Delta x)^4}{15}$

(b) $\frac{2(\Delta x)^4}{15} < 10^{-3}$

$\Delta x < 0.294$

$\Delta x = \frac{b-a}{n} = \frac{2-1}{n} < 0.294$

$n > \frac{1}{0.294} \approx 3.4$

Hence, $n = 4$

Exercise 7.6

1. **a.** Use the rectangular method with $n = 12$ to approximate the value of the definite integral $\int_0^{2\pi} x \sin x dx$.

b. Use integration by parts to calculate the exact value of the definite integral $\int_0^{2\pi} x \sin x dx$.

c. What is the percentage error when using the result from part a. to approximate the value of the definite integral $\int_0^{2\pi} x \sin x dx$?

2. Use the trapezoidal method with the given value of n to approximate the value of each of the following definite integrals.

(i) Give your answers correct to five decimal places.

(ii) Estimate the maximum error with the approximation.

a. $\int_0^1 \frac{dx}{1+x}$ $n = 8$ **b.** $\int_0^1 \frac{dx}{1+x^3}$ $n = 12$

c. $\int_0^{\pi} \sin x dx$ $n = 6$ **d.** $\int_0^1 e^{-x^2} dx$ $n = 4$

3. Use Simpson's method with the given value of n to approximate the value of each of the following definite integrals. Give your answers correct to five decimal places.

a. $\int_1^9 \sqrt{x} dx$ $n = 4$ **b.** $\int_0^{\pi} \sqrt{3 + \cos x} dx$ $n = 6$

c. $\int_0^{\frac{\pi}{2}} \frac{\sin x}{x} dx$ $n = 10$ **d.** $\int_0^1 \frac{x dx}{\ln(1+x)}$ $n = 6$

4. Using the error estimate for the trapezoidal method, determine the minimum number of sub-intervals, n, required to ensure that the approximation of each integral is correct to the specified degree of accuracy.

a. $\int_0^3 \sqrt{x+1}\,dx$ correct to three decimal places

b. $\int_1^3 e^{-x}\,dx$ Error is less than 10^{-6}.

5. Using the error estimate for Simpson's method, determine the minimum number of sub-intervals, n, required to ensure that the approximation of each integral is correct to five decimal places.

a. $\int_0^1 x^4\,dx$ **b.** $\int_{-1}^1 e^x\,dx$

Chapter 8 First order differential equations

8.1. Some concepts of first order differential equations

In the study of many practical problems in the fields of science, engineering, economics, it is often necessary to establish the functional relationship between variables. However, it is not always easy to establish such a relationship directly; instead, it may be easier to establish the relationship between the unknown function and its derivatives or differentials. Equations that contain derivatives or differentials of unknown functions are called **differential equations**. By solving differential equations, we can determine the functional relationship between the variables.

If the unknown function is a function of one single independent variable, the differential equation is called an **ordinary differential equation (ODE)**. For example,

$$\frac{dy}{dx}+x=1 \qquad\qquad \frac{d^2y}{dx^2}+x\frac{dy}{dx}+x^2=y+1$$

If the unknown function is a function of several variables, the differential equation is called a **partial differential equation (PDE)**. For example,

$$\frac{\partial^2 u}{\partial x^2}+\frac{\partial u}{\partial y}=1$$

The **order** of a differential equation is defined as the order of the highest derivative (or differential) present within the differential equation. For instance,

$$\frac{dy}{dx}+x=1$$ is a first order differential equation.

$$\frac{d^2y}{dx^2}+x\frac{dy}{dx}+x^2=y+1$$ is a second order differential equation.

This chapter provides a brief introduction to the fundamental concepts of ordinary differential equations, as well as a few methods for solving simple first order differential equations.

If a function $y = f(x)$ and its derivatives are substituted into a differential equation and satisfy it, then the function is called a solution of the differential equation.

If the solution of a differential equation contains arbitrary constants, and the number of arbitrary constants is equal to the order of the differential equation, then this solution is called the **general solution** of the differential equation. For example,

$$y = x^2 + C$$

is the general solution of the differential equation

$$\frac{dy}{dx} - 2x = 0$$

If initial conditions are provided, specific values for the arbitrary constants can be determined; the resulting function is then called the **particular solution** of the differential equation. For example, given the conditions $x = 1, y = 3$, substituting these into $y = x^2 + C$ gives

$$3 = 1 + \mathrm{C}$$

$$\mathrm{C} = 2$$

Hence, $y = x^2 + 2$ is the particular solution of differential equation $\frac{dy}{dx} - 2x = 0$.

Exercise 8.1

1. Verify that

a. $y = x^5 - 1$ is a solution of $\frac{dy}{dx} = 5x^4$;

b. $y = e^{-x}(x+4)$ is a solution of $\frac{dy}{dx} + y = e^{-x}$;

c. $y = \frac{x^2}{3} + \frac{7}{x}$ is a solution of $x\frac{dy}{dx} + y = x^2$;

d. $y = \sqrt[3]{3x^2 + C}$ is a solution of $y^2 dy = 2x dx$;

e. $y = 3\sin x - 4\cos x$ is a solution of $y'' + y = 0$;

f. $xy + y^2 - x^2 = C$ is a solution of $(x + 2y)y' = 2x - y$.

8.2. Separable first order differential equations

A first order differential equation in the form

$$h(y)\frac{dy}{dx} = g(x)$$

is called a separable first order differential equation. This is because the equation can be rearranged to the following form:

$$h(y)dy = g(x)dx$$

That is, it can be rewritten such that one side contains expressions involving only y and dy, while the other side contains expressions involving only x and dx.

Integrate both sides.

$$\int h(y)dy = \int g(x)dx$$

This will yield the general solution of the given differential equation.

Note:

- After the variables have been separated, dy and dx must be positioned in the numerator on both sides of the equation to allow for integration.
- After integrating both sides, it is only necessary to add a single arbitrary constant C to one side of the equation. This is because if separate arbitrary constants C_1 and C_2 are added to each side of the equation, they can be combined into a single constant C.

Example 1. Find the general solution of the differential equation $\frac{dy}{dx} = -\frac{y}{x}$.

Solution. First, separate the variables.

$$\frac{dy}{y} = -\frac{dx}{x}$$

Integrating both sides gives

$$\int \frac{dy}{y} = \int -\frac{dx}{x}$$

$$\ln y = -\ln x + C$$

$$\ln y + \ln x = C$$

$$\ln(yx) = C$$

$$yx = e^C$$

$$yx = A \qquad \text{where} \qquad A = e^C$$

Hence, either $yx = A$ or $y = \frac{A}{x}$ is the general solution of the given differential equation.

Example 2. Find the particular solution of the differential equation

$$\frac{dy}{dx} + \cos^2 y \sin x = 0 \quad \text{given that } x = 0, y = -\frac{\pi}{4}.$$

Solution. First, separate the variables.

$$\frac{dy}{dx} = -\cos^2 y \sin x$$

$$\frac{dy}{\cos^2 y} = -\sin x dx$$

Integrating both sides gives

$$\int \sec^2 y dy = \int -\sin x dx$$

$$\tan y = \cos x + C$$

Substituting $x = 0,\ y = -\frac{\pi}{4}$ gives

$$\tan(-\frac{\pi}{4}) = \cos 0 + C$$

$$-1 = 1 + C$$

$$C = -2$$

Therefore, the particular solution of the differential equation is

$$\tan y = \cos x - 2$$

Example 3. The net profit, L, of a product is known to be related to the advertising expenditure, x. The net profit increases as the advertising expenditure increases, and its rate of growth is proportional to the difference between a positive constant, G, and the net profit, L. Given that the net profit is L_0 ($L_0 < G$) when the advertising expenditure is zero, find the net profit as a function of x, $L = L(x)$.

Solution. From the information provided, we obtain

$$\frac{dL}{dx} = k(G - L)$$ where k is a positive constant proportionality.

Separating the variables gives

$$\frac{dL}{G - L} = kdx$$

Integrating both sides yields

$$\int \frac{dL}{G - L} = \int kdx$$

$$-\ln(G - L) = kx + C$$

$$\ln(G - L) = -kx - C$$

$$G - L = e^{-kx - C}$$

$$L = G - e^{-kx} \cdot e^{-C}$$

$$L = G - Ae^{-kx} \qquad \text{where} \qquad A = e^{-C}$$

Substituting $x = 0, L = L_0$ gives

$$L_0 = G - Ae^{-k \cdot 0}$$

$$A = G - L_0$$

Therefore, $L = G - (G - L_0)e^{-kx}$

Example 4. According to Newton's Law of Cooling, the rate at which the temperature of a body falls is proportional to the difference between the body's temperature and that of its surroundings. A room is at a constant temperature of 20°C. An object with a temperature of 90°C is brought into the room. After 7 minutes, its temperature has fallen to 70°C. How long will it take for the body's temperature to drop to 30°C?

Solution. Since the temperature of a body is falling, the differential equation must reflect this decrease with a negative sign. Therefore, we have

$$\frac{dT}{dt} = -k(T - 20)$$

$$\int \frac{dT}{T - 20} = \int -kdt$$

$$\ln(T - 20) = -kt + C$$

$$T - 20 = e^{-kt+C}$$

$$T - 20 = Ae^{-kt} \qquad \text{where} \qquad A = e^{C}$$

When $t = 0, T = 90$, we have

$$90 - 20 = Ae^{-k \cdot 0}$$

$$A = 70$$

When $t = 7, T = 70$, we have

$$70 - 20 = 70e^{-7k}$$

$$50 = 70e^{-7k}$$

$$\frac{5}{7} = e^{-7k}$$

$$\ln\frac{5}{7} = -7k$$

$$k = -\frac{1}{7}\ln\frac{5}{7}$$

Therefore, $\quad T - 20 = 70e^{\frac{t}{7}\ln\frac{5}{7}}$

When $T = 30$

$$30 - 20 = 70e^{\frac{t}{7}\ln\frac{5}{7}}$$

$$\frac{1}{7} = e^{\frac{t}{7}\ln\frac{5}{7}}$$

$$\ln\frac{1}{7} = \frac{t}{7}\ln\frac{5}{7}$$

$$t = \frac{7\ln\frac{1}{7}}{\ln\frac{5}{7}} \approx 40.5$$

Therefore, it will take approximately 40.5 minutes for the body's temperature to fall to 30°C.

Example 5. The gradient of the tangent at any point (x, y) on a curve is $\frac{y(1-x)}{x}$.

Given that the curve passes through the point $(1, \frac{3}{e})$, find the equation of the curve.

Solution. $\quad \frac{dy}{dx} = \frac{y(1-x)}{x}$

Separate the variables.

$$\frac{dy}{y} = \frac{1-x}{x}dx$$

Integrate both sides.

$$\int \frac{dy}{y} = \int (\frac{1}{x} - 1)dx$$

$$\ln y = \ln x - x + C$$

$$\ln y - \ln x = -x + C$$

$$\ln \frac{y}{x} = -x + C$$

$$\frac{y}{x} = e^{-x+C}$$

$$y = Axe^{-x} \qquad \text{where} \qquad A = e^C$$

Given that the curve passes through the point $(1, \frac{3}{e})$

$$\frac{3}{e} = Ae^{-1}$$

$$A = 3$$

Therefore, the equation of the curve is

$$y = 3xe^{-x}$$

Example 6. The normal at the point P(x, y) on a curve intersects the x-axis at Q. The y-axis bisects the line segment PQ. Find a differential equation that describes this curve.

Solution. Let the equation of the curve be $y = y(x)$. The equation of the normal at the point P(x, y) is

$$Y - y = -\frac{1}{y'}(X - x)$$

When $Y = 0$

$$X = x + yy'$$

Therefore, the coordinates of Q are $(x + yy', 0)$.

Since the midpoint of line segment PQ is on the y-axis, it follows that

$$\frac{x + x + yy'}{2} = 0$$

Therefore, the differential equation is

$$2x + yy' = 0$$

Example 7. The tangent to a point A(x, y) on a curve in the first quadrant meets the x-axis at the point B($\frac{x}{2}$, 0). Given that the curve passes through the point (9, 81), find the equation of the curve.

Solution. We can use the point A(x, y) and the point B($\frac{x}{2}$, 0) to determine the gradient of the tangent at the point A.

$$\frac{dy}{dx} = \frac{y - 0}{x - \frac{x}{2}}$$

$$\frac{dy}{dx} = \frac{2y}{x}$$

Separating the variables and integrating both sides gives

$$\int \frac{dy}{y} = 2\int \frac{dx}{x}$$

$$\ln y = 2\ln x + \ln C$$

$$y = x^2 C$$

Since the curve passes through the point (9, 81)

$$81 = 9^2 C$$

$$C = 1$$

hence, the equation of the curve is

$$y = x^2$$

Example 8. An object moves in a straight line with uniform acceleration, *a*. The initial conditions are that at time $t = 0$, the speed is $V = V_0$ and the distance is $S = 0$. Determine the physics formulae for the instantaneous speed V, uniform acceleration a and the distance S after time *t*.

Solution. Based on the conditions of the object's linear motion, we can establish the following differential equations.

The instantaneous speed is $$V = \frac{dS}{dt}$$

The instantaneous acceleration is $$a = \frac{dV}{dt}$$

where S is the distance, t is the time.

When the acceleration, a, is uniform (that is, a is a constant), solving the differential equations yields the kinematic formulae for an object's linear motion.

From $$a = \frac{dV}{dt}$$

cross-multiplying and integrating

$$\int_0^t a dt = \int_{V_0}^{V} dV$$

gives $$at = V - V_0$$

That is, $$a = \frac{V - V_0}{t} \quad (1)$$

Hence, $$V = V_0 + at \quad (2)$$

Substitute $V = V_0 + at$ into

$$V = \frac{dS}{dt}$$

Cross-multiplying and integrating

$$\int_0^t (V_0 + at)dt = \int_0^S dS$$

gives $$S = V_0 t + \frac{1}{2}at^2 \qquad (3)$$

Moreover, $$V = \frac{dS}{dt} = \frac{dS}{dt} \cdot \frac{dV}{dV} = \frac{dS}{dV} \cdot \frac{dV}{dt} = \frac{dS}{dV} \cdot a$$

Cross-multiplying and integrating

$$\int_{V_0}^V VdV = \int_0^S adS$$

gives $$\frac{V^2 - V_0^2}{2} = aS$$

That is, $$S = \frac{V^2 - V_0^2}{2a} \qquad (4)$$

Therefore, by solving the differential equations, we obtain a set of physics formulae.

$$a = \frac{V - V_0}{t}$$

$$V = V_0 + at$$

$$S = V_0 t + \frac{1}{2}at^2$$

$$S = \frac{V^2 - V_0^2}{2a}$$

Exercise 8.2

1. Find the general solution of each of the following differential equations.

a. $(1+x^2)\frac{dy}{dx}+xy=0$

b. $y^2dy-(1-x)dx=0$

c. $xydx+\sqrt{1-x^2}\,dy=0$

d. $\frac{dy}{dx}=y\cos x$

e. $\frac{dy}{dx}=e^{x-y}+x^2e^{-y}$

f. $y-x\frac{dy}{dx}=k(y^2+\frac{dy}{dx})$

g. $x\cos^2 ydx=y\cos^2 xdy$

h. $\frac{dy}{dx}=\frac{1+y^2}{xy+x^3y}$

2. Find the particular solution of each of the following differential equations.

a. $3y^4-1+12xy^3\frac{dy}{dx}=0$ $\quad y(1)=2$

b. $(1+e^x)\sin ydy+e^x\cos ydx=0$ $\quad y(0)=\pi$

c. $(1+x^2)y'=\arctan x$ $\quad y(0)=0$

d. $\frac{x}{1+y}dx-\frac{y}{1+x}dy=0$ $\quad y(0)=1$

e. $y\ln xdx+x\ln ydy=0$ $\quad y(1)=1$

f. $3e^x\tan ydx+(1-e^x)\sec^2 ydy=0$ $\quad y(1)=\frac{\pi}{4}$

g. $\frac{dy}{dx}=(1-y^2)\tan x$ $\quad y(0)=2$

h. $(x+1)(x-2)\frac{dy}{dx}+y=0$ $\quad y(1)=-3$

3. Find the equation of the curve that passes through the point (1, 2), given that the gradient of the curve being $\frac{y}{x}$.

4. The tangent at the point (x,y) on a curve intersects the x-axis at $(x-2,0)$. Given that the curve passes through the point (0, 3), find the equation of the curve.

5. The gradient of the tangent at the point P(x, y) on a curve is equal to the square of the x-coordinate of the point P. Given that the curve passes through the point (3, 15), find the equation of the curve.

6. According to Newton's Law of Cooling, the rate at which the temperature of a body falls is proportional to the difference between the body's temperature and that of its surroundings. A room is at a constant temperature of 21°C. A cake has a temperature of 150°C when it is brought from the oven into the room. After 3 minutes, its temperature has fallen to 90°C. How long will it take for the cake's temperature to drop 22°C?

7. A city's population increases at a rate proportional to the total population at any given time. Let $P(t)$ be the population at time t. Given that in 2000, the population was 200,000, and by 2010, it had increased to 250,000. Calculate the population in the year 2023. Use the year 2000 as the starting point.

8. The decay rate of a radioactive substance is proportional to the amount of the substance present at time t. The half-life (the time taken for half of the substance to decay) of the substance is 5,730 years. A piece of animal fossil is found to contain 73% of the normal amount of the radioactive substance. Estimate how many years ago the animal lived.

8.3. Homogeneous differential equations

If a differential equation can be written in the form

$$\frac{dy}{dx} = f(\frac{y}{x})$$

where $\frac{dy}{dx}$ is a function of $\frac{y}{x}$, then it is called a homogeneous differential equation.

Steps to solve a homogeneous differential equation:

Let $u = \frac{y}{x}$ then $y = ux$

Substituting gives $\frac{d(ux)}{dx} = f(u)$

Since $\frac{d(ux)}{dx} = \frac{du}{dx}x + u$

then $\frac{du}{dx}x + u = f(u)$

Rearrange. $\frac{du}{dx}x = f(u) - u$

Cross-multiply. $\frac{du}{f(u) - u} = \frac{dx}{x}$

This is a separable differential equation.

Integrate both sides. $\int \frac{du}{f(u) - u} = \int \frac{dx}{x}$

Finally, substitute $u = \frac{y}{x}$ into the solution to the above equation to obtain the solution of the original equation $\frac{dy}{dx} = f(\frac{y}{x})$.

Example 1. $\frac{dy}{dx} = \frac{y^2}{xy - x^2}$

(a) Show that this equation is a homogeneous differential equation.

(b) Find the general solution.

Solution.

(a) On the right-hand side of the equation, divide both the numerator and the denominator by x^2.

$$\frac{dy}{dx} = \frac{(\frac{y}{x})^2}{\frac{y}{x} - 1}$$

It can be seen that $\frac{dy}{dx}$ is a function of $\frac{y}{x}$; therefore, it is a homogeneous differential equation.

(b) Since $$\frac{dy}{dx} = \frac{(\frac{y}{x})^2}{\frac{y}{x} - 1}$$

let $u = \frac{y}{x}$ then $y = ux$

Substituting gives

$$\frac{d(ux)}{dx} = \frac{u^2}{u-1}$$

Thus $$\frac{du}{dx}x + u = \frac{u^2}{u-1}$$

$$\frac{du}{dx}x = \frac{u^2}{u-1} - u = \frac{u}{u-1}$$

Separating the variables and integrating both sides gives

$$\int \frac{u-1}{u} du = \int \frac{dx}{x}$$

$$u - \ln u + C = \ln x$$

$$\ln(ux) = u + C$$

Substituting $u = \frac{y}{x}$ gives

$$\ln y = \frac{y}{x} + C$$

$$y = Ae^{\frac{y}{x}} \qquad \text{where} \qquad A = e^C$$

Example 2. $(xy - y^2)dx - (x^2 - 2xy)dy = 0$

(a) Show that this equation is a homogeneous differential equation.

(b) Given that $x = 1$ and $y = 1$, find the particular solution of this equation.

Solution.

(a)
$$(xy - y^2)dx - (x^2 - 2xy)dy = 0$$

$$(xy - y^2)dx = (x^2 - 2xy)dy$$

$$\frac{dy}{dx} = \frac{xy - y^2}{x^2 - 2xy}$$

Divide both the numerator and the denominator on the right-hand side of the equation by x^2.

$$\frac{dy}{dx} = \frac{\frac{y}{x} - (\frac{y}{x})^2}{1 - 2\frac{y}{x}}$$

It can be seen that $\frac{dy}{dx}$ is a function of $\frac{y}{x}$; therefore, it is a homogeneous differential equation.

(b) Since
$$\frac{dy}{dx} = \frac{\frac{y}{x} - (\frac{y}{x})^2}{1 - 2\frac{y}{x}}$$

let $u = \frac{y}{x}$ then $y = ux$

Substituting gives

$$\frac{d(ux)}{dx} = \frac{u-u^2}{1-2u}$$

Hence, $$\frac{du}{dx}x + u = \frac{u-u^2}{1-2u}$$

$$\frac{du}{dx}x = \frac{u-u^2}{1-2u} - u = \frac{u^2}{1-2u}$$

Separating the variables and integrating both sides gives

$$\int \frac{1-2u}{u^2} du = \int \frac{dx}{x}$$

$$-\frac{1}{u} - 2\ln u = \ln x + \ln C$$

$$-\frac{1}{u} = \ln(xu^2C)$$

$$xu^2C = e^{-\frac{1}{u}}$$

Substituting $u = \frac{y}{x}$ yields

$$x(\frac{y}{x})^2 C = e^{-\frac{x}{y}}$$

$$\frac{y^2C}{x} = e^{-\frac{x}{y}}$$

Substituting $x = 1$ and $y = 1$ gives

$$C = e^{-1}$$

Therefore, the particular solution of the equation is

$$\frac{y^2 e^{-1}}{x} = e^{-\frac{x}{y}}$$

Example 3. The function $y = f(x)$ satisfies the following differential equation

$$x\frac{dy}{dx} - \sqrt{x^2 - y^2} = y \qquad x > 0$$

(a) Show that this is a homogeneous differential equation.

(b) Use the substitution $y = ux$ to show that $\frac{du}{dx} = \frac{\sqrt{1-u^2}}{x}$.

(c) Hence, prove that if $y = 0$ when $x = 1$, the solution of the original differential equation is $y = x\sin(\ln x)$ for $x > 0$.

Solution.

(a) $$x\frac{dy}{dx} - \sqrt{x^2 - y^2} = y$$

Dividing both sides by x gives

$$\frac{dy}{dx} - \frac{1}{x}\sqrt{x^2 - y^2} = \frac{y}{x}$$

$$\frac{dy}{dx} - \sqrt{1-(\frac{y}{x})^2} = \frac{y}{x}$$

$$\frac{dy}{dx} = \sqrt{1-(\frac{y}{x})^2} + \frac{y}{x}$$

It can be seen that $\frac{dy}{dx}$ is a function of $\frac{y}{x}$; therefore, it is a homogeneous differential equation.

(b) Since $$\frac{dy}{dx} = \sqrt{1-(\frac{y}{x})^2} + \frac{y}{x}$$

substituting $y = ux$ gives

$$\frac{d(ux)}{dx} = \sqrt{1-u^2} + u$$

Thus $$\frac{du}{dx}x + u = \sqrt{1-u^2} + u$$

Hence, $$\frac{du}{dx} = \frac{\sqrt{1-u^2}}{x}$$

(c) Separating the variables and integrating both sides gives

$$\int \frac{du}{\sqrt{1-u^2}} = \int \frac{dx}{x}$$

Then $\qquad \arcsin u = \ln x + \ln C$

$$\arcsin u = \ln(xC)$$

$$u = \sin[\ln(xC)]$$

Substituting $u = \frac{y}{x}$ gives

$$\frac{y}{x} = \sin[\ln(xC)]$$

Substituting $x = 1$ and $y = 0$ gives

$$0 = \sin(\ln C)$$

$$\ln C = 0$$

$$C = 1$$

Therefore, the solution of the original differential equation is $y = x\sin(\ln x), x > 0$.

Example 4. (a) Use the substitutions $x = X + 3$ and $y = Y - 2$ to show that the differential equation

$$(2x + y - 4)dx + (x + y - 1)dy = 0$$

can be transformed into a homogeneous differential equation.

(b) Find the general solution.

Solution.

(a) Since $\qquad (2x + y - 4)dx + (x + y - 1)dy = 0$

let $\qquad x = X + 3, \qquad y = Y - 2$

then $\qquad dx = dX, \qquad dy = dY$

Substituting gives

$$[2(X+3) + Y - 2 - 4]dX + (X + 3 + Y - 2 - 1)dY = 0$$

$$(2X+Y)dX+(X+Y)dY=0$$

$$\frac{dY}{dX}=-\frac{2X+Y}{X+Y}=-\frac{2+\frac{Y}{X}}{1+\frac{Y}{X}}=f(\frac{Y}{X})$$

This is a homogeneous differential equation.

(b) Since $\quad \frac{dY}{dX}=-\frac{2+\frac{Y}{X}}{1+\frac{Y}{X}}$

let $\quad u=\frac{Y}{X} \quad$ then $\quad Y=uX$

Substituting gives

$$\frac{d(uX)}{dX}=-\frac{2+u}{1+u}$$

hence, $$\frac{du}{dX}X+u=-\frac{2+u}{1+u}$$

$$\frac{du}{dX}X=-\frac{2+u}{1+u}-u= \; -\frac{2+2u+u^2}{1+u}$$

Separating the variables and integrating both sides gives

$$\int\frac{(1+u)du}{2+2u+u^2}=-\int\frac{dX}{X}$$

For $\int\frac{(1+u)du}{2+2u+u^2}$

let $\quad T=2+2u+u^2 \quad$ then $\quad dT=(2+2u)du=2(1+u)du\,, \quad du=\frac{dT}{2(1+u)}$

Substituting gives

$$\int\frac{(1+u)du}{2+2u+u^2}=\int\frac{1+u}{T}\cdot\frac{dT}{2(1+u)}=\frac{1}{2}\int\frac{dT}{T}=\frac{1}{2}\ln T=\ln\sqrt{2+2u+u^2}$$

For $-\int\frac{dX}{X}$

$$-\int \frac{dX}{X} = -\ln X$$

then $$\ln\sqrt{2+2u+u^2} = -\ln X + \ln C$$

$$\sqrt{2+2u+u^2} = \frac{C}{X}$$

Substituting $u = \frac{Y}{X} = \frac{y+2}{x-3}$ gives

$$\sqrt{2+2\cdot\frac{y+2}{x-3}+(\frac{y+2}{x-3})^2} = \frac{C}{x-3}$$

Exercise 8.3

1. Find the general solution of each of the following differential equations.

a. $\frac{dy}{dx} = \frac{y}{y+x}$

b. $\frac{dy}{dx} = \frac{y-2x}{x}$

c. $(x^2-2xy)\frac{dy}{dx} = y^2-2xy-3x^2$

d. $x\frac{dy}{dx} = y+\sqrt{x^2-y^2}$

e. $\frac{dy}{dx} = \frac{y^2}{xy-x^2}$

f. $\frac{dy}{dx} = e^{\frac{y}{x}} + \frac{y}{x}$

g. $xy' = y + 2x^3\sec^2\frac{y}{x}$

h. $\frac{dy}{dx} = \frac{y-4x}{x-y}$

2. Find the particular solution of the differential equation $\frac{dy}{dx} = \frac{x}{y} + \frac{y}{x}$ given that $y = 2$ when $x = -1$.

3. Find the particular solution of the differential equation $\frac{dy}{dx} = \frac{y^3+3x^2y}{x^3+3xy^2}$ given that

$y = 2$ when $x = 1$.

4. Solve the initial value problem: $x^2 \frac{dy}{dx} = y^2 - xy$, given that $y(1) = 1$.

5. Solve the initial value problem: $y - x\frac{dy}{dx} = 2x + 2y\frac{dy}{dx}$, given that $y(1) = 1$.

6. Use the substitutions $X = x + 2$ and $Y = y + 3$ to show that the differential equation

$$\frac{dy}{dx} = \frac{y - x + 1}{y + x + 5}$$

can be transformed into a homogeneous differential equation. Hence, show that the general solution of the equation is $\ln\left[(x+2)^2 + (y+3)^2\right] + 2\arctan\left(\frac{y+3}{x+2}\right) = C$.

7. Use the substitution $x = u^2$ to show that the differential equation

$$y^3 dx + 2(x^2 - xy^2)dy = 0$$

can be transformed into a homogeneous differential equation. Hence, show that the general solution of the equation is $y^2 = x(\ln y^2 + C)$.

8. Use the substitution $y = u - x$ to show that the differential equation

$$(x + y)dx + (3x + 3y - 4)dy = 0$$

can be transformed into

$$\frac{3u - 4}{2u - 4} du = dx$$

Hence, show that the general solution of the equation is $x + 3y + 2\ln|x + y - 2| = C$.

8.4. First order linear differential equations

A differential equation in the form

$$\frac{dy}{dx} + P(x)y = Q(x)$$

is called a first order linear differential equation. Its defining characteristic is that the equation contains terms of y and y' to the first power only.

(1) If $Q(x) = 0$, it is called a **first order linear homogeneous differential equation**. It can be solved by separating the variables.

$$\frac{dy}{dx} + P(x)y = 0$$

Separating the variables gives

$$\frac{dy}{y} = -P(x)dx$$

Integrating both sides yields

$$\ln y = -\int P(x)dx + \ln C$$

The general solution formula is given by

$$y = Ce^{-\int P(x)dx}$$

(2) If $Q(x) \neq 0$, it is called a **first order linear nonhomogeneous differential equation**. It can be solved by using an integrating factor.

$$\frac{dy}{dx} + P(x)y = Q(x)$$

The integrating factor is given by

$$u = e^{\int P(x)dx}$$

The general solution formula is then given by

$$uy = \int uQ(x)dx + C$$

Proof. Since the integrating factor is $u = e^{\int P(x)dx}$

thus $$\frac{du}{dx} = \frac{de^{\int P(x)dx}}{dx} = \frac{de^{\int P(x)dx}}{d\int P(x)dx} \cdot \frac{d\int P(x)dx}{dx} = e^{\int P(x)dx} P(x) = uP(x)$$

Since $$\frac{d(uy)}{dx} = u\frac{dy}{dx} + \frac{du}{dx}y = u\frac{dy}{dx} + uP(x)y$$

$$= u(\frac{dy}{dx} + P(x)y) = uQ(x)$$

that is, $$\frac{d(uy)}{dx} = uQ(x)$$

Separating the variables gives

$$d(uy) = uQ(x)dx$$

Integrating both sides gives the general solution formula.

$$uy = \int uQ(x)dx + \text{C}$$ □

When using an integrating factor to solve a differential equation, the differential equation must be first rearranged to the following standard form:

$$\frac{dy}{dx} + P(x)y = Q(x)$$

- The coefficient of $\frac{dy}{dx}$ must be 1.
- $P(x)$ is the coefficient of y.
- $Q(x)$ is positioned on the right-hand side of the equation, and it is a function of x only.
- There is no need to add a constant of integration to the integration result of the integrating factor.
- A frequently used logarithm formula is $a^{\log_a N} = N$. When $a = e$, we have $e^{\ln N} = N$.

Example 1. Find the general solution of $\frac{dy}{dx} + y\tan x = \cos x$.

Solution. Here $P(x) = \tan x$, $Q(x) = \cos x$

The integrating factor is

$$u = e^{\int P(x)dx} = e^{\int \tan x dx} = e^{\int \frac{\sin x}{\cos x}dx} = e^{-\ln\cos x} = e^{\ln\sec x} = \sec x$$

The general solution of the differential equation is

$$uy = \int uQ(x)dx + C$$

$$\sec x \cdot y = \int \sec x \cdot \cos x dx + C$$

$$\sec x \cdot y = x + C$$

$$y = (x + C)\cos x$$

Example 2. Find the particular solution of $x\frac{dy}{dx} + y = \sin x$ which satisfies the initial condition $y\big|_{x=\frac{\pi}{2}} = 1$.

Solution. First, divide both sides of the given equation by x.

$$\frac{dy}{dx} + \frac{y}{x} = \frac{\sin x}{x}$$

Thus $P(x) = \frac{1}{x}$, $Q(x) = \frac{\sin x}{x}$

Hence, the integrating factor is

$$u = e^{\int P(x)dx} = e^{\int \frac{1}{x}dx} = e^{\ln x} = x$$

The general solution of the differential equation is

$$uy = \int uQ(x)dx + C$$

$$xy = \int x \cdot \frac{\sin x}{x} dx + C$$

$$xy = -\cos x + C$$

Substituting the initial condition $y\big|_{x=\frac{\pi}{2}} = 1$ gives

$$\frac{\pi}{2} = -\cos\frac{\pi}{2} + C$$

$$C = \frac{\pi}{2}$$

Therefore, the particular solution of the differential equation is

$$xy = -\cos x + \frac{\pi}{2}$$

$$y = \frac{1}{x}(-\cos x + \frac{\pi}{2})$$

Example 3. Solve the initial value problem $(x^2 - 1)\frac{dy}{dx} + 2xy - \cos x = 0$, given that

$$y\big|_{x=0} = 1.$$

Solution. Rearranging gives

$$(x^2 - 1)\frac{dy}{dx} + 2xy = \cos x$$

Dividing both sides by $x^2 - 1$ yields

$$\frac{dy}{dx} + \frac{2xy}{x^2 - 1} = \frac{\cos x}{x^2 - 1}$$

Thus $\quad P(x) = \dfrac{2x}{x^2 - 1}, \qquad Q(x) = \dfrac{\cos x}{x^2 - 1}$

Hence, the integrating factor is

$$u = e^{\int P(x)dx} = e^{\int \frac{2x}{x^2-1}dx} = e^{\ln(x^2-1)} = x^2 - 1$$

The general solution of the differential equation is

$$uy = \int uQ(x)dx + \mathrm{C}$$

$$(x^2-1)y=\int (x^2-1)\cdot\frac{\cos x}{x^2-1}dx+C$$

$$(x^2-1)y=\sin x+C$$

Substituting the initial condition $y|_{x=0}=1$ gives

$$-1=\sin 0+C$$

$$C=-1$$

Therefore, the particular solution of the differential equation is

$$(x^2-1)y=\sin x-1$$

$$y=\frac{\sin x-1}{x^2-1}$$

Example 4. Solve the initial value problem $\frac{dy}{dx}+\frac{xy}{9-x^2}=2$, given that $y|_{x=0}=3$.

Solution. Here $P(x)=\frac{x}{9-x^2}$, $Q(x)=2$

The integrating factor is

$$u=e^{\int P(x)dx}=e^{\int \frac{x}{9-x^2}dx}$$

Let $T=9-x^2$ then $dT=-2xdx$, $dx=\frac{dT}{-2x}$

Thus $$\int\frac{x}{9-x^2}dx=\int\frac{x}{T}\cdot\frac{dT}{-2x}=-\frac{1}{2}\ln T=\ln(9-x^2)^{-\frac{1}{2}}$$

The integrating factor is

$$u=e^{\int \frac{x}{9-x^2}dx}=e^{\ln(9-x^2)^{-\frac{1}{2}}}=\frac{1}{\sqrt{9-x^2}}$$

The general solution of the differential equation is

$$uy=\int uQ(x)dx+\mathrm{C}$$

$$\frac{1}{\sqrt{9-x^2}}y=\int\frac{2}{\sqrt{9-x^2}}dx+C$$

$$=2\arcsin(\frac{x}{3})+C$$

Hence, $$y=\sqrt{9-x^2}\left(2\arcsin(\frac{x}{3})+C\right)$$

Substituting the initial condition $y|_{x=0}=3$ yields

$$3=3(0+C)$$

$$C=1$$

Therefore, the particular solution of the differential equation is

$$y=\sqrt{9-x^2}\left(2\arcsin(\frac{x}{3})+1\right)$$

The previous example problems were solved using a general formula. In practice, however, this formula can be set aside and the method of variation of an arbitrary constant can be used directly to solve the problem.

Example 5. Find the general solution of the differential equation

$$\frac{dy}{dx}+3y\tan 3x=\sin 6x$$

Solution. Step(1) First find the solution of the corresponding homogeneous equation.

$$\frac{dy}{dx}+3y\tan 3x=0$$

Separating the variables gives

$$\frac{dy}{y}=-3\tan 3x dx$$

Integrating both sides yields

$$\int \frac{dy}{y} = \int -3\tan 3x dx$$

$$\ln y = \ln\cos 3x + \ln C$$

That is, the solution of the homogeneous equation is

$$y = C\cos 3x$$

Step (2) Replace the arbitrary constant C with $u = u(x)$.

That is, let $y = u\cos 3x$. u is a function to be determined.

Substituting this into the original equation gives

$$\frac{d(u\cos 3x)}{dx} + 3u\cos 3x\tan 3x = \sin 6x$$

The first term on the left-hand side is

$$\frac{d(u\cos 3x)}{dx} = \frac{du}{dx}\cdot\cos 3x + u\cdot\frac{d\cos 3x}{dx}$$

$$= \cos 3x\frac{du}{dx} - 3u\sin 3x$$

The second term on the left-hand side is

$$3u\cos 3x\tan 3x = 3u\sin 3x$$

Hence, the left-hand side of the equation becomes

$$\frac{d(u\cos 3x)}{dx} + 3u\cos 3x\tan 3x = \cos 3x\frac{du}{dx}$$

The right-hand side of the equation is

$$\sin 6x = 2\sin 3x\cos 3x$$

From left-hand side = right-hand side, we obtain

$$\cos 3x\frac{du}{dx} = 2\sin 3x\cos 3x$$

Separating the variables gives

$$du = 2\sin 3x dx$$

Integrating both sides yields

$$\int du = \int 2\sin 3x dx$$

$$u = -\frac{2}{3}\cos 3x + C$$

Hence, the general solution of this first order linear nonhomogeneous differential equation is

$$y = u\cos 3x$$

$$= (-\frac{2}{3}\cos 3x + C)\cos 3x$$

Exercise 8.4

1. Find the general solution of each of the following differential equations.

a. $\frac{dy}{dx} + y = e^{-x}$

b. $\frac{dy}{dx} - \frac{2y}{x+1} = (x+1)^3$

c. $(x^2+1)\frac{dy}{dx} + 2xy = 4x^2$

d. $x\frac{dy}{dx} - 2y = x^3e^x$

e. $\frac{dy}{dx} + \frac{y}{x} = \frac{\sin x}{x}$

f. $x\frac{dy}{dx} = y + x^4e^{-2x}$

2. Solve the initial value problem $\frac{dy}{dx} + y\sin x = e^{\cos x}$, given that $y(0) = e$.

3. Solve the initial value problem $(\cos^2 x\sin x)\frac{dy}{dx} + (\cos^3 x)y = 1$, given that $y(\frac{\pi}{3}) = 1$.

4. Solve the initial value problem $\frac{dy}{dx}\cos x = y\sin x + \cos^2 x$, given that $y(\pi) = 1$.

5. Solve the initial value problem $3\frac{dy}{dx}+\frac{6y}{10-x}=1$, given that $y(0)=-20$.

6. Solve the initial value problem $(x^2-5)\frac{dy}{dx}-2xy=-2x(x^2-5)$, given that $y(2)=7$.

7. Solve the initial value problem $\frac{dy}{dx}+2y=x^3e^{-2x}$, given that $y(0)=1$.

8.5. Euler's method

Previous sections introduced several methods for finding the exact solutions of differential equations. However, in many practical applications, various differential equations cannot be solved using these analytical methods. In such cases, Euler's method can be used to find an approximate solution to the differential equation.

The principle of Euler's method is as follows:

From the definition of derivative

$$y'=f'(x)=\lim_{\Delta x\to 0}\frac{\Delta y}{\Delta x}=\lim_{\Delta x\to 0}\frac{f(x+\Delta x)-f(x)}{\Delta x}$$

when Δx is sufficiently small, we have the approximation

$$y'\approx\frac{\Delta y}{\Delta x}=\frac{f(x+\Delta x)-f(x)}{\Delta x}$$

The differential equation to be solved is

$$y'=\frac{dy}{dx}=f(x,y)$$

Hence, $$y' = f(x,y) \approx \frac{f(x+\Delta x) - f(x)}{\Delta x}$$

Cross-multiplying and rearranging gives

$$f(x+\Delta x) \approx f(x) + \Delta x \cdot f(x,y)$$

Let $x = x_n$, $y = y_n$, $\Delta x = x_{n+1} - x_n$ (that is, $x_{n+1} = x_n + \Delta x$), Δx is called the step size. It follows that

$$f(x_n + \Delta x) \approx f(x_n) + \Delta x \cdot f(x_n, y_n)$$

That is, the **formula for Euler's method** is

$$y_{n+1} \approx y_n + \Delta x \cdot f(x_n, y_n) \qquad \text{where} \qquad f(x_n, y_n) = \frac{dy_n}{dx_n}$$

- Condition for Euler's formula: Δx is sufficiently small.

Geometric significance of Euler's method

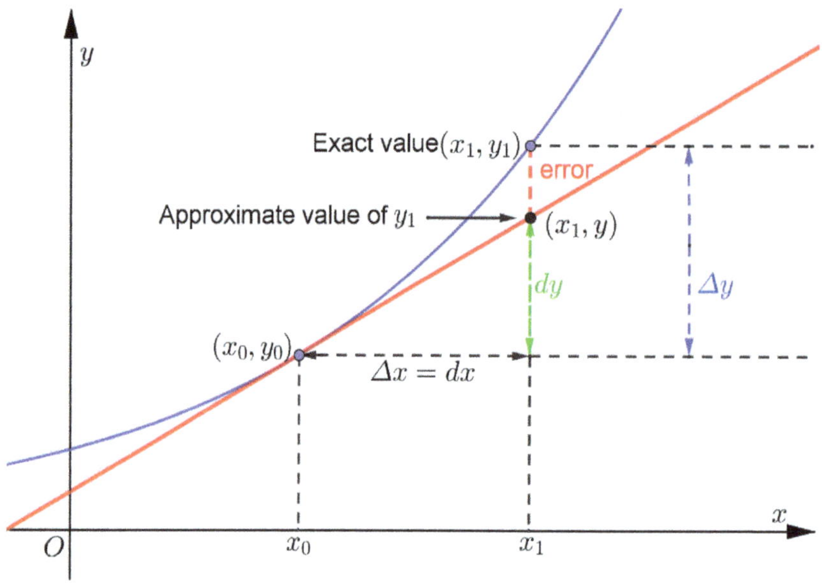

Figure 8.1

As shown in Figure 8.1, the blue curve represents the exact solution of the differential equation. The red line is the tangent to the curve. The initial conditions provided for the differential equation include a point (x_0, y_0), through which the curve passes, and the equation of the tangent of the curve (that is, the differential equation itself). When $x = x_1$, the exact value of y_1 lies on the blue curve. If it is impossible to determine the exact value of y_1, the value of y at the point $x = x_1$ on the tangent passing through the point (x_0, y_0) is used as an approximation of y_1. Since the gradient of the tangent is given by

$$\tan\theta = \frac{dy}{dx} \approx \frac{\Delta y}{\Delta x}$$

thus $$\Delta y \approx \Delta x \cdot \frac{dy}{dx}$$

Hence, $$y_1 = y_0 + \Delta y \approx y_0 + \Delta x \cdot \frac{dy}{dx}$$

As shown in Figure 8.1, only when Δx is sufficiently small will the error between the approximate and exact values be relatively small. Usually, an initial point (x_0, y_0) is given, and we are required to determine the value of y at $x = x$. If Euler's method is applied in a single step, the number of iterations is $N = 1$ and the step size is $\Delta x = x - x_0$.

To obtain a better approximation, we can increase the number of iterations, N, and use a smaller step size. As shown in Figure 8.2

$$\Delta x_n = \Delta x, \quad n = 0, 1, 2, 3, \ldots, N$$

$$\Delta x = \frac{x - x_0}{N} \quad \text{or} \quad N = \frac{x - x_0}{\Delta x}$$ Δx is a small interval of equal length.

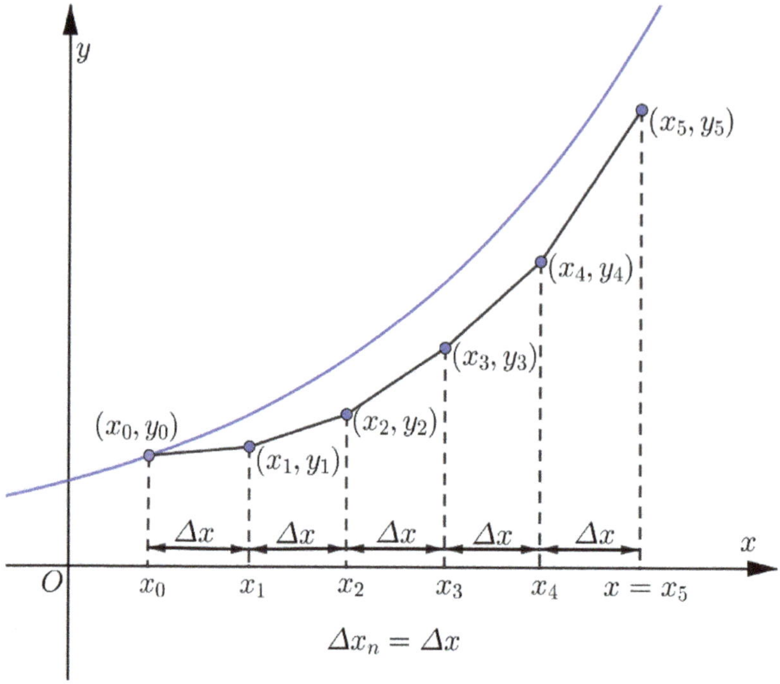

Figure 8.2

Use the Euler method to find an approximate value y_N of y. The number of iterations is determined by N.

In general, if the second derivative is positive ($\frac{d^2y}{dx^2} > 0$), the original function is concave up and its tangent lies below the function's graph; in this case, the approximate value is an underestimate (smaller than the exact value). Conversely, if $\frac{d^2y}{dx^2} < 0$, the original function is concave down and its tangent lies above the function's graph, then the approximate value is an overestimate (larger than the exact value).

Example 1. Consider the differential equation $\frac{dy}{dx} - \frac{y}{x} = \sqrt{x}$ with the initial condition $y|_{x=4} = 0$.

(a) Use Euler's method with a step size $\Delta x = 0.25$ to find an approximate value for y when $x = 5$. Give your answer correct to two decimal places.

(b) Solve the differential equation and find the exact value of y when $x = 5$.

(c) Calculate the percentage error between the approximate value obtained from part (a) and the exact value found in part (b).

(d) Explain why the approximate value is smaller than the exact value.

(e) Use Euler's method with a step size $\Delta x = 0.1$ to find an approximate value for y when $x = 5$. Give your answer correct to two decimal places. What is the percentage error between the approximate value and the exact value from (b)?

Solution.

(a) Since $$\frac{dy}{dx} - \frac{y}{x} = \sqrt{x}$$

thus $$\frac{dy}{dx} = \sqrt{x} + \frac{y}{x}$$

Hence, $$f(x_n, y_n) = \frac{dy_n}{dx_n} = \sqrt{x_n} + \frac{y_n}{x_n}$$

From $\Delta x = 0.25$ we obtain $N = \dfrac{x - x_0}{\Delta x} = \dfrac{5-4}{0.25} = 4$

We can tabulate the values as follows:

n	x_n	y_n
0	4	0
1	4.25	y_1
2	4.5	y_2
3	4.75	y_3
4	5	y_4

In the following, we apply Euler's method to find the approximate value, y_4, of y when $x = 5$ $(N = 4)$.

$$y_{n+1} \approx y_n + \Delta x \cdot f(x_n, y_n) \quad \text{where} \quad f(x_n, y_n) = \frac{dy_n}{dx_n} = \sqrt{x_n} + \frac{y_n}{x_n}$$

$$y_1 = y_0 + \Delta x \cdot f(x_0, y_0) = y_0 + 0.25(\sqrt{x_0} + \frac{y_0}{x_0}) = 0 + 0.25(\sqrt{4} + \frac{0}{4}) = 0.5$$

$$y_2 = y_1 + \Delta x \cdot f(x_1, y_1) = y_1 + 0.25(\sqrt{x_1} + \frac{y_1}{x_1}) = 0.5 + 0.25(\sqrt{4.25} + \frac{0.5}{4.25})$$

$$\approx 1.044799968$$

$$y_3 = y_2 + \Delta x \cdot f(x_2, y_2) = y_2 + 0.25(\sqrt{x_2} + \frac{y_2}{x_2})$$

$$= 1.044799968 + 0.25(\sqrt{4.50} + \frac{1.044799968}{4.50}) \approx 1.633174496$$

$$y_4 = y_3 + \Delta x \cdot f(x_3, y_3) = y_3 + 0.25(\sqrt{x_3} + \frac{y_3}{x_3})$$

$$= 1.633174496 + 0.25(\sqrt{4.75} + \frac{1.633174496}{4.75}) \approx 2.263993417$$

Therefore, when $x = 5$, $y \approx 2.26$ (correct to two decimal places).

Note: A graphical calculator may be used to perform these calculations.

- If using a TI-84 calculator, the 'ANS' button can be used to store the value of y_n.

  For example,

  $$y_3 = y_2 + 0.25(\sqrt{x_2} + \frac{y_2}{x_2}) = \text{ANS} + 0.25(\sqrt{4.50} + \frac{\text{ANS}}{4.50}) = 1.633... \qquad \text{ANS} = y_2$$

  $$y_4 = y_3 + 0.25(\sqrt{x_3} + \frac{y_3}{x_3}) = \text{ANS} + 0.25(\sqrt{4.75} + \frac{\text{ANS}}{4.75}) = 2.263... \qquad \text{ANS} = y_3$$

- If a CASIO FX-CG50 calculator is used, the recurrence relation can be employed to generate a table of values. For example, let a_n represent the value of x_n, and b_n represent the value of y_n.

  $$a_{n+1} = a_n + 0.25$$

$$b_{n+1} = b_n + 0.25(\sqrt{a_n} + \frac{b_n}{a_n})$$

Set the initials values for a_n and b_n then generate the table to obtain all subsequent values.

(b) $$\frac{dy}{dx} - \frac{y}{x} = \sqrt{x}$$

$$P(x) = -\frac{1}{x}, \qquad Q(x) = \sqrt{x}$$

The integrating factor is

$$u = e^{\int P(x)dx} = e^{\int -\frac{1}{x}dx} = e^{-\ln x} = \frac{1}{x}$$

The general solution of the differential equation is

$$uy = \int uQ(x)dx + \mathrm{C}$$

$$\frac{1}{x} \cdot y = \int \frac{1}{x} \cdot \sqrt{x}dx + C$$

$$\frac{1}{x} \cdot y = 2\sqrt{x} + C$$

Substituting the initial condition $x = 4, y = 0$ gives

$$C = -4$$

Therefore, the particular solution of the differential equation is

$$y = 2x^{\frac{3}{2}} - 4x$$

When $x = 5$, the exact value of y is

$$y = 2 \cdot 5^{\frac{3}{2}} - 20 \approx 2.360679775$$

(c) $$\text{percentage error} = \left| \frac{2.263993417 - (2 \cdot 5^{\frac{3}{2}} - 20)}{2 \cdot 5^{\frac{3}{2}} - 20} \right| \times 100\% \approx 4.10\%$$

(d) The derivative of the function is

$$\frac{dy}{dx} = \sqrt{x} + \frac{y}{x}$$

Differentiating both sides with respect to x yields

$$\frac{d^2y}{dx^2} = \frac{1}{2\sqrt{x}} + \frac{y'}{x} - \frac{y}{x^2}$$

Substituting $y' = \frac{dy}{dx} = \sqrt{x} + \frac{y}{x}$ gives

$$\frac{d^2y}{dx^2} = \frac{1}{2\sqrt{x}} + \frac{\sqrt{x} + \frac{y}{x}}{x} - \frac{y}{x^2} = \frac{1}{2\sqrt{x}} + \frac{1}{\sqrt{x}} + \frac{y}{x^2} - \frac{y}{x^2} = \frac{3}{2\sqrt{x}}$$

For $x > 4$, it is found that $\frac{d^2y}{dx^2} > 0$. Therefore, the function is concave up, and its tangent lies below the graph of the function. It follows that the approximate value is smaller than the exact value.

(e) For a step size of $\Delta x = 0.1$, the calculations are performed using a graphical calculator. When $x = 5$, $N = \frac{5-4}{0.1} = 10$, we obtain $y \approx 2.321458672 \approx 2.32$.

$$\text{percentage error} = \left| \frac{2.321458672 - (2 \cdot 5^{\frac{3}{2}} - 20)}{2 \cdot 5^{\frac{3}{2}} - 20} \right| \times 100\% \approx 1.66\%$$

It is evident that by reducing the step size, we obtain a more accurate approximation.

Example 2. Consider the differential equation $x^2y' + xy = y^2$ with the initial condition $y|_{x=1} = 1$.

(a) Use Euler's method with a step size $\Delta x = 0.1$ to find an approximate value for y when $x = 1.5$. Give your answer correct to eight decimal places.

(b) Solve the differential equation and find the exact value of y when $x = 1.5$.

(c) Calculate the percentage error between the approximate value obtained from part (a) and the exact value found in part (b).

(d) Use Euler's method with a step size $\Delta x = 0.025$ to find an approximate value for y when $x = 1.5$. Give your answer correct to eight decimal places. What is the percentage error between the approximate value and the exact value found in (b)?

(e) With reference to the graph of $\frac{d^2 y}{dx^2}$, explain why the approximate value is greater than the exact value.

Solution. (a) Since $\quad x^2 y' + xy = y^2$

thus $$\frac{dy}{dx} + \frac{y}{x} = (\frac{y}{x})^2$$

Therefore, $$f(x_n, y_n) = \frac{dy_n}{dx_n} = (\frac{y_n}{x_n})^2 - \frac{y_n}{x_n} = \frac{y_n^2 - x_n y_n}{x_n^2}$$

From $\Delta x = 0.1$ we obtain $$N = \frac{x - x_0}{\Delta x} = \frac{1.5 - 1}{0.1} = 5$$

We can tabulate the information as follows:

n	x_n	y_n
0	1	1
1	1.1	y_1
2	1.2	y_2
3	1.3	y_3
4	1.4	y_4
5	1.5	y_5

In the following, we apply Euler's method to find the approximate value, y_5, of y when $x = 1.5$ $(N = 5)$.

$$y_{n+1} \approx y_n + \Delta x \cdot f(x_n, y_n) \quad \text{where} \quad f(x_n, y_n) = \frac{dy_n}{dx_n} = \frac{y_n^2 - x_n y_n}{x_n^2}$$

$$y_1 = y_0 + \Delta x \cdot f(x_0, y_0) = y_0 + 0.1(\frac{y_0^{\ 2} - x_0 y_0}{x_0^{\ 2}}) = 1 + 0.1(\frac{1^2 - 1}{1^2}) = 1$$

$$y_2 = y_1 + \Delta x \cdot f(x_1, y_1) = y_1 + 0.1(\frac{y_1^{\ 2} - x_1 y_1}{x_1^{\ 2}}) = 1 + 0.1(\frac{1^2 - 1.1 \cdot 1}{1.1^2}) \approx 0.9917355372$$

$$y_3 = y_2 + \Delta x \cdot f(x_2, y_2) = y_2 + 0.1(\frac{y_2^{\ 2} - x_2 y_2}{x_2^{\ 2}})$$

$$= 0.9917355372 + 0.1(\frac{0.9917355372^2 - 1.2 \cdot 0.9917355372}{1.2^2}) \approx 0.9773922546$$

$$y_4 = y_3 + \Delta x \cdot f(x_3, y_3) = y_3 + 0.1(\frac{y_3^{\ 2} - x_3 y_3}{x_3^{\ 2}})$$

$$= 0.9773922546 + 0.1(\frac{0.9773922546^2 - 1.3 \cdot 0.9773922546}{1.3^2}) \approx 0.9587346031$$

$$y_5 = y_4 + \Delta x \cdot f(x_4, y_4) = y_4 + 0.1(\frac{y_4^{\ 2} - x_4 y_4}{x_4^{\ 2}})$$

$$= 0.9587346031 + 0.1(\frac{0.9587346031^2 - 1.4 \cdot 0.9587346031}{1.4^2}) \approx 0.9371500926$$

Therefore, when $x = 1.5$, $y \approx 0.93715009$ (correct to eight decimal places).

(b) Since $\quad x^2 y' + xy = y^2$

thus $\quad \frac{dy}{dx} + \frac{y}{x} = (\frac{y}{x})^2$

It can be seen that $\frac{dy}{dx}$ is a function of $\frac{y}{x}$; therefore, this is a homogeneous differential equation.

Let $\quad u = \frac{y}{x} \quad$ then $\quad y = ux$

Substituting $y = ux$ gives

$$\frac{d(ux)}{dx} = u^2 - u$$

thus $$\frac{du}{dx}x + u = u^2 - u$$

Hence, $$\frac{du}{dx}x = u^2 - 2u$$

Separating the variables and integrating both sides gives

$$\int \frac{du}{u^2 - 2u} = \int \frac{dx}{x} \qquad (1)$$

For $\int \frac{du}{u^2 - 2u}$ on the left-hand side

first, express $\frac{1}{u^2 - 2u}$ in partial fractions.

$$\frac{1}{u^2 - 2u} = \frac{1}{u(u-2)} = \frac{A}{u} + \frac{B}{u-2}$$

Multiplying both sides by $u(u-2)$ yields

$$1 = A(u-2) + Bu$$

Let $u = 0$, we have $A = -\frac{1}{2}$.

Let $u = 2$, we have $B = \frac{1}{2}$.

Therefore, $$\int \frac{du}{u^2 - 2u} = \frac{1}{2}\int \left(\frac{1}{u-2} - \frac{1}{u}\right)du = \frac{1}{2}\left[\ln(u-2) - \ln u\right]$$

Substituting into (1) gives

$$\ln\frac{u-2}{u} = \ln(xC)^2$$

$$\frac{u-2}{u} = Ax^2 \qquad \text{where} \qquad A = C^2$$

Substituting $u = \frac{y}{x}$ yields

$$\frac{\frac{y}{x}-2}{\frac{y}{x}} = Ax^2$$

$$\frac{y-2x}{y} = Ax^2$$

Substituting $y|_{x=1} = 1$ gives

$$A = -1$$

Hence, $$\frac{y-2x}{y} = -x^2$$

$$y = \frac{2x}{1+x^2}$$

When $x = 1.5$, $y = \frac{2 \cdot 1.5}{1+1.5^2} = \frac{12}{13}$ (≈ 0.9230769231).

(c) percentage error $= \left| \frac{0.9371500926 - \frac{12}{13}}{\frac{12}{13}} \right| \times 100\% \approx 1.52\%$

(d) For a step size of $\Delta x = 0.025$, the calculations are performed using a graphical calculator. When $x = 1.5$, $N = \frac{1.5-1}{0.025} = 20$, we obtain $y \approx 0.92648094$ (correct to eight decimal places).

percentage error $= \left| \frac{0.92648094 - \frac{12}{13}}{\frac{12}{13}} \right| \times 100\% \approx 0.369\%$

It is evident that by reducing the step size, we obtain a more accurate approximation.

(e) As can be seen from the graph of $\frac{d^2y}{dx^2}$, on the interval [1, 1.5], $\frac{d^2y}{dx^2} < 0$; thus, the original function is concave down and its tangent lies above the curve of the original function. Therefore, the approximate value is greater than the exact value.

Exercise 8.5

1. Consider the initial value problem $\frac{dy}{dx} + 2y = x^3e^{-2x}$, given that $y(0) = 1$.

   a. Use Euler's method with a step size $\Delta x = 0.1$ to find an approximate value for y when $x = 0.5$. Give your answer correct to eight decimal places.

   b. Solve the differential equation and find the exact value of y when $x = 0.5$.

   c. Calculate the percentage error between the approximate value obtained from part a. and the exact value found in part b.

   d. Explain why the approximate value is less than the exact value.

   e. Use Euler's method with a step size $\Delta x = 0.01$ to find an approximate value for y when $x = 0.5$. Give your answer correct to eight decimal places. What is the percentage error between the approximate value and the exact value from part b?

2. Consider the initial value problem $\frac{dy}{dx} = x + 2xy$, given that $y(0) = \frac{5}{2}$.

   a. Use Euler's method with a step size $\Delta x = 0.1$ to find an approximate value for y when $x = 1$. Give your answer correct to seven decimal places.

   b. Solve the differential equation and find the exact value of y when $x = 1$.

   c. Calculate the percentage error between the approximate value obtained from part a. and the exact value found in part b.

d. Use Euler's method with a step size $\Delta x = 0.02$ to find an approximate value for y when $x = 1$. Give your answer correct to seven decimal places. What is the percentage error between the approximate value and the exact value from part b?

e. Explain why the approximate value is less than the exact value.

3. Consider the initial value problem $\frac{dy}{dx} + 2xy^2 = 0$, given that $y(0) = 1$.

a. Use Euler's method with a step size $\Delta x = 0.1$ to find an approximate value for y when $x = 0.5$. Give your answer correct to seven decimal places.

b. Solve the differential equation and find the exact value of y when $x = 0.5$.

c. Calculate the percentage error between the approximate value obtained from part a. and the exact value found in part b.

d. Use Euler's method with a step size $\Delta x = 0.025$ to find an approximate value for y when $x = 0.5$. Give your answer correct to seven decimal places. What is the percentage error between the approximate value and the exact value from part b?

e. With reference to the graph of $\frac{d^2y}{dx^2}$, explain why the approximate value is greater than the exact value.

8.6.* Solving differential equations using Picard's method

Both Euler's method and Picard's method are used to find approximate solutions to differential equations. While Euler's formula requires the step size, Δx, is relatively small, Picard's method requires that Δy is relatively small. Both are iterative approximation methods. In this section, we introduce Picard's method and derive Picard's formula.

A differential equation is given.

$$\begin{cases} y' = \dfrac{dy}{dx} = f(x, y) \\ y(x_0) = y_0 \, (initial \quad condition) \end{cases}$$

Separating the variables and integrating both sides gives

$$\int_{y_0}^{y} dy = \int_{x_0}^{x} f(x, y)dx$$

That is, $$y - y_0 = \int_{x_0}^{x} f(x, y)dx$$

Let $y = y_n$. If $\Delta y = y_n - y_{n-1}$ is sufficiently small, then we have $y_n \approx y_{n-1}$.

Therefore, we have

$$y_n - y_0 = \int_{x_0}^{x} f(x, y_n)dx \approx \int_{x_0}^{x} f(x, y_{n-1})dx$$

That is, **Picard's formula** is an integral equation.

$$y_n = y_0 + \int_{x_0}^{x} f(x, y_{n-1})dx \qquad x \in [x_0 - \delta, x_0 + \delta], \quad n = 1, 2, 3, \ldots$$

We can obtain a series of approximate solutions using Picard's formula.

$$y_1(x) = y_0 + \int_{x_0}^{x} f(x, y_0)dx$$

$$y_2(x) = y_0 + \int_{x_0}^{x} f(x, y_1)dx$$

$$y_3(x) = y_0 + \int_{x_0}^{x} f(x, y_2)dx$$

$$\vdots$$

Picard's iterative method provides a means of finding approximate solutions. Each term in the Picard sequence represents an approximate solution; as the number of terms, *n*, increases, a higher degree of accuracy is achieved.

As $n \to \infty$, the limit function, $g(x)$, of the Picard sequence exists, that is,

$$\lim_{n\to\infty} y_n = g(x) \qquad x \in [x_0 - \delta, x_0 + \delta]$$

hence,
$$\lim_{n\to\infty} y_n = \lim_{n\to\infty}\left[y_0 + \int_{x_0}^{x} f(x, y_{n-1})dx \right]$$

That is,
$$g(x) = y_0 + \int_{x_0}^{x} f(x, g(x))dx$$

where $g(x)$ is the exact solution of the differential equation.

Example. Find the first three terms of Picard's sequence that satisfy the following differential equation and initial condition

$$\begin{cases} y' = x^2 + y^2 \\ y(0) = y_0 = 0 \end{cases}$$

Solution. From the differential equation, we obtain the integral equation.

$$y = 0 + \int_0^x (x^2 + y^2)dx$$

By applying Picard's formula

$$y_n = y_0 + \int_{x_0}^{x} f(x, y_{n-1})dx$$

we find
$$y_1 = \int_0^x x^2 dx = \frac{1}{3}x^3$$

$$y_2 = \int_0^x [x^2 + (\frac{1}{3}x^3)^2]dx = \frac{1}{3}x^3 + \frac{1}{63}x^7$$

$$y_3 = \int_0^x [x^2 + (\frac{1}{3}x^3 + \frac{1}{63}x^7)^2]dx = \frac{1}{3}x^3 + \frac{1}{63}x^7 + \frac{2}{2079}x^{11} + \frac{1}{59535}x^{15}$$

It can be proven that this Picard sequence converges to a limit. However, the limit is not an elementary function. In this case, the function y_3 here may be used as an approximate solution of the differential equation, satisfying the given initial condition.

Exercise 8.6

1. Find the first three terms of the Picard sequence that satisfy the following differential equation and the initial condition.

$$y' = x + y, \qquad y(0) = y_0 = 0$$

Chapter 9 Infinite series

9.1. Concept of infinite series

If it is given a sequence $u_1,\ u_2,\ ...,\ u_n,\ ...$, then the expression

$$u_1 + u_2 + ... + u_n + ... \tag{9.1}$$

can be denoted simply by $\sum_{n=1}^{\infty} u_n$, and it is called an infinite series (or simply series), where the nth term u_n is called the general term of the series.

The sum of the first n terms of the series

$$S_n = u_1 + u_2 + ... + u_n = \sum_{i=1}^{n} u_i$$

is called the ***n*th partial sum**. Partial sums

$$S_1,\ S_2,\ ...,\ S_n,\ ...$$

form a sequence.

If as $n \to \infty$, the limit of the sequence of partial sums S_n exists, that is,

$$\lim_{n\to\infty} S_n = S \qquad (S \text{ is a finite constant.})$$

then we say that the series (9.1) **converges**. S is its sum, denoted by

$$S = \sum_{n=1}^{\infty} u_n = u_1 + u_2 + ... + u_n + ...$$

If S_n does not have a limit – that is, $\lim_{n\to\infty} S_n = \sum_{n=1}^{\infty} u_n = \infty$ – then the series $S = \infty$ **diverges,** and it has no sum.

When the series converges, the difference between its sum and the nth partial sum

$$R_n = S - S_n = u_{n+1} + u_{n+2} + \dots$$

is called the **remainder** of the series.

If S_n is used as an approximation of S, then the absolute error is the absolute value of the remainder, $|R_n|$.

Example 1. The infinite series

$$\sum_{n=1}^{\infty} u_1 r^{n-1} = u_1 + u_1 r + u_1 r^2 + \dots + u_1 r^{n-1} + \dots$$

is called a geometric series, where $u_1 \neq 0$. r is called the common ratio of the series, and its general term is $u_n = u_1 r^{n-1}$.

Now, we discuss its convergence.

(1) If $|r| \neq 1$, since

$$S_n = u_1 + u_1 r + u_1 r^2 + \dots + u_1 r^{n-1} = \frac{u_1(1-r^n)}{1-r} = \frac{u_1}{1-r} - \frac{u_1 r^n}{1-r}$$

if $|r| < 1$, then

$$\lim_{n\to\infty} S_n = \frac{u_1}{1-r}$$

Therefore, when $|r| < 1$, the geometric series converges. Its sum is $\frac{u_1}{1-r}$ and the remainder is $\frac{u_1 r^n}{1-r}$.

If $|r| > 1$, then

$$\lim_{n\to\infty} S_n = \infty.$$

Hence, when $|r| > 1$, the geometric series diverges, and it has no sum.

(2) If $r = 1$, then the series becomes

$$u_1 + u_1 + ... + u_1 + ...$$

Because

$$S_n = nu_1$$

then $$\lim_{n\to\infty} S_n = \infty$$

Hence, it diverges.

(3) If $r = -1$, then the series becomes

$$u_1 - u_1 + u_1 - u_1 ... + u_1 - u_1 ...$$

When n is an even number

$$S_n = 0$$

When n is an odd number

$$S_n = u_1$$

As $n \to \infty$

$$\lim_{n\to\infty} S_n = 0 \text{ or } u_1$$ Sometimes it takes 0, sometimes it takes u_1.

We say that the series oscillates without a limit and is therefore divergent.

Based on the above discussion, we can conclude that

the geometric series $\sum_{n=1}^{\infty} u_1 r^{n-1}$ converges if and only if $|r| < 1$, in which case its sum is $\frac{u_1}{1-r}$; The series diverges when $|r| \geq 1$.

For example, the series

$$\sum_{n=1}^{\infty} \frac{(-1)^{n-1}}{2^{n-1}} = 1 - \frac{1}{2} + \frac{1}{4} - \frac{1}{8} + ... + \frac{(-1)^{n-1}}{2^{n-1}} + ...$$

has the common ratio $r = -\frac{1}{2}$. Since $|r| = \frac{1}{2} < 1$, the series converges and its sum is $\frac{2}{3}$.

For the series

$$\sum_{n=1}^{\infty} 2^{n-1} = 1 + 2 + 4 + \ldots + 2^{n-1} + \ldots$$

the common ratio is $r = 2$. Since $|r| > 1$, the series diverges.

Example 2. Determine whether the following series is convergent or divergent.

$$\sum_{n=1}^{\infty} \frac{1}{n(n+1)} = \frac{1}{1 \cdot 2} + \frac{1}{2 \cdot 3} + \frac{1}{3 \cdot 4} + \ldots + \frac{1}{n(n+1)} + \ldots$$

Solution. Using partial fractions, we have

$$\frac{1}{n(n+1)} = \frac{1}{n} - \frac{1}{n+1} \qquad (n = 1, 2, \ldots)$$

Thus $$S_n = \frac{1}{1 \cdot 2} + \frac{1}{2 \cdot 3} + \frac{1}{3 \cdot 4} + \ldots + \frac{1}{n(n+1)}$$

$$= (1 - \frac{1}{2}) + (\frac{1}{2} - \frac{1}{3}) + (\frac{1}{3} - \frac{1}{4}) + \ldots + (\frac{1}{n} - \frac{1}{n+1}) = 1 - \frac{1}{n+1}$$

Hence, $$\lim_{n \to \infty} S_n = \lim_{n \to \infty} (1 - \frac{1}{n+1}) = 1$$

Therefore, the series converges, and its sum is 1.

Example 3. Determine whether the following series is convergent or divergent.

$$\sum_{n=1}^{\infty} \ln \frac{n+1}{n} = \ln \frac{2}{1} + \ln \frac{3}{2} + \ln \frac{4}{3} + \ldots + \ln \frac{n+1}{n} + \ldots$$

Solution. Since

$$\ln\frac{n+1}{n} = \ln(n+1) - \ln n \qquad (n = 1, 2, \ldots)$$

thus

$$S_n = \ln\frac{2}{1} + \ln\frac{3}{2} + \ln\frac{4}{3} + \ldots + \ln\frac{n+1}{n}$$

$$= (\ln 2 - \ln 1) + (\ln 3 - \ln 2) + (\ln 4 - \ln 3) + \ldots + (\ln(n+1) - \ln n)$$

$$= \ln(n+1)$$

Hence,

$$\lim_{n\to\infty} S_n = \lim_{n\to\infty} \ln(n+1) = \infty$$

Therefore, the series is divergent.

Exercise 9.1

1. Determine whether the following series converge or diverge by using the partial sum of the series or other methods. If a series converges, find its sum.

a. $\sum_{n=1}^{\infty}\frac{(-1)^{n-1}}{3^{n-1}}$ **b.** $\sum_{n=1}^{\infty}(\frac{1}{2^n}+\frac{1}{3^n})$ **c.** $\sum_{n=1}^{\infty}(\frac{3}{4^n}-\frac{5}{9^{n-1}})$

d. $\sum_{n=1}^{\infty}\frac{1}{n(n+2)}$ **e.** $\sum_{n=1}^{\infty}\frac{1}{(3n-2)(3n+1)}$ **f.** $\sum_{n=1}^{\infty}\frac{5^{n+1}-2^{n-2}}{3^{2n}}$

g. $\sum_{n=1}^{\infty}(3\cdot 2^{n+1})$ **h.** $\sum_{n=1}^{\infty}\ln\frac{2n+3}{2n+1}$ **i.** $\sum_{n=1}^{\infty}(\sqrt{n+1}-\sqrt{n})$

j. $\sum_{n=1}^{\infty}\ln(1+\frac{1}{n})$ **k.** $\sum_{n=2}^{\infty}\frac{n^2}{n^2-1}$ **l.** $\sum_{n=1}^{\infty}\frac{(2\pi)^n}{5^{n+1}}$

m. $\sum_{n=1}^{\infty}\frac{n}{(n+1)!}$

9.2. Fundamental properties of infinite series

Theorem 9.1

If both series

$$\sum_{n=1}^{\infty} u_n = u_1 + u_2 + \ldots + u_n + \ldots \tag{9.1}$$

and

$$\sum_{n=1}^{\infty} v_n = v_1 + v_2 + \ldots + v_n + \ldots \tag{9.2}$$

converge, and their sums are S and W, respectively, then the series

$$\sum_{n=1}^{\infty} (u_n \pm v_n) = (u_1 \pm v_1) + (u_2 \pm v_2) + \ldots + (u_n \pm v_n) + \ldots \tag{9.3}$$

will also converges, and its sum is $S \pm W$.

Proof. Let

$$S_n = u_1 + u_2 + \ldots + u_n \quad \text{and} \quad \lim_{n\to\infty} S_n = S$$

$$W_n = v_1 + v_2 + \ldots + v_n \quad \text{and} \quad \lim_{n\to\infty} W_n = W$$

Let the nth partial sum of the series (9.3) be T_n, then

$$\begin{aligned} T_n &= (u_1 \pm v_1) + (u_2 \pm v_2) + \ldots + (u_n \pm v_n) \\ &= (u_1 + u_2 + \ldots + u_n) \pm (v_1 + v_2 + \ldots + v_n) \\ &= S_n \pm W_n \end{aligned}$$

Hence,

$$\lim_{n\to\infty} T_n = \lim_{n\to\infty} (S_n \pm W_n) = S \pm W$$

Therefore,

$$\sum_{n=1}^{\infty} (u_n \pm v_n) = S \pm W = \sum_{n=1}^{\infty} u_n \pm \sum_{n=1}^{\infty} v_n \qquad \square$$

Theorem 9.2

If the series

$$\sum_{n=1}^{\infty} u_n = u_1 + u_2 + \ldots + u_n + \ldots \tag{9.1}$$

converges, and its sum is S, then the series formed by multiplying each term by a non-zero constant a,

$$\sum_{n=1}^{\infty} au_n = au_1 + au_2 + \ldots + au_n + \ldots \tag{9.4}$$

also converges, and its sum is aS.

Proof. Let the nth partial sum of the series (9.1) be S_n, that is,

$$S_n = u_1 + u_2 + \ldots + u_n$$

And $$\lim_{n\to\infty} S_n = S$$

Let the nth partial sum of the series (9.4) be W_n, that is,

$$W_n = au_1 + au_2 + \ldots + au_n = a(u_1 + u_2 + \ldots + u_n) = aS_n$$

Hence, $$\lim_{n\to\infty} W_n = \lim_{n\to\infty} aS_n = aS$$

Thus $$\sum_{n=1}^{\infty} au_n = aS = a\sum_{n=1}^{\infty} u_n$$

Since $W_n = aS_n$, if S_n does not have a limit, then W_n does not have a limit either. Therefore, multiplying each term of a series by a non-zero constant does not change its convergence or divergence. □

Theorem 9.3

Adding or removing a finite number of terms at the beginning of a series does not affect its convergence.

Proof. Consider the following two series:

$$u_1 + u_2 + \ldots + u_k + u_{k+1} + u_{k+2} + \ldots + u_{k+m} + \ldots \tag{9.1}$$

And $$u_{k+1} + u_{k+2} + \ldots + u_{k+m} + \ldots \tag{9.5}$$

Observe that series (9.1) is formed by adding k terms to the beginning of series (9.5) (or, conversely, series (9.5) is formed by removing the first k terms from the beginning of series (9.1)).

The sum of the first k terms is denoted by

$$A = u_1 + u_2 + \ldots + u_k$$

The mth partial sum of the series (9.5) is denoted by

$$W_m = u_{k+1} + u_{k+2} + \ldots + u_{k+m}$$

Then, the $(k+m)$th partial sum of the series (9.1) is

$$S_{k+m} = u_1 + u_2 + \ldots + u_k + u_{k+1} + u_{k+2} + \ldots + u_{k+m} = A + W_m$$

Therefore, as $m \to \infty$, the series S_{k+m} and the series W_m will exhibit the same convergence behaviour. Hence, the series $\sum_{n=1}^{\infty} u_n$ and the series $\sum_{m=1}^{\infty} u_{k+m}$ will either both converge or both diverge.

If both the series $\sum_{n=1}^{\infty} u_n$ and the series $\sum_{m=1}^{\infty} u_{k+m}$ converge, and

$$\lim_{m\to\infty} W_m = W \qquad \lim_{m\to\infty} S_{k+m} = S$$

then $$S = A + W \quad \text{or} \quad W = S - A$$ □

Theorem 9.4

If a series converges, the series obtained by adding parentheses will also converge and have the same sum as the original series.

Proof. Let the series $\sum_{n=1}^{\infty} u_n$ converge, and its sum be S, that is,

$$u_1 + u_2 + \ldots + u_n + \ldots = S$$

Assume that after adding parentheses according to a certain rule, the series becomes

$$u_1 + (u_2 + u_3) + (u_4 + u_5 + u_6) + \ldots \tag{9.6}$$

Use W_m to represent the mth partial sum of the series (9.6). Use S_n to represent the nth partial sum of $\sum_{n=1}^{\infty} u_n$ which corresponds to W_m, that is,

$$W_1 = S_1,\ W_2 = S_3,\ W_3 = S_6,\ \ldots,\ W_m = S_n,\ \ldots$$

Obviously, $m < n$, when $m \to \infty$, $n \to \infty$, so

$$\lim_{m\to\infty} W_m = \lim_{n\to\infty} S_n = S \qquad \square$$

On the contrary, if the series obtained after parentheses are added diverges, then the original series must also diverge. If the original series converges, according to Theorem 9.4, the series obtained after parentheses are added will also converge, which contradicts the presupposition.

Note: Divergent series may converge after parentheses are added. That is, the series obtained after parentheses are added converges, but the original series may not converge.

For example, a divergent series is

$$a - a + a - a + a - a + \ldots + (-1)^{n-1}a + \ldots$$

Adding parentheses to adjacent two terms gives

$$(a - a) + (a - a) + \ldots + (a - a) + \ldots = 0 \quad \text{which is convergent.}$$

But the original series does not converge.

As for the series of nonnegative terms (i.e. all terms $u_n \geq 0$, read Section 9.3), adding or deleting parentheses does not affect its convergence nature.

Theorem 9.5 (Necessary condition for convergence)

If the series

$$\sum_{n=1}^{\infty} u_n = u_1 + u_2 + \ldots + u_n + \ldots$$

converges, then $\lim_{n\to\infty} u_n = 0$.

Proof. Since $u_n = S_n - S_{n-1}$

it follows that $\lim_{n\to\infty} u_n = \lim_{n\to\infty} (S_n - S_{n-1}) = \lim_{n\to\infty} S_n - \lim_{n\to\infty} S_{n-1} = S - S = 0$ □

Thus, if the general term of a series does not approach zero, then the series diverges.

Example. For the series

$$\sum_{n=1}^{\infty} n^s = 1^s + 2^s + 3^s + \ldots + n^s + \ldots \qquad (s > 0)$$

since $\lim_{n\to\infty} n^s = \infty$ $(s > 0)$, it follows that the series diverges.

Note:

A series whose general term approaches zero does not necessarily converge.

For example, the series $\sum_{n=1}^{\infty} \ln\frac{n+1}{n}$ satisfies

$$\lim_{n\to\infty} \ln\frac{n+1}{n} = \lim_{n\to\infty} \ln(1 + \frac{1}{n}) = 0$$

However, in Example 3 in Section 9.1 of this chapter, we have proved that it is a divergent series.

From this, we can obtain the Divergence Test.

Theorem 9.6 (The Divergence Test)

For the series

$$\sum_{n=1}^{\infty} u_n = u_1 + u_2 + ... + u_n + ...$$

if $$\lim_{n\to\infty} u_n \neq 0$$

then the series must diverge.

For example, for the series $\sum_{n=1}^{\infty}(3\cdot 2^n)$

$$\lim_{n\to\infty}(3\cdot 2^n) = \infty \neq 0$$

$\sum_{n=1}^{\infty}(3\cdot 2^n)$ is a geometric series. The common ratio is 2, therefore, it diverges.

Exercise 9.2

1. Apply the divergence test to determine whether the following series converge.

a. $\sum_{n=1}^{\infty} \frac{5n}{7n+2}$ **b.** $\sum_{n=1}^{\infty} (-1)^n$ **c.** $\sum_{n=1}^{\infty} \sin(n\pi)$

d. $\sum_{n=1}^{\infty} \ln n$ **e.** $\sum_{n=1}^{\infty} \frac{1}{\sqrt{n}}$ **f.** $\sum_{n=1}^{\infty} \frac{1}{n^3}$

g. $\sum_{n=1}^{\infty} \frac{1}{\sqrt[n]{5}}$ **h.** $\sum_{n=1}^{\infty} \frac{7^n}{3^n}$ **i.** $\sum_{n=1}^{\infty} (\frac{1}{4^n} + \frac{1}{6^n})$

9.3. Series of nonnegative terms

If the series

$$\sum_{n=1}^{\infty} u_n = u_1 + u_2 + \ldots + u_n + \ldots$$

satisfies the condition that $u_n \geq 0$, ($n = 1, 2, \ldots$), then it is called a series of nonnegative terms.

Obviously, the sequence of partial sums S_n of a series of nonnegative terms is a monotonically increasing sequence. That is,

$$S_1 \leq S_2 \leq \ldots \leq S_{n-1} \leq S_n \leq \ldots$$

We know from the principle of the existence of the limit of a sequence that if the sequence S_n is bounded above, it converges. Otherwise, it diverges. From this, we obtain:

Theorem 9.7

The necessary and sufficient condition for a series of nonnegative terms to converge is that its sequence of partial sums S_n is bounded.

According to Theorem 9.7, a commonly used Comparison Test for determining whether a series of nonnegative terms converges can be established.

Theorem 9.8 (The Comparison Test)

If two series of nonnegative terms

$$\sum_{n=1}^{\infty} u_n = u_1 + u_2 + \ldots + u_n + \ldots \tag{9.1}$$

and

$$\sum_{n=1}^{\infty} v_n = v_1 + v_2 + \ldots + v_n + \ldots \tag{9.7}$$

satisfy the relation

$$u_n \leq cv_n \quad (n = 1, 2, 3, \ldots; \text{ where } c \text{ is a constant greater than zero.})$$

then

(1) if series (9.7) converges, series (9.1) also converges;

(2) if series (9.1) diverges, series (9.7) also diverges.

Proof. Let

$$S_n = u_1 + u_2 + ... + u_n$$

$$W_n = v_1 + v_2 + ... + v_n$$

Since $\quad u_n \le cv_n$

it follows that $\quad S_n \le cW_n$

From Theorem 9.7, it follows that

(1) if series (9.7) converges, then W_n is bounded. Hence, S_n is also bounded. Therefore, series (9.1) also converges;

(2) if series (9.1) diverges, then S_n is unbounded. Hence, W_n is unbounded as well. Therefore, series (9.7) also diverges. □

For the convenience of memory, since

$$u_n \le cv_n$$

we can call $\sum_{n=1}^{\infty} u_n$ the 'smaller' series and $\sum_{n=1}^{\infty} v_n$ the 'larger' series:

(1) if the 'larger' series converges, the 'smaller' series also converges;

(2) if the 'smaller' series diverges, the 'larger' series also diverges.

Example 1. Determine the convergence of the harmonic series.

$$\sum_{n=1}^{\infty} \frac{1}{n} = 1 + \frac{1}{2} + \frac{1}{3} + ... + \frac{1}{n} + ...$$

Solution. $$\sum_{n=1}^{\infty} \frac{1}{n} = 1 + \frac{1}{2} + \frac{1}{3} + ... + \frac{1}{n} + ...$$

$$= (1 + \frac{1}{2}) + (\frac{1}{3} + \frac{1}{4}) + (\frac{1}{5} + \frac{1}{6} + \frac{1}{7} + \frac{1}{8}) + \ldots$$

Each of its terms is greater than the corresponding term in the following series.

$$\frac{1}{2} + (\frac{1}{4} + \frac{1}{4}) + (\frac{1}{8} + \frac{1}{8} + \frac{1}{8} + \frac{1}{8}) + \ldots$$

$$= \frac{1}{2} + \frac{1}{2} + \frac{1}{2} + \ldots$$

As the latter series diverges, it follows from the Comparison Test that the harmonic series

$$\sum_{n=1}^{\infty} \frac{1}{n} = 1 + \frac{1}{2} + \frac{1}{3} + \ldots + \frac{1}{n} + \ldots \qquad \text{diverges.}$$

Example 2. Determine the convergence of the following series.

$$\sum_{n=1}^{\infty} \frac{1}{n^p} = 1 + \frac{1}{2^p} + \frac{1}{3^p} + \ldots + \frac{1}{n^p} + \ldots$$

Solution. When $p \leq 1$

$$\frac{1}{n^p} \geq \frac{1}{n}$$

We know from Example 1 above that $\sum_{n=1}^{\infty} \frac{1}{n}$ diverges, thus, $\sum_{n=1}^{\infty} \frac{1}{n^p}$ also diverges.

When $p > 1$

$$\sum_{n=1}^{\infty} \frac{1}{n^p} = 1 + (\frac{1}{2^p} + \frac{1}{3^p}) + (\frac{1}{4^p} + \frac{1}{5^p} + \frac{1}{6^p} + \frac{1}{7^p}) + (\frac{1}{8^p} + \ldots + \frac{1}{15^p}) + \ldots$$

Each of its terms is smaller than the corresponding term in the following series.

$$1 + (\frac{1}{2^p} + \frac{1}{2^p}) + (\frac{1}{4^p} + \frac{1}{4^p} + \frac{1}{4^p} + \frac{1}{4^p}) + (\frac{1}{8^p} + \ldots + \frac{1}{8^p}) + \ldots$$

The latter series is a geometric series with common ratio r $= \frac{1}{2^{p-1}} < 1$, thus, the geometric series converges. Hence, the series $\sum_{n=1}^{\infty}\frac{1}{n^p}$ converges.

Therefore, when $p \le 1$, $\sum_{n=1}^{\infty}\frac{1}{n^p}$ diverges; when $p > 1$, $\sum_{n=1}^{\infty}\frac{1}{n^p}$ converges.

Example 3. Determine the convergence of the following series.

$$\sum_{n=1}^{\infty}\frac{1}{n^n} = 1 + \frac{1}{2^2} + \frac{1}{3^3} + \ldots + \frac{1}{n^n} + \ldots$$

Solution. Since $\quad \frac{1}{n^n} \le \frac{1}{2^{n-1}}$

and the series $\sum_{n=1}^{\infty}\frac{1}{2^{n-1}}$ is a geometric series that converges to 2, it follows that the given series converges, and its sum is less than 2.

Theorem 9.9 (The Limit Comparison Test)

If both $\sum_{n=1}^{\infty} u_n$ and $\sum_{n=1}^{\infty} v_n$ are series of nonnegative terms and

$$\text{L} = \lim_{n\to\infty}\frac{u_n}{v_n} \quad \text{is a positive, finite and non-zero constant}$$

then either both series converge or both diverge.

When using this test, we usually need to determine the convergence of $\sum_{n=1}^{\infty} u_n$. We choose a series $\sum_{n=1}^{\infty} v_n$ whose convergence property is known to compare it with.

Common series for comparison are:

(1) the harmonic series $\sum_{n=1}^{\infty} \frac{1}{n}$, which diverges;

(2) the p-series $\sum_{n=1}^{\infty} \frac{1}{n^p}$, which converges when $p > 1$ and diverges when $p \le 1$;

(3) the geometric series $\sum_{n=1}^{\infty} u_1 r^{n-1}$, which converges when $|r| < 1$ and diverges when $|r| \ge 1$;

(4) the series $\sum_{n=1}^{\infty} \frac{1}{n!}$, which converges.

Example 4. Determine the convergence of the following series.

a) $\sum_{n=1}^{\infty} \frac{5n}{n^2+3}$

b) $\sum_{n=1}^{\infty} \frac{10}{n^2+6}$

c) $\sum_{n=1}^{\infty} \frac{3}{n+\sqrt{n}}$

Solution.

a) Let $u_n = \frac{5n}{n^2+3}$, $v_n = \frac{1}{n}$, then

$$\lim_{n\to\infty} \frac{u_n}{v_n} = \lim_{n\to\infty} \frac{\frac{5n}{n^2+3}}{\frac{1}{n}} = \lim_{n\to\infty} \frac{5n^2}{n^2+3} = \lim_{n\to\infty} \frac{5}{1+\frac{3}{n^2}} = 5$$

Since $\sum_{n=1}^{\infty} \frac{1}{n}$ diverges, by the Limit Comparison Test, $\sum_{n=1}^{\infty} \frac{5n}{n^2+3}$ also diverges.

b) Let $u_n = \frac{10}{n^2+6}$, $v_n = \frac{1}{n^2}$, then

$$\lim_{n\to\infty}\frac{u_n}{v_n}=\lim_{n\to\infty}\frac{\frac{10}{n^2+6}}{\frac{1}{n^2}}=\lim_{n\to\infty}\frac{10n^2}{n^2+6}=\lim_{n\to\infty}\frac{10}{1+\frac{6}{n^2}}=10$$

Since $\sum_{n=1}^{\infty}\frac{1}{n^2}$ converges, by the Limit Comparison Test, $\sum_{n=1}^{\infty}\frac{10}{n^2+6}$ also converges.

c) Let $u_n=\frac{3}{n+\sqrt{n}}$, $v_n=\frac{1}{n}$, then

$$\lim_{n\to\infty}\frac{u_n}{v_n}=\lim_{n\to\infty}\frac{\frac{3}{n+\sqrt{n}}}{\frac{1}{n}}=\lim_{n\to\infty}\frac{3n}{n+\sqrt{n}}=\lim_{n\to\infty}\frac{3}{1+\frac{1}{\sqrt{n}}}=3$$

Since $\sum_{n=1}^{\infty}\frac{1}{n}$ diverges, by the Limit Comparison Test, $\sum_{n=1}^{\infty}\frac{3}{n+\sqrt{n}}$ also diverges.

Below, we introduce the Ratio Test, which is very convenient for practical applications.

Theorem 9.10 (The Ratio Test)

If the series of nonnegative terms

$$\sum_{n=1}^{\infty}u_n=u_1+u_2+\ldots+u_n+\ldots\quad(u_n>0,\ n=1,2,\ldots)$$

satisfies the condition $\lim_{n\to\infty}\frac{u_{n+1}}{u_n}=\mathrm{L}$, then

(1) when L < 1, the series converges;

(2) when L > 1, the series diverges;

(3) when L = 1, this test cannot be used to determine the convergence of the series. The series may converge or diverge.

For example, the series $\sum_{n=1}^{\infty}\frac{1}{n(n+1)}$ satisfies

$$\lim_{n\to\infty}\frac{u_{n+1}}{u_n}=\lim_{n\to\infty}\frac{\frac{1}{(n+1)(n+2)}}{\frac{1}{n(n+1)}}=\lim_{n\to\infty}\frac{n}{n+2}=\lim_{n\to\infty}\frac{1}{1+\frac{2}{n}}=1$$

We have already proved that it is a convergent series in Example 2 of Section 9.1 of this chapter.

For example, the harmonic series $\sum_{n=1}^{\infty}\frac{1}{n}$ also satisfies

$$\lim_{n\to\infty}\frac{u_{n+1}}{u_n}=\lim_{n\to\infty}\frac{\frac{1}{n+1}}{\frac{1}{n}}=\lim_{n\to\infty}\frac{n}{n+1}=\lim_{n\to\infty}\frac{1}{1+\frac{1}{n}}=1$$

But it diverges.

Therefore, when L = 1, this test cannot be used to determine the convergence of a series.

Example 5. Determine the convergence of the series $\sum_{n=1}^{\infty}\frac{x^n}{n}$ $(x>0)$.

Solution.
$$\lim_{n\to\infty}\frac{u_{n+1}}{u_n}=\lim_{n\to\infty}\frac{\frac{x^{n+1}}{n+1}}{\frac{x^n}{n}}=\lim_{n\to\infty}\frac{n}{n+1}x=\lim_{n\to\infty}\frac{1}{1+\frac{1}{n}}x=x$$

Hence, when $0<x<1$, the series converges. When $x\ge 1$, the series diverges.

Example 6. Determine the convergence of the series $\sum_{n=1}^{\infty}\frac{n\cos^2\frac{n}{3}\pi}{2^n}$.

Solution. Because

$$\frac{n\cos^2\frac{n}{3}\pi}{2^n} \le \frac{n}{2^n} \qquad (\cos^2\frac{n}{3}\pi \le 1)$$

and the series $\sum_{n=1}^{\infty}\frac{n}{2^n}$ satisfies

$$\lim_{n\to\infty}\frac{u_{n+1}}{u_n} = \lim_{n\to\infty}\frac{\frac{n+1}{2^{n+1}}}{\frac{n}{2^n}} = \lim_{n\to\infty}\frac{1}{2}\cdot\frac{n+1}{n} = \frac{1}{2}$$

thus, it converges. Therefore, the series $\sum_{n=1}^{\infty}\frac{n\cos^2\frac{n}{3}\pi}{2^n}$ also converges.

Theorem 9.11 (The *n*th Root Test)

Let $\sum_{n=1}^{\infty}u_n$ be a series of nonnegative terms. Let $\lim_{n\to\infty}\sqrt[n]{u_n} = \mathrm{L}$, then

(1) when L < 1, the series converges;

(2) when L > 1, the series diverges;

(3) when L = 1, this test cannot be used to determine the convergence of the series.

Usually, if the general term of the series contains *n*th power, we can consider using this test. When using this test, a very useful result is

$$\lim_{n\to\infty}\sqrt[n]{n} = 1$$

Proof. Because

$$\lim_{x\to\infty}\sqrt[x]{x} = \lim_{x\to\infty}x^{\frac{1}{x}}$$

by taking the logarithm on both sides, we have

$$\ln \lim_{x\to\infty} \sqrt[x]{x} = \lim_{x\to\infty} \ln x^{\frac{1}{x}}$$

$$= \lim_{x\to\infty} \frac{\ln x}{x} \qquad \frac{\infty}{\infty} \text{ type}$$

$$= \lim_{x\to\infty} \frac{1}{x} = 0$$

That is, $\ln \lim_{x\to\infty} \sqrt[x]{x} = 0$

Hence, $\lim_{x\to\infty} \sqrt[x]{x} = 1$

Let $x = n$ and we have

$$\lim_{n\to\infty} \sqrt[n]{n} = 1$$

□

If we apply this test to examine the series $\sum_{n=1}^{\infty} \frac{1}{n}$ and $\sum_{n=1}^{\infty} \frac{1}{n^2}$

$$\lim_{n\to\infty} \sqrt[n]{u_n} = \lim_{n\to\infty} \sqrt[n]{\frac{1}{n}} = \lim_{n\to\infty} \frac{1}{\sqrt[n]{n}} = 1$$

$$\lim_{n\to\infty} \sqrt[n]{u_n} = \lim_{n\to\infty} \sqrt[n]{\frac{1}{n^2}} = \lim_{n\to\infty} \frac{1}{\left(\sqrt[n]{n}\right)^2} = 1$$

we know that $\sum_{n=1}^{\infty} \frac{1}{n}$ diverges and $\sum_{n=1}^{\infty} \frac{1}{n^2}$ converges. Therefore, when L = 1, we cannot use this test to determine the convergence of the series.

Example 7. Determine the convergence of the series $\sum_{n=1}^{\infty} \frac{n}{2^n}$.

Solution. Since

$$\lim_{n\to\infty} \sqrt[n]{\frac{n}{2^n}} = \lim_{n\to\infty} \frac{\sqrt[n]{n}}{2} = \frac{1}{2} < 1$$

it follows that the series $\sum_{n=1}^{\infty} \frac{n}{2^n}$ converges.

Example 8. Determine the convergence of the series $\sum_{n=1}^{\infty} \frac{9^n}{n^3 4^n}$.

Solution. Since

$$\lim_{n\to\infty} \sqrt[n]{\frac{9^n}{n^3 4^n}} = \lim_{n\to\infty} \frac{9}{\left(\sqrt[n]{n}\right)^3 4} = \frac{9}{4} > 1$$

thus, the series diverges.

Theorem 9.12 (The Integral Test)

According to the approximate formula of the rectangular method of definite integral

$$\int_0^{\infty} f(x)dx = \Delta x \sum_{i=0}^{\infty} f(x_i) + R \qquad (R \text{ is an error.})$$

let $\quad x_i = n, \quad \Delta x = x_{i+1} - x_i = n + 1 - n = 1$

then $$\int_0^{\infty} f(x)dx = \sum_{n=0}^{\infty} f(n) + R = \sum_{n=0}^{\infty} u_n + R$$

That is, $$\int_0^{\infty} f(x)dx = \sum_{n=0}^{\infty} u_n + R \qquad \text{where} \qquad f(n) = u_n$$

When the integral $\int_0^{\infty} f(x)dx$ converges, that is,

$$\int_0^{\infty} f(x)dx = \text{a finite number}$$

then the series

$$\sum_{n=0}^{\infty} u_n = \int_0^{\infty} f(x)dx - R = \text{a finite number} - R = \text{a finite number}$$

Therefore, the series $\sum_{n=0}^{\infty} u_n$ converges.

On the contrary, if $\int_0^{\infty} f(x)dx$ diverges, that is,

$$\int_0^{\infty} f(x)dx = \infty$$

then the series

$$\sum_{n=0}^{\infty} u_n = \int_0^{\infty} f(x)dx - R = \infty - R = \infty$$

Therefore, the series $\sum_{n=0}^{\infty} u_n$ diverges.

When the Integral Test is used, we only need to change the n in the general term to x then we will obtain $f(x)$.

Note:

- **The Integral Test applies only to the series with nonnegative terms, where $f(x)$ is positive and continuous on the interval [0, +∞[.**

- Consider using this test only when it is easy to integrate $f(x)$.

Example 9. Determine the convergence of the series $\sum_{n=1}^{\infty} \frac{4}{3n-2}$.

Solution. Let $u_n = f(n) = \frac{4}{3n-2}$, then $f(x) = \frac{4}{3x-2}$

On the interval $[1, +\infty[$, $f(x)$ is positive and continuous; therefore, the Integral Test can be applied.

$$\int_1^{\infty} \frac{4}{3x-2}dx$$

Let $u = 3x - 2$ then $du = 3dx$

when $x = 1$, $u_1 = 1$

when $x = \infty$, $u_2 = \infty$

$$\int_1^{\infty} \frac{4}{3x-2}dx = \int_1^{\infty} \frac{4}{u}\cdot\frac{du}{3} = \frac{4}{3}\lim_{b\to\infty}\int_1^{b}\frac{1}{u}du = \frac{4}{3}\lim_{b\to\infty}\left[\ln u\right]_1^b = \frac{4}{3}\lim_{b\to\infty}\left[\ln b - \ln 1\right] = \frac{4}{3}\ln\infty = \infty$$

Therefore, the series $\sum_{n=1}^{\infty}\frac{4}{3n-2}$ diverges.

Example 10. Determine the convergence of the series $\sum_{n=1}^{\infty} ne^{-n^2}$.

Solution. Let $u_n = f(n) = ne^{-n^2}$ then $f(x) = xe^{-x^2}$

On the interval $[1, +\infty[$, $f(x)$ is positive and continuous; therefore, the Integral Test can be applied.

$$\int_1^{\infty} xe^{-x^2}dx$$

Let $u = -x^2$ then $du = -2xdx$

when $x = 1$, $u_1 = -1$

when $x = \infty$, $u_2 = -\infty$

$$\int_1^{\infty} xe^{-x^2}dx = \int_{-1}^{-\infty} xe^{u}\cdot -\frac{du}{2x} = -\frac{1}{2}\int_{-1}^{-\infty} e^{u}du = -\frac{1}{2}\lim_{b\to-\infty}\int_{-1}^{b} e^{u}du = -\frac{1}{2}\lim_{b\to-\infty}\left[e^{u}\right]_{-1}^{b}$$

$$= -\frac{1}{2}\left[e^{-\infty} - e^{-1}\right] = -\frac{1}{2}\left[0 - e^{-1}\right] = \frac{1}{2}e^{-1}$$

Therefore, the series $\sum_{n=1}^{\infty} ne^{-n^2}$ converges.

Exercise 9.3

1. Use the Comparison Test or the Limit Comparison Test to determine whether the following series converge.

a. $\sum_{n=1}^{\infty} \frac{1}{2n+1}$ **b.** $\sum_{n=1}^{\infty} \frac{2}{n^2+1}$ **c.** $\sum_{n=1}^{\infty} \frac{4n}{n^2+1}$ **d.** $\sum_{n=1}^{\infty} \frac{5}{\sqrt{n}}$

e. $\sum_{n=1}^{\infty} \frac{3n^2}{n^3-2}$ **f.** $\sum_{n=1}^{\infty} \frac{3}{2n^2+n}$ **g.** $\sum_{n=1}^{\infty} \frac{7}{n^3+n}$ **h.** $\sum_{n=1}^{\infty} \frac{6\sin^2 n}{n!}$

i. $\sum_{n=1}^{\infty} \frac{5}{3^n+4}$ **j.** $\sum_{n=1}^{\infty} \frac{n}{n^{\frac{3}{2}} - \frac{3}{2}}$ **k.** $\sum_{n=1}^{\infty} \frac{1}{n\sqrt{n+1}}$ **l.** $\sum_{n=1}^{\infty} \frac{\ln n}{n}$

m. $\sum_{n=1}^{\infty} \frac{1}{\sqrt{(2n+1)(2n-1)}}$ **n.** $\sum_{n=1}^{\infty} \frac{2+n}{3+n^3}$ **o.** $\sum_{n=1}^{\infty} \frac{n+1}{n^2+1}$

2. Use the Ratio Test or the *n*th Root Test to determine whether the following series converge.

a. $\sum_{n=1}^{\infty} \frac{2n+1}{3^n}$ **b.** $\sum_{n=1}^{\infty} \frac{n^2}{5^n}$ **c.** $\sum_{n=1}^{\infty} \frac{n!}{6^n}$ **d.** $\sum_{n=1}^{\infty} \frac{3^{2n}}{4^{3n}}$

e. $\sum_{n=1}^{\infty} \frac{1}{(2n+3)!}$ **f.** $\sum_{n=1}^{\infty} \frac{(n!)^2}{(2n)!}$ **g.** $\sum_{n=1}^{\infty} (\frac{n}{2n+1})^n$ **h.** $\sum_{n=1}^{\infty} \frac{n!}{n^5}$

i. $\sum_{n=1}^{\infty} \frac{(n+1)!9^n}{7^n}$ **j.** $\sum_{n=1}^{\infty} \frac{500^n}{(n+2)!}$ **k.** $\sum_{n=1}^{\infty} \frac{(n!)^2}{n^n}$ **l.** $\sum_{n=1}^{\infty} \frac{(n!)^2}{2^{n^2}}$

m. $\sum_{n=1}^{\infty} \frac{n^2}{(2+\frac{1}{n})^n}$ **n.** $\sum_{n=1}^{\infty} \frac{2+(-1)^n}{2^n}$ **o.** $\sum_{n=1}^{\infty} \frac{n^3\left[\sqrt{2}+(-1)^n\right]^n}{4^n}$

p. $\sum_{n=1}^{\infty}\frac{n!}{e^{n^2}}$ **q.** $\sum_{n=1}^{\infty}\frac{(n+3)!}{3!n!3^n}$ **r.** $\sum_{n=1}^{\infty}\frac{(n!)^3 3^n}{(3n+1)!}$

s. $\sum_{n=1}^{\infty}\frac{(n+2)(n+3)}{5^n}$ **t.** $\sum_{n=1}^{\infty}\frac{n^5 5^n}{7^n}$ **u.** $\sum_{n=1}^{\infty}\frac{5^n n!}{n^n}$

3. Use the Integral Test to determine whether the following series converge.

a. $\sum_{n=1}^{\infty}\frac{1}{3n+2}$ **b.** $\sum_{n=1}^{\infty}\frac{1}{1+4n^2}$ **c.** $\sum_{n=1}^{\infty}\frac{n}{\sqrt{n^2+16}}$ **d.** $\sum_{n=1}^{\infty}\frac{n^2+1}{n^2+3}$

e. $\sum_{n=1}^{\infty}\frac{\arctan n}{1+n^2}$ **f.** $\sum_{n=1}^{\infty}\frac{\ln n}{n}$ **g.** $\sum_{n=1}^{\infty}\frac{1}{n(n+2)}$ **h.** $\sum_{n=1}^{\infty}\frac{1}{(2+3n)^{\frac{3}{2}}}$

i. $\sum_{n=1}^{\infty}n^2 e^{-n^3}$ **j.** $\sum_{n=2}^{\infty}\frac{1}{n(\ln n)^4}$ **k.** $\sum_{n=1}^{\infty}\frac{n+6}{n+2}$ **l.** $\sum_{n=1}^{\infty}e^{-2n}$

4. Use any method to determine whether the following series converge.

a. $\sum_{n=1}^{\infty}\frac{n^{\frac{5}{4}}}{n-1}$ **b.** $\sum_{n=1}^{\infty}\frac{1}{\sqrt{n+6}}$ **c.** $\sum_{n=1}^{\infty}\frac{2^n}{2n-1}$ **d.** $\sum_{n=1}^{\infty}\frac{n^2}{n!}$

e. $\sum_{n=1}^{\infty}\frac{n^2}{5^n}$ **f.** $\sum_{n=1}^{\infty}\frac{\pi}{2^n}$ **g.** $\sum_{n=1}^{\infty}\frac{2+(-1)^n}{4^n}$ **h.** $\sum_{n=1}^{\infty}\frac{1}{n\sqrt[n]{n}}$

i. $\sum_{n=1}^{\infty}\frac{(2n-1)!}{3^n\cdot n!}$ **j.** $\sum_{n=1}^{\infty}\frac{2n\cdot n!}{n^n}$ **k.** $\sum_{n=1}^{\infty}\frac{1}{(n+3)(n+9)}$ **l.** $\sum_{n=1}^{\infty}n\left(\frac{2}{7}\right)^n$

m. $\sum_{n=1}^{\infty}\frac{n^6}{(n+1)!}$ **n.** $\sum_{n=1}^{\infty}\frac{n+4}{n(n+8)}$ **o.** $\sum_{n=1}^{\infty}\sqrt{\frac{n+7}{n}}$

9.4. Series of arbitrary terms, absolute convergence

Now, we shall discuss series with both positive and negative terms. A special case shall be discussed first.

A series that alternates between positive and negative terms is called an **alternating series**. It can be expressed in the following form:

$$\sum_{n=1}^{\infty}(-1)^{n-1}u_n = u_1 - u_2 + u_3 - u_4 + \ldots + u_{2k-1} - u_{2k} + \ldots \qquad (9.8)$$

where $u_n > 0$ $(n = 1, 2, \ldots)$.

We can use the following theorem to determine the convergence of an alternating series.

Theorem 9.13 (Leibniz Theorem)

If the alternating series (9.8) satisfies the following conditions:

(1) $u_n > u_{n+1}$ $(n = 1, 2, \ldots)$;

(2) $\lim_{n\to\infty} u_n = 0$;

then the series converges, and its sum satisfies $S < u_1$.

Proof. Write the first $2k$ terms of the series in the following two forms.

$$S_{2k} = (u_1 - u_2) + (u_3 - u_4) + \ldots + (u_{2k-1} - u_{2k})$$

and

$$S_{2k} = u_1 - (u_2 - u_3) - (u_4 - u_5) - \ldots - (u_{2k-2} - u_{2k-1}) - u_{2k}$$

From condition (1), we note that all the differences in parentheses in the above two equations are nonnegative.

From the first equation, it follows that S_{2k} increases as k increases.

From the second equation, we observe that $S_{2k} < u_1$.

According to the existence principle of limit, we have

$$\lim_{k\to\infty} S_{2k} = S < u_1$$

Furthermore, from $S_{2k+1} = S_{2k} + u_{2k+1}$ and condition (2), we have

$$\lim_{k\to\infty} S_{2k+1} = \lim_{k\to\infty} S_{2k} + \lim_{k\to\infty} u_{2k+1} = S + 0 = S$$

Therefore, regardless of whether n is odd or even, as n grows infinitely, S_n always approaches the same limit S. Consequently, the alternating series (9.8) converges, and its sum satisfies $S < u_1$.

If S_n is used as an approximation of the sum of the series, S, then the absolute error

$$|R_n| < u_{n+1}$$

Since $$|R_n| = u_{n+1} - u_{n+2} + \ldots$$

is also an alternating series, which satisfies the conditions for convergence, it follows that its sum is less than the first term of the remainder series. That is,

$$|R_n| < u_{n+1}$$

Example 1. The alternating series

$$\sum_{n=1}^{\infty} (-1)^{n-1} \frac{1}{n} = 1 - \frac{1}{2} + \frac{1}{3} - \frac{1}{4} + \ldots + \frac{(-1)^{n-1}}{n} + \ldots$$

satisfies $$1 > \frac{1}{2} > \frac{1}{3} > \frac{1}{4} > \ldots > \frac{1}{n} > \frac{1}{n+1} > \ldots$$

and $$\lim_{n\to\infty} \frac{1}{n} = 0$$

Hence, this series converges, and its sum is less than the first term, $u_1 = 1$. If we use the sum of the first n terms $S_n = \sum_{k=1}^{n} \frac{(-1)^{k-1}}{k}$ as an approximation of S, then the absolute error satisfies $|R_n| < \frac{1}{n+1}$.

Example 2. If the first seven terms of the following series are used to approximate the sum S of the series, find the upper limit of the error.

$$\sum_{n=1}^{\infty}(-1)^n \frac{3}{n^4}$$

Solution. Since

$$|R_n| < u_{n+1}$$

thus
$$|R_7| < \frac{3}{(7+1)^4}$$

$$|R_7| < 7.32 \times 10^{-4}$$ □

For a general series containing both positive and negative terms, we have the following Theorem.

Theorem 9.14

For a general series with positive and negative terms

$$\sum_{n=1}^{\infty} u_n = u_1 + u_2 + \ldots + u_n + \ldots$$

if the series formed by the absolute values of all terms of the above series

$$\sum_{n=1}^{\infty} |u_n| = |u_1| + |u_2| + \ldots + |u_n| + \ldots$$

converges, then the original series also converges.

Proof. Since $u_n \le |u_n| \quad (n = 1, 2, \ldots)$

thus
$$\sum_{n=1}^{\infty} u_n \le \sum_{n=1}^{\infty} |u_n|$$

When $\sum_{n=1}^{\infty} |u_n| = S$ (a finite number), that is, $\sum_{n=1}^{\infty} |u_n|$ is convergent

$$\sum_{n=1}^{\infty} u_n = \text{W} \le S \text{ (a finite number)} \qquad \text{(by the Comparison Test)}$$

Therefore, the original series $\sum_{n=1}^{\infty} u_n$ is also convergent. That is, its sum W is also a finite number.

Note:

• If the series formed by the absolute values of all terms of a series is convergent, then we say that the original series converges absolutely.

If a series converges, but the series formed by the absolute values of all its terms diverges, then the series is said to converge conditionally.

For example, the series $\sum_{n=1}^{\infty} \frac{(-1)^{n-1}}{n}$ converges, but the series formed by the absolute values of all of its terms diverges. This is because

$$\sum_{n=1}^{\infty} \frac{(-1)^{n-1}}{n} \le \sum_{n=1}^{\infty} \left| \frac{(-1)^{n-1}}{n} \right| = \sum_{n=1}^{\infty} \frac{1}{n} = \infty \qquad \text{which diverges.}$$

Hence, the series $\sum_{n=1}^{\infty} \frac{(-1)^{n-1}}{n}$ converges conditionally.

Obviously, the series $\sum_{n=1}^{\infty} (-1)^{n-1} r^{n-1}$ $(0 < r < 1)$ converges absolutely. We know from the inequality relationship that the series $\sum_{n=1}^{\infty} (-1)^{n-1} r^{n-1}$ also converges.

• For a series containing both positive and negative terms, because the series formed by the absolute values of its terms is a series of nonnegative terms, all tests used to determine the convergence of a series of nonnegative terms can also be used to determine its convergence.

• For a series that contains both positive and negative terms $\sum_{n=1}^{\infty} u_n$, if $\sum_{n=1}^{\infty} |u_n|$ converges, then $\sum_{n=1}^{\infty} u_n$ must also converge. However, when $\sum_{n=1}^{\infty} |u_n|$ diverges, we can only conclude that $\sum_{n=1}^{\infty} u_n$ does not converge absolutely. We cannot determine that it definitely diverges.

Theorem 9.15 (The Ratio Test for Absolute Convergence)

If a series that contains both positive and negative terms

$$\sum_{n=1}^{\infty} u_n = u_1 + u_2 + \ldots + u_n + \ldots$$

satisfies the condition

$$\lim_{n\to\infty} \left| \frac{u_{n+1}}{u_n} \right| = \mathrm{L}$$

then, by considering the inequality $\sum_{n=1}^{\infty} u_n \le \sum_{n=1}^{\infty} |u_n|$, we observe the following:

when L < 1, the series converges absolutely. $\Rightarrow$ The series $\sum_{n=1}^{\infty} u_n$ converges;

when L > 1, the general term $\lim_{n\to\infty} |u_n| \neq 0$. $\Rightarrow$ The series $\sum_{n=1}^{\infty} u_n$ diverges;

when L = 1, the test is inconclusive.

Proof. According to Theorem 9.8 (the Comparison Test)

when L < 1, $\sum_{n=1}^{\infty} |u_n|$ converges; therefore, $\sum_{n=1}^{\infty} u_n$ also converges.

When $L > 1$, as $n \to \infty$, it is impossible for $|u_n|$ to approach zero. It follows that it is impossible for u_n to approach zero. Therefore, according to Theorem 9.6 (the Divergence Test), the series $\sum_{n=1}^{\infty} u_n$ diverges.

Example 3. Show that the series

$$\sum_{n=1}^{\infty}(-1)^n \frac{n!}{n^n} = -1 + \frac{2!}{2^2} - \frac{3!}{3^3} + \ldots + (-1)^n \frac{n!}{n^n} + \ldots$$

converges absolutely.

Proof.
$$\lim_{n\to\infty}\left|\frac{u_{n+1}}{u_n}\right| = \lim_{n\to\infty}\frac{\frac{(n+1)!}{(n+1)^{n+1}}}{\frac{n!}{n^n}} = \lim_{n\to\infty}\left(\frac{n}{n+1}\right)^n = \lim_{n\to\infty}\frac{1}{\left(1+\frac{1}{n}\right)^n} = \frac{1}{e} < 1$$

Therefore, the series converges absolutely.

Example 4. Determine the convergence of the series $\sum_{n=1}^{\infty}\frac{x^n}{n!}$.

Solution.
$$\lim_{n\to\infty}\left|\frac{u_{n+1}}{u_n}\right| = \lim_{n\to\infty}\frac{\frac{|x|^{n+1}}{(n+1)!}}{\frac{|x|^n}{n!}} = \lim_{n\to\infty}\frac{|x|}{n+1} = 0$$

Therefore, this series converges absolutely for all x ($x \in R$).

According to the necessary condition for convergence, and based on Examples 3 and 4 above, we have

$$\lim_{n\to\infty}\frac{n!}{n^n} = 0 \qquad \lim_{n\to\infty}\frac{x^n}{n!} = 0$$

Example 5. Determine the convergence of the series $\sum_{n=1}^{\infty}\frac{x^n}{n}$.

Solution. $$\lim_{n\to\infty}\left|\frac{u_{n+1}}{u_n}\right| = \lim_{n\to\infty}\frac{\frac{|x|^{n+1}}{n+1}}{\frac{|x|^n}{n}} = \lim_{n\to\infty}\frac{n}{n+1}\cdot|x| = |x|$$

Therefore, when $|x|<1$, the series converges absolutely; when $|x|>1$, the series diverges. When $x=1$, the series becomes the harmonic series, which diverges; when $x=-1$, the series becomes

$$\sum_{n=1}^{\infty}\frac{(-1)^n}{n} = -\sum_{n=1}^{\infty}\frac{(-1)^{n-1}}{n}$$

This series converges conditionally (refer to Example 1 in this section).

Example 6. Determine the convergence of the series $\sum_{n=1}^{\infty} nx^{n-1}$.

Solution. $$\lim_{n\to\infty}\left|\frac{u_{n+1}}{u_n}\right| = \lim_{n\to\infty}\frac{(n+1)|x|^n}{n|x|^{n-1}} = \lim_{n\to\infty}\left(1+\frac{1}{n}\right)|x| = |x|$$

Hence, when $|x|<1$, the series converges absolutely;

when $|x|\geq 1$, the series diverges.

(when $|x|=1$, the general term of the series does not approach zero.)

Exercise 9.4

1. Determine which of the following alternating series converge.

a. $\sum_{n=1}^{\infty}\frac{(-1)^n}{\sqrt{n}}$ **b.** $\sum_{n=1}^{\infty}\frac{(-1)^n}{n!}$ **c.** $\sum_{n=1}^{\infty}\frac{(-1)^{n+1}n}{2n+1}$

d. $\sum_{n=3}^{\infty} \frac{(-1)^n \ln n}{n}$ e. $\sum_{n=1}^{\infty} \frac{(-1)^{n+1} n}{3^n}$ f. $\sum_{n=1}^{\infty} \frac{(-1)^n (2n+3)}{n^2}$

g. $\sum_{n=1}^{\infty} \frac{(-1)^n n}{e^n}$ h. $\sum_{n=1}^{\infty} (-1)^n e^{-n}$ i. $\sum_{n=1}^{\infty} \frac{\cos(n\pi)}{n^{\frac{2}{5}}}$

2. Determine which of the following series converge absolutely, which converge conditionally, and which diverge.

a. $\sum_{n=1}^{\infty} \frac{(-1)^{n+1}}{(2n+1)^2}$ b. $\sum_{n=1}^{\infty} \frac{(-1)^{n+1}}{\ln(n+1)}$ c. $\sum_{n=1}^{\infty} \frac{(-1)^n}{2^n}$

d. $\sum_{n=1}^{\infty} \frac{(-1)^n n^3}{\pi^n}$ e. $\sum_{n=1}^{\infty} \frac{n\cos(n\pi)}{1+n^2}$ f. $\sum_{n=3}^{\infty} \frac{(-1)^n n}{n+\ln n}$

g. $\sum_{n=1}^{\infty} (-1)^n \ln\frac{n+1}{n}$ h. $\sum_{n=1}^{\infty} \frac{(-1)^{n+1} 2^{n^2}}{n!}$ i. $\sum_{n=1}^{\infty} \frac{(-1)^n}{n\ln n}$

j. $\sum_{n=1}^{\infty} \frac{(-1)^n (n^2+1)}{n^3+2}$ k. $\sum_{n=1}^{\infty} \frac{(-1)^{n+1} n!}{(2n+1)!}$ l. $\sum_{n=1}^{\infty} \frac{(-1)^{n+1} 3^{2n+1}}{n^2+1}$

m. $\sum_{n=1}^{\infty} (-1)^n \left(\frac{n+3}{3n}\right)^n$ n. $\sum_{n=1}^{\infty} \frac{(-1)^n \arctan n}{1+n^2}$ o. $\sum_{n=1}^{\infty} \frac{(-1)^n n!}{3^{n^2}}$

3. For each of the following, find an upper bound for the error if the given nth partial sum is used as an approximation of the sum of the series.

a. $\sum_{n=1}^{\infty} \frac{(-1)^{n+1} 4}{n}$ $n=7$ b. $\sum_{n=1}^{\infty} \frac{(-1)^n}{(n+2)(n+3)}$ $n=5$

c. $\sum_{n=1}^{\infty} (-1)^n 5n^{-2}$ $n=4$ d. $\sum_{n=1}^{\infty} \frac{(-1)^n}{\sqrt{n}}$ $n=99$

e. $\sum_{n=1}^{\infty} \frac{(-1)^{n+1} 5}{3^n}$ $n=6$ f. $\sum_{n=1}^{\infty} \frac{(-1)^n}{(n+1)^2}$ $n=3$

4. Find a value of n for each of the following series to give a sum correct to four decimal places.

a. $\sum_{n=1}^{\infty}\frac{(-1)^n}{n^5}$ **b.** $\sum_{n=1}^{\infty}(-1)^n\left(\frac{4}{5}\right)^n$ **c.** $\sum_{n=1}^{\infty}\frac{(-1)^n}{(n+1)!}$

d. $\sum_{n=1}^{\infty}\frac{(-1)^n}{(n+1)^2\ln(n+1)}$ **e.** $\sum_{n=1}^{\infty}\frac{(-1)^{n+1}}{n\cdot 3^n}$ **f.** $\sum_{n=1}^{\infty}\frac{(-1)^n}{2n^3}$

9.5. Power series

A series in the form

$$\sum_{n=0}^{\infty}a_n(x-x_0)^n = a_0 + a_1(x-x_0) + a_2(x-x_0)^2 + \ldots + a_n(x-x_0)^n + \ldots \tag{9.9}$$

is called a power series in $(x-x_0)$, where $a_0,\ a_1,\ a_2,\ \ldots,\ a_n,\ \ldots$ are all constants known as the coefficients of the power series.

When $x_0 = 0$, equation (9.9) becomes

$$\sum_{n=0}^{\infty}a_n x^n = a_0 + a_1 x + a_2 x^2 + \ldots + a_n x^n + \ldots \tag{9.10}$$

which is called a power series in x. Each of its terms is a power function of x. If we change $(x-x_0)$ to x in series (9.9), then the series (9.9) becomes the series (9.10). Therefore, in the following, we shall mainly discuss the power series of the form (9.10).

First, we shall discuss the interval of convergence of a power series. By taking the absolute value of each term of series (9.10), we obtain a series of nonnegative terms:

$$\sum_{n=0}^{\infty}\left|a_n x^n\right| = \left|a_0\right| + \left|a_1 x\right| + \left|a_2 x^2\right| + \ldots + \left|a_n x^n\right| + \ldots \tag{9.11}$$

If we let $\lim_{n\to\infty}\left|\frac{a_{n+1}}{a_n}\right| = \mathrm{L}$

then $\lim_{n\to\infty}\left|\frac{u_{n+1}}{u_n}\right| = \lim_{n\to\infty}\left|\frac{a_{n+1}x^{n+1}}{a_n x^n}\right| = \mathrm{L}\,|x|$

From Theorem 9.15 (The Ratio Test for Absolute Convergence), we observe that:

(1) If $\mathrm{L}\,|x| < 1$ ($\mathrm{L} \neq 0$), that is, $|x| < \frac{1}{\mathrm{L}} = \mathrm{R}$, then series (9.10) converges absolutely;

(2) If $\mathrm{L}\,|x| > 1$, that is, $|x| > \frac{1}{\mathrm{L}} = \mathrm{R}$, then series (9.10) diverges;

(3) If $\mathrm{L}\,|x| = 1$, that is, $|x| = \frac{1}{\mathrm{L}} = \mathrm{R}$, the Ratio Test is inconclusive, and further testing is needed;

(4) If $\mathrm{L} = 0$, then $\mathrm{L}\,|x| = 0 < 1$. In this case, series (9.10) converges for all x.

The above analysis shows that the interval of convergence of a power series is an interval centred at the origin, extending from –R to R. That is, $|x| < \mathrm{R} \Rightarrow -\mathrm{R} < x < \mathrm{R}$, which is called the **interval of convergence** of power series (9.10), where $\mathrm{R} = \frac{1}{L}$ is called the **radius of convergence**.

- If a power series diverges for all $x \neq 0$, that is, $|x| > \mathrm{R} = 0$, the radius of convergence is $\mathrm{R} = 0$. In this case, the interval of convergence of the power series reduces to the point $x = 0$. For example, consider the series

$$\sum_{n=0}^{\infty} n!x^n = 1 + 1!x + 2!x^2 + \ldots + n!x^n + \ldots$$

Since $$\lim_{n\to\infty}\left|\frac{u_{n+1}}{u_n}\right| = \lim_{n\to\infty}\frac{(n+1)!|x|^{n+1}}{n!|x|^n} = \lim_{n\to\infty}(n+1)|x| > 1$$

it follows that this series diverges for all x ($\neq 0$).

- If power series (9.10) converges for all x, that is, $-\infty = -R < x < R = \infty$, then we denote $R = +\infty$. In this instance, the interval of convergence of power series (9.10) is $x \in (-\infty, +\infty)$.

  For example, the series mentioned previously $\sum_{n=0}^{\infty}\frac{x^n}{n!}$ converges absolutely for all x. Therefore, the interval of convergence of this series is $x \in (-\infty, +\infty)$.

- When $0 < R < +\infty$, we need to investigate the convergence at the points $x = \pm R$ separately, in order to determine whether the interval of convergence is an open interval, a closed interval, or a half-open interval.

- The steps to find the interval of convergence of power series (9.10) are as follows:
  (1) First, determine the radius of convergence, R;
  (2) If $0 < R < +\infty$, investigate the convergence at the points $x = \pm R$;
  (3) Finally, state the interval of convergence.

In summary, we can state the theorem for finding the radius of convergence of power series (9.10).

Theorem 9.16

If the coefficients of the power series

$$\sum_{n=0}^{\infty} a_n x^n = a_0 + a_1 x + a_2 x^2 + \ldots + a_n x^n + \ldots$$

satisfy the condition

$$\lim_{n\to\infty}\left|\frac{a_{n+1}}{a_n}\right| = \mathrm{L}$$

then (1) when $0 < \mathrm{L} < +\infty$, $\mathrm{R} = \dfrac{1}{\mathrm{L}}$;

(2) when $\mathrm{L} = 0$, $\mathrm{R} = +\infty$;

(3) when $\mathrm{L} = +\infty$, $\mathrm{R} = 0$.

Example 1. Determine the interval of convergence of the following series.

$$\sum_{n=1}^{\infty}\frac{(-1)^{n-1}x^n}{n} = x - \frac{x^2}{2} + \frac{x^3}{3} - \ldots + (-1)^{n-1}\frac{x^n}{n} + \ldots$$

Solution. Because $\displaystyle\lim_{n\to\infty}\left|\frac{a_{n+1}}{a_n}\right| = \lim_{n\to\infty}\frac{\frac{1}{n+1}}{\frac{1}{n}} = \lim_{n\to\infty}\frac{n}{n+1} = \lim_{n\to\infty}\frac{1}{1+\frac{1}{n}} = 1$

we obtain the radius of convergence, $\mathrm{R} = 1$.

When $x = -1$, it becomes the harmonic series.

$$\sum_{n=1}^{\infty}\frac{(-1)^{2n-1}}{n} = -\sum_{n=1}^{\infty}\frac{1}{n}$$ which diverges.

When $x = 1$, it becomes an alternating series.

$$\sum_{n=1}^{\infty}\frac{(-1)^{n-1}}{n}$$ which converges.

Therefore, the interval of convergence is $x \in\ ]-1, 1]$.

Example 2. Determine the interval of convergence of the following series.

$$\sum_{n=1}^{\infty}(-1)^{n-1}x^{n-1} = 1 - x + x^2 - \ldots + (-1)^{n-1}x^{n-1} + \ldots$$

Solution. $\lim_{n\to\infty}\left|\frac{a_{n+1}}{a_n}\right| = 1$

the radius of convergence is R = 1.

When $x = -1$, the given series becomes

$$\sum_{n=1}^{\infty}(-1)^{n-1}x^{n-1} = \sum_{n=1}^{\infty}(-1)^{n-1}(-1)^{n-1} = \sum_{n=1}^{\infty}(-1)^{2n-2}$$

$$= (-1)^0 + (-1)^2 + (-1)^4 + (-1)^6 + \ldots$$

$$= 1 + 1 + 1 + 1 + \ldots \qquad \text{which diverges.}$$

When $x = 1$, the given series becomes

$$\sum_{n=1}^{\infty}(-1)^{n-1}x^{n-1} = \sum_{n=1}^{\infty}(-1)^{n-1}1^{n-1} = \sum_{n=1}^{\infty}(-1)^{n-1}$$

$$= (-1)^0 + (-1)^1 + (-1)^2 + (-1)^3 + \ldots$$

$$= 1 - 1 + 1 - 1 + \ldots \qquad \text{It oscillates and thus diverges.}$$

Therefore, the interval of convergence is $x \in (-1, 1)$.

Example 3. Determine the interval of convergence of the series $\sum_{n=1}^{\infty}\frac{x^n}{n^n}$.

Solution. Since $\lim_{n\to\infty}\left|\frac{a_{n+1}}{a_n}\right| = \lim_{n\to\infty}\frac{\frac{1}{(n+1)^{n+1}}}{\frac{1}{n^n}} = \lim_{n\to\infty}\frac{1}{n+1}\cdot\frac{1}{(1+\frac{1}{n})^n} = 0\cdot\frac{1}{e} = 0$

it follows that the radius of convergence R = +∞, and the interval of convergence is $x \in (-\infty, +\infty)$.

Example 4. Determine the interval of convergence of the series $\sum_{n=1}^{\infty}\frac{(2x+1)^n}{n}$.

Solution. $$\lim_{n\to\infty}\left|\frac{u_{n+1}}{u_n}\right| = \lim_{n\to\infty}\left|\frac{\frac{(2x+1)^{n+1}}{n+1}}{\frac{(2x+1)^n}{n}}\right| = \lim_{n\to\infty}|2x+1|\frac{n}{n+1} = |2x+1|$$

When $|2x + 1| < 1$, the series converges absolutely. That is,

$$\left|x+\frac{1}{2}\right| < \frac{1}{2}$$

Thus, the radius of convergence is $R = \frac{1}{2}$.

Hence, $$-\frac{1}{2} < x + \frac{1}{2} < \frac{1}{2}$$

$$-1 < x < 0$$

When $x = -1$, it becomes an alternating series.

$$\sum_{n=1}^{\infty}\frac{(-1)^n}{n} \qquad \text{which converges.}$$

When $x = 0$, it becomes the harmonic series.

$$\sum_{n=1}^{\infty}\frac{1}{n} \qquad \text{which diverges.}$$

Therefore, the interval of convergence is $x \in [-1, 0[$.

Note:

It can be seen from Example 4 when each term of the power series is a power function of $ax + b$ ($a \neq 1$), use $\lim_{n\to\infty}\left|\frac{u_{n+1}}{u_n}\right|$ to obtain

$$\lim_{n\to\infty}\left|\frac{u_{n+1}}{u_n}\right| = |ax+b| \qquad \text{the coefficient of } x \quad a \neq 1$$

We need to rearrange this expression so that the coefficient of x becomes 1 before we can determine the radius of convergence.

$$|ax+b| < 1$$

Dividing both sides by the absolute value of a, we obtain

$$\left|x+\frac{b}{a}\right|<\frac{1}{|a|}$$

Therefore, the radius of convergence is $\dfrac{1}{|a|}$.

For ease of reference, all tests are summarised in the table below.

Summary of Convergence Tests

Name of convergence test	Statement	Comments
Divergence Test	If $\lim_{n\to\infty} u_n \neq 0$, then $\sum_{n=1}^{\infty} u_n$ diverges. If $\lim_{n\to\infty} u_n = 0$, then $\sum_{n=1}^{\infty} u_n$ may either converge or diverge. If it is known that $\sum_{n=1}^{\infty} u_n$ converges, then we must have $\lim_{n\to\infty} u_n = 0$.	
Comparison Test	Let $\sum_{n=1}^{\infty} u_n$ and $\sum_{n=1}^{\infty} v_n$ be series of nonnegative terms and they satisfy $u_n \leq cv_n$ (1) If the 'larger' series converges, the 'smaller' series must also converge. (2) If the 'smaller' series diverges, the 'larger' series must also diverge.	This test applies only to the series with nonnegative terms.

Limit Comparison Test	If $\sum_{n=1}^{\infty} u_n$ and $\sum_{n=1}^{\infty} v_n$ are series of nonnegative terms and $L = \lim_{n\to\infty} \frac{u_n}{v_n}$ is a positive, finite, and non-zero constant then both series either converge or both diverge together.	Some skills are needed to select $\sum_{n=1}^{\infty} v_n$ whose convergence properties are already known for comparison.
Ratio Test	Let $\sum_{n=1}^{\infty} u_n$ be a series of nonnegative terms. $$L = \lim_{n\to\infty} \frac{u_{n+1}}{u_n}$$ (1) When $L < 1$, the series converges. (2) When $L > 1$, the series diverges. (3) When $L = 1$, the test is inconclusive.	
***n*th Root Test**	Let $\sum_{n=1}^{\infty} u_n$ be a series of nonnegative terms. $$L = \lim_{n\to\infty} \sqrt[n]{u_n}$$ (1) When $L < 1$, the series converges. (2) When $L > 1$, the series diverges. (3) When $L = 1$, the test is inconclusive.	If the general term of the series contains an *n*th power, we can consider using this test.
Integral Test	If $f(x)$ is a positive and continuous function and to all positive numbers $u_n = f(n)$, then the series $\sum_{n=1}^{\infty} u_n$ and the integral $\int_1^{\infty} f(x)dx$ either both converge or both diverge.	This test applies only to a series of nonnegative terms. $f(x)$ is positive and continuous.

		Consider using this test only when it is easy to integrate $f(x)$.
Alternating Series Test	For the series, $\sum_{n=1}^{\infty}(-1)^{n-1}u_n$, the series converges if $u_n > 0$, and the series also satisfies the following conditions: (1) $u_n > u_{n+1}$ $(n = 1, 2, 3, \ldots)$; (2) $\lim_{n\to\infty} u_n = 0$.	This test applies only to alternating series.
Test For Absolute Convergence	For a series $\sum_{n=1}^{\infty} u_n$ with positive and negative terms, if the series formed by the absolute values of all its terms, $\sum_{n=1}^{\infty}\lvert u_n\rvert$, converges, then the series $\sum_{n=1}^{\infty} u_n$ must also converge.	If $\sum_{n=1}^{\infty}\lvert u_n\rvert$ diverges, this test is inconclusive.
Ratio Test For Absolute Convergence	For any series $\sum_{n=1}^{\infty} u_n$, if it satisfies the condition $L = \lim_{n\to\infty}\left\lvert\frac{u_{n+1}}{u_n}\right\rvert$ then (1) when $L < 1$, the series converges absolutely. $\Rightarrow$ The series $\sum_{n=1}^{\infty} u_n$ converges; (2) when $L > 1$, the general term $\lim_{n\to\infty}\lvert u_n\rvert \neq 0$. $\Rightarrow$ The series $\sum_{n=1}^{\infty} u_n$ diverges; (3) when $L = 1$, the test is inconclusive.	This test can be applied to any series.

We know the convergence properties of the following series:

(1) harmonic series: $\sum_{n=1}^{\infty}\frac{1}{n}$ diverges;

(2) p-series: $\sum_{n=1}^{\infty}\frac{1}{n^p}$ converges when $p>1$, and diverges when $p\le 1$;

(3) geometric series: $\sum_{n=1}^{\infty}u_1r^n$ converges when $|r|<1$, and diverges when $|r|\ge 1$;

(4) $\sum_{n=1}^{\infty}\frac{1}{n!}$ diverges.

- For the power series

$$\sum_{n=0}^{\infty}a_nx^n = a_0+a_1x+a_2x^2+\ldots+a_nx^n+\ldots$$

if the coefficients of the power series satisfy the condition

$$\lim_{n\to\infty}\left|\frac{a_{n+1}}{a_n}\right| = \mathrm{L}$$

then (1) when $0<\mathrm{L}<+\infty$, $\mathrm{R}=\frac{1}{\mathrm{L}}$;

(2) when $\mathrm{L}=0$, $\mathrm{R}=+\infty$;

(3) when $\mathrm{L}=+\infty$, $\mathrm{R}=0$.

($\mathrm{R}=\frac{1}{\mathrm{L}}$ is the radius of convergence; -R to R is the interval of convergence.)

Exercise 9.5

1. Determine the radius of convergence and the interval of convergence for each of the following power series.

a. $\sum_{n=1}^{\infty}(x+3)^n$ b. $\sum_{n=1}^{\infty}\frac{x^n}{n+2}$ c. $\sum_{n=1}^{\infty}4^n(x-2)^n$

d. $\sum_{n=1}^{\infty}\frac{x^n}{(2n)!}$ e. $\sum_{n=1}^{\infty}\frac{(3x+2)^n}{5^n}$ f. $\sum_{n=1}^{\infty}\frac{x^n}{(2n+1)(2n-1)}$

g. $\sum_{n=1}^{\infty}\frac{(-1)^n x^n}{3^n\sqrt{n}}$ h. $\sum_{n=1}^{\infty}\frac{(-1)^n x^{3n}}{n^{\frac{3}{2}}}$ i. $\sum_{n=1}^{\infty}\frac{(x-4)^n}{n^2}$

j. $\sum_{n=1}^{\infty}\frac{(-1)^{n+1}(2x+3)^n}{2n-5}$ k. $\sum_{n=1}^{\infty}\frac{(-1)^n(x+5)^{2n+1}}{(2n+1)!}$ l. $\sum_{n=1}^{\infty}n!(x+1)^n$

2. Determine the radius of convergence and the interval of convergence for each of the following power series.

a. $\sum_{n=1}^{\infty}\frac{x^n}{\sqrt{n^2+4}}$ b. $\sum_{n=1}^{\infty}\frac{(x+1)^n}{n\ln n}$ c. $\sum_{n=1}^{\infty}\frac{x^n}{n\cdot 2^n}$

d. $\sum_{n=1}^{\infty}\frac{(2x+5)^n}{n(n+1)}$ e. $\sum_{n=1}^{\infty}(\ln x)^n$ f. $\sum_{n=1}^{\infty}\frac{x^{2n}}{(2n)!}$

g. $\sum_{n=1}^{\infty}n^5x^n$ h. $\sum_{n=1}^{\infty}\left(\frac{5}{6}\right)^n(x+8)^n$ i. $\sum_{n=1}^{\infty}\frac{3^n x^n}{n!}$

j. $\sum_{n=1}^{\infty}\frac{(-1)^{n+1}(x+2)^{2n+3}}{n^2+9}$ k. $\sum_{n=1}^{\infty}\frac{(2n+1)!(3x-7)^n}{n^3}$ l. $\sum_{n=1}^{\infty}\frac{e^n(x+7)^n}{(2n+1)!}$

9.6. Taylor polynomial, Taylor formula, Taylor series and Maclaurin series

9.6.1. Taylor polynomial, Taylor formula

Taylor's Mean Value Theorem

If a function $f(x)$ has continuous derivatives up to order $n+1$ on an interval (a, b) containing the point x_0 ($x_0 \in (a, b)$), then for any x in (a, b), $f(x)$ can be expanded as a power series in $(x-x_0)$ as follows:

$$f(x) = a_0 + a_1(x-x_0) + a_2(x-x_0)^2 + a_3(x-x_0)^3 + \ldots$$
$$+ a_n(x-x_0)^n + a_{n+1}(x-x_0)^{n+1} + a_{n+2}(x-x_0)^{n+2} + \ldots$$

where $a_0, a_1, a_2, \ldots, a_n, \ldots$ are the coefficients to be determined.

Since the function $f(x)$ has continuous derivatives of all orders, we can differentiate both sides of the equation above up to the nth order. By setting $x = x_0$ we obtain:

$$f(x_0) = a_0 \qquad f'(x_0) = a_1 \qquad f''(x_0) = 2!a_2 \qquad \ldots \qquad f^{(n)}(x_0) = n!a_n$$

Consequently, $f(x)$ can be written in the following form:

$$f(x) = f(x_0) + f'(x_0)(x-x_0) + \frac{f''(x_0)}{2!}(x-x_0)^2 + \frac{f'''(x_0)}{3!}(x-x_0)^3 + \ldots$$
$$+ \frac{f^{(n)}(x_0)}{n!}(x-x_0)^n + a_{n+1}(x-x_0)^{n+1} + a_{n+2}(x-x_0)^{n+2} + \ldots$$

Let

$$P_n(x) = a_0 + a_1(x-x_0) + a_2(x-x_0)^2 + a_3(x-x_0)^3 + \ldots + a_n(x-x_0)^n$$

$$= f(x_0) + f'(x_0)(x-x_0) + \frac{f''(x_0)}{2!}(x-x_0)^2 + \frac{f'''(x_0)}{3!}(x-x_0)^3 + \ldots + \frac{f^{(n)}(x_0)}{n!}(x-x_0)^n$$

$$= \sum_{i=0}^{n} \frac{f^{(i)}(x_0)}{i!}(x-x_0)^i$$

$P_n(x)$ is known as the Taylor polynomial.

Therefore, the absolute error is

$$|R_n| = |f(x) - P_n(x)| = |a_{n+1}(x-x_0)^{n+1} + a_{n+2}(x-x_0)^{n+2} + \ldots|$$

The error R_n is also known as the **Lagrange remainder of order *n*** or simply the **remainder**.

Let $\quad R_n = a_{n+1}(x-x_0)^{n+1} + a_{n+2}(x-x_0)^{n+2} + \ldots = \dfrac{k}{(n+1)!}(x-x_0)^{n+1}$

That is, let $\quad R_n = \dfrac{k}{(n+1)!}(x-x_0)^{n+1}$ (the result of the sum)

If we can determine the value of k, then we can determine the value of $R_n(x)$.

We can write $f(x)$ in the following form:

$$f(x) = \sum_{i=0}^{\infty} \frac{f^{(i)}(x_0)}{i!}(x-x_0)^i = P_n(x) + R_n(x)$$

$$= \sum_{i=0}^{n} \frac{f^{(i)}(x_0)}{i!}(x-x_0)^i + R_n(x)$$

where $\quad R_n(x) = \dfrac{f^{(n+1)}(c)}{(n+1)!}(x-x_0)^{n+1}$ (c is a number between x_0 and x.)

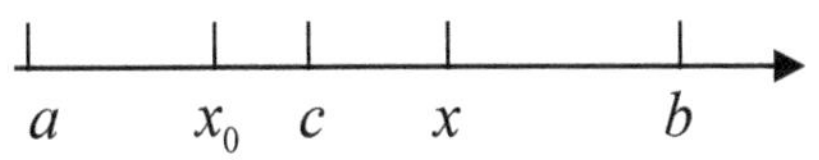

or

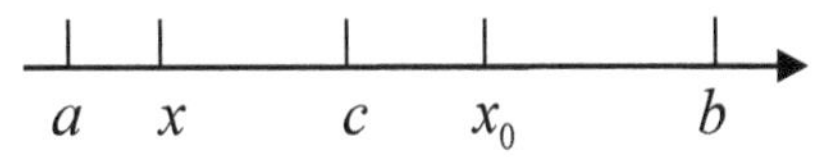

In this form, the expansion of $f(x)$ is known as the **Taylor formula**.

We shall now show that in the expression for $R_n(x)$, $k = f^{(n+1)}(c)$.

Proof. Given that $f(x)$ has continuous derivatives up to the nth order on the interval (a, b) – that is, $f'(x)$, $f''(x)$, …, $f^{(n)}(x)$ exist – then for any $x_0 \in (a, b)$, we can write $f(x)$ as follows:

$$f(x) = f(x_0) + f'(x_0)(x-x_0) + \frac{f''(x_0)}{2!}(x-x_0)^2 + \frac{f'''(x_0)}{3!}(x-x_0)^3 + \ldots$$

$$+ \frac{f^{(n)}(x_0)}{n!}(x-x_0)^n + \frac{k}{(n+1)!}(x-x_0)^{n+1}$$

To find the value of k

let $\phi(t) = f(x) - [P_n(x) + R_n(x)]$ where change x_0 to t, $x_0 \le t \le x$

That is,

$$\phi(t) = f(x) - f(t) - f'(t)(x-t) - \frac{f''(t)}{2!}(x-t)^2 - \frac{f'''(t)}{3!}(x-t)^3 - \ldots$$

$$- \frac{f^{(n)}(t)}{n!}(x-t)^n - \frac{k}{(n+1)!}(x-t)^{n+1}$$

When $t = x_0$ and $t = x$, we have $\phi(x_0) = \phi(x) = 0$.

$\phi(t)$ is continuous on the closed interval $[x_0, x]$ and differentiable on the open interval (x_0, x). By Rolle's Theorem, there exists at least one point $c \in (x_0, x)$ that satisfies $\phi'(c) = 0$.

Since

$$\phi'(t) = \frac{d\phi(t)}{dt}$$

$$= -f'(t) - [f''(t)(x-t) - f'(t)] - [\frac{f^{(3)}(t)}{2!}(x-t)^2 - f''(t)(x-t)] - [\frac{f^{(4)}(t)}{3!}(x-t)^3$$

$$- \frac{f^{(3)}(t)}{2!}(x-t)^2] - \ldots - [\frac{f^{(n+1)}(t)}{n!}(x-t)^n - \frac{f^{(n)}(t)}{(n-1)!}(x-t)^{n-1}] + \frac{k}{n!}(x-t)^n$$

after simplification, we have

$$\phi'(t) = \frac{(x-t)^n}{n!}[k - f^{(n+1)}(t)]$$

Hence, $$\phi'(c) = \frac{(x-c)^n}{n!}[k - f^{(n+1)}(c)] = 0$$

That is, $\dfrac{(x-c)^n}{n!}[k - f^{(n+1)}(c)] = 0$

Since $(x-c)^n \neq 0$, it follows that $k - f^{(n+1)}(c) = 0$.

That is, $k = f^{(n+1)}(c)$

Hence, we have

$$R_n(x) = \frac{k}{(n+1)!}(x-x_0)^{n+1} = \frac{f^{(n+1)}(c)}{(n+1)!}(x-x_0)^{n+1}$$

where $R_n(x)$ is the sum of all terms after the nth term. c is a number between x_0 and x. □

In Taylor formula, when $n = 0$,

$$P_0(x) = f(x_0), \quad R_0(x) = f'(c)(x-x_0)$$

Thus, $f(x) = P_0(x) + R_0(x) = f(x_0) + f'(c)(x-x_0)$

That is, $f(x) = f(x_0) + f'(c)(x-x_0)$

This is Lagrange's Mean Value Theorem. Therefore, Taylor's Mean Value Theorem is a generalisation of Lagrange's Mean Value Theorem.

In particular, when $x_0 = 0$, the Taylor formula becomes

$$f(x) = P_n(x) + R_n(x) = \sum_{i=0}^{n} \frac{f^{(i)}(0)}{i!}x^i + R_n(x)$$

where $R_n(x) = \dfrac{f^{(n+1)}(c)}{(n+1)!}x^{n+1}$ (c is a number between 0 and x.)

When $x_0 = 0$, the Taylor formula is known as the **Maclaurin formula**.

9.6.2. Taylor series, Maclaurin series

The Taylor formula is

$$f(x) = P_n(x) + R_n(x)$$

Taking limits on both sides of the formula yields

$$\lim_{n\to\infty} f(x) = \lim_{n\to\infty} [P_n(x) + R_n(x)]$$

$$f(x) = \lim_{n\to\infty} P_n(x) + \lim_{n\to\infty} R_n(x)$$

$$= \lim_{n\to\infty} \sum_{i=0}^{n} \frac{f^{(i)}(x_0)}{i!}(x - x_0)^i + \lim_{n\to\infty} R_n(x)$$

$$= \sum_{i=0}^{\infty} \frac{f^{(i)}(x_0)}{i!}(x - x_0)^i + \lim_{n\to\infty} R_n(x)$$

When $\lim_{n\to\infty} R_n(x) = 0$

$$f(x) = \sum_{i=0}^{\infty} \frac{f^{(i)}(x_0)}{i!}(x - x_0)^i$$ This is called the **Taylor series.**

If $f(x)$ is a finite number and the Taylor series on the right of the equation above converges, then the sum of the Taylor series is $f(x)$.

If $\lim_{n\to\infty} R_n(x) \neq 0$, then the Taylor series diverges.

$$f(x) \neq \sum_{i=0}^{\infty} \frac{f^{(i)}(x_0)}{i!}(x - x_0)^i$$

In particular, when $x_0 = 0$, if $\lim_{n\to\infty} R_n(x) = 0$, then

$$f(x) = \sum_{i=0}^{\infty} \frac{f^{(i)}(0)}{i!} x^i$$

Taylor series at $x_0 = 0$ is generally known as the **Maclaurin series**.

Note:

1. From Taylor expansion

$$f(x) = \sum_{i=0}^{\infty} \frac{f^{(i)}(x_0)}{i!}(x - x_0)^i$$

$$= \sum_{i=0}^{n} \frac{f^{(i)}(x_0)}{i!}(x-x_0)^i + R_n(x)$$

we see that whether a function $f(x)$ can be expanded as a power series at a specific value x_0 in the interval (a, b) depends on whether the function has derivatives of all orders at $x = x_0$.

For example,

$$f(x) = \frac{1}{x}, \qquad f'(x) = -\frac{1}{x^2}, \qquad f''(x) = \frac{2}{x^3}, \qquad \ldots$$

When $x = x_0 = 0$,

$$f'(0) = -\frac{1}{0^2} = -\infty, \qquad f''(0) = \infty, \qquad \ldots$$

since the derivatives are

$$f^{(i)}(0) = \pm\infty$$

derivatives of all orders do not exist. Consequently, the function $f(x)$ does not have a Taylor expansion at $x = x_0 = 0$.

However, when $x = x_0 = 1$,

$$f'(1) = -\frac{1}{1^2} = -1, \qquad f''(1) = 2, \qquad \ldots$$

by induction, we have

$$f^{(i)}(1) = (-1)^i i!$$

Hence, $$f(x) = \frac{1}{x} = \sum_{i=0}^{n} \frac{(-1)^i i!}{i!}(x-1)^i + R_n(x) = \sum_{i=0}^{n} (-1)^i (x-1)^i + R_n(x)$$

When $0 < x < 2$, $\lim_{n\to\infty} R_n(x) = 0$ (The proof is omitted.)

Therefore, we have

$$f(x) = \frac{1}{x} = \sum_{i=0}^{\infty} (-1)^i (x-1)^i \qquad (0 < x < 2)$$

We can deduce that certain functions possess a Taylor expansion and series at suitable points x_0. Otherwise, the functions cannot be represented by a Taylor expansion and series.

2. When the error of the Taylor polynomial, $R_n(x) \approx 0$

$$f(x) = \sum_{i=0}^{n} \frac{f^{(i)}(x_0)}{i!}(x - x_0)^i + R_n(x) \approx \sum_{i=0}^{n} \frac{f^{(i)}(x_0)}{i!}(x - x_0)^i$$

that is,
$$f(x) \approx \sum_{i=0}^{n} \frac{f^{(i)}(x_0)}{i!}(x - x_0)^i$$

However, in many cases, the exact value of c cannot be identified. Consequently, it is not always possible to determine whether $R_n(x) \approx 0$. In general, we can only determine whether $|R_n| < a$.

If a is a very small positive number, then

$$|R_n| < a \qquad \Rightarrow \qquad R_n \approx 0$$

3. Since we must first determine whether the Taylor series is convergent to determine whether $f(x)$ can be represented by a Taylor series, we need to use $\lim_{n\to\infty}|R_n(x)| < a$ to determine.

When $a = 0$, we have

$$\lim_{n\to\infty}|R_n(x)| < 0 \qquad \Rightarrow \qquad \lim_{n\to\infty}|R_n(x)| = 0$$

Since
$$\lim_{n\to\infty}|R_n(x)| = 0$$

it follows that the Taylor series converges.

On the contrary, when $a \neq 0$, we have

$$\lim_{n\to\infty}|R_n(x)| < a \neq 0 \quad \Rightarrow \quad \lim_{n\to\infty}|R_n(x)| \neq 0$$

Since $\lim_{n\to\infty}|R_n(x)| \neq 0$

it follows that the Taylor series diverges.

In many cases, if a cannot be determined, the Ratio Test must be used to determine whether the Taylor series converges.

$$\lim_{n\to\infty}\left|\frac{u_{n+1}}{u_n}\right| = L$$

where $u_n = \dfrac{f^{(n)}(x_0)}{n!}(x-x_0)^n$ and $u_{n+1} = \dfrac{f^{(n+1)}(x_0)}{(n+1)!}(x-x_0)^{n+1}$.

Depending on whether $L < 1$ or $L > 1$, we can determine whether the Taylor series converges. When $L < 1$, the Taylor series converges. Since if a series converges, then we definitely have

$$\lim_{n\to\infty}|u_n| = 0.$$

Hence, the error of the Taylor series $\lim_{n\to\infty}|R_n(x)|$ is definitely equal to zero.

That is, $\lim_{n\to\infty}|R_n(x)| = 0$

On the contrary, when $L > 1$, the Taylor series diverges and the error of the Taylor series is

$$\lim_{n\to\infty}|R_n(x)| \neq 0$$

Example 1. Find the Maclaurin series for $f(x) = e^x$.

Solution. Since

$$
\begin{aligned}
f(x) &= e^x & f(0) &= e^0 = 1 \\
f'(x) &= e^x & f'(0) &= e^0 = 1 \\
f''(x) &= e^x & f''(0) &= e^0 = 1 \\
&\vdots & &\vdots \\
f^{(n)}(x) &= e^x & f^{(n)}(0) &= e^0 = 1
\end{aligned}
$$

hence, we have

$$\sum_{n=0}^{\infty} \frac{f^{(n)}(0)}{n!}(x-0)^n = \sum_{n=0}^{\infty} \frac{x^n}{n!}$$

Since $$\lim_{n\to\infty}\left|\frac{u_{n+1}}{u_n}\right| = \lim_{n\to\infty}\left|\frac{\dfrac{x^{n+1}}{(n+1)!}}{\dfrac{x^n}{n!}}\right| = \lim_{n\to\infty}\left|\frac{x^{n+1}n!}{x^n(n+1)!}\right| = \lim_{n\to\infty}\left|\frac{x}{n+1}\right|$$

when $-\infty < x < \infty$, we have

$$\lim_{n\to\infty}\left|\frac{x}{n+1}\right| = 0 < 1$$

The series $\sum_{n=0}^{\infty} \frac{x^n}{n!}$ converges; it therefore necessarily follows that

$$\lim_{n\to\infty}|u_n| = \lim_{n\to\infty}\left|\frac{x^n}{n!}\right| = 0$$

Thus $$\lim_{n\to\infty}|R_n(x)| = \lim_{n\to\infty}\left|\frac{e^c}{(n+1)!}x^{n+1}\right| < \lim_{n\to\infty} e^{|x|}\frac{|x|^{n+1}}{(n+1)!} = 0$$

Therefore, for all values of x, we have

$$e^x = \sum_{n=0}^{\infty} \frac{x^n}{n!} \qquad (-\infty < x < \infty)$$

Example 2. Find the Taylor series for $f(x) = e^x$ at $x_0 = 3$.

Solution. Since

$$f(x) = e^x \qquad f(3) = e^3$$

$$f'(x) = e^x \qquad f'(3) = e^3$$

$$f''(x) = e^x \qquad f''(3) = e^3$$

$$\vdots \qquad \vdots$$

$$f^{(n)}(x) = e^x \qquad f^{(n)}(3) = e^3$$

hence, we have

$$\sum_{n=0}^{\infty} \frac{f^{(n)}(3)}{n!}(x-3)^n = \sum_{n=0}^{\infty} \frac{e^3}{n!}(x-3)^n$$

Similarly to Example 1, it can be shown that

$$\lim_{n\to\infty} |R_n(x)| = 0$$

For all values of x, we have

$$e^x = \sum_{n=0}^{\infty} \frac{e^3}{n!}(x-3)^n \qquad (-\infty < x < \infty)$$

Example 3. Find the Maclaurin series for $\sin x$.

Solution. Since

$$f(x) = \sin x \qquad f(0) = \sin 0 = 0$$

$$f'(x) = (\sin x)' = \cos x \qquad f'(0) = \cos 0 = 1$$

$$f''(x) = (\cos x)' = -\sin x \qquad f''(0) = -\sin 0 = 0$$

$$f^{(3)}(x) = (-\sin x)' = -\cos x \qquad f^{(3)}(0) = -\cos 0 = -1$$

$$f^{(4)}(x) = (-\cos x)' = \sin x \qquad f^{(4)}(0) = \sin 0 = 0$$

$$f^{(5)}(x) = (\sin x)' = \cos x \qquad f^{(5)}(0) = \cos 0 = 1$$

$$\vdots \qquad \vdots$$

$$f^{(n)}(x) = \sin(x + \frac{n}{2}\pi) \qquad f^{(2n)}(0) = 0$$

$$f^{(2n+1)}(0) = (-1)^n$$

we have

$$\sum_{n=0}^{\infty} \frac{f^{(n)}(0)}{n!} x^n = f(0) + f'(0)x + \frac{f''(0)}{2!}x^2 + \frac{f^{(3)}(0)}{3!}x^3 + \frac{f^{(4)}(0)}{4!}x^4 + \frac{f^{(5)}(0)}{5!}x^5 + \ldots$$

$$= 0 + 1 \cdot x + \frac{0}{2!}x^2 + \frac{(-1)}{3!}x^3 + \frac{0}{4!}x^4 + \frac{1}{5!}x^5 + \ldots$$

$$= x - \frac{x^3}{3!} + \frac{x^5}{5!} - \ldots + (-1)^n \frac{x^{2n+1}}{(2n+1)!} + \ldots = \sum_{n=0}^{\infty} (-1)^n \frac{x^{2n+1}}{(2n+1)!}$$

Since $\lim\limits_{n\to\infty} \left| \frac{u_{n+1}}{u_n} \right| = \lim\limits_{n\to\infty} \frac{(2n+1)!}{(2n+3)!} |x|^2 = \lim\limits_{n\to\infty} \frac{1}{(2n+3)(2n+2)} |x|^2 = 0 \cdot |x|^2 = 0 < 1$

its interval of convergence is $x \in (-\infty, \infty)$.

Because $\lim\limits_{n\to\infty} |u_n| = \lim\limits_{n\to\infty} \left| (-1)^n \frac{x^{2n+1}}{(2n+1)!} \right| = 0$

it follows that

$$\lim_{n\to\infty} |R_n(x)| = \lim_{n\to\infty} \left| \sin(c + \frac{(2n+3)\pi}{2}) \frac{x^{2n+3}}{(2n+3)!} \right| \le \lim_{n\to\infty} \frac{|x|^{2n+3}}{(2n+3)!} = 0$$

Therefore, for all values of x, we have

$$\sin x = \sum_{n=0}^{\infty} (-1)^n \frac{x^{2n+1}}{(2n+1)!} \qquad (-\infty < x < \infty)$$

Example 4. Find the Maclaurin series for $\cos x$.

Solution. Since

$$f(x) = \cos x \qquad f(0) = \cos 0 = 1$$

$$f'(x) = (\cos x)' = -\sin x \qquad f'(0) = -\sin 0 = 0$$

$$f''(x) = (-\sin x)' = -\cos x \qquad f''(0) = -\cos 0 = -1$$

$$f^{(3)}(x) = (-\cos x)' = \sin x \qquad f^{(3)}(0) = \sin 0 = 0$$

$$f^{(4)}(x) = (\sin x)' = \cos x \qquad f^{(4)}(0) = \cos 0 = 1$$

$$f^{(5)}(x) = (\cos x)' = -\sin x \qquad f^{(5)}(0) = -\sin 0 = 0$$

$$\vdots \qquad \vdots$$

$$f^{(n)}(x) = \cos(x + \frac{n\pi}{2})$$

we have

$$\sum_{n=0}^{\infty} \frac{f^{(n)}(0)}{n!} x^n = f(0) + f'(0)x + \frac{f''(0)}{2!}x^2 + \frac{f^{(3)}(0)}{3!}x^3 + \frac{f^{(4)}(0)}{4!}x^4 + \frac{f^{(5)}(0)}{5!}x^5 + \dots$$

$$= 1 + 0 \cdot x + \frac{(-1)}{2!}x^2 + \frac{0}{3!}x^3 + \frac{1}{4!}x^4 + \frac{0}{5!}x^5 + \dots$$

$$= 1 - \frac{x^2}{2!} + \frac{x^4}{4!} + \dots + (-1)^n \frac{x^{2n}}{(2n)!} + \dots = \sum_{n=0}^{\infty} (-1)^n \frac{x^{2n}}{(2n)!}$$

Because $\lim_{n\to\infty} \left|\frac{u_{n+1}}{u_n}\right| = \lim_{n\to\infty} \frac{(2n)!}{(2n+2)!}|x|^2 = \lim_{n\to\infty} \frac{1}{(2n+2)(2n+1)}|x|^2 = 0 \cdot |x|^2 = 0 < 1$

its interval of convergence is $x \in (-\infty, \infty)$.

Since $\lim_{n\to\infty} |u_n| = \lim_{n\to\infty} \left|(-1)^n \frac{x^{2n}}{(2n)!}\right| = 0$

it follows that

$$\lim_{n\to\infty} |R_n(x)| = \lim_{n\to\infty} \left|\cos(c + \frac{(2n+2)\pi}{2}) \frac{x^{2n+2}}{(2n+2)!}\right| \le \lim_{n\to\infty} \frac{|x|^{2n+2}}{(2n+2)!} = 0$$

Therefore, for all values of x, we have

$$\cos x = \sum_{n=0}^{\infty} (-1)^n \frac{x^{2n}}{(2n)!} \qquad (-\infty < x < \infty)$$

Example 5. Find the Maclaurin series for $\ln(1 + x)$.

Solution. Since $f(x) = \ln(1+x)$

$$f'(x) = [\ln(1+x)]' = \frac{d\ln(1+x)}{d(1+x)} \cdot \frac{d(1+x)}{dx} = \frac{1}{1+x}$$

$$f''(x) = [\frac{1}{1+x}]' = \frac{d(1+x)^{-1}}{d(1+x)} \cdot \frac{d(1+x)}{dx} = -\frac{1}{(1+x)^2}$$

$$f^{(3)}(x) = [-\frac{1}{(1+x)^2}]' = -\frac{d(1+x)^{-2}}{d(1+x)} \cdot \frac{d(1+x)}{dx} = \frac{2}{(1+x)^3} = \frac{2!}{(1+x)^3}$$

$$f^{(4)}(x) = [\frac{2!}{(1+x)^3}]' = \frac{2!d(1+x)^{-3}}{d(1+x)} \cdot \frac{d(1+x)}{dx} = \frac{-3!}{(1+x)^4}$$

$$f^{(5)}(x) = [\frac{-3!}{(1+x)^4}]' = \frac{-3!d(1+x)^{-4}}{d(1+x)} \cdot \frac{d(1+x)}{dx} = \frac{4!}{(1+x)^5}$$

$$\vdots$$

$$f^{(n)}(x) = (-1)^{n-1}\frac{(n-1)!}{(1+x)^n}$$

thus $f(0) = \ln(1+0) = 0$

$$f'(0) = \frac{1}{1+0} = 1$$

$$f''(0) = -\frac{1}{(1+0)^2} = -1$$

$$f^{(3)}(0) = \frac{2!}{(1+0)^3} = 2!$$

$$f^{(4)}(0) = \frac{-3!}{(1+0)^4} = -3!$$

$$f^{(5)}(0) = \frac{4!}{(1+0)^5} = 4!$$

$$\vdots$$

We have

$$\sum_{n=0}^{\infty}\frac{f^{(n)}(0)}{n!}x^n = f(0)+f'(0)x+\frac{f''(0)}{2!}x^2+\frac{f^{(3)}(0)}{3!}x^3+\frac{f^{(4)}(0)}{4!}x^4+\frac{f^{(5)}(0)}{5!}x^5+...$$

$$=0+1\cdot x+\frac{(-1)}{2!}x^2+\frac{2!}{3!}x^3+\frac{(-3!)}{4!}x^4+\frac{4!}{5!}x^5+...$$

$$=x-\frac{x^2}{2}+\frac{x^3}{3}-\frac{x^4}{4}+\frac{x^5}{5}+...+(-1)^{n+1}\frac{x^n}{n}+...=\sum_{n=1}^{\infty}(-1)^{n+1}\frac{x^n}{n}$$

Because $\displaystyle\lim_{n\to\infty}\left|\frac{u_{n+1}}{u_n}\right|=\lim_{n\to\infty}\frac{n}{n+1}|x|=|x|$

When $|x| < 1$, the series converges. When $x = -1$, the series diverges. When $x = 1$, the series converges. Therefore, its interval of convergence is $x \in]-1, 1]$.

Because $\displaystyle\lim_{n\to\infty}|u_n|=\lim_{n\to\infty}\left|(-1)^{n+1}\frac{x^n}{n}\right|=0$

it follows that $\displaystyle\lim_{n\to\infty}|R_n(x)|=\lim_{n\to\infty}\left|\frac{(-1)^n x^{n+1}}{(1+c)^{n+1}(n+1)}\right|\le\lim_{n\to\infty}\frac{|x|^{n+1}}{n+1}=0$

Hence, for $-1 < x \le 1$, we have

$$\ln(1+x)=\sum_{n=1}^{\infty}(-1)^{n+1}\frac{x^n}{n} \qquad (-1 < x \le 1)$$

Exercise 9.6

1. Find the first four non-zero terms of the Maclaurin series for each of the following.

a. $f(x)=(1+x)^{\frac{2}{3}}$ **b.** $f(x)=(2-x)^{-3}$ **c.** $f(x)=e^{\frac{x}{5}}$

d. $f(x)=\dfrac{1}{4-x}$ **e.** $f(x)=1-3x+2x^2$ **f.** $f(x)=\ln(1-x)$

2. Find the Taylor series up to the term in x^3 at $x = x_0$ for each of the following.

a. $f(x)=\frac{1}{x^2}$, $x_0 = -1$ **b.** $f(x)=2-x-x^2$, $x_0 = 1$

c. $f(x)=e^x$, $x_0 = 2$ **d.** $f(x)=\sqrt{x}$, $x_0 = 4$

9.7. Power series expansion of functions

In the previous section, we used Taylor's or Maclaurin's formulae to expand the function $f(x)$ into a power series. This is known as the **direct expansion method.** In this section, we will learn about the **indirect expansion method**. The indirect expansion method is based on the existing power series expansions of certain functions, and uses the four arithmetic operations, substitution, differentiation, integration, and other methods to find the power series expansion of a given function. Some commonly used Maclaurin series formulae are

$$e^x = \sum_{n=0}^{\infty}\frac{x^n}{n!} = 1+x+\frac{x^2}{2!}+... \qquad (-\infty < x < \infty)$$

$$\ln(1+x) = \sum_{n=1}^{\infty}(-1)^{n+1}\frac{x^n}{n} = x-\frac{x^2}{2}+\frac{x^3}{3}-... \qquad (-1 < x \leq 1)$$

$$\sin x = \sum_{n=0}^{\infty}(-1)^n\frac{x^{2n+1}}{(2n+1)!} = x-\frac{x^3}{3!}+\frac{x^5}{5!}-... \qquad (-\infty < x < \infty)$$

$$\cos x = \sum_{n=0}^{\infty}(-1)^n\frac{x^{2n}}{(2n)!} = 1-\frac{x^2}{2!}+\frac{x^4}{4!}-... \qquad (-\infty < x < \infty)$$

Example 1. Use the Maclaurin series to prove Euler's formula.

$$e^{ix} = \cos x + i\sin x$$

Proof. We know that $\cos x = \sum_{n=0}^{\infty}(-1)^n \frac{x^{2n}}{(2n)!}$, $\sin x = \sum_{n=0}^{\infty}(-1)^n \frac{x^{2n+1}}{(2n+1)!}$, $e^x = \sum_{n=0}^{\infty}\frac{x^n}{n!}$

thus
$$\cos x = \sum_{n=0}^{\infty}(-1)^n \frac{x^{2n}}{(2n)!} = 1 - \frac{x^2}{2!} + \frac{x^4}{4!} - \ldots \qquad (1)$$

$$i\sin x = \sum_{n=0}^{\infty}(-1)^n \frac{ix^{2n+1}}{(2n+1)!} = ix - \frac{ix^3}{3!} + \frac{ix^5}{5!} - \ldots \qquad (2)$$

Adding equations (1) and (2) gives

$$\cos x + i\sin x = 1 + ix - \frac{x^2}{2!} - \frac{ix^3}{3!} + \frac{x^4}{4!} + \frac{ix^5}{5!} - \ldots = \sum_{n=0}^{\infty}\frac{(ix)^n}{n!} = e^{ix}$$

Hence, we obtain Euler's formula:

$$e^{ix} = \cos x + i\sin x$$

Example 2. Find the Maclaurin series for e^{3x}.

Solution. We can use the Maclaurin series for e^x and substitute $3x$ for x to obtain

$$e^{3x} = \sum_{n=0}^{\infty}\frac{(3x)^n}{n!}.$$

Example 3. Find the Maclaurin series for $\sin(2x^2)$.

Solution. We can use the Maclaurin series of $\sin x$ and substitute $2x^2$ for x to obtain

$$\sin(2x^2) = \sum_{n=0}^{\infty}(-1)^n \frac{(2x^2)^{2n+1}}{(2n+1)!}.$$

Example 4. Find the Maclaurin series for $x\cos(x^3)$.

Solution. We can use the Maclaurin series for $\cos x$ and substitute x^3 for x to obtain

$$\cos(x^3) = \sum_{n=0}^{\infty}(-1)^n \frac{(x^3)^{2n}}{(2n)!} = \sum_{n=0}^{\infty}(-1)^n \frac{x^{6n}}{(2n)!}$$

Thus $$x\cos(x^3) = x\sum_{n=0}^{\infty}(-1)^n \frac{(x^3)^{2n}}{(2n)!} = \sum_{n=0}^{\infty}(-1)^n \frac{x^{6n+1}}{(2n)!}$$

Example 5. Find the terms up to x^4 of the Maclaurin series for $e^{3x}\cos(2x)$.

Solution. From Example 2, we have

$$e^{3x} = \sum_{n=0}^{\infty}\frac{(3x)^n}{n!} = 1+3x+\frac{(3x)^2}{2!}+\frac{(3x)^3}{3!}+\frac{(3x)^4}{4!}+\ldots$$

$$= 1+3x+\frac{9}{2}x^2+\frac{9}{2}x^3+\frac{27}{8}x^4+\ldots$$

And $$\cos(2x) = \sum_{n=0}^{\infty}(-1)^n \frac{(2x)^{2n}}{(2n)!} = 1-\frac{(2x)^2}{2!}+\frac{(2x)^4}{4!}-\ldots = 1-2x^2+\frac{2}{3}x^4-\ldots$$

Thus $$e^{3x}\cos(2x) = \left(1+3x+\frac{9}{2}x^2+\frac{9}{2}x^3+\frac{27}{8}x^4+\ldots\right)\cdot\left(1-2x^2+\frac{2}{3}x^4-\ldots\right)$$

By only taking terms whose product is of a degree less than or equal to four, we have

$$e^{3x}\cos(2x) = 1-2x^2+\frac{2}{3}x^4+3x-6x^3+\frac{9}{2}x^2-9x^4+\frac{9}{2}x^3+\frac{27}{8}x^4+\ldots$$

$$= 1+3x+\frac{5}{2}x^2-\frac{3}{2}x^3-\frac{119}{24}x^4+\ldots$$

Example 6. Find the Maclaurin series for $\sin^2 x$.

Solution. From the double angle theorem $\cos 2x = 1 - 2\sin^2 x$ we have

$$\sin^2 x = \frac{1}{2}(1-\cos 2x)$$

We can use the Maclaurin series for $\cos x$ and substitute $2x$ for x to obtain

$$\sin^2 x = \frac{1}{2}(1-\cos 2x) = \frac{1}{2} - \frac{1}{2}\cos 2x = \frac{1}{2} - \frac{1}{2}\sum_{n=0}^{\infty}(-1)^n \frac{(2x)^{2n}}{(2n)!}$$

Example 7. Find the Maclaurin series for $e^{\sin x}$ up to the term in x^4.

Solution. **Method 1** Substitute the Maclaurin series for $\sin x$.

$$e^{\sin x} = e^{x-\frac{x^3}{3!}+\dots} = e^x \cdot e^{-\frac{x^3}{3!}} \cdot \dots$$

$$= \left(1+x+\frac{x^2}{2!}+\frac{x^3}{3!}+\frac{x^4}{4!}+\dots\right)\cdot\left(1-\frac{x^3}{3!}+\dots\right)\cdot(\dots)$$

$$= 1-\frac{1}{6}x^3+x-\frac{1}{6}x^4+\frac{1}{2}x^2+\frac{1}{6}x^3+\frac{1}{24}x^4+\dots$$

$$= 1+x+\frac{1}{2}x^2-\frac{1}{8}x^4+\dots$$

Method 2 We can use the Maclaurin series for e^x and substitute $\sin x$ for x to obtain

$$e^{\sin x} = 1+\sin x+\frac{(\sin x)^2}{2!}+\frac{(\sin x)^3}{3!}+\frac{(\sin x)^4}{4!}+\dots$$

From Example 6, we have

$$\sin^2 x = \frac{1}{2}(1-\cos 2x) = \frac{1}{2}\left[1-\sum_{n=0}^{\infty}(-1)^n\frac{(2x)^{2n}}{(2n)!}\right]$$

$$= \frac{1}{2}\left\{1-\left[1-\frac{(2x)^2}{2!}+\frac{(2x)^4}{4!}-\dots\right]\right\}$$

$$= \frac{1}{2}\left[\frac{(2x)^2}{2!}-\frac{(2x)^4}{4!}+\dots\right] = \frac{1}{2}\left(2x^2-\frac{2}{3}x^4+\dots\right)$$

$$= x^2-\frac{1}{3}x^4+\dots$$

Thus $\quad \sin^3 x = \sin x \cdot \sin^2 x = \left(x - \frac{x^3}{3!} + \frac{x^5}{5!} - \ldots\right)\cdot\left(x^2 - \frac{1}{3}x^4 + \ldots\right) = x^3 + \ldots$

$$\sin^4 x = \sin^2 x \cdot \sin^2 x = \left(x^2 - \frac{1}{3}x^4 + \ldots\right)\cdot\left(x^2 - \frac{1}{3}x^4 + \ldots\right) = x^4 + \ldots$$

Therefore,

$$e^{\sin x} = 1 + \left(x - \frac{x^3}{3!} + \ldots\right) + \frac{1}{2!}\left(x^2 - \frac{1}{3}x^4 + \ldots\right) + \frac{1}{3!}\left(x^3 + \ldots\right) + \frac{1}{4!}\left(x^4 + \ldots\right) + \ldots$$

$$= 1 + x - \frac{1}{6}x^3 + \ldots + \frac{1}{2}x^2 - \frac{1}{6}x^4 + \ldots + \frac{1}{6}x^3 + \ldots + \frac{1}{24}x^4 + \ldots$$

$$= 1 + x + \frac{1}{2}x^2 - \frac{1}{8}x^4 + \ldots$$

Example 8. Find the Maclaurin series for $\ln\frac{\sin x}{x}$ $(x \neq 0)$ up to the term in x^4.

Solution. Substituting $\sin x = \sum_{n=0}^{\infty}(-1)^n \frac{x^{2n+1}}{(2n+1)!} = x - \frac{x^3}{3!} + \frac{x^5}{5!} - \ldots$, we have

$$\ln\frac{\sin x}{x} = \ln\left[\frac{1}{x}\left(x - \frac{x^3}{3!} + \frac{x^5}{5!} - \ldots\right)\right] = \ln\left(1 - \frac{1}{6}x^2 + \frac{1}{120}x^4 - \ldots\right)$$

Using $\ln(1 + x) = \sum_{n=1}^{\infty}(-1)^{n+1}\frac{x^n}{n} = x - \frac{x^2}{2} + \frac{x^3}{3} - \ldots$ gives

$$\ln\frac{\sin x}{x} = \left(-\frac{1}{6}x^2 + \frac{1}{120}x^4\right) - \frac{1}{2}\left(-\frac{1}{6}x^2\right)^2 + \ldots = -\frac{1}{6}x^2 - \frac{1}{180}x^4 + \ldots$$

Example 9. Find the Maclaurin series for $\ln\left[(5-2x)\sqrt{3+4x}\right]$ up to the term in x^3.

Also, find the interval of convergence of this series.

Solution. $\ln\left[(5-2x)\sqrt{3+4x}\right]=\ln(5-2x)+\frac{1}{2}\ln(3+4x)$

$$=\ln\left[5\left(1+\left(-\frac{2}{5}x\right)\right)\right]+\frac{1}{2}\ln\left[3\left(1+\frac{4}{3}x\right)\right]$$

$$\ln\left[5\left(1+\left(-\frac{2}{5}x\right)\right)\right]=\ln 5+\ln\left(1+\left(-\frac{2}{5}x\right)\right)$$

$$=\ln 5+\left(-\frac{2}{5}x\right)-\frac{1}{2}\left(-\frac{2}{5}x\right)^2+\frac{1}{3}\left(-\frac{2}{5}x\right)^3-\ldots$$

$$=\ln 5-\frac{2}{5}x-\frac{2}{25}x^2-\frac{8}{375}x^3-\ldots$$

$$\frac{1}{2}\ln\left[3\left(1+\frac{4}{3}x\right)\right]=\frac{1}{2}\left[\ln 3+\ln\left(1+\frac{4}{3}x\right)\right]$$

$$=\frac{1}{2}\left[\ln 3+\frac{4}{3}x-\frac{1}{2}\left(\frac{4}{3}x\right)^2+\frac{1}{3}\left(\frac{4}{3}x\right)^3-\ldots\right]$$

$$=\frac{1}{2}\left(\ln 3+\frac{4}{3}x-\frac{8}{9}x^2+\frac{64}{81}x^3-\ldots\right)$$

$$=\frac{1}{2}\ln 3+\frac{2}{3}x-\frac{4}{9}x^2+\frac{32}{81}x^3-\ldots$$

Thus $\ln\left[(5-2x)\sqrt{3+4x}\right]=\ln 5-\frac{2}{5}x-\frac{2}{25}x^2-\frac{8}{375}x^3-\ldots$

$$+\frac{1}{2}\ln 3+\frac{2}{3}x-\frac{4}{9}x^2+\frac{32}{81}x^3-\ldots$$

$$=\ln 5\sqrt{3}+\frac{4}{15}x-\frac{118}{225}x^2+\frac{3784}{10125}x^3-\ldots$$

Because the interval of convergence for $\ln(1+x)$ is $-1<x\leq 1$

thus, the interval of convergence for $\ln\left(1+\left(-\frac{2}{5}x\right)\right)$ is $-1<-\frac{2}{5}x\leq 1$.

That is, $-\frac{5}{2}\leq x<\frac{5}{2}$

The interval of convergence for $\ln\left(1+\frac{4}{3}x\right)$ is $-1 < \frac{4}{3}x \leq 1$.

That is, $-\frac{3}{4} < x \leq \frac{3}{4}$

Hence, the interval of convergence for $\ln\left[(5-2x)\sqrt{3+4x}\right]$ is $-\frac{3}{4} < x \leq \frac{3}{4}$.

Example 10.

(a) Find the power series for $\frac{1}{1+x}$ and $\frac{1}{1+x^2}$ in x.

(b) Hence find the power series for $\ln(1+x)$ and $\arctan x$. Also, find the interval of convergence of the series.

(c) Find the power series for $\frac{1}{(1-x)^2}$.

Solution. (a) We know that

$$\frac{1}{1-r} = 1 + r + r^2 + r^3 + r^4 + \ldots + r^{n-1} + \ldots \qquad -1 < r < 1$$

Let $r = -x$ and $r = -x^2$ respectively. We have

$$\frac{1}{1+x} = 1 - x + x^2 - x^3 + x^4 + \ldots + (-1)^{n-1}x^{n-1} + \ldots \qquad -1 < x < 1$$

$$\frac{1}{1+x^2} = 1 - x^2 + x^4 - x^6 + x^8 + \ldots + (-1)^{n-1}x^{2n-2} + \ldots \qquad -1 < x < 1$$

(b) Integrating both sides of the results from (a) from 0 to x term by term, we have

$$\int_0^x \frac{1}{1+x}dx = \int_0^x \left(1 - x + x^2 - x^3 + x^4 + \ldots + (-1)^{n-1}x^{n-1} + \ldots\right)dx$$

$$\ln(1+x) = x - \frac{1}{2}x^2 + \frac{1}{3}x^3 - \frac{1}{4}x^4 + \frac{1}{5}x^5 + \ldots + \frac{(-1)^{n-1}}{n}x^n + \ldots$$

When $x = 1$, it becomes an alternating series $\sum_{n=1}^{\infty}\frac{(-1)^{n-1}}{n}$, which converges.

When $x = -1$, it becomes the harmonic series $\sum_{n=1}^{\infty}\frac{(-1)^{2n-1}}{n} = -\sum_{n=1}^{\infty}\frac{1}{n}$, which diverges.

Therefore, its interval of convergence is $-1 < x \le 1$, which is the same as the result obtained in the previous section.

Similarly, $$\int_0^x \frac{1}{1+x^2}dx = \int_0^x \left(1 - x^2 + x^4 - x^6 + x^8 + \ldots + (-1)^{n-1}x^{2n-2} + \ldots\right)dx$$

$$\arctan x = x - \frac{1}{3}x^3 + \frac{1}{5}x^5 - \frac{1}{7}x^7 + \frac{1}{9}x^9 + \ldots + \frac{(-1)^{n-1}}{2n-1}x^{2n-1} + \ldots$$

When $x = 1$, it becomes an alternating series $\sum_{n=1}^{\infty}\frac{(-1)^{n-1}}{2n-1}$, which converges.

When $x = -1$, it becomes an alternating series $\sum_{n=1}^{\infty}\frac{(-1)^{n}}{2n-1}$, which converges.

Therefore, its interval of convergence is $-1 \le x \le 1$.

(c) We know that

$$\frac{1}{1-x} = 1 + x + x^2 + x^3 + x^4 + \ldots + x^{n-1} + \ldots \qquad -1 < x < 1$$

Differentiating both sides term by term, we have

$$\frac{1}{(1-x)^2} = 1 + 2x + 3x^2 + 4x^3 + \ldots + nx^{n-1} + \ldots \qquad -1 < x < 1$$

Example 11. Find the power series for the function $f(x) = \frac{x}{2+x-x^2}$.

Solution. First, express $\frac{x}{2+x-x^2}$ in partial fractions.

$$\frac{x}{2+x-x^2}=\frac{A}{1+x}+\frac{B}{2-x}$$

Multiplying both sides by $(1+x)(2-x)$ gives

$$x=A(2-x)+B(1+x)$$

Let $x=2$ and we have $B=\frac{2}{3}$

Let $x=-1$ and we have $A=-\frac{1}{3}$

Thus $$\frac{x}{2+x-x^2}=-\frac{1}{3}\cdot\frac{1}{1+x}+\frac{2}{3}\cdot\frac{1}{2-x}=\frac{1}{3}\left(\frac{2}{2-x}-\frac{1}{1+x}\right)$$

We know that $\frac{1}{1-x}=1+x+x^2+x^3+x^4+\ldots+x^{n-1}+\ldots$ $\quad -1<x<1$

Therefore,

$$\frac{2}{2-x}=2\cdot\frac{1}{2-x}=2\cdot\frac{1}{2\left(1-\frac{x}{2}\right)}=\frac{1}{1-\frac{x}{2}}=1+\frac{x}{2}+\left(\frac{x}{2}\right)^2+\left(\frac{x}{2}\right)^3+\ldots=\sum_{n=0}^{\infty}\left(\frac{x}{2}\right)^n$$

$$-2<x<2$$

$$\frac{1}{1+x}=1-x+x^2-x^3+x^4+\ldots+(-1)^{n-1}x^{n-1}+\ldots=\sum_{n=0}^{\infty}(-1)^n x^n \qquad -1<x<1$$

Hence,

$$\frac{x}{2+x-x^2}=\frac{1}{3}\left(\frac{2}{2-x}-\frac{1}{1+x}\right)=\frac{1}{3}\left(\sum_{n=0}^{\infty}\left(\frac{x}{2}\right)^n-\sum_{n=0}^{\infty}(-1)^n x^n\right)=\frac{1}{3}\sum_{n=0}^{\infty}\left[\left(\frac{x}{2}\right)^n-(-1)^n x^n\right]$$

$$-1<x<1$$

Example 12. Find the power series for the following functions.

(a) $\int_0^x e^{-x^2}dx$

(b) $\int_0^x \frac{\ln(1+x)}{x}dx$

(c) $\int_0^x \cos x^2 dx$

Solution. (a) We can use the Maclaurin series for e^x and substitute $-x^2$ for x to obtain

$$e^{-x^2} = 1 + \left(-x^2\right) + \frac{\left(-x^2\right)^2}{2!} + \frac{\left(-x^2\right)^3}{3!} + \ldots = \sum_{n=0}^{\infty} \frac{(-1)^n x^{2n}}{n!}$$

Thus $$\int_0^x e^{-x^2} dx = \int_0^x \sum_{n=0}^{\infty} \frac{(-1)^n x^{2n}}{n!} dx = \sum_{n=0}^{\infty} \frac{(-1)^n}{n!} \int_0^x x^{2n} dx = \sum_{n=0}^{\infty} \frac{(-1)^n x^{2n+1}}{n!(2n+1)} \qquad -\infty < x < \infty$$

(b) Substituting $\ln(1+x) = x - \frac{1}{2}x^2 + \frac{1}{3}x^3 - \frac{1}{4}x^4 + \frac{1}{5}x^5 + \ldots + \frac{(-1)^{n-1}}{n}x^n + \ldots$ gives

$$\int_0^x \frac{\ln(1+x)}{x} dx = \int_0^x \frac{1}{x}\left(x - \frac{1}{2}x^2 + \frac{1}{3}x^3 - \frac{1}{4}x^4 + \frac{1}{5}x^5 + \ldots + \frac{(-1)^{n-1}}{n}x^n + \ldots \right) dx$$

$$= \int_0^x \left(1 - \frac{1}{2}x + \frac{1}{3}x^2 - \frac{1}{4}x^3 + \ldots \right) dx = \int_0^x \sum_{n=0}^{\infty} \frac{(-1)^n x^n}{n+1} dx$$

$$= \sum_{n=0}^{\infty} \frac{(-1)^n}{n+1} \int_0^x x^n dx = \sum_{n=0}^{\infty} \frac{(-1)^n x^{n+1}}{(n+1)^2} \qquad -1 < x < 1$$

(c) We can use the Maclaurin series for $\cos x$ and substitute x^2 for x to obtain

$$\cos x^2 = \sum_{n=0}^{\infty} (-1)^n \frac{(x^2)^{2n}}{(2n)!} = \sum_{n=0}^{\infty} (-1)^n \frac{x^{4n}}{(2n)!}$$

Thus $$\int_0^x \cos x^2 dx = \int_0^x \sum_{n=0}^{\infty} (-1)^n \frac{x^{4n}}{(2n)!} dx = \sum_{n=0}^{\infty} \frac{(-1)^n}{(2n)!} \int_0^x x^{4n} dx = \sum_{n=0}^{\infty} \frac{(-1)^n x^{4n+1}}{(2n)!(4n+1)}$$

$$-\infty < x < \infty$$

Example 13. Expand $\frac{d}{dx}\left(\frac{e^x-1}{x}\right)$ in a power series in x.

Solution. Because $e^x = \sum_{n=0}^{\infty}\frac{x^n}{n!} = 1+x+\frac{x^2}{2!}+...$

thus $$\frac{e^x-1}{x} = 1+\frac{x}{2!}+\frac{x^2}{3!}+\frac{x^3}{4!}+...+\frac{x^{n-1}}{n!}+...$$

Differentiating both sides term by term, we have

$$\frac{d}{dx}\left(\frac{e^x-1}{x}\right) = \frac{1}{2!}+\frac{2}{3!}x+\frac{3}{4!}x^2+...+\frac{n-1}{n!}x^{n-2}+... = \sum_{n=2}^{\infty}\frac{n-1}{n!}x^{n-2} \qquad x \neq 0$$

Exercise 9.7

1. Using the Maclaurin series for e^x, $\ln(1+x)$, $\sin x$ and $\cos x$, find the Maclaurin series for the following.

a. $f(x)=e^{-x^2}$ **b.** $f(x)=x^3e^{-x}$ **c.** $f(x)=\cos^2 x$

d. $f(x)=\ln(1+3x)$ **e.** $f(x)=x\cos(2x)$ **f.** $f(x)=\ln\sqrt{\frac{1+x}{1-x}}$

g. $f(x)=\sin(5x^3)$ **h.** $f(x)=\cos(4\sqrt{x})$ **i.** $f(x)=e^{\frac{x}{5}}$

2. Find the Maclaurin series up to the term in x^4 for the following.

a. $f(x)=e^{\cos x}$ **b.** $f(x)=\ln(1+\sin x)$ **c.** $f(x)=\ln(2+5x)\cos(3x)$

d. $f(x)=e^{4x}\sin(3x)$ **e.** $f(x)=\frac{1}{3-x}$ **f.** $f(x)=\frac{x}{x^2-2x-3}$

3. **a.** Find the Maclaurin series for $f(x) = \tan x$ up to the term in x^4.

b. Hence, find the Maclaurin series for $f(x) = e^{\tan x}$ up to and including the term in x^4.

4. **a.** Use the Maclaurin series for $\sin x$ to find the Maclaurin series for $f(x) = \ln(\sin x + 1)$ up to the term in x^4.

b. Hence, find the Maclaurin series for $f(x) = \ln(secx)$ up to the term in x^4.

5. Find a power series in x for the following.

a. $\int_0^x e^{t^2}\,dt$ **b.** $\int_0^x \frac{\sin t}{t}\,dt$ **c.** $\int_0^x \frac{\arctan t}{t}\,dt$

d. $\int_0^x \frac{1}{3-t}\,dt$ **e.** $\int_0^x \cos\sqrt{t}\,dt$ **f.** $\int_0^x \sin t^2\,dt$

6. **a.** Show that

$$1 + x^2 + x^4 + \ldots = \frac{1}{1-x^2}$$

b. Show that

$$\int_0^x \frac{1}{1-t^2}\,dt = \frac{1}{2}\ln\frac{1+x}{1-x}$$

c. Use the results from above to find the sum of the series:

$$x + \frac{x^3}{3} + \frac{x^5}{5} + \ldots$$

7. Express $y = \ln(1 - x - 2x^2)$ as a power series in x. Also, find its interval of convergence.

9.8. Applications of power series

The expansion of a power series can be used to perform approximate calculations, find limits, determine approximate solutions to differential equations, and so on.

I. <u>Use power series expansions to calculate approximations</u>

When performing approximate calculations using power series expansions, we need to determine the value of n in the expansion to achieve the required accuracy. To estimate the size of the error, we can use:

(1) The Lagrange remainder of order n. This method works in all cases.

$$R_n(x) = \frac{f^{(n+1)}(c)}{(n+1)!}(x-x_0)^{n+1} \qquad (c \text{ is a number between } x_0 \text{ and } x.)$$

(2) If the series happens to be an alternating series, we can use the error estimate for an alternating series.

$$|R_n| < u_{n+1}$$

The following examples illustrate these methods.

Example 1. Calculate the value of e^{-1}, correct to four decimal places.

Solution. The Maclaurin series for e^x is

$$e^x = \sum_{n=0}^{\infty}\frac{x^n}{n!} = 1 + x + \frac{x^2}{2!} + \ldots$$

To be correct to four decimal places, we need the error to be less than $5 \times 10^{-5} = 0.00005$. The Lagrange remainder of order n is

$$R_n(x) = \frac{e^c}{(n+1)!}x^{n+1} \qquad c \in (0, x)$$

When $x = -1$

$$|R_n(-1)| = \left|\frac{e^c}{(n+1)!}x^{n+1}\right| = \left|\frac{e^c}{(n+1)!}(-1)^{n+1}\right| < \frac{1}{(n+1)!} \qquad c \in (-1, 0)$$

Because $\quad \frac{1}{8!} \approx 2.48 \times 10^{-5}$

it follows that $\quad n + 1 = 8 \quad$ that is, $\quad n = 7$

Therefore,

$$e^{-1} \approx \sum_{n=0}^{7} \frac{(-1)^n}{n!} = \frac{(-1)^0}{0!} + \frac{(-1)^1}{1!} + \frac{(-1)^2}{2!} + \frac{(-1)^3}{3!} + \frac{(-1)^4}{4!} + \frac{(-1)^5}{5!} + \frac{(-1)^6}{6!} + \frac{(-1)^7}{7!}$$

$$= 1 - 1 + \frac{1}{2} - \frac{1}{6} + \frac{1}{24} - \frac{1}{120} + \frac{1}{720} - \frac{1}{5040} \approx 0.3679$$

Example 2. Calculate the value of $\ln 1.1$, correct to five decimal places.

Solution. The Maclaurin series for $\ln(1 + x)$ is

$$\ln(1 + x) = \sum_{n=1}^{\infty} (-1)^{n+1} \frac{x^n}{n} = x - \frac{x^2}{2} + \frac{x^3}{3} - \ldots$$

To be correct to five decimal places, we need the error to be less than 5×10^{-6}. As this is an alternating series, we can use the error estimate for the alternating series.

$$|R_n| < u_{n+1}$$

Given that $\; 1 + x = 1.1, \;$ thus $\; x = 0.1.$

$$|R_4| < u_5 = \frac{0.1^5}{5} = 2 \times 10^{-6}$$

We find that $\quad n = 4.$

Therefore,

$\ln 1.1 = \ln(1 + 0.1)$

$$\approx \sum_{n=1}^{4}(-1)^{n+1}\frac{0.1^n}{n} = (-1)^2\frac{0.1^1}{1} + (-1)^3\frac{0.1^2}{2} + (-1)^4\frac{0.1^3}{3} + (-1)^5\frac{0.1^4}{4} \approx 0.09531$$

Example 3. Calculate the value of $\int_0^{0.2} e^{-x^2}\,dx$, correct to five decimal places.

Solution. To be correct to five decimal places, we need the error to be less than 5×10^{-6}. From Example 12 of the previous section, we have

$$\int_0^x e^{-x^2}\,dx = \int_0^x \sum_{n=0}^{\infty}\frac{(-1)^n x^{2n}}{n!}dx = \sum_{n=0}^{\infty}\frac{(-1)^n}{n!}\int_0^x x^{2n}\,dx = \sum_{n=0}^{\infty}\frac{(-1)^n x^{2n+1}}{n!(2n+1)}$$

Thus $\quad \int_0^{0.2} e^{-x^2}\,dx = \sum_{n=0}^{\infty}\frac{(-1)^n 0.2^{2n+1}}{n!(2n+1)}$

As this is an alternating series, we can use the error estimate for an alternating series.

$$|R_n| < u_{n+1}$$

Because $\quad |R_2| < u_3 = \dfrac{0.2^7}{3!\cdot 7} \approx 3\times 10^{-7}$

we find that $\quad n = 2$.

Therefore, $\quad \int_0^{0.2} e^{-x^2}\,dx \approx \sum_{n=0}^{2}\frac{(-1)^n 0.2^{2n+1}}{n!(2n+1)} = \frac{0.2}{1} - \frac{0.2^3}{3} + \frac{0.2^5}{2!\cdot 5} \approx 0.19737$

Example 4. (a) Find the Taylor series for $\sin x$ at $x_0 = \dfrac{\pi}{6}$.

(b) Find the value of $\sin 31°$, correct to six decimal places.

Solution. (a) Because

$f(x) = \sin x \qquad f(\frac{\pi}{6}) = \sin\frac{\pi}{6} = \frac{1}{2}$

$$f'(x) = (\sin x)' = \cos x \qquad f'(\frac{\pi}{6}) = \cos\frac{\pi}{6} = \frac{\sqrt{3}}{2}$$

$$f''(x) = (\cos x)' = -\sin x \qquad f''(\frac{\pi}{6}) = -\sin\frac{\pi}{6} = -\frac{1}{2}$$

$$f^{(3)}(x) = (-\sin x)' = -\cos x \qquad f^{(3)}(\frac{\pi}{6}) = -\cos\frac{\pi}{6} = -\frac{\sqrt{3}}{2}$$

$$f^{(4)}(x) = (-\cos x)' = \sin x \qquad f^{(4)}(\frac{\pi}{6}) = \sin\frac{\pi}{6} = \frac{1}{2}$$

$$f^{(5)}(x) = (\sin x)' = \cos x \qquad f^{(5)}(\frac{\pi}{6}) = \cos\frac{\pi}{6} = \frac{\sqrt{3}}{2}$$

$$\vdots \qquad \vdots$$

$$f^{(n)}(x) = \sin(x + \frac{n}{2}\pi) \qquad f^{(n)}(\frac{\pi}{6}) = \sin(\frac{\pi}{6} + \frac{n}{2}\pi)$$

we have

$$\sin x = \sum_{n=0}^{\infty} \frac{f^{(n)}(\frac{\pi}{6})}{n!}(x - \frac{\pi}{6})^n$$

$$= f(\frac{\pi}{6}) + f'(\frac{\pi}{6})\left(x - \frac{\pi}{6}\right) + \frac{f''(\frac{\pi}{6})}{2!}\left(x - \frac{\pi}{6}\right)^2 + \frac{f^{(3)}(\frac{\pi}{6})}{3!}\left(x - \frac{\pi}{6}\right)^3 + \dots$$

$$+ \frac{\sin\left(\frac{\pi}{6} + \frac{n}{2}\pi\right)}{n!}\left(x - \frac{\pi}{6}\right)^n + \dots$$

$$= \frac{1}{2} + \frac{\sqrt{3}}{2}\left(x - \frac{\pi}{6}\right) - \frac{1}{4}\left(x - \frac{\pi}{6}\right)^2 - \frac{\sqrt{3}}{12}\left(x - \frac{\pi}{6}\right)^3 + \dots + \frac{\sin\left(\frac{\pi}{6} + \frac{n}{2}\pi\right)}{n!}\left(x - \frac{\pi}{6}\right)^n + \dots$$

$-\infty < x < \infty$ (Proof is omitted.)

(b) $31° = \frac{31\pi}{180}$

$$\sin\frac{31\pi}{180}=\frac{1}{2}+\frac{\sqrt{3}}{2}\left(\frac{31\pi}{180}-\frac{\pi}{6}\right)-\frac{1}{4}\left(\frac{31\pi}{180}-\frac{\pi}{6}\right)^2-\frac{\sqrt{3}}{12}\left(\frac{31\pi}{180}-\frac{\pi}{6}\right)^3+\ldots$$

$$+\frac{\sin\left(\frac{\pi}{6}+\frac{n}{2}\pi\right)}{n!}\left(\frac{31\pi}{180}-\frac{\pi}{6}\right)^n+\ldots$$

$$=\frac{1}{2}+\frac{\sqrt{3}}{2}\left(\frac{\pi}{180}\right)-\frac{1}{4}\left(\frac{\pi}{180}\right)^2-\frac{\sqrt{3}}{12}\left(\frac{\pi}{180}\right)^3+\ldots+\frac{\sin\left(\frac{\pi}{6}+\frac{n}{2}\pi\right)}{n!}\left(\frac{\pi}{180}\right)^n+\ldots$$

Now, we need to use the Lagrange remainder of order n to determine the value of n required for the approximation to be correct to six decimal places.

$$|R_n(x)|=\left|\frac{f^{(n+1)}(c)}{(n+1)!}(x-\frac{\pi}{6})^{n+1}\right| \qquad c\in(\frac{\pi}{6},\frac{31\pi}{180})$$

When $x=\dfrac{31\pi}{180}$

$$\left|R_n(\frac{31\pi}{180})\right|=\left|\frac{\sin(c+\frac{n+1}{2}\pi)}{(n+1)!}(\frac{31\pi}{180}-\frac{\pi}{6})^{n+1}\right|=\left|\frac{\sin(c+\frac{n+1}{2}\pi)}{(n+1)!}(\frac{\pi}{180})^{n+1}\right|$$

$$<\left|\frac{1}{(n+1)!}(\frac{\pi}{180})^{n+1}\right|$$

To be correct to six decimal places, we need the error to be less than 5×10^{-7}.

Since $\qquad \dfrac{1}{(3+1)!}(\dfrac{\pi}{180})^{3+1}\approx 3.9\times10^{-9}$

thus $\qquad n=3$

Therefore, $\qquad \sin\dfrac{31\pi}{180}\approx\dfrac{1}{2}+\dfrac{\sqrt{3}}{2}\left(\dfrac{\pi}{180}\right)-\dfrac{1}{4}\left(\dfrac{\pi}{180}\right)^2-\dfrac{\sqrt{3}}{12}\left(\dfrac{\pi}{180}\right)^3\approx 0.515038$

Example 5. (a) Find the Taylor series for $f(x)=\sqrt[3]{x}$ at $x_0=27$ up to the term in x^2.

(b) Use this series to evaluate the value of $\sqrt[3]{28}$.

(c) Estimate the size of the error.

Solution. (a) Because

$$f(x) = \sqrt[3]{x} \qquad f(27) = \sqrt[3]{27} = 3$$

$$f'(x) = \frac{1}{3}x^{-\frac{2}{3}} \qquad f'(27) = \frac{1}{3}\cdot 27^{-\frac{2}{3}} = \frac{1}{27}$$

$$f''(x) = -\frac{2}{9}x^{-\frac{5}{3}} \qquad f''(27) = -\frac{2}{9}\cdot 27^{-\frac{5}{3}} = -\frac{2}{2187}$$

$$f'''(x) = \frac{10}{27}x^{-\frac{8}{3}}$$

thus
$$f(x) = \sqrt[3]{x} \approx f(27) + f'(27)(x-27) + \frac{f''(27)}{2!}(x-27)^2$$

$$= 3 + \frac{1}{27}(x-27) - \frac{1}{2187}(x-27)^2$$

(b)
$$\sqrt[3]{28} \approx 3 + \frac{1}{27}(28-27) - \frac{1}{2187}(28-27)^2 = \frac{6641}{2187} \approx 3.03658$$

(c)
$$|R_2(x)| = \left|\frac{f'''(c)}{3!}(x-27)^3\right|$$

Substituting $f'''(x) = \frac{10}{27}x^{-\frac{8}{3}}$ gives

$$|R_2(x)| = \left|\frac{\frac{10}{27}c^{-\frac{8}{3}}}{3!}(x-27)^3\right| = \left|\frac{5(x-27)^3}{81c^{\frac{8}{3}}}\right|$$

Since $x = 28$, $c \in (27, 28)$

thus $$|R_2(28)| = \left|\frac{5(28-27)^3}{81c^{\frac{8}{3}}}\right| < \left|\frac{5(28-27)^3}{81\cdot 27^{\frac{8}{3}}}\right| \approx 9.41\times 10^{-6}$$

This is the maximum error. After calculation, we find the absolute error is

$$\left|\frac{6641}{2187} - \sqrt[3]{28}\right| \approx 9.18\times 10^{-6}$$

Example 6. (a) Find the Maclaurin series for $f(x) = \ln\cos x$ up to the term in x^4.

(b) Use this series to calculate the value of $\ln\sqrt{2}$.

Solution. (a) Substituting the Maclaurin series for $\cos x$

$$\cos x = \sum_{n=0}^{\infty}(-1)^n \frac{x^{2n}}{(2n)!} = 1 - \frac{x^2}{2!} + \frac{x^4}{4!} - \ldots$$

gives $$\ln\cos x = \ln(1 - \frac{x^2}{2!} + \frac{x^4}{4!} - \ldots) = \ln[1 + (-\frac{x^2}{2!} + \frac{x^4}{4!} - \ldots)]$$

Then use the Maclaurin series for $\ln(1+x)$

$$\ln(1+x) = \sum_{n=1}^{\infty}(-1)^{n+1}\frac{x^n}{n} = x - \frac{x^2}{2} + \frac{x^3}{3} - \ldots$$

we have $$\ln\cos x = (-\frac{x^2}{2!} + \frac{x^4}{4!} - \ldots) - \frac{1}{2}(-\frac{x^2}{2!} + \frac{x^4}{4!} - \ldots)^2 + \ldots$$

$$= -\frac{x^2}{2} + \frac{x^4}{24} - \frac{x^4}{8} + \ldots = -\frac{x^2}{2} - \frac{x^4}{12} + \ldots$$

That is, $$\ln\cos x = -\frac{x^2}{2} - \frac{x^4}{12} + \ldots$$

(b) Let $$\ln\cos x = \ln\frac{1}{\sqrt{2}}$$

then $$\cos x = \frac{1}{\sqrt{2}}$$

$$x = \frac{\pi}{4}$$

Hence, $$\ln\frac{1}{\sqrt{2}} = \ln\cos\frac{\pi}{4} \approx -\frac{1}{2}\left(\frac{\pi}{4}\right)^2 - \frac{1}{12}\left(\frac{\pi}{4}\right)^4$$

$$\ln 1 - \ln\sqrt{2} \approx -\frac{1}{2}\left(\frac{\pi}{4}\right)^2 - \frac{1}{12}\left(\frac{\pi}{4}\right)^4$$

$$\ln\sqrt{2} \approx \frac{1}{2}\left(\frac{\pi}{4}\right)^2 + \frac{1}{12}\left(\frac{\pi}{4}\right)^4 = \frac{96\pi^2 + \pi^4}{3072}$$

That is, $$\ln\sqrt{2} \approx \frac{96\pi^2 + \pi^4}{3072}$$

II. Use power series expansions to find limits

Earlier, we learned to use L'Hôpital's rule to find limits; however, for some problems, we need to use L'Hôpital's rule repeatedly to obtain the result. After differentiating several times, the function can become increasingly complex without a result being reached. We can now use a power series expansion to an appropriate order, as determined by the problem, to find the limit.

Example 7. Evaluate $\lim\limits_{x\to 0}\dfrac{x\arctan x - x^2\cos x}{x^4}$.

Solution. From Example 10 in the previous section, we get the Maclaurin series for $\arctan x$.

$$\arctan x = x - \frac{1}{3}x^3 + \frac{1}{5}x^5 - \frac{1}{7}x^7 + \frac{1}{9}x^9 + \ldots + \frac{(-1)^{n-1}}{2n-1}x^{2n-1} + \ldots$$

And the Maclaurin series for $\cos x$ is

$$\cos x = \sum_{n=0}^{\infty}(-1)^n\frac{x^{2n}}{(2n)!} = 1 - \frac{x^2}{2!} + \frac{x^4}{4!} - \ldots$$

Thus $$x\arctan x - x^2\cos x = \left(x^2 - \frac{1}{3}x^4 + \frac{1}{5}x^6 - \ldots\right) - \left(x^2 - \frac{x^4}{2!} + \frac{x^6}{4!} - \ldots\right)$$

$$= \frac{1}{6}x^4 + \frac{19}{120}x^6 + \ldots$$

Therefore, $$\lim_{x\to 0}\frac{x\arctan x - x^2\cos x}{x^4} = \lim_{x\to 0}\left(\frac{1}{6} + \frac{19}{120}x^2 + \ldots\right) = \frac{1}{6}$$

Example 8. Evaluate $\lim_{x\to 0}\frac{e^x\sin x - x(1+x)}{x^3}$.

Solution. Substituting the Maclaurin series for e^x and $\sin x$ gives

$$\lim_{x\to 0}\frac{e^x\sin x - x(1+x)}{x^3} = \lim_{x\to 0}\frac{(1+x+\frac{x^2}{2!}+\ldots)(x-\frac{x^3}{3!}+\ldots) - x - x^2}{x^3}$$

$$= \lim_{x\to 0}\frac{\frac{1}{3}x^3 - \frac{1}{6}x^4 - \frac{1}{12}x^5 + \ldots}{x^3} = \frac{1}{3}$$

III. Use power series expansions to find approximate solutions to differential equations

In the previous chapter, we learned about different methods for finding exact solutions to differential equations. However, in practical applications, many differential equations cannot be solved by these methods. In such cases, in addition to Euler's and Picard's methods, we can also use power series expansions to find approximate solutions to differential equations.

Example 9. The differential equation $y\frac{dy}{dx} = e^{2y}$ satisfies the condition $y = 1$ when $x = 0$.

(a) Find the first three terms of the Maclaurin series solution to the differential equation.

(b) Hence, estimate the value of y when $x = 0.1$, correct to two decimal places.

(c) Solve the differential equation exactly and calculate the value of y when $x = 0.1$, correct to two decimal places.

Solution. (a) The Maclaurin series is

$$f(x) = f(0) + f'(0)x + \frac{f''(0)}{2!}x^2 + \ldots$$

We first need to find the values of $f'(0)$ and $f''(0)$.

Substituting $x = 0$ and $y = 1$ into $y\frac{dy}{dx} = e^{2y}$, we have

$$1 \cdot \left.\frac{dy}{dx}\right|_{x=0} = e^{2 \cdot 1}$$

$$\left.\frac{dy}{dx}\right|_{x=0} = e^2$$

That is, $f'(0) = e^2$

To obtain $\frac{d^2y}{dx^2}$, differentiate both sides of $y\frac{dy}{dx} = e^{2y}$ with respect to x.

$$(y\frac{dy}{dx})' = (e^{2y})'$$

$$\frac{dy}{dx} \cdot \frac{dy}{dx} + y \cdot \frac{d^2y}{dx^2} = \frac{de^{2y}}{d2y} \cdot \frac{d2y}{dx}$$

$$(\frac{dy}{dx})^2 + y \cdot \frac{d^2y}{dx^2} = 2e^{2y}\frac{dy}{dx}$$

Substituting $x = 0$, $y = 1$, and $\left.\frac{dy}{dx}\right|_{x=0} = e^2$, we have

$$(e^2)^2 + 1 \cdot \left. \frac{d^2 y}{dx^2} \right|_{x=0} = 2e^{2 \cdot 1} \cdot e^2$$

$$\left. \frac{d^2 y}{dx^2} \right|_{x=0} = e^4$$

That is, $f''(0) = e^4$

Therefore, $y = f(0) + f'(0)x + \frac{f''(0)}{2!}x^2 + \ldots = 1 + e^2 x + \frac{e^4}{2}x^2 + \ldots$

(b) When $x = 0.1$

$$y = 1 + e^2 \cdot 0.1 + \frac{e^4}{2}0.1^2 \approx 2.01$$

(c) $y\frac{dy}{dx} = e^{2y}$

Separating the variables and integrating both sides gives

$$\int ye^{-2y} dy = \int dx \qquad (1)$$

For $\int ye^{-2y} dy$

let $u = y$ $\quad dv = e^{-2y} dy$

then $du = dy$ $\quad v = \int e^{-2y} dy = -\frac{1}{2}e^{-2y}$

thus $\int ye^{-2y} dy = -\frac{1}{2}ye^{-2y} + \int \frac{1}{2}e^{-2y} dy = -\frac{1}{2}ye^{-2y} - \frac{1}{4}e^{-2y}$

Hence, (1) changes to

$$-\frac{1}{2}ye^{-2y} - \frac{1}{4}e^{-2y} = x + C$$

Substituting $x = 0$ and $y = 1$, we have

$$-\frac{1}{2}e^{-2} - \frac{1}{4}e^{-2} = C$$

$$C = -\frac{3}{4}e^{-2}$$

Therefore, the particular solution to the differential equation is

$$-\frac{1}{2}ye^{-2y} - \frac{1}{4}e^{-2y} = x - \frac{3}{4}e^{-2}$$

Multiplying both sides by -4 gives

$$2ye^{-2y} + e^{-2y} = -4x + 3e^{-2}$$

When $x = 0.1$, solving with a calculator gives

$$y \approx 3.61$$

Exercise 9.8

1. Use power series to estimate the value of the following integrals correct to four decimal places. Give your answers correct to four decimal places.

a. $\int_0^1 e^{-x^2} dx$ **b.** $\int_2^4 x \sin x dx$ **c.** $\int_0^1 \cos x^2 dx$

d. $\int_0^2 \frac{\sin x}{x} dx$ **e.** $\int_0^{\frac{1}{2}} \frac{\arctan x}{x} dx$ **f.** $\int_0^{\frac{1}{10}} \frac{\ln(1+x)}{x} dx$

2. Use suitable power series to calculate the value of the following to the required accuracy. Give your answers correct to six decimal places.

a. $\sin 18°$, correct to 10^{-5} **b.** $\cos 1°$, correct to 10^{-6}

c. $\ln 1.2$, correct to 10^{-4} **d.** $\sqrt{e}$, correct to 10^{-7}

3. Find the following limits using Maclaurin series.

a. $\lim_{x \to 0} \frac{\sin x - xe^x}{x^2}$ **b.** $\lim_{x \to 0} \frac{x\cos(2x) + \arctan x}{x}$

c. $\lim_{x \to 0} \frac{\cos x - e^{-x}}{2x}$

d. $\lim_{x \to 0} \frac{\ln\sqrt{1+x} - \sin(2x)}{x}$

4. $f(x) = \sqrt[5]{1+x}$

   a. Find the Maclaurin series for $f(x)$ up to the term in x^4.

   b. Use this series to find an approximate value for $\sqrt[5]{1.3}$. Give your answer correct to five decimal places.

   c. Using the Lagrange error term, find an upper bound for the error in this approximation.

5. Find the Maclaurin series for $f(x) = \sin(\sin x)$ up to the term in x^3. Hence, or otherwise, find the limit $\lim_{x \to 0} \frac{\sin(\sin x) - x}{x^3}$.

6. $f(x) = \ln(1 + \cos x)$

   a. Find the Maclaurin series for $f(x)$ up to the term in x^4.

   b. Use this series to find an approximation for $\ln\frac{3}{2}$.

   c. Find an upper bound for the error in this approximation.

7. $f(x) = \ln(\sin x)$

   a. Find the Taylor series for $f(x)$ at $x_0 = \frac{\pi}{6}$ up to the term in x^3.

   b. Hence, estimate the value of $\ln\frac{\sqrt{2}}{2}$ and give your answer correct to six decimal places.

   c. Give an upper bound for the error in your estimate.

   d. Find the actual error.

e. Using term-by-term differentiation of the series from a., find a power series for $\cot x$.

8. The differential equation

$$\frac{dy}{dx} = -4xy^2$$

satisfies the initial condition $y = 1$ when $x = 0$.

a. Find the first three non-zero terms of the Maclaurin series solution to the differential equation.

b. Hence, estimate the value of y when $x = 0.5$.

c. Solve the differential equation exactly and estimate the value of y when $x = 0.5$.

9. The differential equation

$$(x^2 + 1)\frac{dy}{dx} + xy = 0$$

satisfies the initial condition $y = 2$ when $x = 0$.

a. Find the first three non-zero terms of the Maclaurin series solution to the differential equation.

b. Hence, estimate the value of y when $x = 0.1$. Give your answer correct to six decimal places.

c. Solve the differential equation exactly and calculate the value of y when $x = 0.1$. Give your answer correct to six decimal places.

10. The differential equation

$$(1 + y^2)\frac{dy}{dx} = e^x y$$

satisfies the initial condition $y = 1$ when $x = 0$.

a. Find the first three terms of the Maclaurin series solution to the differential equation.

b. Hence, estimate the value of y when $x = 0.2$.

c. Solve the differential equation exactly and calculate the value of y when $x = 0.2$. Give your answer correct to six decimal places.

11. The differential equation

$$\frac{dy}{dx} - xe^{y} = 2e^{y}$$

satisfies the initial condition $x = 0, y = -1$.

a. Find the terms of the Maclaurin series solution to the differential equation up to the term in x^3.

b. Hence, estimate the value of y when $x = 0.5$. Give your answer correct to six decimal places.

c. Solve the differential equation exactly and calculate the value of y when $x = 0.5$. Give your answer correct to six decimal places.

9.9. Extension of the Binomial Theorem for any real number λ

Find the Maclaurin series for $f(x) = (1 + x)^{\lambda}, \lambda \in R$.

Because

$$f(x) = (1 + x)^{\lambda} \qquad f(0) = 1$$

$$f'(x) = \lambda(1 + x)^{\lambda - 1} \qquad f'(0) = \lambda$$

$$f''(x) = \lambda(\lambda - 1)(1 + x)^{\lambda - 2} \qquad f''(0) = \lambda(\lambda - 1)$$

$$f'''(x) = \lambda(\lambda - 1)(\lambda - 2)(1 + x)^{\lambda - 3} \qquad f'''(0) = \lambda(\lambda - 1)(\lambda - 2)$$

$$\vdots \qquad \qquad \vdots$$

it follows that the Maclaurin series for $f(x) = (1+x)^{\lambda}$ is

$$(1+x)^{\lambda} = f(0) + f'(0)x + \frac{f''(0)}{2!}x^2 + \frac{f^{(3)}(0)}{3!}x^3 + \ldots + \frac{f^{(k)}(0)}{k!}x^k + \ldots$$

$$= 1 + \lambda x + \frac{\lambda(\lambda-1)}{2!}x^2 + \frac{\lambda(\lambda-1)(\lambda-2)}{3!}x^3 + \ldots$$

$$+ \frac{\lambda(\lambda-1)(\lambda-2)\cdots(\lambda-k+1)}{k!}x^k + \ldots \qquad |x| < 1$$

Thus, the Binomial Theorem can be extended to cases where λ is a real number.

$$(1+x)^{\lambda} = 1 + \lambda x + \frac{\lambda(\lambda-1)}{2!}x^2 + \frac{\lambda(\lambda-1)(\lambda-2)}{3!}x^3 + \ldots$$

$$+ \frac{\lambda(\lambda-1)(\lambda-2)\cdots(\lambda-k+1)}{k!}x^k + \ldots$$

$$= 1 + \sum_{k=1}^{\infty} \frac{\lambda(\lambda-1)(\lambda-2)\cdots(\lambda-k+1)}{k!}x^k \qquad \lambda \in \mathrm{R}$$

This series converges for $|x| < 1$.

When $\lambda = n \in \mathrm{Z}^+$, it is our familiar Binomial Theorem, and it can also be written as

$$(1+x)^n = 1 + \binom{n}{1}x + \binom{n}{2}x^2 + \binom{n}{3}x^3 + \ldots + \binom{n}{k}x^k + \ldots = \sum_{k=0}^{n}\binom{n}{k}x^k$$

The result above can be applied to the expansion of $(a+bx)^{\lambda}$ ($\lambda \in \mathrm{R}$), but we need to rearrange the terms inside the brackets to be in the form of 1 **plus** a second term.

$$(a+bx)^{\lambda} = [a(1 + \frac{bx}{a})]^{\lambda} = a^{\lambda}(1 + \frac{bx}{a})^{\lambda}$$

$$= a^{\lambda}[1 + \lambda\frac{bx}{a} + \frac{\lambda(\lambda-1)}{2!}\left(\frac{bx}{a}\right)^2 + \frac{\lambda(\lambda-1)(\lambda-2)}{3!}\left(\frac{bx}{a}\right)^3 + \ldots]$$

$$= a^{\lambda}[1+\sum_{k=1}^{\infty}\frac{\lambda(\lambda-1)(\lambda-2)\cdots(\lambda-k+1)}{k!}\left(\frac{bx}{a}\right)^{k}]$$

This expansion converges for $\left|\frac{bx}{a}\right| < 1$.

Example 1. Expand $\frac{1}{(1+x)^2}$ up to the term in x^4.

Solution. $\frac{1}{(1+x)^2} = (1+x)^{-2}$

$$= 1 + (-2)x + \frac{(-2)(-3)}{2!}x^2 + \frac{(-2)(-3)(-4)}{3!}x^3 + \frac{(-2)(-3)(-4)(-5)}{4!}x^4 + \ldots$$

$$= 1 - 2x + 3x^2 - 4x^3 + 5x^4 + \ldots$$

Example 2. Expand $\frac{1}{\sqrt{1+x}}$ up to the term in x^3.

Solution. $\frac{1}{\sqrt{1+x}} = (1+x)^{-\frac{1}{2}}$

$$= 1 + \left(-\frac{1}{2}\right)x + \frac{\left(-\frac{1}{2}\right)\left(-\frac{1}{2}-1\right)}{2!}x^2 + \frac{\left(-\frac{1}{2}\right)\left(-\frac{1}{2}-1\right)\left(-\frac{1}{2}-2\right)}{3!}x^3 + \ldots$$

$$= 1 - \frac{1}{2}x + \frac{3}{8}x^2 - \frac{5}{16}x^3 + \ldots$$

Example 3.

(a) Find the first three terms in ascending powers of x in the binomial expansion of $\frac{1}{2-3x}$.

(b) Find the range of values of x for which this expansion is valid.

Solution. (a) $\dfrac{1}{2-3x} = (2-3x)^{-1} = \{2[1 + \left(-\dfrac{3}{2}x\right)]\}^{-1}$

$$= 2^{-1}[1 + (-1)\left(-\frac{3}{2}x\right) + \frac{(-1)(-1-1)}{2!}\left(-\frac{3}{2}x\right)^2 + \ldots]$$

$$= \frac{1}{2}(1 + \frac{3}{2}x + \frac{9}{4}x^2 + \ldots)$$

(b) The expansion converges when

$$\left|-\frac{3}{2}x\right| < 1$$

That is, $-2 < 3x < 2$

$$-\frac{2}{3} < x < \frac{2}{3}$$

Example 4. Find the coefficient of x^2 in the binomial expansion of $\sqrt{\dfrac{1-x}{1+x}}$ for $|x| < 1$.

Solution. $\sqrt{\dfrac{1-x}{1+x}} = (1-x)^{\frac{1}{2}}(1+x)^{-\frac{1}{2}} = \left[1+(-x)\right]^{\frac{1}{2}}(1+x)^{-\frac{1}{2}}$

$$= [1 + \frac{1}{2}(-x) + \frac{\left(\frac{1}{2}\right)\left(\frac{1}{2}-1\right)}{2!}(-x)^2 + \ldots][1 + \left(-\frac{1}{2}\right)x + \frac{\left(-\frac{1}{2}\right)\left(-\frac{1}{2}-1\right)}{2!}x^2 + \ldots]$$

$$= (1 - \frac{1}{2}x - \frac{1}{8}x^2 + \ldots)(1 - \frac{1}{2}x + \frac{3}{8}x^2 + \ldots)$$

Using selective multiplication, we obtain the terms with x^2.

$$1\cdot(\frac{3}{8}x^2) + \left(-\frac{1}{2}x\right)\cdot\left(-\frac{1}{2}x\right) + \left(-\frac{1}{8}x^2\right)\cdot 1 = \frac{1}{2}x^2$$

Therefore, the coefficient of x^2 is $\dfrac{1}{2}$.

Example 5.

(a) Find the first three terms in ascending powers of x in the binomial expansion of $\sqrt{\dfrac{4-3x}{9-4x}}$.

(b) Find the interval of convergence of this expansion.

(c) By substituting $x = 0.05$, find the approximate value of $\sqrt{7}$, correct to four decimal places.

Solution. (a) $\sqrt{\dfrac{4-3x}{9-4x}} = (4-3x)^{\frac{1}{2}}(9-4x)^{-\frac{1}{2}}$

$$(4-3x)^{\frac{1}{2}} = \left[4(1+(-\frac{3}{4}x))\right]^{\frac{1}{2}} = 4^{\frac{1}{2}}[1+\frac{1}{2}(-\frac{3}{4}x)+\frac{\frac{1}{2}\left(\frac{1}{2}-1\right)}{2!}(-\frac{3}{4}x)^2+...]$$

$$= 2(1-\frac{3}{8}x-\frac{9}{128}x^2+...) = 2-\frac{3}{4}x-\frac{9}{64}x^2+...$$

$$(9-4x)^{-\frac{1}{2}} = \left[9(1+(-\frac{4}{9}x))\right]^{-\frac{1}{2}} = 9^{-\frac{1}{2}}[1+(-\frac{1}{2})(-\frac{4}{9}x)+\frac{(-\frac{1}{2})\left(-\frac{1}{2}-1\right)}{2!}(-\frac{4}{9}x)^2+...]$$

$$= \frac{1}{3}(1+\frac{2}{9}x+\frac{2}{27}x^2+...) = \frac{1}{3}+\frac{2}{27}x+\frac{2}{81}x^2+...$$

Thus $\sqrt{\dfrac{4-3x}{9-4x}} = (4-3x)^{\frac{1}{2}}(9-4x)^{-\frac{1}{2}} = (2-\frac{3}{4}x-\frac{9}{64}x^2+...)(\frac{1}{3}+\frac{2}{27}x+\frac{2}{81}x^2+...)$

$$= \frac{2}{3}-\frac{11}{108}x-\frac{275}{5184}x^2+...$$

(b) The interval of convergence for $(4-3x)^{\frac{1}{2}}$ is

$$\left|-\frac{3}{4}x\right| < 1 \quad \Rightarrow \quad |x| < \frac{4}{3}$$

The interval of convergence for $(9-4x)^{-\frac{1}{2}}$ is

$$\left|-\frac{4}{9}x\right|<1 \qquad \Rightarrow \qquad |x|<\frac{9}{4}$$

Hence, the interval of convergence of this expansion is

$$|x|<\frac{4}{3}$$

(c) Substituting $x=0.05$ gives

$$\sqrt{\frac{4-3\cdot 0.05}{9-4\cdot 0.05}}=\sqrt{\frac{7}{16}}=\frac{\sqrt{7}}{4}=\frac{2}{3}-\frac{11}{108}\cdot(0.05)-\frac{275}{5184}(0.05)^2+\ldots$$

$$\sqrt{7}=4[\frac{2}{3}-\frac{11}{108}\cdot(0.05)-\frac{275}{5184}(0.05)^2+\ldots]\approx 2.6458$$

Example 6.

(a) Find the first three terms in ascending powers of x up to the term x^2 in the binomial expansion of $\frac{\left(2-x^2\right)^2}{(8+3x)^{\frac{2}{3}}}$.

(b) Hence, find the approximate value for $\int_{-0.4}^{0.4}\frac{\left(2-x^2\right)^2}{(8+3x)^{\frac{2}{3}}}dx$.

Solution. (a) $\frac{\left(2-x^2\right)^2}{(8+3x)^{\frac{2}{3}}}=\left(2-x^2\right)^2(8+3x)^{-\frac{2}{3}}$

$$\left(2-x^2\right)^2=4-4x^2+x^4$$

$$(8+3x)^{-\frac{2}{3}}=\left[8(1+\frac{3}{8}x)\right]^{-\frac{2}{3}}=8^{-\frac{2}{3}}[1+(-\frac{2}{3})(\frac{3}{8}x)+\frac{(-\frac{2}{3})(-\frac{2}{3}-1)}{2!}(\frac{3}{8}x)^2+\ldots]$$

$$=\frac{1}{4}(1-\frac{1}{4}x+\frac{5}{64}x^2+\ldots)=\frac{1}{4}-\frac{1}{16}x+\frac{5}{256}x^2+\ldots$$

Therefore,

$$\frac{\left(2-x^2\right)^2}{(8+3x)^{\frac{2}{3}}}=\left(2-x^2\right)^2(8+3x)^{-\frac{2}{3}}=(4-4x^2+x^4)(\frac{1}{4}-\frac{1}{16}x+\frac{5}{256}x^2+...)$$

$$=1-\frac{1}{4}x-\frac{59}{64}x^2+...$$

(b) $$\int_{-0.4}^{0.4}\frac{\left(2-x^2\right)^2}{(8+3x)^{\frac{2}{3}}}dx=\int_{-0.4}^{0.4}(1-\frac{1}{4}x-\frac{59}{64}x^2+...)dx\approx\frac{1141}{1500}$$

Exercise 9.9

1. Expand each of the following in ascending powers of x, up to the term in x^3. State the range of values of x for which each expansion is valid.

a. $(1+x)^{-3}$ **b.** $\frac{1}{(1-2x)^2}$ **c.** $(1-4x)^{\frac{2}{3}}$

d. $\sqrt{1+5x}$ **e.** $\frac{1}{(1+x^2)^2}$ **f.** $\frac{1}{\sqrt[5]{(1-7x)^3}}$

2. Expand each of the following in ascending powers of x, up to the term in x^3. State the range of values of x for which each expansion is valid.

a. $(3+x)^{-1}$ **b.** $\sqrt{4+x}$ **c.** $(6+5x)^{\frac{3}{4}}$

d. $(3-2x)^{-\frac{1}{2}}$ **e.** $\frac{2+x}{2-x}$ **f.** $\frac{1+5x}{3+x}$

g. $\sqrt{\frac{1+x}{1-4x}}$ **h.** $\frac{(3-x^2)^2}{(8+3x)^3}$ **i.** $(7-4x)^{\frac{2}{3}}(3+6x)^{\frac{1}{4}}$

3. Given that $(1+bx)^n$ is expanded in ascending powers of x, the first three terms are $1 - 18x + 172x^2$. Find the values of the constants b and n.

4. When $(1+px)^n$ is expanded in ascending powers of x, the first three terms are $1+\frac{x}{8}-\frac{7}{156}x^2$. Find the values of the constants p and n.

5. Expand $(1-3x)^{\frac{1}{5}}$ in ascending powers of x up to and including the term x^2. Hence, evaluate $\sqrt[5]{0.997}$ correct to five decimal places.

6. Expand $(6-4x)^{-\frac{1}{3}}$ in ascending powers of x up to and including the term x^2. Hence, evaluate $\frac{1}{\sqrt[3]{5.96}}$ correct to four decimal places.

7. Find the first three terms in the expansion in ascending powers of x of $\sqrt{\frac{1+4x}{1-2x}}$. State the range of values of x for which the expansion is valid. By substituting $x = 0.01$ in your expansion, find an approximate value for $\sqrt{52}$, correct to two decimal places.

8. **a**. Expand $(1+x)^{-\frac{1}{2}}$ in ascending powers of x up to and including the term x^2.

b. Prove that $\frac{11}{2}(1+\frac{1}{120})^{-\frac{1}{2}} = 30^{\frac{1}{2}}$.

c. Use your expansion of $(1+x)^{-\frac{1}{2}}$ with $x=\frac{1}{120}$ to find an approximate value for $30^{\frac{1}{2}}$, correct to four decimal places.

9. Find the first three terms in the expansion in ascending powers of x of $\dfrac{x^2+3x+2}{\sqrt[3]{1-5x}}$. Hence, find an approximation to $\displaystyle\int_{-0.3}^{0.5}\frac{x^2+3x+2}{\sqrt[3]{1-5x}}dx$. Give your answer correct to six decimal places.

10. **a.** Express $\dfrac{9-8x}{(2x-1)(3-x)}$ in partial fractions.

b. Hence, find the expansion of $\dfrac{9-8x}{(2x-1)(3-x)}$ in ascending powers of x up to the term in x^2.

c. State the set of values of x for which this expansion is valid.

Answers to Exercises

Exercise 1

1. a. $\{x \mid x > 5\}$ **b.** $\{(x, y) \mid x^2 + y^2 < 25\}$ **c.** $\{(x, y) \mid y = x^2$ and $x - y = 0\}$

2. a. $\{3, 4\}$ **b.** $\{(0, 0), (1, 1)\}$ **c.** $\{-4, -3, -2, -1, 0, 1, 2, 3, 4, 5, 6\}$ **3.** B, C

4. a. $A \cup B = \{1, 2, 3, 5\}$ **b.** $A \cap B = \{1, 3\}$ **c.** $A \cup B \cup C = \{1, 2, 3, 4, 5, 6\}$

d. $A \cap B \cap C = \varnothing$ **e.** $A - B = \{2\}$ **5. a.** $A \cup B = \{x \mid x > 3\}$

b. $A \cap B = \{x \mid 4 < x < 5\}$ **c.** $A - B = \{x \mid 3 < x \le 4\}$ **6. a.** $A' = \{4, 5, 6\}$

b. $B' = \{1, 3, 5\}$ **c.** $A' \cup B' = \{1, 3, 4, 5, 6\}$ **d.** $A' \cap B' = \{5\}$

7. $A \times B = \{(a, a), (a, b), (a, c), (b, a), (b, b), (b, c), (c, a), (c, b), (c, c), (d, a), (d, b), (d, c)\}$ **10. a.** $-3 < x < 3$ **b.** $-3 < x < 11$ **c.** $0 < x < 4$ and $x \ne 2$ **11.** $f(0) = 2$, $f(1) = 0$, $f(2) = 0$, $f(-x) = x^2 + 3x + 2$, $f(x + 1) = x^2 - x$. **12. a.** $[-3, 3]$

b. $x \ge -2, x \ne \pm 1$ **c.** $x \in R$ **d.** $[-1, 3]$ **e.** $x > -3$ **f.** $x > 1$

13. $f[f(x)] = \dfrac{x}{1-2x}$, $f\{f[f(x)]\} = \dfrac{x}{1-3x}$

17. a.

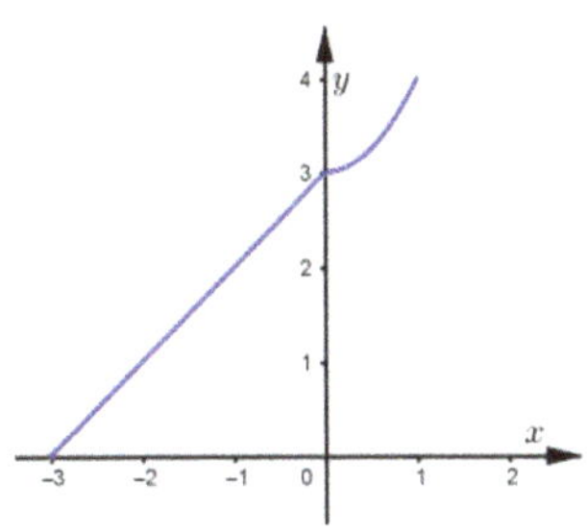

b.

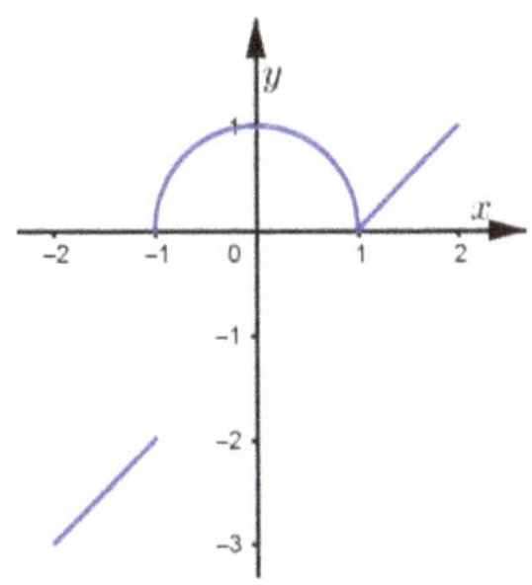

18. a. even **b.** odd **c.** even **d.** odd **e.** neither even nor odd **f.** even

19. a. $y = \dfrac{x-1}{2}$ **b.** $y = \log_2 x$ **c.** $y = \dfrac{2+2x}{x-1}$ **d.** $y = \sqrt[3]{x-2}$

e. $y = 10^{x-1} - 2$ **f.** $y = \dfrac{3-x}{x-4}$ **20.** $y = \log_a^2 x$ **21.** $y = \sqrt{2 + \cos^2 x}$

22. $y = 3\log^3(1+t) + 2\log(1+t)$

23. $a = 2$, it is a composite function. D = $(-\infty, +\infty)$; $a = 0.5$, it is a composite function. D = $\{x \mid 2n\pi - \frac{7\pi}{6} < x < 2n\pi + \frac{\pi}{6}, n = 0, \pm1, \ldots\}$; $a = -2$, it is not a composite function.

24.

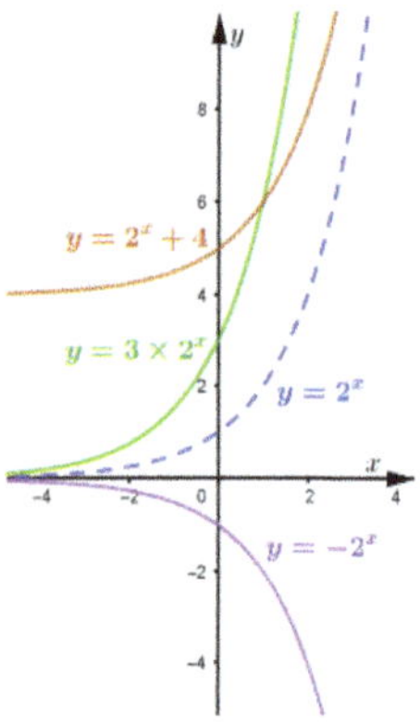

25. a.

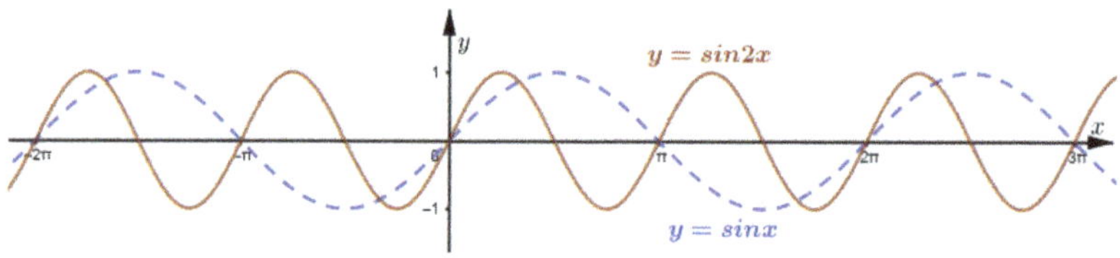

25. b.

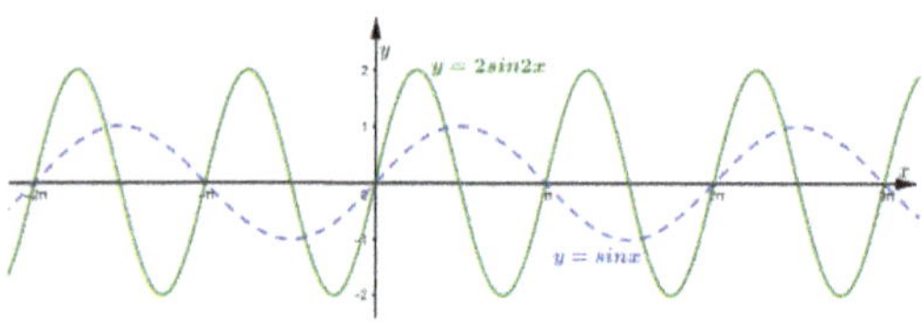

c.

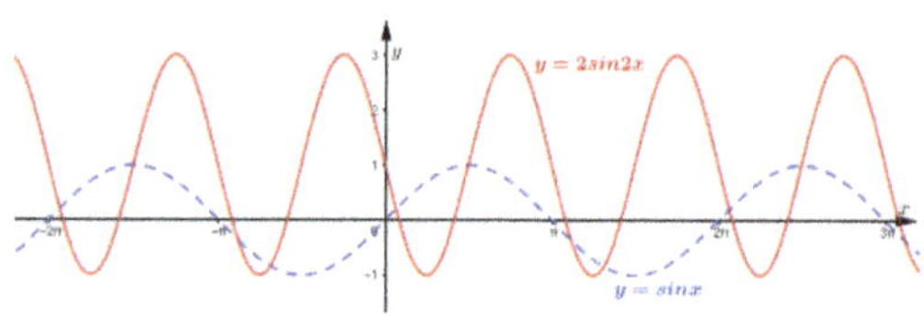

Exercise 2

1. a. 11 **b.** 8 **c.** 0 **d.** -1 **e.** $\frac{1}{6}$ **f.** $\frac{1}{2}$ **g.** 4 **h.** 2 **i.** 4 **j.** 0

k. -1 **l.** 4 **2. a.** 3 **b.** 4 **c.** 0 **d.** 12 **e.** 6 **f.** $\frac{2}{3}$ **g.** ∞ **h.** -2

3. a. 12 **b.** $\frac{32}{15}$ **c.** 4 **d.** 0 **e.** $\frac{1}{5}$ **f.** $\frac{1}{2}$ **g.** $\frac{1}{4}$ **h.** -2

4. a. $\frac{1}{4}$ **b.** 10 **c.** $\sqrt{2}$ **d.** $\frac{3}{4}$ **e.** $\frac{45}{32}$ **f.** $\frac{1}{5}$

5. a. $\frac{2}{5}$ **b.** e^9 **c.** 3 **d.** $\frac{1}{5}$ **e.** 8 **f.** $e^{\frac{2}{5}}$ **g.** e^{-2} **h.** 0

Exercise 3

1. a. $2x$ **b.** $3x^2 + 1$ **c.** $-4x$ **d.** $-\frac{1}{x^2}$ **e.** $\frac{1}{2\sqrt{x}}$ **f.** $\frac{-2}{x^3}$

2. a. $6x^2 - 1$ **b.** $-9x^{-4} + x^2$ **c.** $\frac{5}{2}x^{-\frac{1}{2}} - 2x^{-2} - \frac{3}{5}x^{-\frac{2}{5}}$ **d.** $\frac{2}{(x+1)^2}$

e. $2\sqrt{x} + \frac{1}{2}(2x-3)x^{-\frac{1}{2}}$ **f.** $\frac{x^2\cos x - 2x\sin x}{x^4}$ **g.** $-\sin x\ln x + \frac{\cos x}{x}$

h. $5\sec^2(5x) + 3\cos(3x) - 2\sin x\cos x$ **i.** $-\frac{5}{2}[\cos(5x)]^{-\frac{1}{2}}\sin(5x)$

j. $\frac{5(1+\cos x) + 5x\ln x\sin x}{x(1+\cos x)^2}$ **k.** $\frac{\sin x + \cos x - x(\cos x - \sin x)}{(\sin x + \cos x)^2}$

l. $2\sec(2x-3)\tan(2x-3)\ln(3x+2) + \frac{3\sec(2x-3)}{3x+2}$

3. a. $3\cos(3x) + 6\cos(2x)\sin^2(2x)$ **b.** $3x^2\cos(x^3) + \frac{2x}{(x^2+1)\ln 2}$

c. $-7\sin(7x-5) - 2\cos(2x+3)$ **d.** $-(2+x^3)^2 + 6x^2(2-x)(2+x^3)$

e. $-2\cos x(1-\sin x)$ **f.** $\tan^2\frac{x}{3}\sec^2\frac{x}{3} - \frac{x^{-\frac{1}{2}}}{2}\cos(\cos\sqrt{x})\sin\sqrt{x}$

g. $\frac{(1+x^2)^{-\frac{1}{2}}(2x - x^2 + 1)}{2(1-x)^{\frac{3}{2}}}$ **h.** $5\sec^2(5x) + \frac{6(\ln x^2)^2}{x}$

i. $-\frac{3}{2}[\cos(3x+5)]^{-\frac{1}{2}}\sin(3x+5)$ **j.** $10\sec^5(2x+1)\tan(2x+1)$

k. $\frac{3}{6x+8}\left[1 + (\ln(3x+4))^{-\frac{1}{2}}\right]$ **l.** $\frac{5(\ln x)^4}{x} + \frac{8\sin(2x)}{\cos^5(2x)}$

4. a. $\frac{5}{4}\sec^2\frac{5x}{4}\cos^3\frac{x}{2} - \frac{3}{2}\tan\frac{5x}{4}\cos^2\frac{x}{2}\sin\frac{x}{2}$ **b.** $\frac{1}{3x}\left(\ln\frac{2x}{5}\right)^{-\frac{2}{3}}$ **c.** $\frac{1}{x} - \tan x$

d. $4x^3\sqrt{5-x} - \frac{1}{2}(9+x^4)(5-x)^{-\frac{1}{2}}$ **e.** $6x\sqrt{1+4x^3} + 6x^2(2+3x^2)(1+4x^3)^{-\frac{1}{2}}$

f. $\dfrac{3x^2}{(1+x^3)\ln 3}$ **g.** $-\dfrac{1}{\sqrt{x}(1-x)}$ **h.** $\dfrac{12\tan^2\left[\log_3(4x+5)\right]\sec^2\left[\log_3(4x+5)\right]}{(4x+5)\ln 3}$

i. $\dfrac{6\ln^2\left[\tan(2x+3)\right]\sec^2(2x+3)}{\tan(2x+3)}$ **j.** $-\sin(2x^3\ln x)(6x^2\ln x+2x^2)$

k. $\dfrac{-\sin x\sqrt{1-x^2}+x\cos x(1-x^2)^{-\frac{1}{2}}}{1-x^2}$ **l.** $6x^5\sqrt[5]{x^7-2}+\dfrac{7}{5}x^{12}(x^7-2)^{-\frac{4}{5}}$

5. a. $\dfrac{2x}{3}(1-x^2)^{-\frac{4}{3}}$ **b.** $\dfrac{-6\sec^2(3x)}{\tan^3(3x)}$ **c.** $\left[2x\cos(3x)-3x^2\sin(3x)\right]\cos[x^2\cos(3x)]$

d. $\dfrac{-2\sin(2x)\sin^2 x-2\ln\left[\cos(2x)\right]\sin x\cos x\cos(2x)}{\sin^4 x\cos(2x)}$

e. $3\left[\dfrac{1}{x}\cos(5x)-5\ln x\sin(5x)\right]\sin^2\left[\ln x\cos(5x)\right]\cdot\cos\left[\ln x\cos(5x)\right]$

f. $\dfrac{5\ln^4 x\sec(5x)-5x\ln^5 x\sec(5x)\tan(5x)}{x\sec^2(5x)}$ **g.** $7(x^4-x^3+2x)^6(4x^3-3x^2+2)$

h. $\dfrac{28(x-3)^3}{(2x+1)^5}$ **i.** $\dfrac{3}{4}(x^5+3x^2)^{-\frac{1}{4}}(5x^4+6x)$ **j.** $-\dfrac{5}{2}\left[\cos ec(5x)\right]^{\frac{1}{2}}\cot(5x)$

k. $\dfrac{3-3x^2}{(1+x^2)^2}\sec\left(\dfrac{3x}{1+x^2}\right)\tan\left(\dfrac{3x}{1+x^2}\right)$ **l.** $10\cot(5x+2)$

6. a. $-\dfrac{x^2}{y^2}$ **b.** $\dfrac{2xy-2y}{2x-x^2}$ **c.** $\dfrac{8}{3}-\dfrac{3y}{x}$ **d.** $\dfrac{-e^{xy}yx-y-3x\sin(3x)}{e^{xy}x^2+x\ln x}$

e. $\dfrac{3-2x\cos(x^2+y^2)}{2y\cos(x^2+y^2)}$ **f.** $\dfrac{e^y}{1-xe^y}$ **g.** $\dfrac{-3y^3-2xy}{9xy^2+x^2-2y}$ **h.** $\dfrac{2\sqrt{x+y^2}-1}{2y}$

i. $\dfrac{(x^2+y)(4x^3y+\sin x)-2x}{1-x^6-x^4y}$ **j.** $\dfrac{-8y-2x}{e^y+8x}$ **k.** $\dfrac{xy\ln y-y^2}{xy\ln x-x^2}$ **l.** $-\dfrac{2y^2}{x^3}$

7. a. $\dfrac{\cos x-5\sin x}{e^{5x}}$ **b.** $\dfrac{1}{2}3^{\sqrt{x}}\ln 3\cdot x^{-\frac{1}{2}}\arctan x^2+\dfrac{3^{\sqrt{x}}\cdot 2x}{1+x^4}$

c. $-5^{\sqrt{\cos(2x)}}\cdot\ln 5\left[\cos(2x)\right]^{-\frac{1}{2}}\cdot\sin(2x)$

d. $\frac{2(\sin x+\cos x)-\arcsin(2x)(\cos x-\sin x)\sqrt{1-4x^2}}{(\sin x+\cos x)^2\sqrt{1-4x^2}}$ **e.** $\frac{1+3x}{x}$

f. $-\frac{4(\arccos\frac{x}{3})^3}{\sqrt{9-x^2}}$ **g.** $4e^{4x+3}\sin(e^{x^2})+2xe^{4x+3}\cos(e^{x^2})e^{x^2}$

h. $-(6x^2e^{-x}-2x^3e^{-x})\sin(2x^3e^{-x})$ **i.** $\frac{x\arccos x-\sqrt{1-x^2}}{(1-x^2)^{\frac{3}{2}}}$

j. $3e^{\sin^2(3x)}\sin(6x)\cos(e^{-2x})+2e^{\sin^2(3x)-2x}\sin(e^{-2x})$ **k.** $\frac{-4\arctan\left(\frac{1}{2x}\right)}{4x^2+1}$

l. $3\left[2e^{2x}\cos(5x)-5e^{2x}\sin(5x)\right]\sin^2\left[e^{2x}\cos(5x)\right]\cos\left[e^{2x}\cos(5x)\right]$

m. $\frac{5\ln^4 x-2x\ln^5 x}{xe^{2x+3}}$ **n.** $2\cos(2^{\sin(2x)})\cdot 2^{\sin(2x)}\ln 2\cos(2x)$

o. $9e^{2-\cos^3(3x+5)}\cos^2(3x+5)\sin(3x+5)$ **p.** $\frac{1}{2^{\sqrt{x}}-e^{-2x}}(\frac{1}{2\sqrt{x}}2^{\sqrt{x}}\ln 2+2e^{-2x})$

q. $\frac{1}{2(1+x^5-x)}(x^5-x)^{-\frac{1}{2}}(5x^4-1)$ **r.** $-\frac{3}{(x+2)\sqrt{3(2x+1)}}$ **s.** $\frac{e^{\arcsin\sqrt{2x}}}{\sqrt{(1-2x)2x}}$

t. $-\sin\left(\frac{\arcsin x}{5}\right)\frac{1}{5\sqrt{1-x^2}}$ **u.** $3^{4^{x^3}}\ln 3\cdot 4^{x^3}\cdot\ln 4\cdot 3x^2$ **v.** $10^{3+x^4}\ln 10\cdot 4x^3$

w. $\frac{1}{9(x\arctan\sqrt[3]{x})^{\frac{2}{3}}\left[1+(\sqrt[3]{x})^2\right]}$ **z.** $\frac{-\sin(2x)}{\sqrt{1-\cos^4 x}}$ **8. a.** $-x^{\cos x}\sin x\cdot\ln x+\frac{x^{\cos x}\cos x}{x}$

b. $x\sqrt{\frac{1+x}{1-x}}(\frac{1}{x}+\frac{1}{2(1+x)}+\frac{1}{2(1-x)})$ **c.** $(\ln x)^x\left[\ln(\ln x)+\frac{1}{\ln x}\right]$

d. $\frac{1}{3}\sqrt[3]{\frac{x^2+3}{x+2}}(\frac{2x}{x^2+3}-\frac{1}{x+2})$ **e.** $(\tan x)^{\sin x}\left[\cos x\cdot\ln(\tan x)+\frac{1}{\cos x}\right]$

f. $(x^3+2)^{\frac{1}{4}}\sin^5 x\cdot 3^{x+2}\left[\frac{3x^2}{4(x^3+2)}+5\cot x+\ln 3\right]$

g. $x^{x^{x^3}} \cdot x^{x^3} \ln x(3x^2 \ln x + x^2 + \frac{1}{x \ln x})$ **h.** $(\sin x)^{\cos^2 x}\left[-\sin(2x)\cdot\ln(\sin x)+\frac{\cos^3 x}{\sin x}\right]$

i. $\frac{\sqrt{4x+3}}{5^{x^3}\cos(2x)}\left[\frac{2}{4x+3}-3x^2\ln 5+2\tan(2x)\right]$

j. $(2x+3)^5(4x-1)^7(x^3+2)^8(\frac{10}{2x+3}+\frac{28}{4x-1}+\frac{24x^2}{x^3+2})$

9. a. $y'=\frac{3x^2}{1+x^3}$, $y''=\frac{6x-3x^4}{(1+x^3)^2}$ **b.** $y'=4x^3+5$, $y''=12x^2$

c. $y'=e^{-2x}-2xe^{-2x}$, $y''=-4e^{-2x}+4xe^{-2x}$ **d.** $y'=\frac{-y\cos x-\sin(x-y)}{\sin x-\sin(x-y)}$,

$$y''=\frac{-\cos(x-y)(\sin x+y\cos x)^2+2\cos x\left[y\cos x+\sin(x-y)\right]\left[\sin x-\sin(x-y)\right]}{\left[\sin x-\sin(x-y)\right]^3}$$

$$+\frac{y\sin x\left[\sin x-\sin(x-y)\right]^2}{\left[\sin x-\sin(x-y)\right]^3}$$

e. $y'=e^x+2x\cos(x^2)$, $y''=e^x+2\cos(x^2)-4x^2\sin(x^2)$

f. $y'=\frac{3x^2y}{3y^3+2}$, $y''=\frac{6xy+3x^2y'-9y^2(y')^2}{3y^3+2}$ **g.** $y'=y^{-2}+1$, $y''=-2y^{-3}(y^{-2}+1)$

h. $y'=\frac{-2x-2xy^2}{2y+2y^3-e^{\arctan y}}$,

$$y''=\frac{\left[-2-2y^2-4xyy'-2(y')^2-6y^2(y')^2\right](1+y^2)+e^{\arctan y}(y')^2}{(2y+2y^3-e^{\arctan y})(1+y^2)}$$

10. $y'=-\frac{1}{2\pi}$, $y''=\frac{1}{4\pi^2}$ **11. a.** $dy=8x^3dx$ **b.** $dy=\frac{1+x^2}{(1-x^2)^2}dx$

c. $dy=-e^{-x}(\cos x+\sin x)dx$ **d.** $dy=e^{\sin x}\cos x dx$ **e.** $dy=-\frac{y}{x}dx$

f. $dy=\frac{e^y}{1-xe^y}dx$ **12. a.** 0.99 **b.** 2.0017 **c.** 0.01 **d.** 1.05

13. a. $3^x(\ln 3)^n$ **b.** $(-1)^{n+1}(n-1)!(1+x)^{-n}$ **c.** $5^n e^{5x}$

d. $m(m-1)\cdots(m-n+1)(2+x)^{m-n} = \dfrac{m!}{(m-n)!}(2+x)^{m-n}$

Exercise 4.1

1. a. $\frac{5}{4}$ **b.** 0 **c.** 0 **d.** π **2. a.** 2.10 **b.** 0.523 **c.** 0 **d.** 8.92

4. $-38 \le f(x) \le 26$ **9.** $2.96202 < \sqrt[3]{26} < 2.96296$ **10.** $2.82322 < \sqrt{8} < 2.83333$

Exercise 4.2

1. a. 5 **b.** $-\infty$ **c.** 2 **d.** $\frac{1}{4}$ **e.** 0 **f.** $\dfrac{\ln 7 - \ln 2}{3}$

2. a. $-\frac{3}{5}$ **b.** $-\infty$ **c.** 1 **d.** 0 **e.** $\frac{1}{2}$ **f.** $\frac{4}{3}$

3. a. $-\frac{1}{2}$ **b.** 0 **c.** 1 **d.** 0 **e.** 0 **f.** 0 **4. a.** 1 **b.** 1

Exercise 4.3

1. (a) continuous **(b)** not differentiable **2.** continuous and differentiable

3. At $x = 0$, continuous and not differentiable. At $x = 1$, continuous and differentiable; At $x = 2$, not continuous and not differentiable.

4. $f(x)$ is continuous for $x \in$ R, $x \neq -1$. $f(x)$ is differentiable for $x \in$ R, $x \neq -1$, $x \neq 2$, $x \neq 3$. **5.** $a = -12$, b = 16 **6.** $a = -1$, b = 0 **7.** $a = 1$

10. (a) $a = -5$, b = 6 **(b)** $f(x)$ is differentiable for $x \in$ R, $x \neq 2$ and $x \neq 4$.

Exercise 5.1

1. a. tangent: $y = 13x - 12$ normal: $y = -\dfrac{1}{13}x + \dfrac{184}{13}$

b. tangent: $y = -x + 2$ normal: $y = x$

c. tangent: $y=-2e^{-3}x+9e^{-3}$ normal: $y=\frac{e^3}{2}x+\frac{6-3e^6}{2e^3}$

d. tangent: $y=-2x$ normal: $y=\frac{1}{2}x-\frac{5\pi}{2}$

e. tangent: $y=-\frac{\sqrt{3}}{3}x+\frac{4\sqrt{3}}{3}$ normal: $y=\sqrt{3}x$

f. tangent: $y=(\ln 5+1)x-5$ normal: $y=-\frac{1}{\ln 5+1}x+5\ln 5+\frac{5}{\ln 5+1}$

g. tangent: $y=0.326x+1.43128$ normal: $y=-3.067x+4.82428$

h. tangent: $y=-\frac{1}{2}x+1$ normal: $y=2x+1$

i. tangent: $y=-\frac{13}{10}x+\frac{33}{10}$ normal: $y=\frac{10}{13}x+\frac{16}{13}$

j. tangent: $y=\frac{2-\pi}{4}x+\frac{\pi-1}{2}$ normal: $y=\frac{4}{\pi-2}x+\frac{\pi}{4}-\frac{4}{\pi-2}$

k. tangent: $y=-\frac{4}{5}x+\frac{42}{5}$ normal: $y=\frac{5}{4}x+\frac{9}{4}$

l. tangent: $y=-\frac{1}{6}x+\frac{13}{6}$ normal: $y=6x-4$

m. tangent: $y=\frac{-1}{2e+1}x+\frac{1}{2e+1}$ normal: $y=(2e+1)x-2e-1$

n. tangent: $y=\frac{12}{47}x+\frac{82}{47}$ normal: $y=-\frac{47}{12}x+\frac{71}{12}$

2. $y=21.3628x-91$ **3.** $(7+\sqrt{5},-\sqrt{10}-5\sqrt{2})$ **4.** $a=-1, b=5$

5. $y=0$, $y=\frac{(\sqrt[3]{16})^2}{2}$ **6.** (-1.25, 3.375) **7.** (1, 3) **8.** $y=2x+3$

9. $y=\frac{1}{2}x+1$ **10.** $y=\frac{4}{5}x+\frac{6}{5}$ **11.** tangent: $y=\frac{2\pi}{3}$ normal: $x=\frac{\pi}{3}$ **12.** $\frac{24}{5}$

Exercise 5.2

1. a. interval of increase $-1 < x < 0$, $x > 1$; interval of decrease $x < -1$, $0 < x < 1$

b. interval of increase $0 < x < \frac{\pi}{6}$, $\frac{3\pi}{2} < x < 2\pi$, $\frac{\pi}{2} < x < \frac{5\pi}{6}$;

interval of decrease $\frac{\pi}{6} < x < \frac{\pi}{2}$, $\frac{5\pi}{6} < x < \frac{3\pi}{2}$

c. interval of increase $x < 1, x > 2$; interval of decrease $1 < x < 2$

d. interval of increase $x < -2$, $x > 0$; interval of decrease $-2 < x < -1$, $-1 < x < 0$

e. interval of increase $x > -\frac{5}{2}$; interval of decrease $x < -\frac{5}{2}$

f. interval of increase $x \in R$; interval of decrease none

2. $x > e$ **3.** $\pi < x < 2\pi$ **4.** $-\sqrt{2} < x < 1$, $1 < x < \sqrt{2}$

Exercise 5.3

1. a. A local maximum is -5 at $x = 0$. A local minimum is -9 at $x = 2$.

b. A local maximum is $\frac{4}{27}$ at $x = \frac{7}{3}$. A local minimum is 0 at $x = 3$.

c. A local maximum is 25.1 at $x = -\frac{1}{7}$. A local minimum is 0 at $x = 5$.

d. A local maximum is 10.3 at $x = \frac{-9 + \sqrt{41}}{2}$.

Local minimums are 0 at $x = 0$ and -777 at $x = \frac{-9 - \sqrt{41}}{2}$.

e. A local minimum is -0.404 at $x = -1.16$.

f. A local minimum is 0 at $x = 0$. A local maximum is 0.541 at $x = 2$.

g. A local minimum is -1 at $x = -1$. A local maximum is 1 at $x = 1$.

h. A local maximum is 1.5 at $x = \frac{1}{2}$. **i.** A local minimum is 3.93 at $x = -\sqrt[3]{\frac{3}{2}}$.

j. A local minimum is $2\sqrt{2}$ at $x = \frac{1}{\sqrt{2}}$. A local maximum is $-2\sqrt{2}$ at $x = -\frac{1}{\sqrt{2}}$.

k. A local minimum is -3 at $x = 1$. **l.** A local minimum is -0.25 at $x = \frac{1}{16}$.

2. $(\frac{-2+\sqrt{13}}{9}, 3.90)$ is a local minimum. $(\frac{-2-\sqrt{13}}{9}, 4.67)$ is a local maximum.

3. (2, 0) and (-2, 0) are local minimums. (0, 16) is a local maximum.

4. interval of increase $-1 < x < 1$; interval of decrease $x < -1, x > 1$; A local maximum is 3 at $x = 1$. A local minimum is -1 at $x = -1$.

5. $a = 2$. Local minimums are $\frac{5}{3}$ at $x = \frac{\pi}{2}$, $-\sqrt{3}$ at $x = \frac{5\pi}{3}$ and $-\sqrt{3}$ at $x = \frac{4\pi}{3}$.

Local maximums are $-\frac{5}{3}$ at $x = \frac{3\pi}{2}$, $\sqrt{3}$ at $x = \frac{\pi}{3}$ and $\sqrt{3}$ at $x = \frac{2\pi}{3}$.

6. A local minimum is 1 at $x = 0$. A local maximum is $\frac{\pi}{2}$ at $x = \frac{\pi}{2}$. Global minimum is -1 at $x = \pi$. **7.** $f(x)$ has extreme value at $x = x_0$ and it is local minimum.

8. x_0 is an extreme value of $f(x)$ and it is local maximum.

Exercise 5.4

1. The maximum height is 13.1 m when t = 1.8125s.

2. When 14.2 (thousands) items are sold, the maximum profit is 2720 euros.

3. a. $\frac{80-4x}{8}$ **b.** $\frac{3(80-4x)}{8}$ **c.** $\frac{3(80-4x)^2}{64}$ **d.** $\frac{60}{7} \approx 8.57$ cm

4. 100 m width, 200 in length **5.** 1830 cm^3 **6.** 36 cm^2 **7.** (1.06, 1.18)

8. width is $\frac{32}{3}$ and length is $\frac{8\sqrt{3}}{3}$. **9.** $(1, \frac{3}{2})$ **10.** $\frac{4\sqrt{5}R}{5}$ **11.** $32\sqrt{3}$

12. $a = 45$, $b = -30$ **13.** $\frac{3R}{2}$ **14.** 108 m^2 **15.** 71.4

16. The radius of the base circle is $\sqrt[3]{\frac{150}{\pi}}$ and the height is $\frac{300}{\pi(\frac{150}{\pi})^{\frac{2}{3}}}$.

17. The radius of the semicircle, the width and the length of the rectangle are $\dfrac{20}{\pi+4}$.

18. 60° **19.** $\dfrac{12\sqrt{21}}{5}$ **20.** 7.5 **21.** $r = h = \sqrt[3]{\dfrac{3V}{5\pi}}$ **22.** Both width and length are equal to $\sqrt{A}$. **23.** height $= 2\sqrt{\dfrac{1}{3}}R$, volume $= \dfrac{4\sqrt{3}\pi R^3}{9}$

24. AD = 30, AF = 40 **25.** AB = BC $= \sqrt{2}r$ **26.** 16.2

27. a. $r = \dfrac{h}{2}(\sin x + \cos x - 1)$ **b.** 45° **28.** $P(\dfrac{a}{\sqrt{2}}, \dfrac{b}{\sqrt{2}})$, ab

29. width = length = 10 **30.** $3\sqrt{3}a$ **31.** $\dfrac{\sqrt{2}}{2}a$

Exercise 5.5

1. $x < -\dfrac{\sqrt{3}}{3}$, $x > \dfrac{\sqrt{3}}{3}$ **2.** concave up: $-1 < x < 1$, points of inflexion: (-1, ln2) & (1, ln2). **3. a.** (0, 0), $(\dfrac{3\sqrt{10}}{10}, -1.02)$, $(-\dfrac{3\sqrt{10}}{10}, 1.02)$ **b.** (0, 0) is a point of inflexion. $(\dfrac{3\sqrt{10}}{10}, -1.02)$ is a local minimum. $(-\dfrac{3\sqrt{10}}{10}, 1.02)$ is a local maximum. **c.** (0, 0) is a horizontal point of inflexion. $(\dfrac{3\sqrt{5}}{10}, -0.634)$ & $(-\dfrac{3\sqrt{5}}{10}, 0.634)$ are non-horizontal points of inflexion. **d.** $-\dfrac{3\sqrt{5}}{10} < x < 0$, $x > \dfrac{3\sqrt{5}}{10}$

e.

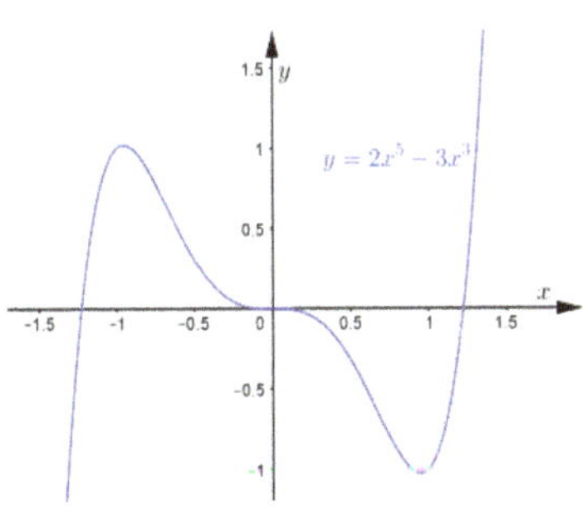

4. a. (1, 1) is a local maximum. (-1, -1) is a local minimum. Points of inflexion are $(-\sqrt{3},-\frac{\sqrt{3}}{2})$, (0, 0) and $(\sqrt{3},\frac{\sqrt{3}}{2})$. The interval the curve is concave down: $x<-\sqrt{3}$, $0<x<\sqrt{3}$.

b. (0, 0) is a local minimum. (-1, ln2) and (1, ln2) are non-horizontal points of inflexion. Concave down intervals: $x<-1,\ x>1$

c. $(2,2\sqrt{2})$ is a local minimum. $(6,\frac{4\sqrt{6}}{3})$ is a non-horizontal point of inflexion. Concave down interval: $x>6$

d. (-1, 0) is a local minimum. $(\frac{5}{3},9.48)$ is a local maximum. $(\frac{1}{3},\frac{128}{27})$ is a non-horizontal point of inflexion. Concave down interval: $x>\frac{1}{3}$

e. (0, 0) is a local minimum. (1, e^{-2}) is a local maximum. (0.293, 0.0478) & (1.71, 0.0959) are non-horizontal points of inflexion. Concave down interval: $0.293<x<1.71$

f. (0, 1) is a local minimum. No point of inflexion. No interval for concave down.

g. No stationary point. (0, 0) is non-horizontal point of inflexion. Concave down interval: $x<0$

h. (0, 1) is a local maximum. $(-\frac{\sqrt{2}}{2},e^{-\frac{1}{2}})$ & $(\frac{\sqrt{2}}{2},e^{-\frac{1}{2}})$ are non-horizontal points of inflexion. Concave down interval: $-\frac{\sqrt{2}}{2}<x<\frac{\sqrt{2}}{2}$

i. (0, -1) is a local maximum. No point of inflexion. The function is concave down for $x\in$ R.

j. (0, 0) is a local minimum. No point of inflexion. No interval for concave down.

5. a. $a=-3$, $b=-5$, $c=-6$ **b.** $-1<x<6$ **c.** No point of inflexion

6. a. $(1+\sqrt{5},0.254)$ is a local maximum. $(1-\sqrt{5},-8.51)$ is a local minimum.

b. $(2+\sqrt{6}, 0.185)$ & $(2-\sqrt{6}, -5.95)$ **c.** $2-\sqrt{6} < x < 2+\sqrt{6}$

7. a. (0, 3), (π, -1) (2π, 3), $(\frac{2\pi}{3}, -1.5)$, $(\frac{4\pi}{3}, -1.5)$

b. (0.936, 0.890), (2.57, -1.26), (3.71, -1.26), (5.35, 0.890) **c.** $0.936 < x < 2.57$, $3.71 < x < 5.35$ **8. a.** $(\sqrt{e^{-2}}, 2\sqrt{e^{-2}})$ is a local maximum. $(-\sqrt{e^{-2}}, -2\sqrt{e^{-2}})$ is a local minimum. **b.** No point of inflexion **c.** $x < 0$

9.

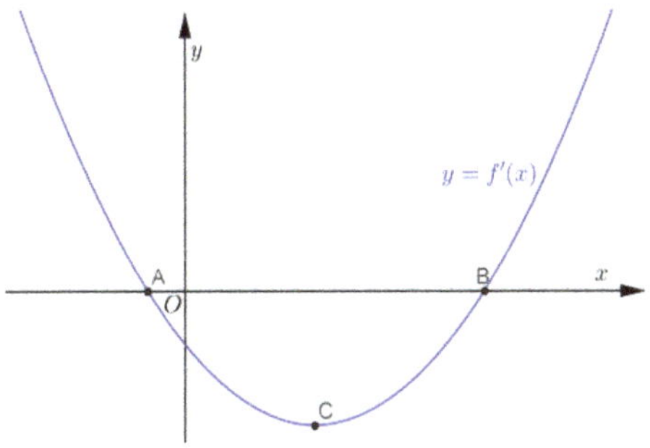

A: local maximum

B: local minimum

C: point of inflexion

10.

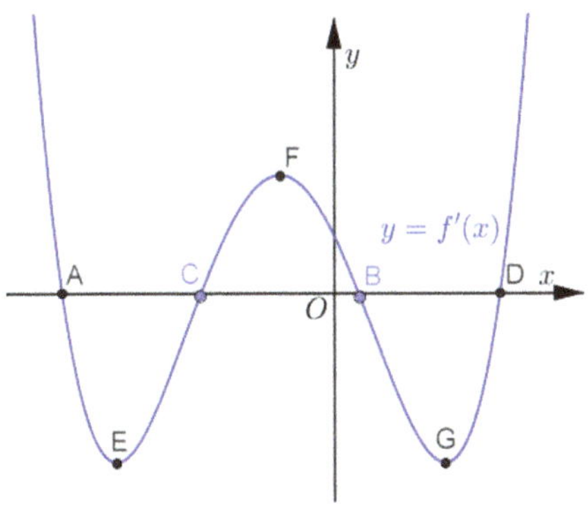

A & B: local maximum

C & D: local minimum

E, F & G: points of inflexion

11.

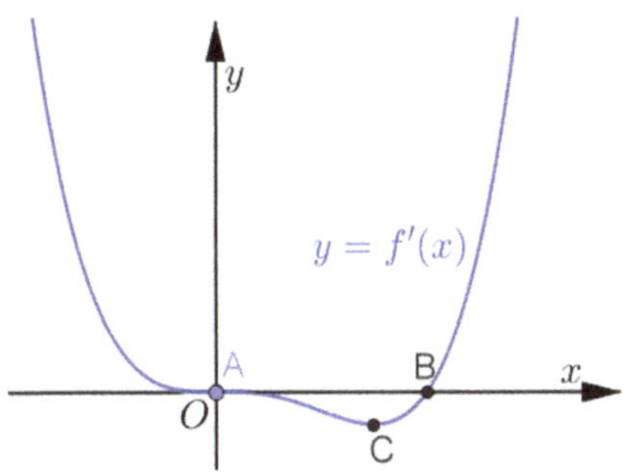

A: local maximum

B: local minimum

C: point of inflexion

Exercise 5.6

1. a. horizontal asymptote $y = 0$ **b.** No asymptote **c.** vertical asymptote $x = -1$

d. horizontal asymptote $y = 1$ **e.** horizontal asymptote $y = 1$, vertical asymptote $x = 4$

f. horizontal asymptote $y = \frac{3}{5}$, vertical asymptote $x = -\frac{1}{5}$

g. vertical asymptote $x = 0$, oblique asymptote $y = 2x$

h. horizontal asymptote $y = 6$, vertical asymptote $x = -3$

i. vertical asymptote $x = 0$, oblique asymptote $y = -x$

j. horizontal asymptote $y = 1$, vertical asymptotes $x = 1$ & $x = 2$

2. a.

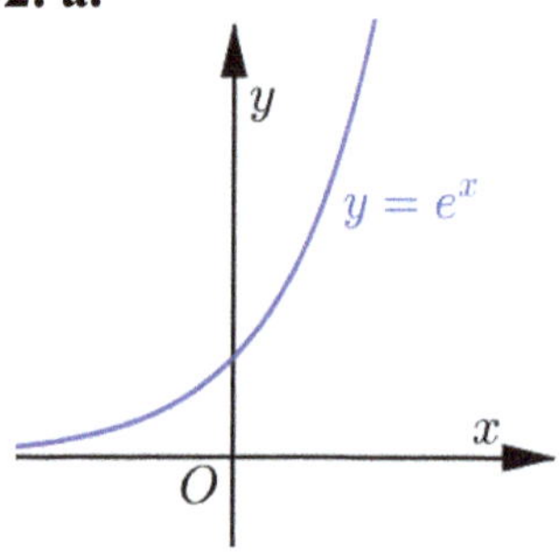

b.

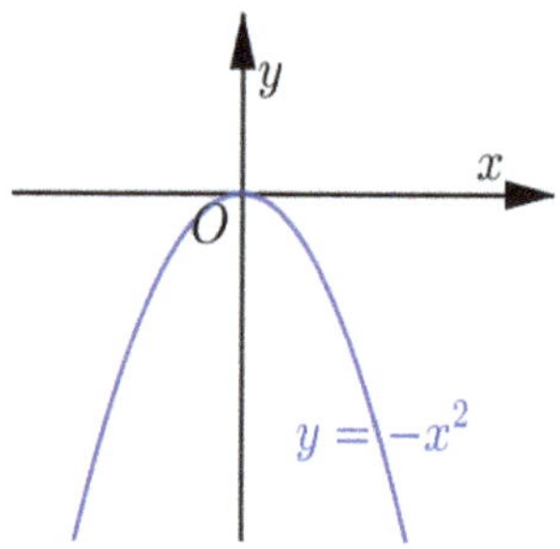

c.

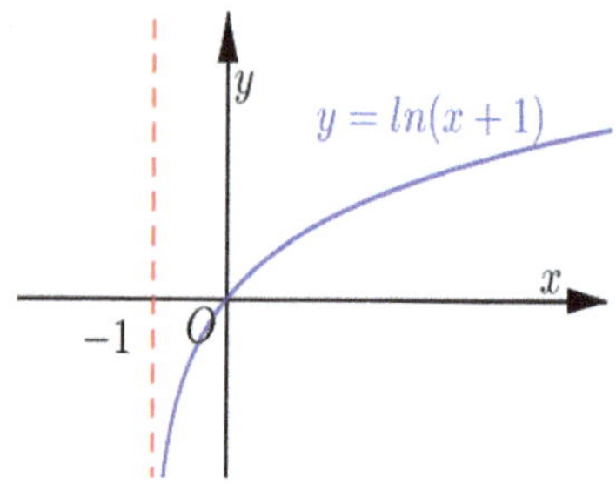

d.

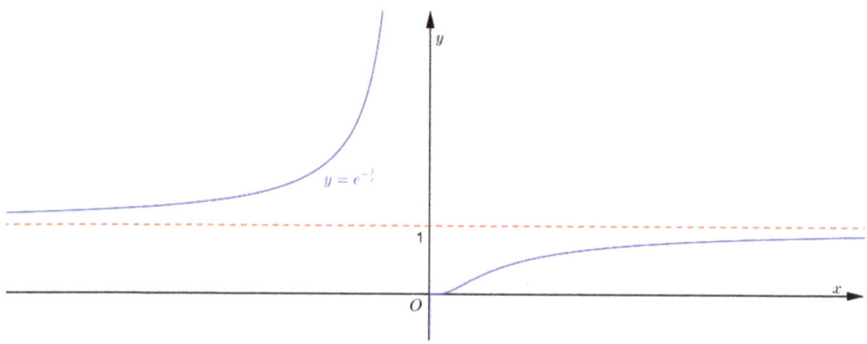

e.

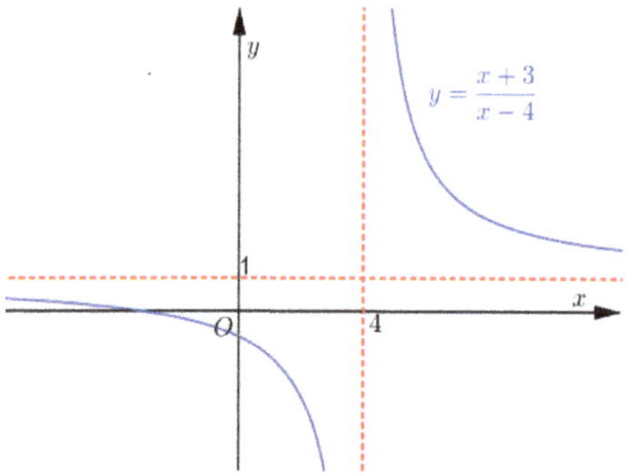

f.

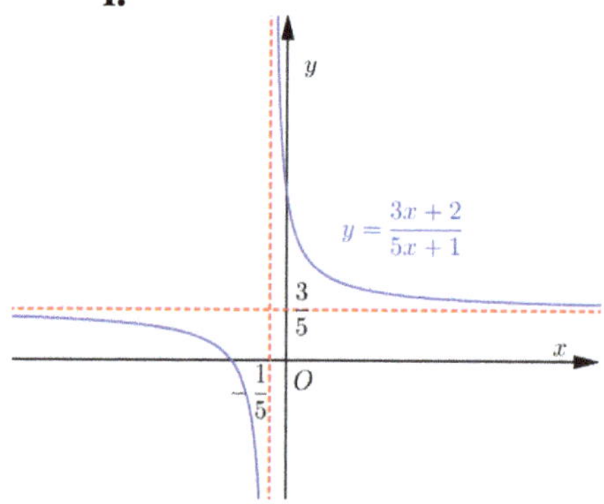

g.

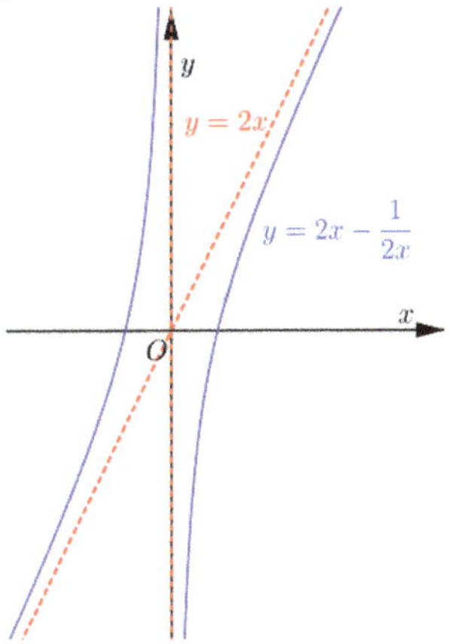

y
y = 2x
$y = 2x - \frac{1}{2x}$
x
O

h.

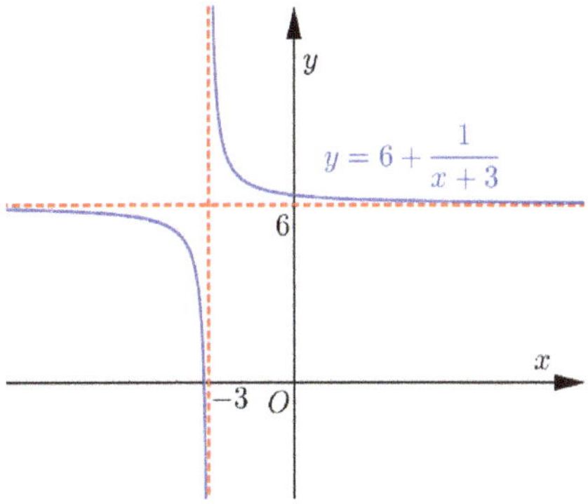

y
$y = 6 + \frac{1}{x+3}$
6
x
−3
O

i.

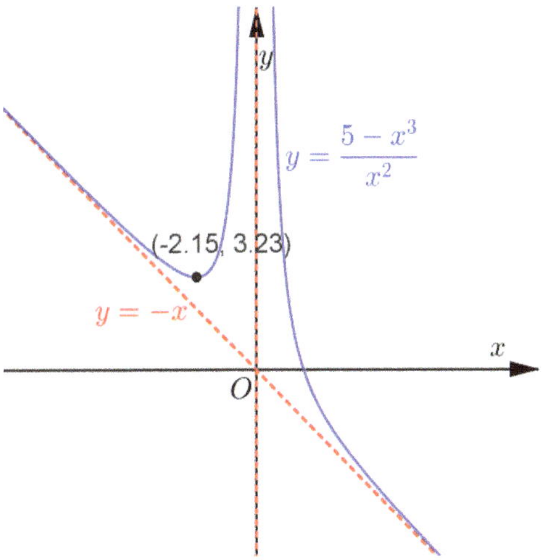

y
$y = \frac{5 - x^3}{x^2}$
(-2.15, 3.23)
y = −x
x
O

j.

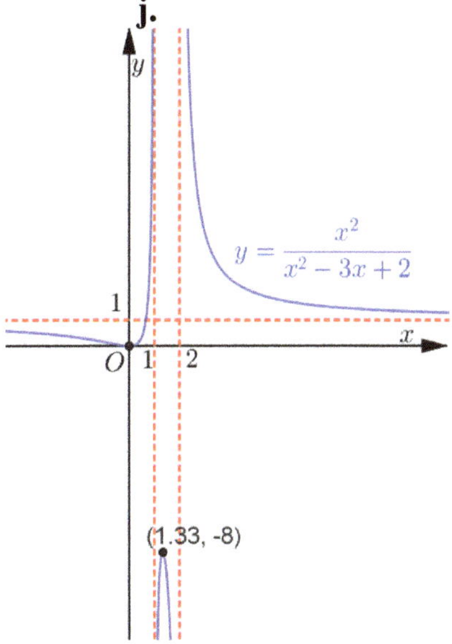

y
$y = \frac{x^2}{x^2 - 3x + 2}$
1
x
O
1
2
(1.33, -8)

3. a.

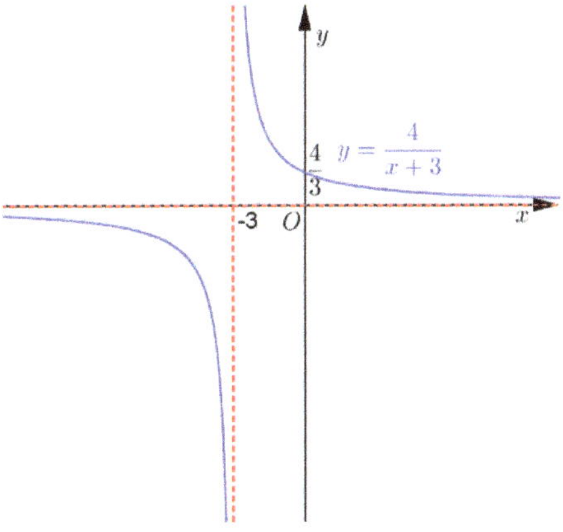

y
$\frac{4}{3}$
$y = \frac{4}{x+3}$
-3
O
x

b.

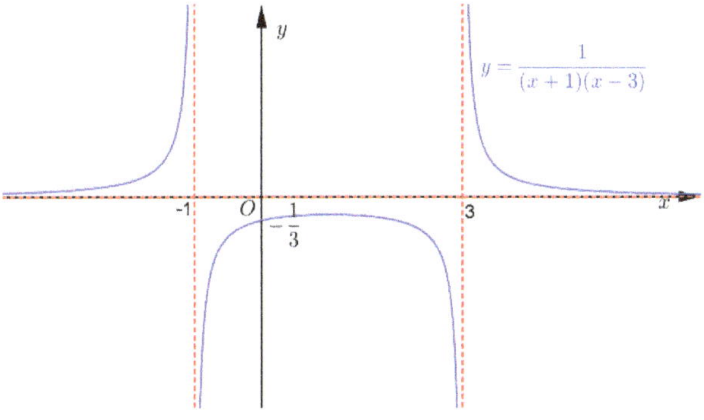

y
$y = \frac{1}{(x+1)(x-3)}$
-1
O
$-\frac{1}{3}$
3
x

c.

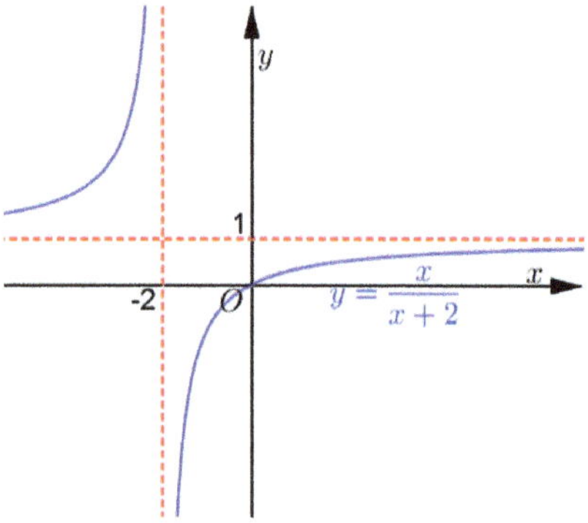
y
1
-2
O
x
$y = \frac{x}{x+2}$

d.

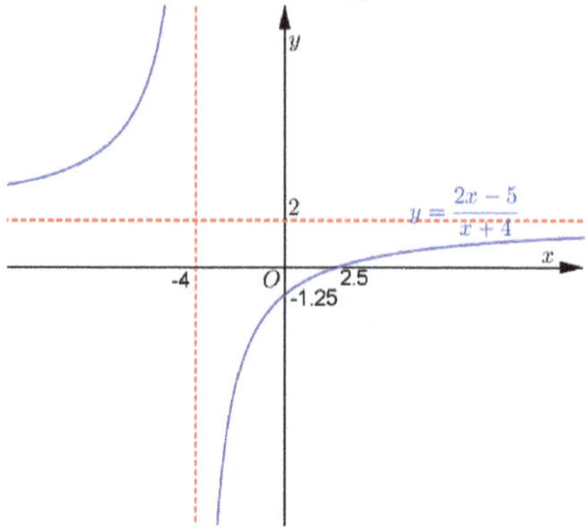
y
2
$y = \frac{2x-5}{x+4}$
-4
O
2.5
-1.25
x

e.

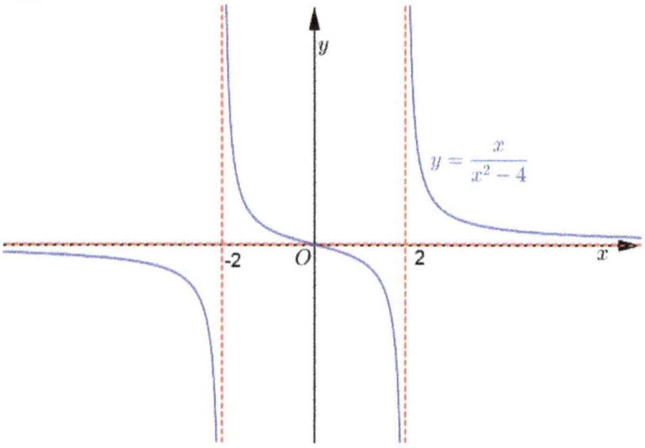
y
$y = \frac{x}{x^2-4}$
-2
O
2
x

f.

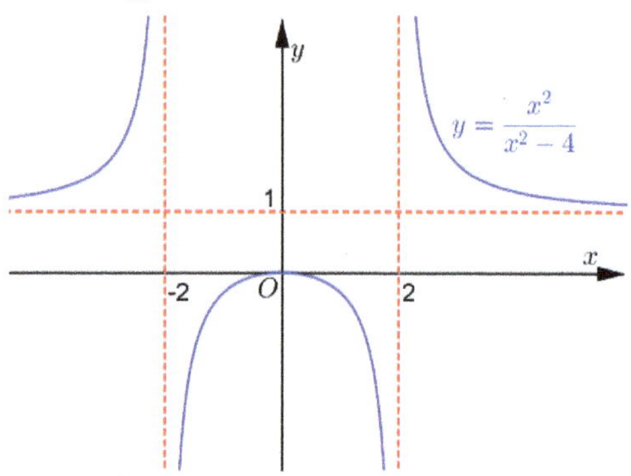
y
$y = \frac{x^2}{x^2-4}$
1
x
-2
O
2

g.

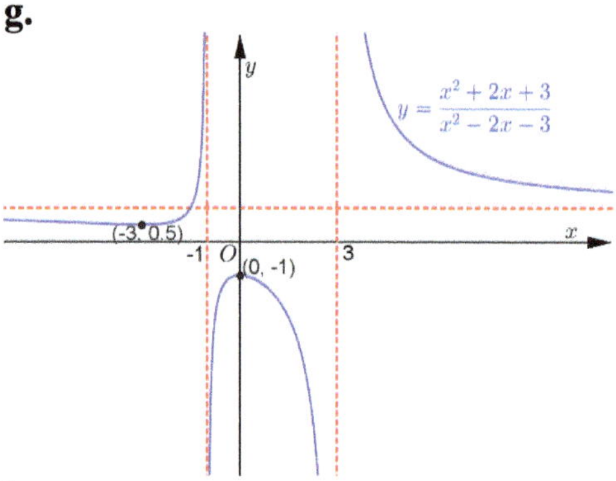
y
$y = \frac{x^2+2x+3}{x^2-2x-3}$
(-3, 0.5)
x
-1
O
(0, -1)
3

h.

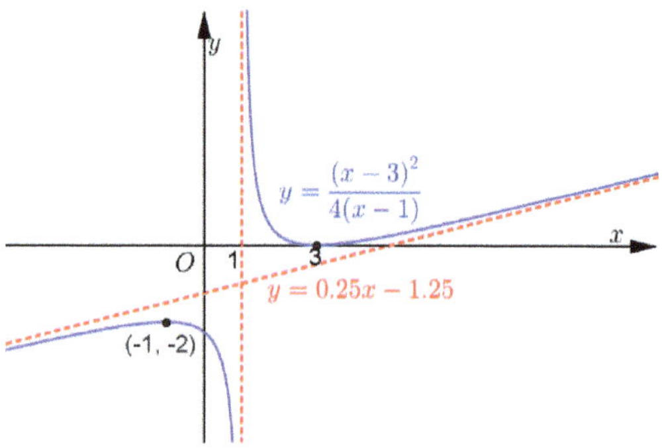
y
$y = \frac{(x-3)^2}{4(x-1)}$
x
O
1
3
$y = 0.25x - 1.25$
(-1, -2)

i.

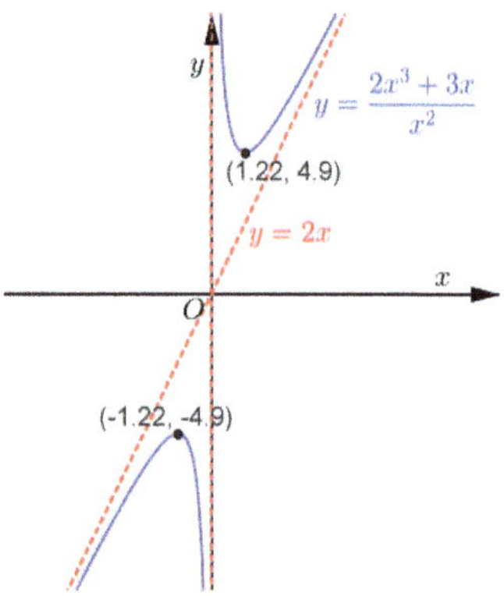
y
$y = \frac{2x^3+3x}{x^2}$
(1.22, 4.9)
$y = 2x$
x
O
(-1.22, -4.9)

j.

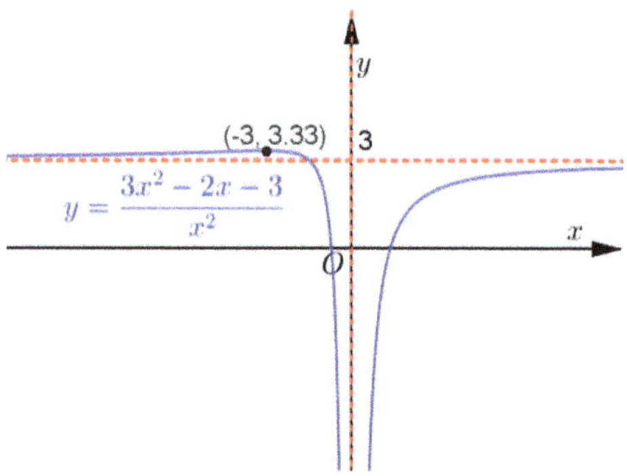
y
(-3, 3.33)
3
$y = \frac{3x^2-2x-3}{x^2}$
x
O

4. $a = 2$, $k = 5$,

5. a. (i) (0, 0) **(ii)** horizontal asymptote $y = 0$, vertical asymptotes $x = 1$ & $x = -1$

(iii) (-0.223, -0.192) **(iv)** $x < -1$, $-0.223 < x < 1$, $x > 1$

b.

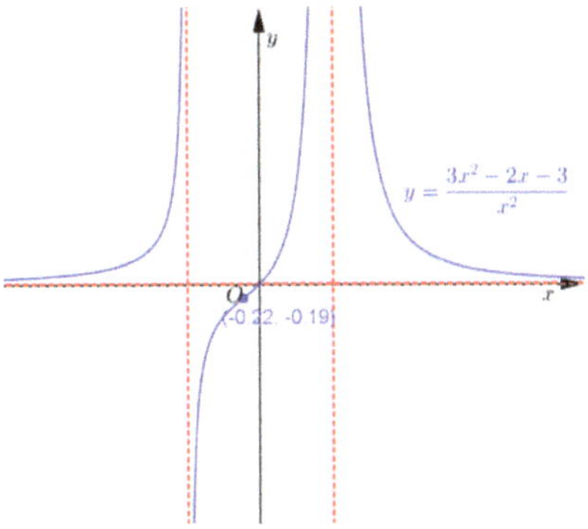

6. a. (i) (0, 0) **(ii)** vertical asymptote $x = 1$, oblique asymptote $y = x + 2$

(iii) (0, 0) is a horizontal point of inflexion, a local minimum is at (3, 6.75).

(iv) (0, 0)

b.

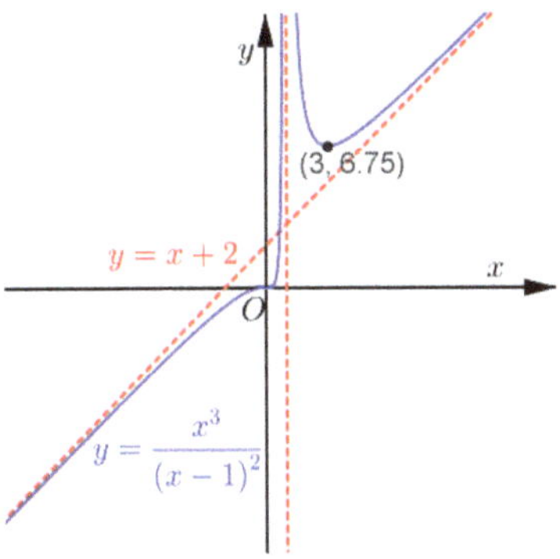

7. a. (i) x-intercepts are (-1, 0) & (2, 0), y-intercept: (0, 4)

(ii) A local maximum is at (0, 4). A local minimum is at (2, 0). **(iii)** (1, 2)

b.

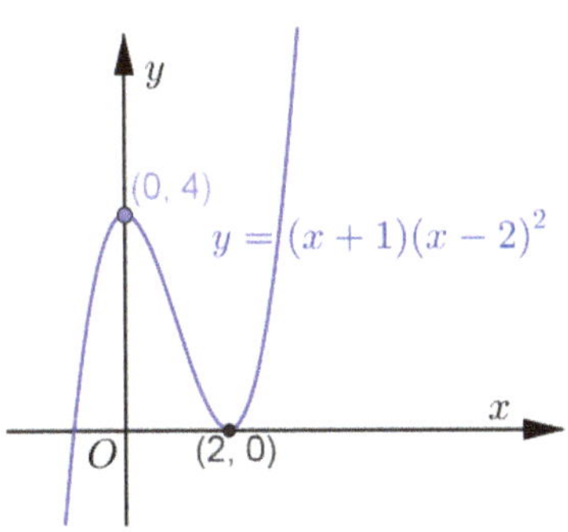

8. a. (i) (0, 0) **(ii)** vertical asymptote $x = -1$, oblique asymptote $y = x - 3$

(iii) A local minimum is at (0, 0). A local maximum is at $(-4, -\frac{256}{27})$.

(iv) no point of inflexion

b.

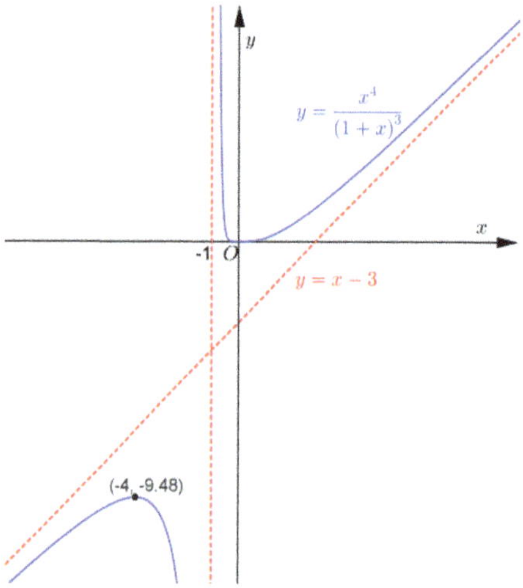

9. a. (i) x-intercepts are (-3, 0) & (3, 0), y-intercept: $(0, \frac{9}{16})$

(ii) horizontal asymptote $y = 1$, vertical asymptotes $x = 4$ & $x = -4$

(iii) A local maximum is at $(0, \frac{9}{16})$ **(iv)** no point of inflexion

b.

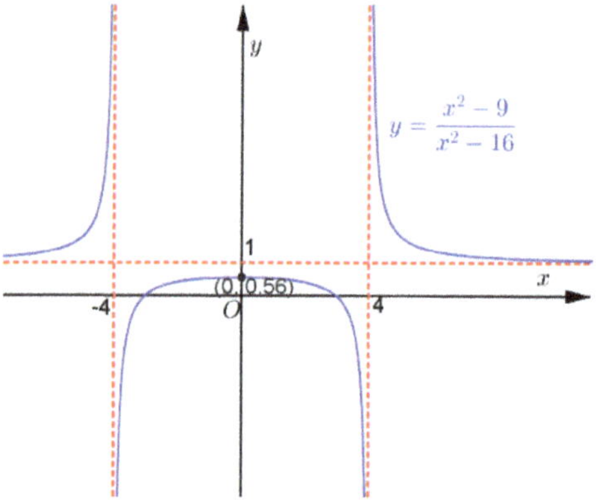

10. a. (i) x-intercepts are (1, 0) & (4, 0), y-intercept: (0, -2)

(ii) vertical asymptote $x = 2$, oblique asymptote $y = x - 3$

(iii) no stationary point **(iv)** no point of inflexion

b.

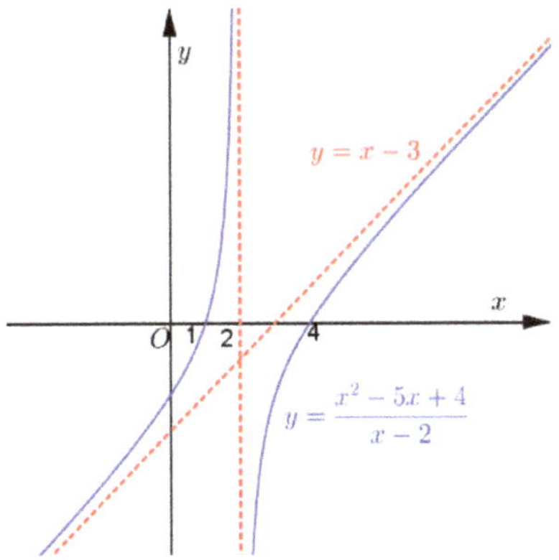

11. a. (i) x-intercepts are (-1, 0) & (0.5, 0), no y-intercept

(ii) vertical asymptote $x = 0$, horizontal asymptote $y = 2$

(iii) A local maximum is at (1, 4). A local minimum is at (-1, 0).

(iv) Points of inflexions are $(\sqrt{2}, 3.77)$, $(-\sqrt{2}, 0.232)$.

b.

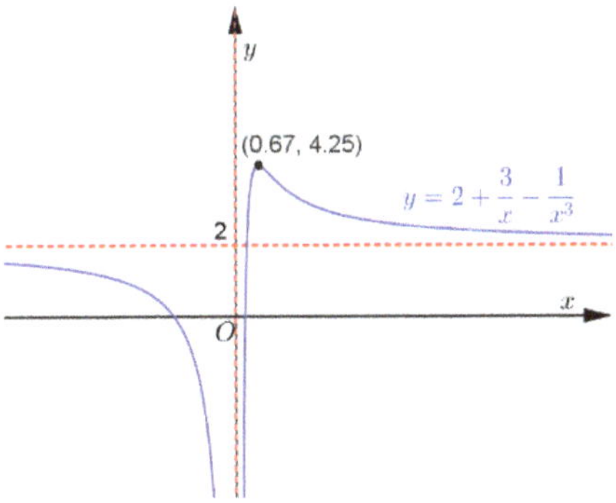

12. a.

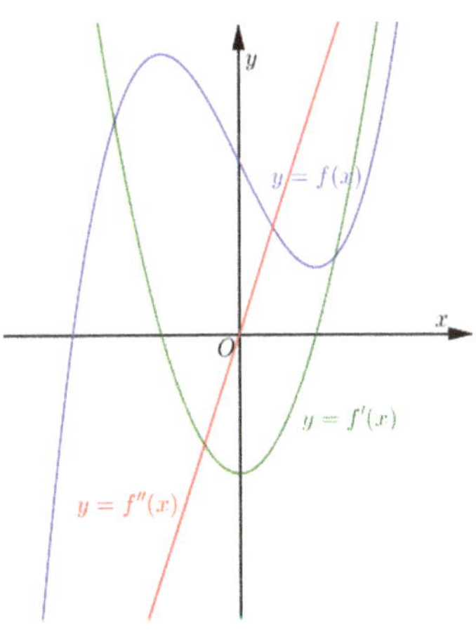

b.

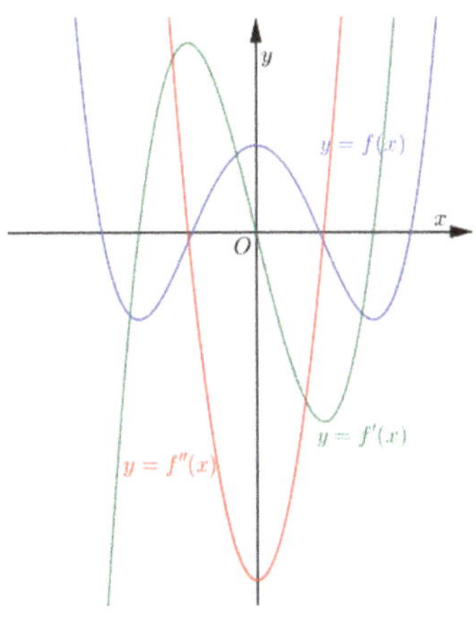

c.

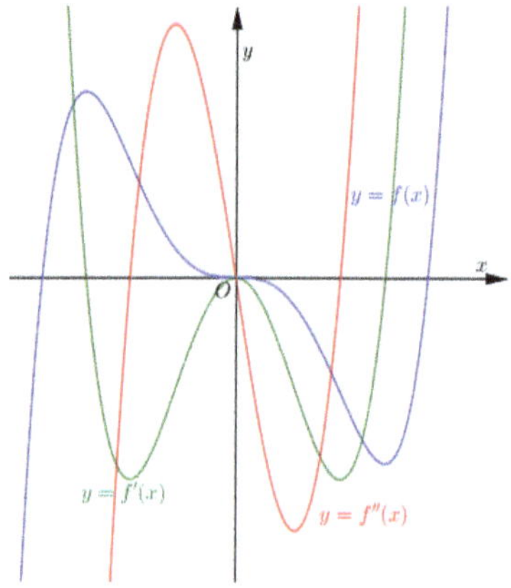

d.

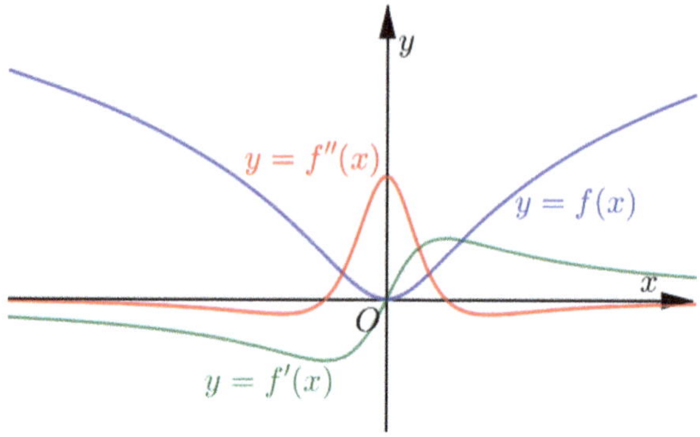

e.

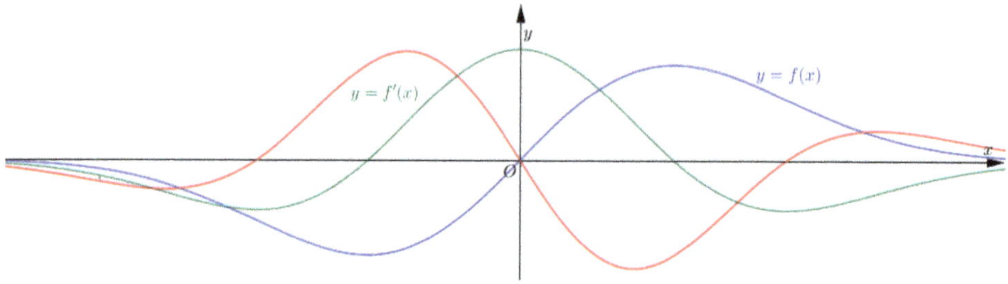

f.

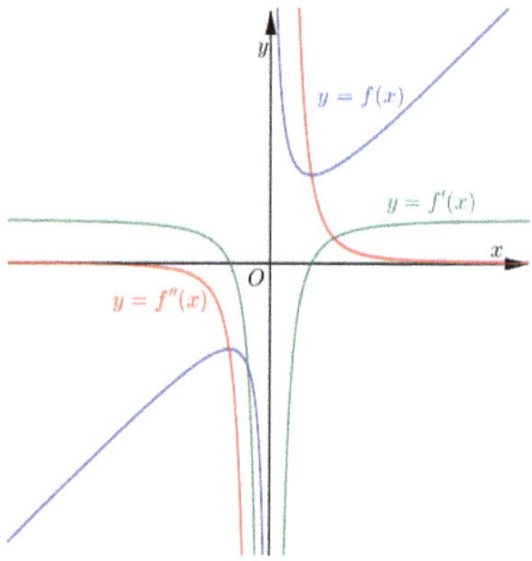

g.

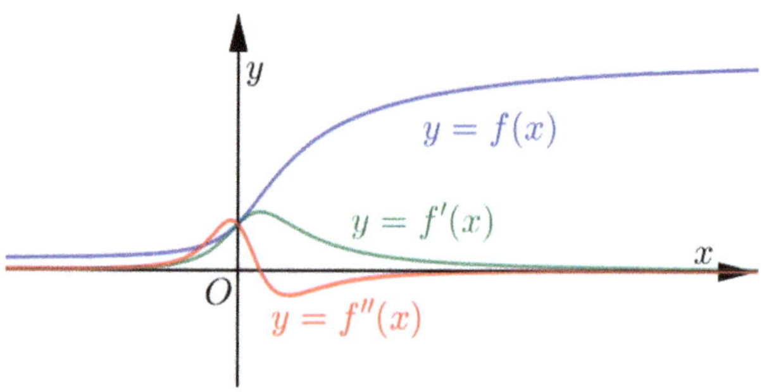

Exercise 5.7

1. a. increasing at 10π cm/s **b**. increasing at 80π cm^2/s

2. a. decreasing at 1200 cm^2/s **b.** decreasing at 384 cm^3/s

c. decreasing at $2\sqrt{3}$ cm/s **3.** increasing at 2.27 cm/s

4. Surface area is decreasing at 160π cm^2/s and volume is decreasing at 1600π cm^3/s.

5. rising at $\dfrac{1}{2\pi}$ m/minute **6.** increasing at 16 rad/hour **7.** increasing at $\dfrac{1}{30}$ km/s

8. 75π m³/minute **9. a.** increasing at $\frac{160}{3}$ m/minute **b.** increasing at $\frac{400}{3}$ m/minute

10. $a = \frac{2\sqrt{3}}{3}$ **11.** $\sqrt{5}$ m/s **12.** $\left.\frac{dy}{dt}\right|_{y\approx-0.647} \approx 2.99$, $\left.\frac{dy}{dt}\right|_{y\approx-1.72} \approx -4.46$

13. increasing at 42 cm²/minute **14.** 548 km/minute

15. The area is increasing at 0.25 m²/s. The diagonal is increasing at 0.004 m/s.

16. The distance is increasing at 50 km/h. **17.** The distance is decreasing at 2.8 km/h.

Exercise 6.1

1. a. $\frac{x^4}{4}+\frac{x^3}{3}-2x+C$ **b.** $\frac{5x^2}{2}+\ln|x|+\frac{1}{2x^2}+C$ **c.** $\frac{x^3}{3}-\frac{3x^2}{2}+2x+C$

d. $6x^{\frac{2}{3}}+C$ **e.** $9x-2x^3+\frac{x^5}{5}+C$ **f.** $-\cos x+\sqrt{3}\sin x+\frac{3}{4}x^{\frac{4}{3}}+C$

g. $3e^x-5\tan x+2\arcsin x+C$ **h.** $8x-12\ln|x|-\frac{6}{x}+\frac{1}{2x^2}+C$ **i.** $\frac{8}{5}x^{\frac{5}{2}}+C$

j. $-\frac{2}{3}x^{-\frac{3}{2}}+C$ **k.** $\frac{4}{13}x^{\frac{13}{4}}+C$ **l.** $\frac{3x^2}{2}+2\ln|x|-\frac{5}{2x^2}+C$

2. a. $-\frac{2}{5x^5}-\frac{3}{4x^4}+C$ **b.** $\frac{6}{5}x^{\frac{5}{2}}+2x^{-\frac{1}{2}}+C$ **c.** $-\frac{2\sqrt{5}}{3}x^{-\frac{3}{2}}+C$

d. $\ln|x|-\frac{2}{3}x^{-\frac{3}{2}}+C$ **e.** $x^2-x-\ln|x|+C$ **f.** $\frac{6}{13}x^{\frac{13}{6}}+\frac{6}{7}x^{\frac{7}{6}}+C$

g. $\frac{x^3}{3}-2x-\frac{1}{x}+C$ **h.** $\frac{10}{13}x^{\frac{13}{10}}-\frac{10}{7}x^{\frac{7}{5}}+\frac{5}{4}x^{\frac{4}{5}}+C$ **i.** $-2x^{-\frac{1}{2}}+6x^{\frac{1}{2}}+2x^{\frac{3}{2}}-\frac{2}{5}x^{\frac{5}{2}}+C$

j. $4e^x+\tan x+\arctan x+C$ **k.** $7x-\frac{3^x}{\ln 3}+C$ **l.** $6\sec x-2\ln|x|+C$

3. $f(x)=\frac{x^2}{2}+\frac{1}{x}-\frac{1}{2}$ **4.** $f(x)=\frac{5}{3}x^3-\frac{3}{4}x^{\frac{4}{3}}+5$ **5.** $f(x)=\frac{3}{2}x^2-2x-44$

6. $f(x)=\frac{1}{2}x^2+2x+5$ **7.** $y=2e^x-\frac{3}{2}x^2+6-e^2$ **8.** $g(x)=-4x^2-2x+5$

9. $h(x)=-\frac{1}{2}x^3+3x^2+\frac{3}{2}x+3$ **10. a.** $f(x)=\frac{1}{6}x^3-\frac{1}{2}x^2-\frac{1}{2}x+\frac{7}{6}$

10. b. $f(x)=\frac{1}{6}x^3-\cos x+2x+2$ **c.** $f(x)=\frac{1}{2}e^x+\sin x+\frac{5}{2}$

d. $f(x)=-3\cos x-\sin x+5x+1-\pi$

Exercise 6.2

1. a. $\frac{(2x+5)^{11}}{22}+C$ **b.** $-\frac{3}{16}(1-4x)^{\frac{4}{3}}+C$ **c.** $-\frac{4}{3}\sqrt{3-6x}+C$

d. $-\frac{1}{10(5x+3)^2}+C$ **e.** $\frac{1}{7}\sin(7x+1)+C$ **f.** $\cos(2-3x)+C$

g. $\ln\left|1+x^2\right|+C$ **h.** $\frac{3}{2}\ln\left|2x+7\right|+C$ **i.** $\frac{1}{4}(1+x^3)^{\frac{4}{3}}+C$ **j.** $-\frac{1}{2}e^{-x^2}+C$

k. $-\frac{2}{25(x^5+3)^5}+C$ **l.** $-e^{\cos x}+C$ **m.** $\frac{\sin^8 x}{8}+C$ **n.** $\frac{2}{3}e^{x^3+6}+C$

o. $\frac{1}{81}(6x+7)^{\frac{9}{4}}-\frac{7}{45}(6x+7)^{\frac{5}{4}}+C$ **p.** $\frac{(1+\ln x)^4}{4}+C$ **q.** $\ln\left|x+\cos x\right|+C$

r. $\frac{1}{2}(\arctan x)^2+C$ **2. a.** $\frac{2}{3}(x-1)^{\frac{3}{2}}+2\sqrt{x-1}+C$ **b.** $\ln\left|x^2-4x+3\right|+C$

c. $-\frac{1}{3}e^{-x^3}+C$ **d.** $\frac{1}{4}(\ln x)^4+C$ **e.** $\frac{1}{3}(x^2+3)^{\frac{3}{2}}+C$ **f.** $-\frac{3}{4}\ln\left|5-2x^2\right|+C$

g. $-\frac{1}{3}e^{-3x}-e^{-x}+C$ **h.** $-\frac{1}{\arcsin x}+C$ **i.** $\frac{2}{5}e^{\sin(5x)}+C$

j. $\frac{1}{2}\ln\left|x^2+1\right|-\arctan x+C$ **k.** $\ln\left|\ln x\right|+C$ **l.** $-\frac{1}{4\sin^2(2x+\frac{\pi}{3})}+C$

m. $\frac{3}{2}\arcsin(\frac{2x}{3})+C$ **n.** $\frac{5}{32}\arctan(\frac{x}{2})+C$

3. a. $-\frac{7}{24}(5-3x)^{\frac{8}{7}}+C$ **b.** $-\frac{3}{4}(4x+6)^{-\frac{2}{3}}+C$ **c.** $-\sqrt{1-x^2}+C$ **d.** $\cos\frac{1}{x}+C$

e. $2\ln\left|3+e^x\right|+C$ **f.** $\frac{2}{3}(\cos x)^{-\frac{3}{2}}+C$ **g.** $\frac{1}{2}e^{\tan(2x)}+C$ **h.** $-\ln\left|1+\cos x\right|+C$

i. $-\frac{5}{16}(1-2x)^{\frac{4}{5}}+\frac{5}{36}(1-2x)^{\frac{9}{5}}+C$ **j.** $-\frac{3}{4}(\cos\frac{x}{3})^4+C$ **k.** $\frac{2}{7}(5-x)^7-\frac{13}{6}(5-x)^6+C$

l. $\frac{2}{5}(x+3)^{\frac{5}{2}}+\frac{2}{3}(x+3)^{\frac{3}{2}}+C$ **m.** $\frac{1}{4}e^{4x-6}-2e^{2x-3}+4x+C$ **n.** $\frac{1}{2}\sqrt{9+x^4}+C$

o. $\frac{1}{3}(x^2+16)^{\frac{3}{2}}-16(x^2+16)^{\frac{1}{2}}+C$ **p.** $\frac{1}{9}\tan^3(3x)+\frac{1}{15}\tan^5(3x)+C$

4. a. $-\frac{2}{27}(3x-1)^{-3}-\frac{11}{36}(3x-1)^{-4}+C$ **b.** $\frac{1}{3}\arctan(3x)+C$

c. $\frac{7}{20}\arctan(\frac{4}{5}x)+C$ **d.** $\arcsin(\frac{x}{4})+C$ **e.** $\frac{1}{4}\ln\left|x^4+4x\right|+C$

f. $x+e^{-2x}-\frac{1}{4}e^{-4x}+C$ **g.** $\frac{1}{2}\sqrt{e^{4x}-2}+C$ **h.** $-\frac{1}{3}e^{-3x}-5e^{-2x}-25e^{-x}+C$

i. $\ln\left|x^2-x\right|+C$ **j.** $\frac{5}{2}\sin(x^2-8)+C$ **k.** $\sqrt{x^2-2\sin x}+C$

l. $\frac{4}{7}(5+\sin x)^7+C$ **m.** $\frac{1}{8}x-\frac{1}{32}\sin(4x)+C$ **n.** $\frac{(\cos x)^8}{8}-\frac{(\cos x)^6}{6}+C$

o. $(\tan\frac{x}{2})^2+\frac{1}{2}(\tan\frac{x}{2})^4+C$ **p.** $-\frac{1}{2\left[1+\tan(2x+5)\right]}+C$

5. a. $\frac{1}{20}\tan(5x^4)+C$ **b.** $-2\cos(\sqrt{x})+C$ **c.** $x-\sin x+C$

d. $-\frac{1}{4}\ln\left|\cos(4x)\right|+C$ **e.** $\frac{1}{3}\ln\left|\sin(3x)\right|+C$ **f.** $\ln\left|\sin x\right|+C$ **g.** $\frac{1}{12}\sin^6(2x)+C$

h. $-\frac{1}{2}\cos(2x)+\frac{1}{6}\cos^3(2x)+C$ **i.** $-\frac{1}{15}\cos^5(3x)+C$ **j.** $-\frac{1}{3}\cos^3 x+C$

k. $\frac{1}{12}\sin(6x)+\frac{1}{2}x+C$ **l.** $\frac{1}{4}\left(\frac{1}{8}\sin(4x)+\frac{3}{2}x+\sin(2x)\right)+C$

m. $-\frac{2}{5}\cos^5 x+C$ **n.** $-\frac{4}{3}\cos^3(\frac{x}{2})+C$ **o.** $\frac{1}{10}\tan^2(5x)+C$

p. $\frac{1}{8}\sin(8x)+x+C$ **6. a.** $\frac{1}{2}\sin(4x-\pi)+2x+C$ **b.** $\frac{\tan^6(x+\frac{\pi}{3})}{6}+C$

c. $\frac{1}{3}\arcsin(3\tan x)+C$ **d.** $\frac{10}{3}\sqrt{\sin(3x)}-\frac{2}{3}[\sin(3x)]^{\frac{5}{2}}+C$ **e.** $\frac{\cos^7 x}{7}-\frac{\cos^5 x}{5}+C$

f. $\frac{3}{4}(\sin x)^{\frac{4}{3}}-\frac{3}{10}(\sin x)^{\frac{10}{3}}+C$ **g.** $-\frac{5}{2}\cos(\frac{2x}{5})-\frac{5}{6}\cos^3(\frac{2x}{5})+C$

h. $2\left[\frac{\sin^3(3x+1)}{3}-\frac{\sin^5(3x+1)}{5}\right]+C$ **i.** $\sec^3 x+C$ **j.** $\cos^2 x-2\ln|\cos x|+C$

k. $3x-\frac{3}{2}\sin(2x)+C$ **l.** $8(\frac{\cos^5(\frac{x}{2})}{5}-\frac{\cos^3(\frac{x}{2})}{3})+C$

7.a. $\frac{x\sqrt{9-x^2}}{2}+\frac{9}{2}\arcsin(\frac{x}{3})+C$ **b.** $\sqrt{x^2+16}+C$ **c.** $\sqrt{x^2-16}-4\arccos(\frac{4}{x})+C$

d. $\frac{x}{2(x^2+1)}+\frac{1}{2}\arctan x+C$ **e.** $-\sqrt{25-x^2}+C$ **f.** $\sqrt{x^2-4}+C$

g. $-\frac{1}{3}\left(\frac{\sqrt{1+x^2}}{x}\right)^3+\frac{\sqrt{1+x^2}}{x}+C$ **h.** $\sqrt{x^2-9}-3\arccos(\frac{3}{x})+C$ **i.** $\frac{x}{\sqrt{1-x^2}}+C$

j. $\frac{x}{a^2\sqrt{x^2+a^2}}+C$ **k.** $\frac{a^2}{2}\arcsin(\frac{x}{a})-\frac{x}{2}\sqrt{a^2-x^2}+C$ **l.** $-\frac{1}{3a^2}\left(\frac{\sqrt{a^2-x^2}}{x}\right)^3+C$

Exercise 6.3

1. a. xe^x-e^x+C **b.** $x\sin x+\cos x+C$ **c.** $-\frac{\ln x}{x}-\frac{1}{x}+C$

d. $\frac{x^2}{2}\ln(1+x)-\frac{x^2}{4}+\frac{x}{2}-\frac{1}{2}\ln|1+x|+C$ **e.** $-\frac{x^2+1}{2}e^{-x^2}+C$

f. $-\frac{x}{2}\cos(2x)+\frac{1}{4}\sin(2x)+C$ **g.** $\frac{5}{34}e^{3x}\sin(5x)+\frac{3}{34}e^{3x}\cos(5x)+C$

h. $-\frac{1}{3}(4x+9)\cos(3x)+\frac{4}{9}\sin(3x)+C$

i. $\frac{2}{3}x^{\frac{3}{2}}\ln^2 x-\frac{8}{9}x^{\frac{3}{2}}\ln x+\frac{16}{27}x^{\frac{3}{2}}+C$

j. $-\frac{1}{3}(2x-1)e^{-3x}-\frac{2}{9}e^{-3x}+C$

k. $\frac{1}{2}(6x+7)\sin(2x)+\frac{3}{2}\cos(2x)+C$

l. $x\ln(4x)-x+C$

2. a. $\frac{1}{3}x\sin(3x)+\frac{1}{9}\cos(3x)+C$

b. $-24x\cos(\frac{x}{6})+144\sin(\frac{x}{6})+C$

c. $\frac{x^2}{2}\ln(3x)-\frac{1}{4}x^2+C$

d. $-x^2e^{-x}-2xe^{-x}-2e^{-x}+C$

e. $-\frac{1}{3}x^2\cos(3x)+\frac{2}{9}x\sin(3x)+\frac{2}{27}\cos(3x)+C$

f. $\frac{x^4}{4}\ln^2 x-\frac{x^4}{8}\ln x+\frac{x^4}{32}+C$

g. $x\ln^3 x-3x\ln^2 x+6x\ln x-6x+C$

h. $x\tan x+\ln|\cos x|+C$

i. $-\frac{1}{8}e^{-4x}\left[\cos(4x)+\sin(4x)\right]+C$

j. $\frac{1}{2}x^3\sin(2x)+\frac{3}{4}x^2\cos(2x)-\frac{3}{4}x\sin(2x)-\frac{3}{8}\cos(2x)+C$

k. $\sin x\ln(\sin x)-\sin x+C$

l. $\frac{x^3}{3}\arctan(\frac{x}{2})-\frac{1}{3}(4+x^2-4\ln|4+x^2|)+C$

3. a. $\frac{2}{3}\sin x\sin(2x)+\frac{1}{3}\cos x\cos(2x)+C$

b. $e^x\sin(e^x)+\cos(e^x)+C$

c. $-\frac{1}{5}x^2\cos(5x)+\frac{2}{25}x\sin(5x)+\frac{2}{125}\cos(5x)+C$

d. $-x\cot x+\ln|\sin x|+C$

e. $\frac{x^3}{3}\arcsin x+\frac{1}{3}(1-x^2)^{\frac{1}{2}}-\frac{1}{9}(1-x^2)^{\frac{3}{2}}+C$

f. $-\frac{x^4}{2}\cos(2x)+x^3\sin(2x)+\frac{3}{2}x^2\cos(2x)-\frac{3}{2}x\sin(2x)-\frac{3}{4}\cos(2x)+C$

g. $\frac{1}{2}e^{2x}\arctan(e^x)-\frac{1}{2}e^x+\frac{1}{2}\arctan(e^x)+C$

h. $x(\arcsin x)^2+2\sqrt{1-x^2}\arcsin x-2x+C$

i. $2\sqrt{x}\arctan\sqrt{x}-\ln|1+x|+C$

j. $-2\sqrt{x+3}\cos\sqrt{x+3}+2\sin\sqrt{x+3}+C$

k. $\frac{2}{3}(\sqrt{3x+5}\sin\sqrt{3x+5}+\cos\sqrt{3x+5})+C$

l. $2(\sqrt{x})^3e^{\sqrt{x}}-6xe^{\sqrt{x}}+12\sqrt{x}e^{\sqrt{x}}-12e^{\sqrt{x}}+C$

m. $x\arccos(2x)-\frac{1}{2}\sqrt{1-4x^2}+C$

n. $\frac{x}{2}\left[\cos(\ln x)+\sin(\ln x)\right]+C$

4. a. $\frac{x^3}{3}\arctan x-\frac{x^2}{6}+\frac{1}{6}\ln\left(1+x^2\right)+C$

b. $\frac{x}{2}\left[\sin(\ln x)-\cos(\ln x)\right]+C$

c. $e^x\sin^2 x-\frac{1}{5}\left[e^x\sin(2x)-2e^x\cos(2x)\right]+C$

d. $\frac{1}{4}x^2+\frac{1}{4}x\sin(2x)+\frac{1}{8}\cos(2x)+C$

e. $-2\sqrt{1-x}\arcsin x+4\sqrt{1+x}+C$

f. $8\arcsin(\frac{x}{4})-\frac{x\sqrt{16-x^2}}{4}+C$

g. $x\tan x+\ln|\cos x|+C$

h. $-\frac{1}{4}x\cos(2x)+\frac{1}{8}\sin(2x)+C$

Exercise 6.4

1. a. $\frac{9}{7}\ln|x-2|+\frac{5}{7}\ln|x+5|+C$

b. $2\ln|x+3|+\ln|x-2|+C$

c. $-\frac{3}{2}\ln|x+4|+\frac{3}{2}\ln|x-4|+C$

d. $-3\ln|x-2|+8\ln|x-5|+C$

e. $-\frac{10}{3}\ln|3x-1|+\frac{7}{2}\ln|2x-1|+C$

f. $\frac{19}{85}\ln|5x-2|+\frac{3}{17}\ln|x+3|+C$

g. $-\frac{2}{3}\ln|x+2|+\frac{1}{2}\ln|x+3|+\frac{1}{6}\ln|x-1|+C$

h. $-\frac{10}{49}\ln|2x+1|-\frac{2}{49}\ln|x-3|+\frac{61}{49}\ln|x+4|+C$

i. $\frac{8}{7}\ln|x^2+3|+\frac{11}{7\sqrt{3}}\arctan\left(\frac{x}{\sqrt{3}}\right)+\frac{12}{7}\ln|x-2|+C$

j. $2\ln|x+3|+\frac{3}{x+3}+C$

k. $5\ln|x-4|-\frac{23}{x-4}+C$

l. $x+\ln|x+4|+2\ln|x-2|+C$

2. a. $\frac{31}{39}\ln|3x-1|-\frac{6}{13}\ln|x+4|+C$ **b.** $\ln|4x+1|-\ln|x+1|+C$

c. $\frac{7}{10}\ln|2x-3|-\frac{7}{10}\ln|4x-1|+C$ **d.** $-\frac{9}{24}\ln|x+3|-\frac{1}{4}\ln|x-3|+\frac{5}{8}\ln|x-1|+C$

e. $\frac{13}{10}\ln|x-3|-\frac{3}{2}\ln|x-1|+\frac{1}{5}\ln|x+2|+C$

f. $\frac{2}{9}\ln|x^2+5|-\frac{37}{9\sqrt{5}}\arctan\left(\frac{x}{\sqrt{5}}\right)+\frac{14}{9}\ln|x+7|+C$

g. $\frac{3}{2}x^2-13x+\frac{20}{9}\ln|x-2|+\frac{880}{9}\ln|x+7|+C$ **h.** $\frac{2}{x+1}+\ln|x+2|+C$

i. $\frac{2}{9}\ln|x|-\frac{2}{9}\ln|x-3|-\frac{17}{3(x-1)}+C$

j. $-\frac{1}{13}\left[\frac{17+37\sqrt{17}}{34}\ln\left|x-\frac{-5+\sqrt{17}}{2}\right|+\frac{17-37\sqrt{17}}{34}\ln\left|x-\frac{-5-\sqrt{17}}{2}\right|\right]+\frac{14}{13}\ln|x-3|+C$

k. $-\frac{1}{6}\ln|x^2+1|+\frac{1}{3}\arctan x+\frac{7}{6}\ln|x^2+4|-\frac{1}{6}\arctan\left(\frac{x}{2}\right)+C$

l. $\ln|x-1|+\ln|x^2+1|-3\arctan x+C$

3. a. $2\ln|1+x|+C$ **b.** $1+x-\ln|1+x|+C$ **c.** $\frac{5}{2}\ln|1+x^2|+C$ **d.** $4\arctan x+C$

e. $3\ln|1+x|+\frac{3}{x+1}+C$ **f.** $\frac{2}{3}(1+x)^{\frac{3}{2}}-2\sqrt{1+x}+C$ **g.** $\frac{4}{1-x}+C$

h. $\frac{5}{2}\ln|1+x|-\frac{5}{2}\ln|1-x|+C$ **i.** $-\frac{7}{2}\ln|1-x^2|+C$ **j.** $3\arcsin x+C$

k. $-2\sqrt{1-x^2}+C$ **l.** $14\sqrt{1+x}+C$

4. a. $\frac{1}{4}x^2-\frac{5}{4}x+\frac{27}{88}\ln|2x-3|+\frac{64}{11}\ln|x+4|+C$

b. $\frac{1}{2}x^2-4x-\frac{3}{4}\ln|x|-\frac{3}{4x}+\frac{51}{4}\ln|x+2|+\frac{29}{4(x+2)}+C$

c. $\frac{3}{4}\ln|x-3|-\frac{5}{x-3}+\frac{1}{4}\ln|x+1|+C$

d. $\frac{2}{5}\ln|x|+\frac{3}{5}\ln|x+5|+C$ **e.** $\frac{4}{3}\ln|x+3|+\frac{5}{3}\ln|x-3|+C$

f. $\frac{1}{9}\ln|x-2|+\frac{44}{9}\ln|x+7|+C$ **g.** $-\frac{15}{7}\ln|x+5|+\frac{15}{7}\ln|x-2|+C$

h. $-\frac{1}{2}\ln|x-3|-\frac{2}{x-3}+\frac{1}{2}\ln|x+1|+C$ **i.** $-3\ln|x+3|+5\ln|x+2|-\ln|x+5|+C$

j. $\ln|x^2+7|+\frac{3}{\sqrt{7}}\arctan\left(\frac{x}{\sqrt{7}}\right)-\ln|x+5|+C$ **k.** $3\ln|x|+\frac{2}{x}+2\ln|x+1|+\frac{1}{x+1}+C$

l. $-31\ln|x-1|+\frac{17}{x-1}+34\ln|x-2|+C$

Exercise 7.1

1. a. 14.1 **b.** $\sqrt{3}\tan^{-1}\left(\frac{\sqrt{3}}{2}\right)$ **c.** $\ln\frac{4}{3}$ **d.** 4 **e.** 65

f. $8\sqrt{3}+6$ **g.** $\frac{11}{6}$ **h.** 4 **2. a.** 8 **b.** -3 **c.** -6 **d.** $\frac{16}{3}$

3. k = 0 or k = 1 **4.** k = 1

Exercise 7.2

1. a. $-\frac{16}{49}$ **b.** $\frac{1}{3}\left[(\ln 4)^3-(\ln 2)^3\right]$ **c.** $\frac{31}{10}$ **d.** $\frac{28\sqrt{5}-20}{3}$

e. $\frac{14\sqrt{14}-10\sqrt{10}}{3}$ **f.** $\frac{7}{204}$ **g.** $\frac{1}{4}\ln\frac{5}{3}$ **h.** $\frac{9}{64}$

i. $-\frac{1}{4}(\cos\frac{81\pi^4+48}{16}-\cos\frac{\pi^4+768}{256})$ **j.** $6+12\ln 2$ **k.** $\frac{1}{12}$ **l.** $\frac{e^4}{2}-\frac{1}{2e^4}+4$

m. $\frac{1}{3}(\sin e^6-\sin e^{-6})$ **n.** $\frac{1}{6}(\arctan 2-\arctan\frac{4}{3})$

2. a. $2(\ln 2)^2 - 2\ln 2 + \frac{3}{4}$ **b.** $\frac{1}{2}\ln\frac{e}{2}$ **c.** π **d.** 4π

e. 1 **f.** $\frac{2\pi}{3} - \frac{\sqrt{3}}{2}$ **g.** $\frac{1}{5}(e^{\pi} - 2)$ **h.** $4\ln 4 - 4$ **i.** $\ln 2$

j. $\sin 1$ **k.** $\frac{2}{3}\ln\frac{5}{4}$ **l.** $\frac{1}{4}e^2 + \frac{1}{4}$ **m.** $\frac{8 - 5\sqrt{2}}{12}$ **n.** $\frac{3e^{\frac{\pi}{2}} - 2}{5}$

3. $\frac{\pi a^4}{16}$ **4.** $\frac{12 - 4\sqrt{3} - \pi}{12}$ **5.** $\frac{2+\pi}{4}$ **6.** $\frac{e}{2}(\sin 1 - \cos 1) + \frac{1}{2}$

7. $\frac{\pi a^2}{4}$ **8.** $\frac{\pi}{4} - \frac{1}{2}$ **9.** $\sqrt{2} - \frac{2\sqrt{3}}{3}$ **10.** $\frac{\pi}{4}$

11. b. $\frac{263}{315} - \frac{\pi}{4}$ **12. c.** $3.00217596 < \pi < 3.30239355$

13. a. $p = -e^{-1}$ **b.** $I_5 = 120 - 326e^{-1}$ **c.** $-82e^{-1}$

Exercise 7.3

1. $]0, 1.82\pi]$ **2. a.** $\sqrt{1+x}$ **b.** $-xe^{-x}$ **c.** $\frac{2x}{\sqrt{1+x^8}}$

d. $2xe^{x^2} - 3x^2e^{x^3}$ **e.** $\sin\sqrt{x}$ **f.** $\frac{2x\cos x^2}{1+\sqrt{\sin x^2}} - \frac{3(x+1)^2}{1+\sqrt{(x+1)^3}}$

3. Maximum is $F(0) = 0$. Minimum is $F(4) = -\frac{32}{3}$.

4. a. $\frac{4\pi^2}{3}$ **b.** $\frac{10\sqrt{2}}{3}$ **c.** 6

Exercise 7.4.1

1. a. $\frac{3}{2} - \ln 2$ **b.** 7.81 **c.** 6.93 **d.** $\frac{13}{3}$ **e.** $\frac{8}{3}$ **f.** $e + \frac{1}{e} - 2$

g. $\frac{\pi}{2} - 1$ **h.** $\frac{28}{3}$ **2. a.** 0.950 **b.** $\frac{4}{3}$ **c.** 1 **d.** 3.08

e. 3.11 **f.** 10.1 **g.** $\frac{40\sqrt{10}}{3}a^3$ **h.** $\frac{11}{6}$ **i.** 1.26 **j.** $\frac{16}{3}$

3. $\frac{1}{2}$ **4.** 6.32 **5.** 4 **6. a.** A (-2, 0), B (0, $\sqrt{2}$)

b. $y=-2\sqrt{2}x+\sqrt{2}$ **c.** C ($\frac{1}{2}$, 0) **d.** 2.24 **7.** $\frac{13}{4}$ **8.** $\frac{27}{4}$ **9.** 5.68

10. 9 **11.** $\frac{243}{32}$ **12.** 18 **13.** $\frac{8}{3}$ **14.** A ($\frac{16}{3}$, $\frac{256}{9}$)

15. $y=\frac{1}{4}x+\ln 4-1$ **16.** 5 **17. a.** 1 **b.** 3.60 **18. a.** $e^{-1}-e^{-3}$

b. $e^{-1}+e^{-3}$ **19. a.** $a=-\frac{4}{5}, b=3$ **b.** $\frac{225}{32}$ **20. a.** $a=e^{-1}$,

$(x_0, y_0)=(e^2,1)$ **b.** $\frac{1}{6}e^2-\frac{1}{2}$

Exercise 7.4.2

1. a. a^2 **b.** $\frac{1}{4}a^2(e^{4\pi}-1)$

Exercise 7.4.3

1. a. **i.** $\frac{8}{3}\pi$ **ii.** $\frac{8}{3}\pi$ **b.** **i.** 7.5π **ii.** $\frac{62}{5}\pi$

c. **i.** $\frac{\pi^2}{4}$ **ii.** π **d.** **i.** $\frac{128}{7}\pi$ **ii.** $\frac{32}{5}\pi$

e. **i.** 4.5π **ii.** 2.25π **f.** **i.** $\frac{88}{15}\pi$ **ii.** $\frac{7}{12}\pi$

2. 3π **3. a.** $\frac{4}{3}\pi ab^2$ **b.** $\frac{4}{3}\pi a^2 b$

4. a. 325π **b.** 63π **c.** 168π **d.** 522π

5. a. A (1, 1) **b.** $y=2x-1$ **c.** $\frac{\pi}{30}$ **6.** 8.27π

7. $\pi\int_a^b [k-g(x)]^2 dx - \pi\int_a^b [k-f(x)]^2 dx$ **8. a.** $a = e^{-1}$, $(x_0, y_0) = (e^2, 1)$ **b.** $\frac{\pi}{2}$

9. $\frac{448\pi}{15}$ **10. a.** $y = \frac{ax}{\sqrt{1+a}}$ **b.** $a = 4$ **11. a.** $\frac{e}{2} - 1$ **b.** $\frac{\pi}{6}(5e^2 - 12e + 3)$

12. $\frac{4\pi}{3}$ **13.** $\frac{\pi}{6}$ **14.** $a = 0$, b = A **15. a.** $\frac{9}{4}$ **b.** $\frac{81\pi}{10}$

16. a. $\frac{32\pi}{5}$ **b.** $\frac{8\pi}{3}$ **c.** 8π **d.** $\frac{176\pi}{15}$ **17. a.** $a = \frac{-4+\sqrt{66}}{5}$ **b.** $\frac{17\pi}{25}$

Exercise 7.4.4

1. $\frac{2}{3}a^2 b \tan\alpha$

Exercise 7.4.5

1. $\frac{8}{27}(10\sqrt{10} - 1)$ **2.** $\frac{\sqrt{10}}{3}(e^6 - 1)$ **3.** $\frac{3\pi a}{2}$ **4.** $6a$

Exercise 7.4.6

1. a. 50 meters in negative direction from the origin.

b. Initial velocity is 60 m/s. Initial acceleration is -30 m/s^2.

c. $t \approx 2.76$ s or $t \approx 7.24$ s **d.** three times **e.** twice

f. $2.76 < t < 5$ or $t > 7.24$ **g.** $t > 5$ **h.** 100 **i.** 189

2. a. 1 meter in negative direction from the origin. **b.** 5 meters

c. 4 meters **d.** 12 m/s **e.** $\frac{\pi}{3}$ seconds **f. (i)** $-36\sin(3t - \pi)$ **(ii)** $-9s - 9$

3. a. (i) $s = -\frac{3}{2}e^{-2t} + \frac{3}{2}$ **(ii)** $a = -6e^{-2t}$

b. (i) $s = 5\ln|t+4| - 5\ln 4$ **(ii)** $a = -\frac{5}{(t+4)^2}$

c. (i) $s = \ln\left|t^2 + 3\right| - \ln 3$ **(ii)** $a = \dfrac{6 - 2t^2}{(t^2 + 3)^2}$

d. (i) $s = \dfrac{2}{3}(t+2)^{\frac{3}{2}} - \dfrac{4\sqrt{2}}{3}$ **(ii)** $a = \dfrac{1}{2}(t+2)^{-\frac{1}{2}}$

4. a. (i) 1.44 **(ii)** 1.44 **b. (i)** 8π **(ii)** 29.6 **c. (i)** $-\dfrac{8}{3}$ **(ii)** $\dfrac{19}{3}$

5. a. $-\dfrac{3}{1024}$ **b.** $-\dfrac{9\sqrt{3}}{2}$ **c.** $-\dfrac{9}{512}$ **d.** $100e^{-\frac{5}{4}}$

6. a. (i) $v = -\dfrac{7}{t+2} + \dfrac{13}{2}$ **(ii)** 18.3

b. (i) $v = \dfrac{2}{3}\cos(3t) - \dfrac{2}{3}$ **(ii)** 2.92

c. (i) $v = \dfrac{1}{2}\sqrt{4t+5} + 2 - \dfrac{\sqrt{5}}{2}$ **(ii)** 7.56 **7.** $v^2 = 2s + 21 - 6\ln\left|s+2\right| + 6\ln 4$

Exercise 7.5

1. a. $\dfrac{\pi}{4}$ **b.** ∞ **c.** $\dfrac{1}{2}\ln 2$ **d.** -1 **e.** no limit **f.** $\dfrac{\pi^2}{8}$

2. a. 0 **b.** $\dfrac{1}{3}\ln 4$ **c.** π **d.** 2 **e.** $\dfrac{\pi}{2}$ **f.** $\dfrac{\pi}{4} + \dfrac{1}{2}\ln 2$

Exercise 7.6

1. a. $\dfrac{\pi}{6}(-2 - \sqrt{3})$ **b.** -2π **c.** 2.30%

2. a. (i) 0.69412 **(ii)** 2.60×10^{-3} **b. (i)** 0.83521 **(ii)** 1.01×10^{-3}

c. (i) 1.95410 **(ii)** 7.18×10^{-2} **d. (i)** 0.74298 **(ii)** 1.04×10^{-2}

3. a. 17.32223 **b.** 5.40258 **c.** 1.37076 **d.** 1.22927

4. a. n = 34 **b.** n = 496 **5. a.** n = 14 **b.** n = 18

Exercise 8.2

1. a. $y=\dfrac{C}{\sqrt{1+x^2}}$ **b.** $\dfrac{y^3}{3}=x-\dfrac{x^2}{2}+C$ **c.** $y=Ae^{\sqrt{1-x^2}}$ **d.** $y=Ae^{\sin x}$

e. $e^y=e^x+\dfrac{x^3}{3}+C$ **f.** $y=A(1-ky)(x+k)$

g. $x\tan x+\ln|\cos x|=y\tan y+\ln|\cos y|+C$ **h.** $\sqrt{1+y^2}=\dfrac{Cx}{\sqrt{1+x^2}}$

2. a. $x(1-3y^4)=-47$ **b.** $\cos y=-\dfrac{1}{2}(1+e^x)$ **c.** $y=\dfrac{1}{2}(\arctan x)^2$

d. $3(x^2-y^2)+2(x^3-y^3)+5=0$ **e.** $(\ln x)^2+(\ln y)^2=0$

f. $(e^x-1)^3=(e-1)^3\tan y$ **g.** $(1+y)\cos^2 x=3(y-1)$ **h.** $y=3\left[\dfrac{x+1}{2(x-2)}\right]^{\frac{1}{3}}$

3. $y=2x$ **4.** $y=3e^{\frac{1}{2}x}$ **5.** $y=\dfrac{x^3}{3}+6$ **6.** 23 **7.** 334136

8. 2602 years ago

Exercise 8.3

1. a. $y=Ce^{\frac{x}{y}}$ **b.** $x^2e^{\frac{y}{x}}=C$ **c.** $y^2x-x^2y-x^3=C$ **d.** $xC=e^{\arcsin\frac{y}{x}}$

e. $y=Ce^{\frac{y}{x}}$ **f.** $\ln|x|+e^{-\frac{y}{x}}+C=0$ **g.** $\dfrac{1}{4}\sin\dfrac{2y}{x}+\dfrac{y}{2x}=x^2+C$

h. $Cx^4(\dfrac{y}{x}+2)^3(\dfrac{y}{x}-2)=1$ **2.** $-e^2x=e^{\frac{1}{2}\left(\frac{y}{x}\right)^2}$ **3.** $y=\dfrac{2(x^2-y^2)^2}{9x}$

4. $y+x^2y-2x=0$ **5.** $\ln(x^2+y^2)+\arctan\dfrac{y}{x}=\ln 2+\dfrac{\pi}{4}$

Exercise 8.4

1. a. $y=e^{-x}(x+C)$ **b.** $y=\dfrac{1}{2}(x+1)^4+C(x+1)^2$ **c.** $y=\dfrac{4x^3+3C}{3(x^2+1)}$

d. $y = x^2(e^x + C)$ **e.** $xy = -\cos x + C$ **f.** $y = -\frac{1}{4}xe^{-2x}(2x^2 + 2x + 1) + Cx$

2. $y = e^{\cos x}(x+1)$ **3.** $y\sin x = \tan x - \frac{\sqrt{3}}{2}$ **4.** $y\cos x = \frac{1}{4}\sin 2x + \frac{1}{2}x - \frac{\pi}{2} - 1$

5. $y = \frac{1}{3}(10-x) - \frac{7}{30}(10-x)^2$ **6.** $y = -(x^2-5)\left[\ln\left|x^2-5\right| + 7\right]$

7. $y = \frac{e^{-2x}}{4}(x^4 + 4)$

Exercise 8.5

1. a. 0.33212626 **b.** $y = \frac{e^{-2x}}{4}(x^4+4)$, $\frac{e^{-1}(0.5^4+4)}{4} = \frac{65}{64}e^{-1}$ **c.** 11.1%

d. On the interval [0, 1], $y'' > 0$. The function is concave up and its tangent lies below the function. **e.** 0.36979097, 1.03%

2. a. 6.5039001 **b.** $y = -\frac{1}{2} + 3e^{x^2}$, $y = -\frac{1}{2} + 3e$ (≈ 7.65484549) **c.** 15.0%

d. 7.392606753 3.43% **e.** $y'' > 0$ hence, the function is concave up and its tangent lies below the function.

3. a. 0.82525035 **b.** $y = \frac{1}{1+x^2}$, 0.8 **c.** 3.16% **d.** 0.80582436, 0.728%

e. On the interval [0, 0.5], $y'' < 0$. The function is concave down and its tangent lies above the function.

Exercise 8.6

1. $y_1 = \frac{x^2}{2}$ $y_2 = \frac{x^2}{2} + \frac{x^3}{6}$ $y_3 = \frac{x^2}{2} + \frac{x^3}{6} + \frac{x^4}{24}$

Exercise 9.1

1. a. converges, $\frac{3}{4}$ **b.** converges, $\frac{3}{2}$ **c.** converges, $-\frac{37}{8}$ **d.** converges, $\frac{3}{4}$

e. converges, $\frac{1}{3}$ **f.** converges, $\frac{173}{28}$ **g.** diverges **h.** diverges

i. diverges **j.** diverges **k.** diverges **l.** diverges **m.** converges, 1

Exercise 9.2

1. a. diverges **b.** diverges **c.** converges **d.** diverges

e. cannot determine. Other test is needed. **f.** cannot determine. Other test is needed. **g.** diverges **h.** diverges **i.** converges

Exercise 9.3

1. a. diverges **b.** converges **c.** diverges **d.** diverges

e. diverges **f.** converges **g.** converges **h.** converges

i. converges **j.** diverges **k.** converges **l.** diverges

m. diverges **n.** converges **o.** diverges

2. a. converges **b.** converges **c.** diverges **d.** converges

e. converges **f.** converges **g.** converges **h.** diverges

i. diverges **j.** converges **k.** diverges **l.** converges

m. converges **n.** converges **o.** converges **p.** converges

q. converges **r.** converges **s.** converges **t.** converges

u. diverges

3. a. diverges **b.** converges **c.** diverges **d.** diverges

e. converges **f.** diverges **g.** converges **h.** converges

i. converges **j.** converges **k.** diverges **l.** converges

4. a. diverges **b.** diverges **c.** diverges **d.** converges

e. converges **f.** converges **g.** converges **h.** diverges

i. diverges **j.** converges **k.** converges **l.** converges

m. converges **n.** diverges **o.** diverges

Exercise 9.4

1. **a.** converges **b.** converges **c.** diverges **d.** converges **e.** converges **f.** converges **g.** converges **h.** converges **i.** converges

2. **a.** converges absolutely **b.** converges conditionally **c.** converges absolutely **d.** converges absolutely **e.** converges conditionally **f.** diverges **g.** converges conditionally **h.** diverges **i.** converges conditionally **j.** converges conditionally **k.** converges absolutely **l.** diverges **m.** converges absolutely **n.** converges absolutely **o.** converges absolutely

3. **a.** $\frac{1}{2}$ **b.** $\frac{1}{72}$ **c.** $\frac{1}{5}$ **d.** $\frac{1}{10}$ **e.** $\frac{5}{6561}$ **f.** $\frac{1}{25}$

4. **a.** n = 7 **b.** n = 44 **c.** n = 6 **d.** n = 8 **e.** n = 7 **f.** n = 21

Exercise 9.5

1. **a.** $R = 1,\ -4 < x < -2$ **b.** $R = 1,\ -1 \le x < 1$ **c.** $R = \frac{1}{4},\ \frac{7}{4} < x < \frac{9}{4}$

d. $R = \infty,\ -\infty < x < \infty$ **e.** $R = \frac{5}{3},\ -\frac{7}{3} < x < 1$ **f.** $R = 1,\ -1 \le x \le 1$

g. $R = 3,\ -3 < x \le 3$ **h.** $R = 1,\ -1 \le x \le 1$ **i.** $R = 1,\ 3 \le x \le 5$

j. $R = \frac{1}{2},\ -2 < x \le -1$ **k.** $R = \infty,\ -\infty < x < \infty$ **l.** $R = 0,\ x = -1$

2. **a.** $R = 1,\ -1 \le x < 1$ **b.** $R = 1,\ -2 \le x < 0$ **c.** $R = 2,\ -3 < x \le 1$

d. $R = \frac{1}{2},\ -3 \le x \le -2$ **e.** $R = 1,\ \frac{1}{e} < x < e$ **f.** $R = \infty,\ -\infty < x < \infty$

g. $R = 1,\ -1 < x < 1$ **h.** $R = \frac{6}{5},\ -\frac{46}{5} < x < -\frac{34}{5}$ **i.** $R = \infty,\ -\infty < x < \infty$

j. $R = 1,\ -3 \le x \le -1$ **k.** $R = 0,\ x = \frac{7}{3}$ **l.** $R = \infty,\ -\infty < x < \infty$

Exercise 9.6

1. a. $1+\frac{2}{3}x-\frac{1}{9}x^2+\frac{4}{81}x^3$ **b.** $\frac{1}{8}+\frac{3}{16}x+\frac{3}{16}x^2+\frac{5}{32}x^3$

c. $1+\frac{1}{5}x+\frac{1}{50}x^2+\frac{1}{750}x^3$ **d.** $\frac{1}{4}+\frac{1}{16}x+\frac{1}{64}x^2+\frac{1}{256}x^3$

e. $1-3x+2x^2$ **f.** $-x-\frac{1}{2}x^2-\frac{1}{3}x^3-\frac{1}{4}x^4$

2. a. $1+2(x+1)+3(x+1)^2+4(x+1)^3$ **b.** $-3(x-1)-(x-1)^2$

c. $e^2+e^2(x-2)+\frac{e^2}{2}(x-2)^2+\frac{e^2}{6}(x-2)^3$ **d.** $2+\frac{1}{4}(x-4)-\frac{1}{64}(x-4)^2+\frac{1}{512}(x-4)^3$

Exercise 9.7

1. a. $\sum_{n=0}^{\infty}\frac{(-1)^n x^{2n}}{n!}$ **b.** $\sum_{n=0}^{\infty}\frac{(-1)^n x^{n+3}}{n!}$ **c.** $\frac{1}{2}+\frac{1}{2}\sum_{n=0}^{\infty}\frac{(-1)^n (2x)^{2n}}{(2n)!}$

d. $\sum_{n=1}^{\infty}\frac{(-1)^{n+1}(3x)^n}{n}$ **e.** $\sum_{n=0}^{\infty}\frac{(-1)^n 2^{2n} x^{2n+1}}{(2n)!}$ **f.** $\sum_{n=0}^{\infty}\frac{x^{2n+1}}{2n+1}$

g. $\sum_{n=0}^{\infty}\frac{(-1)^n (5x^3)^{2n+1}}{(2n+1)!}$ **h.** $\sum_{n=0}^{\infty}\frac{(-1)^n (4\sqrt{x})^{2n}}{(2n)!}$ **i.** $\sum_{n=0}^{\infty}\frac{\left(\frac{x}{5}\right)^n}{n!}$

2. a. $e(1-\frac{x^2}{2}+\frac{x^4}{6}-...)$ **b.** $x-\frac{x^2}{2}+\frac{x^3}{6}-\frac{x^4}{12}+...$

c. $\ln 2+\frac{5}{2}x-(\frac{9}{2}\ln 2+\frac{25}{8})x^2-\frac{145}{24}x^3+(\frac{27}{8}\ln 2+\frac{275}{64})x^4+...$

d. $3x+12x^2+\frac{39}{2}x^3+14x^4+...$ **e.** $\sum_{n=0}^{\infty}\frac{x^n}{3^{n+1}}$, $-3<x<3$ **f.** $\frac{1}{4}\sum_{n=0}^{\infty}[(-1)^n-\frac{1}{3^n}]x^n$

3. a. $x+\frac{x^3}{3}+...$ **b.** $1+x+\frac{x^2}{2}+\frac{x^3}{2}+\frac{3x^4}{8}+...$ **4. a.** $x-\frac{x^2}{2}+\frac{x^3}{6}-\frac{x^4}{12}+...$

b. $\frac{1}{2}x^2+\frac{1}{12}x^4+...$ **5. a.** $\sum_{n=0}^{\infty}\frac{x^{2n+1}}{n!(2n+1)}$ **b.** $\sum_{n=0}^{\infty}\frac{(-1)^n x^{2n+1}}{(2n+1)!(2n+1)}$

c. $\sum_{n=1}^{\infty}\frac{(-1)^{n-1}x^{2n-1}}{(2n-1)^2}$ **d.** $\sum_{n=1}^{\infty}\frac{x^n}{n\cdot 3^n}$ **e.** $\sum_{n=0}^{\infty}\frac{(-1)^n x^{n+1}}{(2n)!(n+1)}$ **f.** $\sum_{n=0}^{\infty}\frac{(-1)^n x^{4n+3}}{(2n+1)!(4n+3)}$

6. c. $\frac{1}{2}\ln\frac{1+x}{1-x}$ **7.** $\sum_{n=1}^{\infty}\frac{(-1)^{n+1}[1+(-2)^n]x^n}{n}, \quad -\frac{1}{2}\le x<\frac{1}{2}$

Exercise 9.8

1. a. 0.7468 **b.** 0.1161 **c.** 0.9045 **d.** 1.6054 **e.** 0.4872 **f.** 0.0976

2. a. 0.309017 **b.** 0.999848 **c.** 0.182267 **d.** 1.648721 **3. a.** -1 **b.** 2

c. $\frac{1}{2}$ **d.** $-\frac{3}{2}$ **4. a.** $1+\frac{1}{5}x-\frac{2}{25}x^2+\frac{6}{125}x^3-\frac{21}{625}x^4+\ldots$ **b.** 1.05382

c. 6.21×10^{-5} **5. a.** $x-\frac{1}{3}x^3+\ldots$ **b.** $-\frac{1}{3}$ **6. a.** $\ln 2-\frac{1}{4}x^2-\frac{1}{96}x^4+\ldots$

b. 0.406465 **c.** 0.0121 **7. a.** $-\ln 2+\sqrt{3}(x-\frac{\pi}{6})-2(x-\frac{\pi}{6})^2+\frac{4\sqrt{3}}{3}(x-\frac{\pi}{6})^3+\ldots$

b. -0.335337 **c.** 0.0156586 **d.** 0.0112367 **e.** $\sqrt{3}-4(x-\frac{\pi}{6})+4\sqrt{3}(x-\frac{\pi}{6})^2+\ldots$

8. a. $1-2x^2+4x^4+\ldots$ **b.** 0.75 **c.** $\frac{2}{3}$ **9. a.** $2-x^2+\frac{3}{4}x^4+\ldots$ **b.** 1.990075

c. 1.990074 **10. a.** $1+\frac{1}{2}x+\frac{1}{4}x^2+\ldots$ **b.** 1.11 **c.** 1.110494

11. a. $-1+\frac{2}{e}x+\frac{4+e}{2e^2}x^2+\frac{8+3e}{3e^3}x^3+\ldots$ **b.** -0.484955 **c.** -0.465796

Exercise 9.9

1. a. $1-3x+6x^2-10x^3+\ldots,\ -1<x<1$ **b.** $1+4x+12x^2+32x^3+\ldots,\ -\frac{1}{2}<x<\frac{1}{2}$

c. $1-\frac{8}{3}x-\frac{16}{9}x^2-\frac{256}{81}x^3+\ldots,\ -\frac{1}{4}<x<\frac{1}{4}$

d. $1+\frac{5}{2}x-\frac{25}{8}x^2+\frac{125}{16}x^3+\ldots,\ -\frac{1}{5}<x<\frac{1}{5}$

e. $1-2x^2+3x^4-4x^6+\dots$, $-1<x<1$

f. $1+\frac{21}{5}x+\frac{588}{25}x^2+\frac{17836}{125}x^3+\dots$, $-\frac{1}{7}<x<\frac{1}{7}$

2. a. $\frac{1}{3}(1-\frac{x}{3}+\frac{x^2}{9}-\frac{x^3}{27}+\dots)$, $-3<x<3$ **b.** $2(1+\frac{x}{8}-\frac{x^2}{128}+\frac{x^3}{1024}+\dots)$, $-4<x<4$

c. $6^{\frac{3}{4}}(1+\frac{5}{8}x-\frac{25}{384}x^2+\frac{625}{27648}x^3+\dots)$, $-\frac{6}{5}<x<\frac{6}{5}$

d. $3^{-\frac{1}{2}}(1+\frac{1}{3}x+\frac{1}{6}x^2+\frac{5}{54}x^3+\dots)$, $-\frac{3}{2}<x<\frac{3}{2}$ **e.** $1+x+\frac{1}{2}x^2+\frac{1}{4}x^3+\dots$, $-2<x<2$

f. $\frac{1}{3}(1+\frac{14}{3}x-\frac{14}{9}x^2+\frac{14}{27}x^3+\dots)$, $-3<x<3$

g. $1+\frac{5}{2}x+\frac{55}{8}x^2+\frac{365}{16}x^3+\dots$, $-\frac{1}{4}<x<\frac{1}{4}$

h. $\frac{1}{512}(9-\frac{81}{8}x+\frac{51}{32}x^2+\frac{513}{256}x^3+\dots)$, $-\frac{8}{3}<x<\frac{8}{3}$

i. $7^{\frac{2}{3}}\times 3^{\frac{1}{4}}(1+\frac{5}{42}x-\frac{2123}{3528}x^2+0.553x^3+\dots)$, $-\frac{1}{2}<x<\frac{1}{2}$

3. $b=\frac{10}{9}$, $n=-16.2$ **4.** $p=\frac{263}{312}$, $n=\frac{39}{263}$ **5.** 0.99940 **6.** 0.5515 **7.** 7.21

8. a. $1-\frac{1}{2}x+\frac{3}{8}x^2+\dots$ **c.** 5.4772 **9.** 2.973630

10. a. $\frac{2}{2x-1}-\frac{3}{3-x}$ **b.** $-3-\frac{13}{3}x-\frac{73}{9}x^2-\dots$ **c.** $-\frac{1}{2}<x<\frac{1}{2}$

www.ingramcontent.com/pod-product-compliance
Ingram Content Group UK Ltd.
Pitfield, Milton Keynes, MK11 3LW, UK
UKHW051956041026
15590UKWH00009B/2338

9 798895 71040